Hayes, C.
Anglian Water
Ambury Road
PE18 6NZ HUNTINGDON
UNITED KINGDOM

Heijden, C.A. van der
National Institute of Public Health
and Environmental Hygiene
P.O. Box 1
3720 BA BILTHOVEN
THE NETHERLANDS

Hofman, M.
Gem. Drinkwaterleiding Rotterdam
Postbus 6610
3002 AP ROTTERDAM
THE NETHERLANDS

Hoigne, J.
EAWAG
Überlandstrasse 133
CH-8600 DÜBENDORF
SWITZERLAND

Hoof, F. van
Antwerpse Waterwerken
Mechelsesteenweg 64
2018 ANTWERPEN
BELGIUM

Hrubec, J.
National Institute of Public Health
and Environmental Hygiene
P.O.Box 150
2260 AD LEIDSCHENDAM
THE NETHERLANDS

Huck, P.
Department of Civil Engineering
University of Alberta
EDMONTON, Alberta
CANADA

Huijgen-Regtien, Mrs.I.J.J.
National Institute of Public Health
and Environmental Hygiene
P.O.Box 150
2260 AD LEIDSCHENDAM
THE NETHERLANDS

Keller, A.
CIBA - GEIGY AG
CH-4002 BASEL
SWITZERLAND

Klijnhout, A.F.
Heineken Technisch Beheer B.V.
P.O. Box 510
2380 BB ZOETERWOUDE
THE NETHERLANDS

Kool, H.J.
National Institute of Public Health
and Environmental Hygiene
P.O.Box 150
2260 AD LEIDSCHENDAM
THE NETHERLANDS

Kooy, D. van der
Netherlands Waterworks Testing
& Research Inst. KIWA N.V.
P.O. Box 1072
3430 BB NIEUWEGEIN
THE NETHERLANDS

Kraus, J.
Deutscher Verein des Gas- und
Wasserfaches
Frankfurter Allee 27
6263 ESCHBORN
FEDERAL REPUBLIC OF GERMANY

Kreijl, C.F. van
National Institute of Public Health and
Environmental Hygiene
P.O.Box 1
3720 BA BILTHOVEN
THE NETHERLANDS

Kronberg, L.
Institute of Organic Chemistry
Abo Akademi
Akademig 9
SF-20500 TURKU
FINLAND

Kruijf, H.A.M. de
National Institute of Public Health and
Environmental Hygiene
P.O.Box 150
2260 AD LEIDSCHENDAM
THE NETHERLANDS

Kruithof, J.C.
Netherlands Waterworks Testing and
Research Institute KIWA N.V.
P.O. Box 1072
3430 BB NIEUWEGEIN
THE NETHERLANDS

Laegreid, Mrs. M.
National Inst. of Public Health
Geithyrsvn. 75
0566 OSLO 5
NORWAY

Leer, E.W.B. De
Lab. Anal. Chem. TH - Delft
Jaffalaan 9
2628 BX DELFT
THE NETHERLANDS

Legube, B.
University of Poitiers
40, Ave. du Recteur Pineau
86022 POITIERS
FRANCE

Leuken, R.G.J. van
Division of Technology TNO
P.O. Box 217
2600 AE DELFT
THE NETHERLANDS

Lewis, W.M.
World Health Organisation
8, Scherfigsvej
DK-2100 COPENHAGEN
DENMARK

Lieshout, J. van
Water Laboratorium Oost
Kastanjelaan 8
7004 AK DOETINCHEM
THE NETHERLANDS

Liimatainen, A.
Dept. of Environmental Hygiene
University of Kuopio
P.O. Box 6
70211 KUOPIO 21
FINLAND

Luijten, J.A.
National Institute of Public Health
and Environmental Hygiene
P.O.Box 150
2260 AD LEIDSCHENDAM
THE NETHERLANDS

Mallevialle, J.
Laboratoire Central
Lyonnaise des Eaux
38, rue du President Wilson
78230 LE PECQ
FRANCE

Masters, Mrs. S.J.
EAWAG - ETH
Überlandstrasse
8600 DÜBENDORF
SWITZERLAND

ORGANIC MICROPOLLUTANTS IN DRINKING WATER AND HEALTH

ORGANIC MICROPOLLUTANTS IN DRINKING WATER AND HEALTH

Proceedings of an International Symposium, Amsterdam, The Netherlands, 11–14 June 1985

edited by

H.A.M. de Kruijf

and

H.J. Kool

National Institute of Public Health and Environmental Hygiene, Bilthoven, The Netherlands

Reprinted from *The Science of the Total Environment,* Vol. 47, 1985

ELSEVIER 1985
Amsterdam — Oxford — New York — Tokyo

ELSEVIER SCIENCE PUBLISHERS B.V.
Sara Burgerhartstraat 25
P.O. Box 211, 1000 AE Amsterdam, The Netherlands

Distributors for the United States and Canada:

ELSEVIER SCIENCE PUBLISHING COMPANY INC.
52, Vanderbilt Avenue
New York, NY 10017, U.S.A.

ISBN 0-444-42583-7

© Elsevier Science Publishers B.V., 1985

All rights reserved. No part of this publication may be reproduced, stored in a retrieval system or transmitted in any form or by any means, electronic, mechanical, photocopying, recording or otherwise, without the prior written permission of the publisher, Elsevier Science Publishers B.V./Science & Technology Division, P.O. Box 330, 1000 AH Amsterdam, The Netherlands.

Special regulations for readers in the USA — This publication has been registered with the Copyright Clearance Center Inc. (CCC), Salem Massachusetts. Information can be obtained from the CCC about conditions under which photocopies of parts of this publication may be made in the USA. All other copyright questions, including photocopying outside of the USA, should be referred to the copyright owner, Elsevier Science Publishers B.V., unless otherwise specified.

Printed in The Netherlands

CONTENTS

Preface ... XI

List of participants ... XIII

Welcome
   H. Cohen (Bilthoven, The Netherlands) ... 1

Opening Address
   F.W.R. Evers (Leidschendam, The Netherlands) ... 3

Organic Micropollutants in Drinking Water: An Overview
   J.A. Cotruvo (Washington, USA) ... 7

Impact of different types of organic micropollutants present on sources of drinking water on the quality of drinking water
   H. Sontheimer, H.-J. Brauch and W. Kühn (Karlsruhe, Federal Republic of Germany) ... 27

The control of organics in drinking water in Canada and the United States (standards, Legislation and practice)
   P. Toft (Ottawa, Canada) ... 45

Legislation and policy for the protection of the drinking water supply in The Netherlands
   A.M. van Dijk-Looijaard and H.A.M. de Kruijf (Leidschendam, The Netherlands) ... 59

World Health Organization, guidelines on organic micropollutants
   W.M. Lewis (Copenhagen, Denmark) ... 83

The effects of a hazardous and a domestic waste landfill on the trace organic quality of chalk groundwater at a site in East Anglia
   K.M. Baxter (Medmenham, United Kingdom) ... 93

Biological processes and organic micropollutants in treatment processes
   B.E. Rittmann (Urbana, USA) ... 99

Influence of microbiological activity in granular activated carbon filters on the removal of organic compounds
   J. De Laat, F. Bouanga and M. Dore (Poitiers, France) ... 115

Developments in biotechnology of relevance to drinking water preparation
   D.B. Janssen and B. Witholt (Groningen, The Netherlands) ... 121

The influence of water treatment processes on the presence of organic surrogates and mutagenic compounds in water
   M.A. Van Der Gaag, J.C. Kruithof and L.M. Puijker (Nieuwegein, The Netherlands) ... 137

Removal of organic micropollutants by coagulation and adsorption
    V.L. Snoeyink and A.S.C. Chen (Urbana, USA) 155

Organic micropollutants and treatment processes: kinetics and final
    effects of ozone and chlorine dioxide
    J. Hoigné (Dübendorf, Switzerland) 169

Formation of linear aldehydes during surface water preozonization and
    their removal in water treatment in relation to mutagenic activity
    and sum parameters
    F. Van Hoof, J. Janssens and H. Van Dijck (Antwerp, Belgium) 187

New directions in oxidant by-product research: identification and
    significance
    R.F. Christman, D.L. Norwood and J.D. Johnson (Chapel Hill, USA) 195

Chloroform production from model compounds of aquatic humic
material. The role of pentachlororesorcinol as an intermediate
    E.W.B. De Leer and C. Erkelens (Delft, The Netherlands) 211

Chlorination of humic substances in aqueous solution: yields of
    volatile and major non-volatile organic halides
    B. Legube, J.P. Croue and M. Dore (Poitiers, France) 217

Chloropicrin formation during oxidative treatments in the preparation
    of drinking water
    N. Merlet, H. Thibaud and M. Dore (Poitiers, France) 223

Evaluation of different treatment processes with respect to mutagenic
    activity in drinking water
    H.J. Kool, J. Hrubec, C.F. van Kreijl and G.J. Piet (Leidschendam,
    The Netherlands)... 229

Mutagenic activity in humic water and alum flocculated humic water
    treated with alternative disinfectants
    P. Backlund, L. Kronberg, G. Pensar (Turku, Finland) and L. Tikkanen
    (Espoo, Finland)... 257

A comparison of methods for concentrating mutagens in drinking water -
    recovery aspects and their implications for the chemical character of
    major unidentified mutagens
    B. Wigilius, H. Borén, A. Grimvall (Linköping, Sweden), G.E. Carlberg
    and M. Möller (Oslo, Norway) 265

The quality of drinking water prepared from bank-filtered river water in
    The Netherlands
    A. Noordsij, L.M. Puyker and M.A. Van Der Gaag (Nieuwegein,
    The Netherlands) 273

Mutagenicity testing of water with fish: a step forward to a reliable assay
    M.A. Van Der Gaag and J.F.J. Van De Kerkhoff (Nieuwegein, The Netherlands) 293

Alternative methods for chlorination
    F. Fiessinger, J.J. Rook and J.P. Duguet (Le Pecq, France) 299

Identification and assessment of hazardous compounds in drinking water
    J.K. Fawell and M. Fielding (Medmenham, United Kingdom) 317

Fractionation of mutagenic compounds formed during chlorination of humic water
    L. Kronberg, B. Holmbom (Turku, Finland) and L. Tikkanen (Espoo, Finland) 343

Liquid chromatographic determination of the fungicide iprodione in surface water, using on-line preconcentration
    C.E. Goewie and E.A. Hogendoorn (Bilthoven, The Netherlands) 349

Progress in the isolation and characterization of non-volatile mutagens in a drinking-water
    P.G. van Rossum (Pretoria, South Africa) 361

Characterization of low molecular weight products desorbed from polyethylene tubings
    C. Anselme, K. N'Guyen, A. Bruchet, J. Mallevialle (Le Pecq, France).. 371

Carcinogenic and mutagenic properties of chemicals in drinking water
    R.J. Bull (Pullman, USA) 385

Effect of trichloroethylene on the exploratory and locomotor activity of rats exposed during development
    D.H. Taylor, K.E. Lagory, D.J. Zaccaro, R.J. Pfohl and R.D. Laurie (Cincinnati, USA) 415

Results of a 90-day toxicity study on 1,2,3- and 1,1,2-trichloropropane administered via the drinking water
    D.C. Villeneuve, I. Chu, V.E. Secours, M.G. Coté, G.L. Plaa and V.E. Valli (Montréal, Canada) 421

Carcinogenicity study in rats with a mixture of eleven volatile halogenated hydrocarbon drinking water contaminants
    P.W. Wester, C.A. van der Heijden, A. Bisschop, G.J. van Esch, R.C.C. Wegman and Th. de Vries (Bilthoven, The Netherlands) 427

Target organ toxicology of halocarbons commonly found contaminating drinking water
    L.W. Condie (Cincinnati, USA) 433

Inhalation exposure in the home to volatile organic contaminants of
    drinking water
    J.B. Andelman (Pittsburgh, USA) 443

Epidemiologic studies of organic micropollutants in drinking water
    G.F. Craun (Cincinnati, USA) 461

Metabolites of chlorinated solvents in blood and urine of subjects
    exposed at environmental level
    G. Ziglio, G. Beltramelli, F. Pregliasco and G. Ferrari (Milan, Italy) 473

Critical considerations on the significance of carcinogenic and mutagenic
    compounds in drinking water
    C.A. Van Der Heijden and C.F. van Kreijl (Bilthoven, The Netherlands). 479

CLOSING REMARKS

Drinking water and health hazards in environmental perspective
    B.C.J. Zoeteman (Bilthoven, The Netherlands) 487

Author Index 505

Subject Index 506

Preface

The occurrence of organic micropollutants in raw water sources presents a serious threat to the quality of drinking water and therefore to human health. Under various conditions before, during, and after treatment processes, hazardous compounds may be formed. The papers in this volume were presented at a Symposium which was specifically directed to exchange and communicate the most recent information and concepts on the fate and behaviour of organic micropollutants in raw water sources before, during and after treatment processes and the potential risks such compounds may present to human health when present in drinking water.

Legislation and standards in some countries are based on toxicity data of individual compounds. As several speakers reported that numerous unknown compounds occur in water, there is a demand for reliable group parameters and methods to selectively identify compounds which are biologically active.

Treatment methods based on (micro)biological techniques tend to be the most promising with regard to the effectiveness of removal of dangerous substances. Chlorination, a widely used disinfection method, is certainly a major cause of the increase of mutagenicity in drinking water. Analytical methods may, within a reasonable timespan, enable identification of such mutagenic compounds, but several carcinogenic and mutagenic substances have already been identified in drinking water.

Disinfection by chlorination should be minimized as much as possible or banned altogether where the bacteriological conditions allow such a drastic measure. Several alternatives exist, chlorine dioxide especially appears to be a promising alternative. An excellent suggestion was made to simply increase the residence time reservoirs, which will then apparently decrease the load of biologically active compounds.

Drinking water is not only used for intake but also for washing, showers etc. In an elegant study it was shown that if a rather volatile compound, such as trichloroethylene, is present in drinking water the inhalation route in the household may be much more important than the actual consumption route via drinking water with respect to health hazards. Studies on the carcinogenicity of compounds in drinking water unequivocally indicate that indeed such compounds are present. Results of epidemiological studies in this respect, however, are not too convincing.

Although the importance of hazardous compounds in drinking water should not at all be neglected, its place in the total burden of hazardous compounds to which man is exposed should not be overrated. This Symposium was only possible through the generous assistance of the National Institute of Public Health and Environmental Hygiene, the United States Environmental Protection Agency, the Ministry of Housing, Physical Planning and Environment, and the Netherlands Waterworks Testing and Research Institute Ltd. Also, the World Health Organization, Copenhagen and the Water Research Center of the United Kingdom cosponsored this Symposium.

It is a pleasure for the editors to acknowledge the help of a great number of persons in preparing this manuscript. We are especially grateful to Mrs. I. Huijgen-Regtien for her invaluable assistance during the whole process of organizing the Symposium and the preparation of the manuscript. We also thank a number of colleagues who helped us at various stages of the Symposium organisation, especially Drs. A. Minderhoud, Drs. W. van de Meent, Ir. J. Hrubec, G. Piet, Dr. B. Haring and others. We thank a number of scientists for reviewing the manuscript and finally we are indebted to Elsevier Science Publishers for their assistance.

We hope that this volume may contribute substantially to the development of methods and understanding in the field of organic micropollutants in drinking water and their significance for public health.

H.A.M. de Kruijf
H.J. Kool

Leidschendam, July 1985.

LIST OF PARTICIPANTS
International Symposium on Organic Micropollutants in Drinking Water and
Health, Amsterdam 11-14th June, 1985

Andelman, J.
University of Pittsburgh
Graduate School of Public Health
PITTSBURGH, PA 15217
USA

Backlund, P.
Institute of Organic Chemistry
Abo Akademi
Akademig. 1
20500 TURKU 50
FINLAND

Bassie, W.
N.V. Watermij. Z.W. Nederland
Postbus 121
4460 AC GOES
THE NETHERLANDS

Baxter, K.
Water Research Centre
Medmenham Laboratory
P.O. Box 16
MEDMENHAM, Marlow, Bucks
UNITED KINGDOM

Böre, M.
NOKIA Chemicals
SF - 32740 AESTA
FINLAND

Brauch, H.J.
Engler-Bunte Institute
University Karlsruhe
R. Willstätter Allee 5
7500 KARLSRUHE
FEDERAL REPUBLIC OF GERMANY

Brumen, S.B.
Zavod ZA
Promajska 1
MARIBOR, 62000
YUGOSLAVIA

Bull, R.J.
College of Pharmacy
State University of Washington
PULLMAN WA 99164
USA

Christman, R.F.
University of North Carolina
Department of Environmental Sciences
and Engineering
Rosenau Hall 201 H
CHAPEL HILL, NC 27514
USA

Condie, L.
U.S. Environmental Protection Agency
26 West St. Clair
CINCINNATI, Ohio 45268
USA

Cotruvo, J.
U.S. Environmental Protection Agency
Office of Drinking Water
401 M. Street SW
WASHINGTON DC 20460
USA

Craun, G.F.
U.S. Environmental Protection Agency
26 West St. Clair
CINCINNATI OH 45268
USA

Dienst, J.A. van
Aramco Overseas Co.
Plesmanlaan 100
Room 3-134
2332 CB LEIDEN
THE NETHERLANDS

Dijk-Looijaard, Mrs. A.M.van
National Institute of Public Health
and Environmental Hygiene
P.O. Box 150
2260 AD LEIDSCHENDAM
THE NETHERLANDS

Dreo, A.
Public Health Institute
Pavomajska 1
62000 MARIBOR
YUGOSLAVIA

Erkelens, Mrs. C.
TH - Delft
Jaffalaan 9
2628 BX DELFT
THE NETHERLANDS

Fawell, J.
Water Research Centre
Medmenham Laboratory
P.O. Box 16
MEDMENHAM, Marlow, Bucks
UNITED KINGDOM

Fayad, N.
Univ. of Petroleum and Minerals
P.O. Box 1479
DHAHRAN 31261
SAUDI ARABIA

Fielding, M.
Water Research Centre
P.O. Box 16
MEDMENHAM, Marlow S17 2HD
UNITED KINGDOM

Fiessinger, F.
Laboratoire Central
Lyonnaise des Eaux
38, rue du President Wilson
78230 LE PECQ
FRANCE

Fletcher, I.J.
The South Staffordshire Waterworks Co.
Central Office
Green Lane, Walsall
WEST MIDLANDS WS2 7PD
UNITED KINGDOM

Gaag, M.A. van der
Netherlands Waterworks Testing and
Research Institute KIWA NV
P.O. Box 1072
3430 BB NIEUWEGEIN
THE NETHERLANDS

Gjessing, E.
Norwegian Institute for Water Research
P.O. Box 333
Brekkeveien 19
OSLO 3
NORWAY

Goewie, C.
National Institute of Public Health
and Environmental Hygiene
P.O. Box 1
3720 BA BILTHOVEN
THE NETHERLANDS

Grimvall, A.
Department of Water in Environment and
Society
Linkoping University
S-58183 LINKOPING
SWEDEN

Haring, B.J.A.M.
Ministry of Housing, Physical Planning
and Environment
P.O.Box 450
2260 MB LEIDSCHENDAM
THE NETHERLANDS

Medved, M.
Inst. for Public Health
Provomajska 1
62000 MARIBOR
YUGOSLAVIA

Meent, W. van de
Netherlands Waterworks Testing
& Research Institute KIWA N.V.
P.O. Box 1072
3430 BB   NIEUWEGEIN
THE NETHERLANDS

Meijers, A.P.
WRK
P.O.Box 10
3430 AA   NIEUWEGEIN
THE NETHERLANDS

Merlet, Mrs. N.
Univ. of Poitiers
Lab. Chimie d L'eau
40, Avenue du Recteur Pineau
86022 POITIERS
FRANCE

Minderhoud A.
National Institute of Public Health
and Environmental Hygiene
P.O.Box 1
3720 BA   BILTHOVEN
THE NETHERLANDS

Naerssen, E.A. van
Municipal Waterworks Dordrecht
Postbus 62
3300 AB   DORDRECHT
THE NETHERLANDS

Noordam, P.C.
Netherlands Waterworks Testing
& Research Institute KIWA N.V.
P.O.Box 1072
3430 BB   NIEUWEGEIN
THE NETHERLANDS

Noordsij, A.
Netherlands Waterworks Testing
& Research Institute KIWA N.V.
P.O. Box 1972
3430 BB   NIEUWEGEIN
THE NETHERLANDS

Norwood, D.
University of North Carolina
Department of Environmental Sciences
and Engineering
Rosenau Hall 201 H
CHAPEL HILL, NC 27514
USA

Piet, G.
National Institute of Public Health
and Environmental Hygiene
P.O.Box 150
2260 AD   LEIDSCHENDAM
THE NETHERLANDS

Praag, F. van
Purewater B.V.
Amperestraat 42
SCHAESBERG
THE NETHERLANDS

Prest, J.
Imperial Chemical Industries
Mond Division
P.O. Box 8
THE HEATH, Runcorn
UNITED KINGDOM

Puffelen, J. van
Duinwaterleiding van 's-Gravenhage
P.O. Box 710
2501 CS   DEN HAAG
THE NETHERLANDS

Rauzy, S.
Ville de Paris - Controle des Eaux
144 Av PV Couturier
75014 PARIS
FRANCE

Reckhow, D.
Compagnie Generale des Eaux
Rue de la Dique
78600 MAISONS LAFITTE
FRANCE

Rismal, R.
Institut za zdraustu. hidrotechniko
Hajdrihoua 28
61000 LJUBLJANA
YUGOSLAVIA

Ritsema, R.
Government Institute for Sewage
Treatment
Herenweg 99a
9721 AA   GRONINGEN
THE NETHERLANDS

Rittman, B.E
University of Illinois
URBANA IL 61801
USA

Rook, J.J.
Laboratoire Central
Lyonnaise des Eaux
38, rue du President Wilson
78230 LE PECQ
FRANCE

Rossum, P. van
National Institute for Water Research
P.O. Box 395
PRETORIA 0001
SOUTH AFRICA

Schraa, G.
Department of Microbiology
Agricultural University
H. v. Suchtelenweg 4
6703 CT   WAGENINGEN
THE NETHERLANDS

Snoeyink, V.L.
University of Illinois
208, N. Romine Street
URBANA IL 61801
USA

So, Mrs. M.L.
Hong Kong Baptist College
224 Waterloo Road
HONG KONG

Sontheimer, H.
Engler Bunte Institute
University of Karlsruhe
KARLSRUHE
FEDERAL REPUBLIC OF GERMANY

Soppe, A.I.A.
Gemeentelijk Waterbedrijf Groningen
Van Kerckhoffstraat 2
9701 BN   GRONINGEN
THE NETHERLANDS

Struijs, J.
National Institute of Public Health
and Environmental Hygiene
P.O. Box 1
3720 BA   BILTHOVEN
THE NETHERLANDS

Suijlekom, G. van
Natronchemie
Westersingel 102
3015 LD ROTTERDAM
THE NETHERLANDS

Taylor, D.H.
University of Miami
Department of Zoology
OXFORD, Ohio 45056
USA

Toft, P.
Environmental Standard Division
Bureau Chemical Hazards
OTTAWA, Ontario
CANADA

Tiovanen, E.
Helsinki City Water and Sewage Works
Water Research Office
Kylasaarenkatu 10
0550 HELSINKI
FINLAND

Trouwborst T.
Ministry of Housing, Physical Planning
and Environment
P.O.Box 450
2260 MB LEIDSCHENDAM
THE NETHERLANDS

Vachon, J.
Environment - Quebec
723, Rue Marly (4th floor)
3900 STRASBOURG - SAINTE FOY
QC-GIX 4E4 FRANCE

Vakkuri, T.
Helsinki City Water and Sewage Works
Water Research Office
Kuninkaantammentie 11
00430 HELSINKI
FINLAND

Vasseur, P.
Centre des Sciences de l'Environment
Universite de Metz
57000 METZ
FRANCE

Veenendaal, G.
Netherlands Waterworks Testing
& Research Institute KIWA N.V.
P.O. Box 1072
3420 BB NIEUWEGEIN
THE NETHERLANDS

Villeneuve, D.
Environmental Contaminants Section
Bureau of Chemical Hazards
Health Protection Branch
Ottawa, K1A OL2
CANADA

Waring, M.
Department of Health and Social Security
Room 907, Hannibal House
Elephant and Castle
LONDON SE1 6TE
UNITED KINGDOM

Warren, S.
Water Research Centre
Medmenham Laboratory
P.O. Box 16 Marlow
BUCKINGHAMSHIRE SL 7 2HD
UNITED KINGDOM

Wigilius, B.
Department of Water in Environment
and Society
Linkoping University
S-8183 LINKOPING
SWEDEN

Witholt, B.
Rijksuniversiteit Groningen
Lab. voor Biochemie
Nijenborgh 16
9747 AG GRONINGEN
THE NETHERLANDS

Ziglio, G.
University of Milano
Istituto di Igiene
Via F. Sforza 35
20122 MILANO
ITALY

Zoeteman, B.C.J.
National Institute of Public Health
and Environmental Hygiene
P.O. Box 1
3720 BA BILTHOVEN
THE NETHERLANDS

WELCOME

H. Cohen

Director-General National Institute of Public Health and Environmental Hygiene, P.O.Box 1, 3720 BA BILTHOVEN, The Netherlands

Ladies and Gentlemen

It is a great honour to heartily welcome all of you to this International Symposium on Organic Micropollutants in Drinking Water and Health, which is co-sponsored by the Environmental Protection Agency of the United States, the National Institute of Public Health and Environmental Hygiene and The Netherlands Waterworks Testing and Research Institute.

I especially welcome you, Mr. Evers, deputy director-general for Environmental Protection, for your willingness to present the Opening Address of Minister Winsemius, who unexpectedly had to be present elsewhere.

I would first like to take the opportunity to put this symposium as well as the National Institute for Health and Environmental Protection in perspective.

In the past years many opportunities have been sought to promote an intensive exchange between drinking water experts on both sides of the Atlantic Ocean. One of the important examples in this context is the NATO-CCMS Drinking Water Pilot Study, which was so successfully carried out under the project leadership of Dr. Jo Cotruvo, who also is one of the major driving forces behind the organisation of this symposium.

Although not explicitly stated as such, this symposium and the earlier symposium on Drinking Water and Health, which was held in Noordwijkerhout in 1980, are good examples of the co-operative activities in environmental protection which is the aim of the Memorandum of Understanding between the United States and The Netherlands. International co-operation is essential, particularly in this time of budget cuts, to exchange experiences and recent research findings as well as to learn from each other's overall approaches to the management of health hazards of the potable water supply.

Human health has been a major subject of our Institute since the early years of this century. RIVM is the National Institute for Public Health and Environmental Hygiene and was founded on 1st January, 1984. It is mainly the result of a fusion between the National Institute for Public Health and the National Institute for Water Supply. The new institute, which employs about 1300 people, provides a better basis for integrated environmental research and for a close co-operation between, for instance, toxicologists, microbiologists and chemists. Its program is financed on a yearly basis by two Ministries, one responsible for Public Health and the other for Environmental Protection.

It is my personal opinion that the roots of the potable water supply industry go back to the medical profession in the past century. The conviction that hygiene circumstances had to be improved, paved the way for the sanitary engineer to construct his pumping stations, sand filters and distribution systems. Since the sanitary engineer has been very successful and water supply has rapidly extended over the world, the medical profession in our industrialized nations has more or less lost its interest in drinking water as a vector of disease. Yet potable water supply is a potential major vehicle of toxic chemicals and particularly of infectious diseases. Bacteria and viruses are living organisms. They adapt themselves and are capable of suddenly creating health risks not known before. A very important example is AIDS, and more closely related to water supply the recent outbreak of Legionnaires disease in the UK. The constant control of these potential risks by disinfection techniques is the primary reason for the existence of the potable water supply industry.

As Dr. Schaeffer, one of our former health inspectors said: "a good public water supply is a blessing, a bad one is a curse!" One minor fault may lead to the distribution of contaminated drinking water to millions of houses. The price for microbiological safety of water is often the presence of by-products of chemical disinfection. These by-products are one of the main issues for the discussions of the coming days. The halogenated organics formed increase the risk of tumor induction in the population to a level which is difficult to quantify and to regulate. And even if this risk is well taken care of, the public may not be satisfied. As Dr. Cotruvo stated in a recent paper which was brought to my attention: "Not only should public water supplies be safe, but it should also be perceived as safe". But here we probably go beyond the scope of this symposium, which will primarily deal with the assessment of the risk of organic chemicals in drinking water. The new administrator of EPA, Dr. Lee Thomas, said three weeks ago at the occasion of the second US-Dutch international symposium on aerosols in Williamsburg, that "it is in the process of risk management that science and policymaking most visibly come together".

Against this background I thank you, Mr. Evers, that you are here to give us on behalf of Minister Winsemius the view of the policymaker on our problem of today.

OPENING ADDRESS

F.W.R. Evers
Deputy of the Minister of Housing, Physical Planning and Environment,
P.O.Box 450, 2260 MB LEIDSCHENDAM, The Netherlands

Ladies and gentlemen,

The positive aspects of the introduction of a central drinking water supply in relation to public health are best illustrated by the successful abatement of water-borne disease outbreaks caused by bacteriological contaminated surface water at the end of the 19th century.

Since the last decades, people became more and more concerned about the increasing deterioration of the quality of surface water and groundwater by contamination with organic and inorganic pollutants. As a result of increased industrialization and intensification of agriculture new hygienic problems arose due to the introduction of many new potentially toxic chemicals in the environment. It became apparent that legal measures have to be taken to control emissions and to protect drinking water resources. In order to guarantee the quality of drinking water it became necessary to extend the legally prescribed systems for quality control and to set standards for a great number of parameters. It must always be kept in mind that the choice of these parameters was based on the scientific knowledge about prevalence of contaminants in drinking water and recognized risks of contamination, available at that time. In the past it has been shown that experience gained from the necessary activities to protect the drinking water quality, has broad implications for environment protection policies as well. Evaluation of health aspects of contaminants in drinking water is an integral part of the necessary activities. But it is a very difficult task to estimate the potential health risks of lifetime consumption of drinking water containing low concentrations of various pollutants.

The scope of this symposium includes the exchange of information on the inventorisation, behaviour and effects of organic micropollutants in drinking water. Therefore I should like to recall some of the achievements made in this field during the last two decades and mention also some new developments in the policy for the protection of water catchment areas. Subsequently I will discuss some major aspects such as the quality of water sources, water purification and drinking water distribution.

Organic micropollutants in drinking water can originate from the raw water source such as ground water or surface water but can also be formed during treatment processes like chlorination or by leaching from distribution materials such as plastic or bitumen lined pipes.

With respect to the contamination of water sources with organic micropollutants one must realise that despite the rapid development of advanced analytical instruments, chemists nowadays are capable of identifying only 10 to 20% of the organic micropollutants present in water.

Although the individual concentrations of most of the about 600 identified organic compounds in drinking water are rather low, concentrates prepared on XAD resins were demonstrated to have mutagenic activity in the Ames test.

Those related responses were observed with concentrates corresponding to 0,5 to 3,0 liters of drinking water.

International negotiations for the prevention of the pollution of surface water by organic micropollutants in view of its use for potable water supply were initiated at the beginning of the seventies and resulted in several directives of the Council of the European Communities.

The positive effects of these directives have been demonstrated by a decrease of the contamination of the river Rhine with a number of organic and inorganic micropollutants. Unfortunately these positive developments are not found to the same extent for certain macro contaminants like chlorides, phosphates and nitrates. Recently, a list of priority pollutants with respect to the pollution of surface water has been proposed which enables member states of the European Communities to initiate concerted actions in order to abate this kind of water pollution.

Groundwater is generally believed to be the most reliable source for the preparation of drinking water, however since the discovery of some major soil pollution events like Lekkerkerk people became aware of the vulnerability of groundwater.

In The Netherlands about 60% of the total amount of the distributed drinking water is derived from groundwater. For a small densely populated country like The Netherlands with a total yearly drinking water demand of over 1000 million m^3 this means that careful planning and protection of the 250 existing groundwater catchment areas is necessary.

Since the end of the seventies the quality of these 250 groundwater pumping stations is regularly investigated by the National Institute of Public Health and Environmental Hygiene. This yearly groundwater survey programme has led to the discovery of a few groundwater catchment areas contaminated with organohalogenated compounds like tri and perchloroethylene. Recently the draft legislation for soil protection was adopted in Parliament. Special regulations based on this law are being prepared at the moment for the protection of groundwater catchment areas. In this respect protection zones, with water residence times of 60 days, 10 years and 25 years, have been proposed. In this way a differentiation can be made for different kinds of activities like application of pesticides or the building of petrol stations in the vicinity of water catchment areas.

The actual human exposure to organic micropollutants present in drinking water is not only determined by the kind of water source (ground water, surface water, bankfiltered surface water) but also by the applied purification techniques.

Some steps in the purification processes, like chlorination for instance, may be responsible for the introduction of organohalogen compounds, which are formed by the reaction of chlorine with humic acids. The most common purification techniques for groundwater include aeration, deacidification and sometimes softening whereas for surface water more complicated techniques are used like coagulation, slow sand filtration, active coal treatment, softening and chlorination. In general many combinations of such purification steps are possible including also advanced and expensive techniques like ultrafiltration.

New techniques which are being developed nowadays include the denitrification of groundwater and deep-well-infiltration which is used for underground storing and purification of water. Progress has also been made in the development of alternatives for chlorine disinfection. In this respect it can be mentioned that the toxicological aspects of the application of chlorine dioxide are being evaluated.

Generally speaking the present available water-treatment techniques enable the waterworks to deliver a good and safe quality of drinking water provided that the internationally accepted standards for the basic quality of surface water used as a source for drinking water are being met.

Apart from the quality aspects of water sources and purification techniques careful attention is needed to prevent the introduction of hazardous organic micropollutants during the distribution of water. In some cases plastic pipes (or materials used to connect them) were shown to leach organics into the water.

A special problem in connection with the application of PVC and polythene pipes is the possibility of permeation of these pipes with organic solvents from accidental spills, waste dumping or the application of disinfectants like for instance methylbromide in horticulture.

Ladies and gentlemen, in addition to the items mentioned and discussed during the symposium it is evident that drinking water preparation and distribution demands constant attention as was recently emphasized again with the discovery of outbreaks of Legionnaires disease. In this respect the distribution of drinking water is back in the phase where it all began : the care of distributing water of safe and hygienically reliable quality.

It is therefore that I look forward with special interest to the recommendations prepared by your symposium.

I hereby open this International Symposium on Organic Micropollutants in Drinking Water and Health and I wish you a very successful meeting.

Organic Micropollutants in Drinking Water: An Overview

Joseph A. Cotruvo, U.S. Environmental Protection Agency
Office of Drinking Water

ABSTRACT

Biological contamination is still the most significant public health risk from drinking water even in industrialized countries. High potential for organic chemical transport to drinking water continues to exist even with source protection because of the multitude of chemical types and quantities. Drinking water is usually not a unique source nor the most significant contributor to total exposure from synthetic organic chemicals but it might be one of the most controllable.

The major public concern with drinking water contamination has been possible contribution to cancer risks from organic micropollutants. Even though the actual risks are probably small in most cases it is clearly within the public interest to prevent adulteration of water supplies and to protect their quality for the future so that these concerns or risks can be avoided.

A risk assessment/management decision model is suggested which may assist the process of making rational assessments of these contamination problems and control decisions that consciously consider all of the available data in a consistent manner.

The U.S. Environmental Protection Agency (1) is pleased to be associated with the Netherlands National Institute of Public Health and Environmental Hygiene, and the Netherlands Waterworks Testing and Research Institute Ltd. and the World Health Organization in organizing this International Symposium on Organic Micropollutants in Drinking Water and Health. This is a subject of immediate and continuing interest especially in industrialized countries and among all persons and organizations that are concerned with the protection of water resources, production of drinking water, and protection of public health.

Within the last ten years the extension of analytical detection methodologies to water has resulted in identification of hundreds of natural and synthetic organic chemical constituents in rivers, lakes, groundwaters and finished drinking waters. These techniques are now capable of detecting some substances at parts per trillion levels. It is likely that many of these recent detections are actually identifying contamination that has existed for some time. The concentration of any given substance is usually quite low (parts per billion or less), but the number and variety of substances detected in a particular water body can be large and generally unpredictable.

Public interest lies in both the protection of the quality of the water environment but also in protection of public health, which may be at risk from excessive exposures to contaminants from inhalation of purged volatile substances from water into the air, residues that may accumulate in the fish, dermal contact with processed water, inhalation of volatiles transported to indoor air from showering and other water uses in the home, and ultimately from direct consumption of contaminated drinking water.

The significance of this source of contaminant exposure relative to human health risks is not completely known, but in general it is not likely to be large in terms of numbers of cases of disease per million persons per lifetime, nor large relative to other risks to public health and safety. Nevertheless there are examples of egregious water contamination that have been associated with improper waste disposal practices. There is great public demand to protect the quality of our water resources for the future and to avoid the unnecessary risks that may be associated with involuntary exposure to environmental contaminants.

As an introduction to the Symposium this paper will briefly describe the types and sources of these organic micropollutants in drinking waters, management and control opportunities, risk assessment issues, and some recent ideas concerning a regulatory management decision model regarding differentiation between substances based upon their possible contributions to human cancer risks.

Sources of Organic Micropollutants in Drinking Water

Contamination of drinking water sources by organic micropollutants may occur from numerous sources including: natural products, by-products of the chemical reaction between natural products and the oxidizing agents used in the treatment of drinking water and waste water, commercial and industrial production and use activities, and waste disposal. (2)

Water is usually neither a unique source nor the most significant contributor to human exposure to these substances, with the principal exception of the by-products of interaction between natural products and treatment chemicals. Occupational exposures to chemicals in production may be three to six orders of magnitude greater than drinking water. Inhalation of ambient air is a continuing and general source of exposure to halogenated and petroleum hydrocarbons often even in rural areas. Residues of many pesticides in foods is an exposure source for the general population.

- Natural Products in Water

These diverse and mostly unidentified components of all natural waters to a greater or lesser degree include broad classes such as humic or fulvic acids which are relatively refractory by-products of decomposition; tannins; terpines; amino acids and proteins and a variety of other nitrogen and sulfur containing substances. Because of their complexity, variety and water solubility these are usually not identified individually but rather in surrogate general parameters such as total organic carbon (TOC), total organic nitrogen (TON), color, or absorbance spectra. Total concentrations range from about 0.1 mg/l in deep groundwaters to 1 to 5 mg/l in many surface waters and 20 to 25 mg/l in some highly colored surface and groundwaters.

- Treatment By-products

The most studied treatment generated products are those derived from chlorination; however, all chemically active treatment chemicals including ozone, chlorine dioxide, iodine and chloramines are known to produce reaction products in water. The most familiar identified substances include chloroform and other trihalomethanes, chlorinated and brominated acetonitriles, chloropicrin, halogenated phenols which often produce undesirable tastes and odors, and numerous acids, aldehydes and ketones. Quantities that have been detected in finished drinking water range from several hundred micrograms per liter for trihalomethanes to a few micrograms per liter or less for most of the other substances.

- Industrial Products

A large number of halogenated hydrocarbon solvents, aerosol propellants and refrigerants including trichloroethylene, tetrachloroethylene, carbon tetrachloride, methylene chloride and freons are produced worldwide at rates of several billion kilograms per year. They tend to be the most frequently found synthetic contaminants in drinking water because of their huge production, highly delocalized use patterns, chemical and biological stability, volatility, and negligible adsorption to soils and sediments. Drinking water contamination ranging from 0.1 ug/l to several milligrams per liter has been reported with groundwater being the most significant and recent concern. Results of the 1982 Ground Water Supply Survey are contained in Table I (3).

Literally hundreds of other synthetic organic chemicals have been reported at least once in drinking water. Individual frequencies and concentrations are usually at parts per billion or less in finished drinking water. Principal contamination potential seems to be associated with surface waters

TABLE 1

Summary of Ground Water Supply Survey Occurrence Data

Random sample, n= 466

Chemical	Quantification Limit (ug/l)	Positives Number	Percent	Median (ug/l)	Maximum (ug/l)
Tetrachloroethylene	0.2	34	7.3	0.5	23
Trichloroethylene	0.2	30	6.4	1	78
1,1,1-Trichloroethane	0.2	27	5.8	0.8	18
1,2-Dichloroethane	0.2	18	3.9	0.5	3.2
1,2-Dichloroethylenes (cis and/or trans)	0.2	16	3.4	1.1	2
Carbon tetrachloride	0.2	15	3.2	0.4	16
1,1-Dichloroethylene	0.2	9	1.9	0.3	6.3
m-Xylene	0.2	8	1.7	0.3	1.5
o+p-Xylene	0.2	8	1.7	0.3	.9
Toluene	0.2	6	1.3	0.8	2.9
1,2-Dichloropropane	0.2	6	1.3	0.9	21
p-Dichlorobenzene	0.5	5	1.1	0.7	1.3
Bromobenzene	0.5	4	.9	1.8	5.8
Ethylbenzene	0.5	3	.6	0.8	1.1
Benzene	0.5	3	3	3	15
1,2-Dichloroethane	0.5	3	.6	0.6	1
Vinyl chloride	1	1	1.1	1.1	1.1
1,2-Dibromo-3-chloropropane	5	1	5.5	5.5	5.5
1,1,2-Trichloroethane	0.2	0	---------	---------	---------
1,1,1,2-Tetrachloroethane	0.2	0	---------	---------	---------
1,1,2,2-Tetrachloroethane	0.5	0	---------	---------	---------
Chlorobenzene	0.5	0	---------	---------	---------
n-Propylbenzene	0.5	0	---------	---------	---------
o-Chlorotoluene	0.5	0	---------	---------	---------
p-Chlorotoluene	0.5	0	---------	---------	---------
m-Dichlorobenzene	0.5	0	---------	---------	---------
o-Dichlorobenzene	0.5	0	---------	---------	---------
Styrene	0.5	0	---------	---------	---------
Isopropylbenzene	0.5	0	---------	---------	---------

receiving industrial discharges and groundwaters in the vicinity of waste disposal sites.

- Petroleum Products

Water contamination from gasoline and other refined fuels and solvents is becoming a more frequent occurrance. These include especially gasoline and its components such as benzene, toluene and xylenes which are enriched in non leaded gasoline, and which tend to be more highly mobile than many other hydrocarbons in the aqueous environment. The most significant problems appear to be during transportation (spills and leaks from pipelines) and storage. In the United States we are just beginning to assess the contamination potential from Leaking Underground Storage Tanks known as LUST. More than 1,000,000 underground gasoline storage tanks are in use and since they are subject to damage from mechanical stress and corrosion the ultimate potential for leaks and environmental contamination is great.

- Pesticides

Hundreds of pesticides totaling hundreds of millions of pounds per year are in use worldwide. Some such as alachlor are applied above ground and they may migrate to surface waters by runoff or by evaporation and deposition in rainfall. Irrigation return waters can contaminate surface and groundwaters. Some herbicides such as atrazine are highly mobile in water and are frequently found in surface waters. Nematocides such as aldicarb, and dibromochloropropane are being found in ground waters. Most of this water contamination appears to be the result of uses that are currently legally acceptable rather than from manufacturing or disposal. Until recently it was mistakenly assumed that many of these chemicals were sufficiently biologically or chemically degraded or immobilized in soils so that surface and groundwater contamination would not occur. Many of those assumptions must now be reexamined.

- In Situ Transformation

Biological and chemical transformations of synthetic chemicals in the environment is a well known phenomenon (eg. DDT DDD and DDE) and often relied upon as a detoxification process. However this is not necessarily always the case. For example aldicarb oxidizes readily to aldicarb sulfoxide and then to aldicarb sulfone. The sulfoxide is more toxic than the parent and the sulfone is somewhat less toxic; all these forms are found in water. Of more widespread interest are the recent indications that vinyl chloride, 1,1-dichloroethylene and cis and trans-1,2-dichloroethylenes when found in groundwaters may be principally the result of degradation of higher chlorinated compounds such as 1,1,1-trichloroethane, trichloroethylene or tetrachloroethylene. This would be a case of generation of a product of significant carcinogenic risk (vinyl chloride) from less toxic precursors.

- Waste Disposal

The disposal of chemical wastes is a mounting problem in industrialized countries. Contamination of groundwaters from improper disposal practices has been identified as the most significant source of their present and future contamination.

Management of Contamination

- Source Protection

The most desirable means of assuring the quality and safety of drinking water would be to prevent contamination of source waters by potentially harmful organic chemicals. However, source protection at a high level may not always be possible for these substances. The problem is attributable to the potentially large number, variety and quantity of the contaminants and ubiquitous sources and uses. Some are capable of migrating to surface waters in solution and others adsorbed to particulates. Others can migrate to groundwater in solution or suspended but not retarded by soils.

- Water Treatment

Drinking water supplies will always need to rely on treatment processes tailored to the potential contaminants in the source to assure the safety of the finished water. Conventional treatment processes normally applied to surface waters, consisting of a particulate removal stage involving coagulation and filtration are reasonably effective for removal of a number of organic chemicals that are readily bound to sediments. These may include polynuclear aromatic hydrocarbons, polychlorinated biphenyls and some pesticides such as chlordane, and even chlorinated dioxins.

Aeration technology is a highly effective technology for removal of volatile synthetic organic chemicals, i.e. those with a high Henry's Law constant. These include such common contaminants as trihalomethanes, benzene, trichloroethylene and vinyl chloride. Packed tower aeration systems can be designed and operated to routinely achieve up to 99% removal of those types of substances.

Aeration results in transport of volatile substances from water into the ambient air so questions could be raised on its desirability and risks from inhalation. Both theoretical calculations and field studies have demonstrated that except for cases of extreme groundwater contamination, aeration results in negligible risks to persons in the vicinity. Air concentrations decline rapidly within only a few meters from the source and initial quantities and total loadings are very small at the same time. For example aeration of 1 million liters per day containing 100 ug/l of trichloroethylene at 99% efficiency results in total emission of only about 100 g/day.

Granular activated carbon (GAC) and to a lesser degree powdered activated carbons (PAC) are being used increasingly in both large and small public water systems. GAC can effectively remove many high and low molecular weight organic chemicals that are hydrophobic, including many volatile chemicals. The GAC must be periodically reactivated to remove accumulated chemicals and prevent breakthrough into finished water. The performance of a GAC system for organic chemicals is a function of numerous factors including contact time and the type of carbon. Carbons can be produced to have particular affinities for certain types of substances. Treatment processes can be simulated in the laboratory using adsorption isotherms to predict the design and expected performance of full scale treatment plants.

Numerous additional treatment techniques can be mentioned which are not currently widely used but which may have future potential for selective use, eg. adsorption on macroreticular resins, and oxidation with ozone and ultra-violet light. Small point-of-use units containing GAC have been shown to be effective in some applications for removal of organic chemicals at the consumer's tap or for a household water supply. Cases exist where with effective external management the point-of-use approach is an effective and efficient control measure.

- Legislative Controls

Legislation and regulations to protect drinking water supplies and their sources and to control waste disposal have been installed in many countries. For example in the USA the Water Pollution Control Act (also called the Clean Water Act) requires control of effluent discharges to surface waters through both technology requirements and permits. Ambient water quality is regulated through criteria related to ultimate use eg. agricultural or drinking water source.

Prevention of groundwater contamination is regulated in the U.S.A. by the Underground Injection Control programs of the Safe Drinking Water Act, and the hazardous waste controls under the Resource Conservation and Recovery Act. The Superfund program in the U.S.A. is a fund provided from a tax on chemical products that is used to mitigate and clean sites that have been contaminated by disposal of chemical wastes. Approximately three fourths of these cases to date have involved actions to recover or protect current or potential groundwater drinking water supplies.

- Drinking Water Standards and Guidelines

Drinking water standards and guidelines have been recently prepared or updated by several international organizations and individual countries. The European Economic Communities' Directive on drinking water quality was proposed in 1975. The World Health Organization in 1984 published Guidelines for Drinking Water Quality for 20 organic chemicals. A WHO guideline represents the level of a constituent that ensures an esthetically pleasing water and does not result in any significant risk to the health of the consumer.

Canada revised its Drinking Water Guidelines in 1978 to include 21 organic chemicals and pesticides (5). It is now considering further revisions with an expected increase in the number of organic chemicals.

The USEPA has interim regulations in effect for ten organic chemicals including six pesticides and four trihalomethanes (6). Revised regulations are under development for up to 60 organic chemicals with emphasis on the volatile synthetic organic chemicals and solvents and pesticides.

- Monitoring

Monitoring for organic chemicals becomes increasingly complicated and costly as the number of chemicals of interest increases and the detectable levels become lower with improvements in analytical and concentration techniques. With this explosion in the quantity of data that can be obtained, it is essential to pay greater attention to the quality of that data. Any laboratory that produces such data must maintain a rigorous quality control program, and adequately document all aspects of the analyses. Accuracy and precision information must be part of every data report.

In addition to numerical limits for a number of organic chemicals, the U.S.A. is also planning to institute monitoring and reporting requirements for all public water systems in two phases. The first phase would include up to 50 volatile synthetic organic chemicals and the second would include an unspecified number of other synthetic organic chemicals with special attention paid to pesticides.

In a separate action EPA's Drinking Water and Pesticide offices are now designing a comprehensive survey of pesticides in public and private groundwaters. This survey will study those pesticides which because of chemistry or use pattern are suspected of migrating to groundwaters. The study to begin in 1986 will examine the environmental conditions that affect the migration as well as drinking water concentrations so that future pesticides registrations will be tailored to prevent such contamination.

Risk Assessment

Having identified the sources and extent of drinking water contamination, public health officials must determine the type and extent of the risks involved and then make judgements on the most reasonable method for alleviating the risk. These two activities are called risk assessment and risk management, respectively.

The principal concerns raised about the contamination of drinking water by trace amounts of synthetic organic chemicals has been the possible increased risk of cancer in the population. There has also been some discussion on the possibility of reproductive or teratogenic risks: however these have been generally treated as threshold effects that occur at chemical doses that are usually higher than those received from drinking water. Public health officials usually assume conservatively that substances which have been shown to cause cancers in animals may operate by non threshold mechanisms. It then follows that some albeit small risk of cancer is theoretically assumed to exist at any dose above zero.

Because the normal concentrations of potential "carcinogens" in drinking water are very small, the postulated risks are largely empirically unquantifiable. If they do infact exist, they are small in magnitude relative to overall cancer risk rates.

Projections of risks from chemicals in drinking water are attempted using two principal means: retrospective epidemiology studies of exposed populations and surrogate toxicology studies using experimental animals.

- Epidemiology

Numerous epidemiology studies have been made on populations exposed to organic chemical contaminants in drinking water. Most of these have examined possible correlations between THMs or some related parameter and incidents of cancer of various types; a few have dealt with cases of water contamination from chemical wastes.

The earlier THM studies sometimes reported correlation between THMs and risks of bladder cancer. Later studies have focused mainly on colon/rectal cancer risk, although no consensus has been reached on whether these studies have discovered a true exposure risk relationship or just fortuitous correlations. The basic problem with attempting to correlate drinking water contamination and cancer risks are that the theoretical risks from drinking water are small relative to the overall cancer risks in the population, thus the confounding factors are always very difficult to distinguish and quantify. Thus epidemiology studies have not generally been effective in helping decision makers to qualitatively and quantitatively assess drinking water risks from organic micropollutants.

- Animal Experimentation

Apart from those few cases (eg. vinyl chloride and benzene) where a substance has been shown to be a human carcinogen, controlled feeding studies (bioasays) using rats and mice exposed to high chemical doses have been the principal means of making judgements of the likelihood that a chemical may be carcinogenic to humans. Groups of about 50 animals per sex and dose with suitable controls are exposed for most of their lifetimes to 2 or 3 doses of the test chemical by food, drinking water, corn oil gavage or by inhalation. The highest dose employed is the Maximum Tolerated dose (MTD); the highest dose at which no other significant adverse health effect should occur. The human risk assessment requires mathematical conversion of the dose/response effect in the high dose animal test to the projected dose/response risk in humans at environmental exposures.

Numerous mathematical models have been developed in attempts to estimate potential risks to humans from low dose exposure to carcinogens; each of them incorporates numerous unverifiable assumptions. Low dose calculations are highly model dependent leading to widely differing risk conclusions and none of the models can be firmly justified on either statistical or biological grounds.

One hit and a multistage model that is linear at low doses are commonly used among the many choices because they tend to be conservative - unlikely to underestimate the actual risk.

A recent series of analyses has been performed which illustrates the range of uncertainties inherent in risk calculations for a number of volatile synthetic organic chemicals. These uncertainties range from the inability to precisely estimate the exposure of the population crossection, to interpretation of the scientific data, to uncertainties in the models. It was suggested that uncertainties in the toxicological data may be in the range of 1 to 3 orders of magnitude; uncertainties in the choice of model may exceed 6 orders of magnitude. Figure I illustrates a comparison of 4 models applied to the same experimental data for trichloroethylene (7).

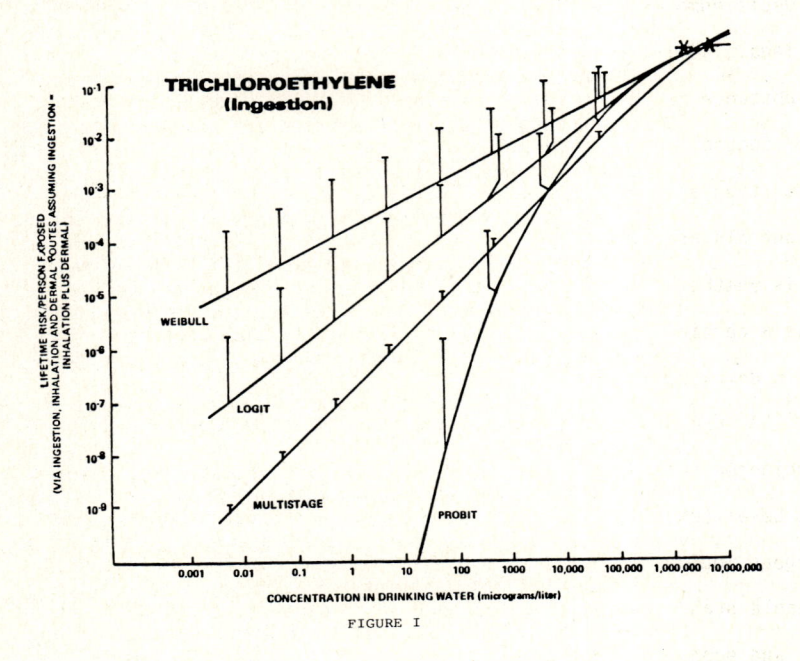

FIGURE I

Note that at a drinking water concentration of about 50 ug/l the projected incremental lifetime risks range from 10^{-9} (probit) to 10^{-3} (Weibull). The decision-maker must as a matter of judgement choose which model(s) to apply, which assumptions to incorporate and which "acceptable" risk to allow. These huge variations aptly demonstrate that these seemingly systematic and sophisticated calculation techniques must be used with great caution and skepticism. They are, however, the only quantitative tools available at this time.

Risk Management Decisions

Risk management is the public process of deciding what actions to take where risk has been determined to exist. It includes integration of risk assessment with consideration of engineering feasibility and a determination of how the decisionmaker may apply imperatives to reduce risk in light of all of the legal, social, economic and political factors (8).

If a substance is considered to be carcinogenic to humans, the most conservative control approaches must be considered, whereas if it is a threshold toxicant, then an acceptable daily intake can be calculated and less stringent controls are possible. Thus the most critical initial judgement to be made is whether or not a sufficient body of evidence exists to treat a substance as a carcinogen (non-threshold toxicant) for the purposes of regulation in drinking water.

In the USA, EPA has concluded that the ideal goal for drinking water is that no carcinogen should be present. This is called a Recommended Maximum Contaminant Level (RMCL) which is not a legally enforceable standard. RMCL for carcinogens would thus be zero (9). Maximum Contaminant Levels (MCLs) are enforceable standards and they must be set as close to the RMCL as is technically and economically feasible.

Identification of Probable Human Carcinogens (2)

The following section is a decision model proposal for systematically determining the appropriate goal level based upon health consideration. It involves use of a ranking system for evaluation of carcinogenicity evidence which has been developed by EPA based upon a system used by the International Agency for Research on Cancer (IARC)(10). Based upon the ranking the substance is placed in one of three regulatory classes and the RMCL is determined from selection of the appropriate data base.

- IARC Classifications

IARC has provided guidelines for assessing the epidemiological and animal toxicology data base leading to a conclusion of the strength of the evidence of the carcinogenicity of numerous substances. Substances are classified into three groups as follows:

IARC Criteria

Group 1 - Chemical is carcinogenic to humans (sufficient evidence from epidemiological studies).

Group 2 - Chemical is probably carcinogenic to humans.

Group 2A - At least limited evidence of carcinogenicity to humans.

Group 2B - Usually a combination of sufficient evidence in animals and inadequate data in humans.

Group 3 - Chemical cannot be classified as to its carcinogenicity to humans.

USEPA has recently proposed a five category scheme that is an expansion of the IARC criteria (11). The primary difference is that the IARC three categories do not distinguish between those chemicals with "inadequate" animal evidence of carcinogenicity and those with no evidence of carcinogenicity.

Group A - Human carcinogen (sufficient evidence from epidemiological studies).

Group B - Probable human carcinogen.

Group B1 - At least limited evidence of carcinogenicity to humans.

Group B2 - Usually a combination of sufficient evidence in animals and inadequate data in humans.

Group C - Possible human carcinogen (limited evidence of carcinogenicity in animals in the absence of human data).

Group D - Not classified (inadequate animal evidence of carcinogenicity).

Group E - No evidence of carcinogenicity for humans (no evidence for carcinogenicity in at least two adequate animal tests in different species or in both epidemiological and animal studies).

Both of these classification schemes are based upon a qualitative review of all available evidence, including: short term tests; long term animal studies; human data; pharmacokinetics; comparative metabolism; and structure/activity.

After the chemicals are examined and ranked based upon the qualitative evidence, the next step in the model process is to establish a 3 level classification for quantitative regulatory decisions as follows:

Three-Category Approach for Setting RMCLs

Category I - Known or probable human carcinogens: Strong evidence of carcinogenicity.

- o EPA Group A or Group B
- o IARC Group 1, 2A or 2B

Category II - Equivocal evidence of carcinogenicity.

- o EPA Group C
- o IARC Group 3

Category III - Non-carcinogens: Inadequate or no evidence of carcinogenicity in animals.

- o EPA Group D or E
- o IARC Group 3

Category I includes those chemicals which have sufficient human or animal evidence of carcinogenicity to warrant their control as known or probable human carcinogens. The carcinogenicity of these compounds has not been demonstrated to exhibit a threshold and thus it can be assumed that any exposure could contribute some finite level of risk.

Category II includes those chemicals for which some limited but insufficient evidence of carcinogenicity exists from animal data. Goals should reflect the fact that some experimental evidence of carcinogenicity in animals has been reported although not verified.

There are two main options for obtaining a quantitative goal for these types of chemicals. The first consists of calculations based upon non-carcinogenic endpoints (ADI) according to the adequacy of the data and toxicological principles. To account for the equivocal evidence of carcinogenicity an additional factor could be applied (e.g., ADI divided by a factor of 10 or some other value). The second option consists of a calculation based upon a lifetime risk in the range of 10^{-5} to 10^{-6} using a conservative method such as the linear multi-stage model. This is intended to be protective in a conservative manner for compounds that are not being treated as carcinogens, but for which questions have been raised concerning potential effects. Either option may be employed depending on the quality of the data.

The goals for Category II chemicals should be less conservative than those for Category I chemicals and more conservative than those for Category III chemicals and that this is reflective of the quantity and quality of the toxicology data that are available.

Category III includes those substances with inadequate or no evidence of carcinogenicity. Goals would be calculated based upon chronic toxicity data using ADIs. This approach is well accepted in the scientific community as a method for determining acceptable exposure levels for threshold toxicants.

VOCs Using Three Category Regulatory Approach

Category I - Known or Probable Human Carcinogens: Strong evidence of carcinogenicity.

Benzene
Vinyl chloride
1,2-Dichloroethane
Trichloroethylene

Category II - Equivocal Evidence of Carcinogenicity.

Tetrachloroethylene
1,1-Dichloroethylene

Category III - Non-Carcinogens: Inadequate or no evidence of carcinogenicity in animals.

1,1,1-Trichloroethane
p-Dichlorobenzene

The foregoing suggested risk assessment/management decision model provides a rationale and methodology for using all of the experimental data and reasonable judgements for control decisions on potentially harmful chemicals in the environment. The quality and quantity of the toxicology data are considered along with technological and economic factors.

REFERENCES

1. The opinions expressed in this paper are those of the author, and they do not necessarily reflect the policies of the U.S. Environmental Protection Agency.

2. Cotruvo, J.A., "Risk Assessment and Control Decisions for Protecting Drinking Water Quality", working paper prepared for the International Agency for Research on Cancer Working Group on Drinking Water and Cancer Risks, Lyon, December 14, 1985. In press.

3. Proposed Regulations: National Primary Drinking Water Regulations: Volatile Synthetic Organic Chemicals, Federal Register Vol. 49, No. 114, June 12, 1984.

4. Guidelines for Drinking Water Quality, Vol. 1, World Health Organization, Geneva, 1984.

5. Guidelines for Canadian Drinking Water Quality, H 48 10/1978, Canadian Government Printing Centre, Hull, Quebec, Canada.

6. National Interim Primary Drinking Water Regulations: Vol. 44(231) pp. 68624-68705, 1979; and 1975.

7. Cothern, C.R., Coniglio, W.A. and Marcus, W.L., in Assessment of Carcinogenic Risk to the U.S. Population due to Selected Volatile Organic Compounds from Drinking Water via the Ingestion, Inhalation and Dermal Routes. Draft Report, USEPA, Officr of Drinking Water, Criteria and Standards Division 1984.

8. Ruckelshaus William D. "Risk in a Free Society", EPA Journal, P. 12, April 1984.

9. Proposed Regulations: National Primary Drinking Water Regulations: Volatile Synthetic Organic Chemicals, Federal Register Vol. 49, No. 114, June 12, 1984.

10. "Evaluation of the Carcinogenic Risk of Chemicals to Humans", IARC Monographs Supplement 4, Lyon, France pp. 1114, October 1982.

11. U.S.E.P.A. 1984, Proposed Guidelines for Carcinogen Risk Assessment; Request for Comments Federal Register 49(227): 46294-46301

IMPACT OF DIFFERENT TYPES OF ORGANIC MICROPOLLUTANTS PRESENT ON SOURCES OF DRINKING WATER ON THE QUALITY OF DRINKING WATER

H. SONTHEIMER, H.-J. BRAUCH and W. KÜHN

INTRODUCTION

Following the observation some years ago, that disinfection with chlorine can lead to the formation of THM-substances (Rook, 1974), the subject of organic micropollutants in drinking water has received a lot of public interest. This is owing to the fact, that some of the micropollutants formed during chlorination and other compounds found in the raw water used for drinking water preparation may have some impact on human health, as these substances may be carcinogenic etc. (Drinking Water and Health, 1980, 1982) Since the first publications from Rook and others (Rook 1974) about the THM-formation many studies have been made on micropollutants in treated drinking water, which have not been removed by the usual treatment processes.

In order to give an overview on the different impacts observed in connection with such pollutants one should consider the following question: How do we define organic micropollutants? A first answer to this can be seen in the following table.

> Micropollutants are identifiable water constituents
> having a concentration in the µg/l range

Fig. 1

There is no doubt that we have to consider all those organics as micropollutants, which can be found in the µg/l range or in still lower concentrations, while the overall organics concentration in most of our drinking waters will be in the mg/l range. This means that micropollutants are defined on the one side through their low concentration. But on the other side, the word itself means, they must be pollutants. And as nobody can decide if a substance has to be counted as a pollutant or not without knowing the exact chemical formula and structure, we usually restrict the word micropollutant to those substances which can be analysed individually as single defined substances. But here some of the problems begin, which will be discussed in the following chapter.

MICROPOLLUTANTS AND ANALYSES

Water works directors and engineers very often say, that only the modern analytical methods like GC-MS, HPLC etc. are responsible for all publicity about the micropollutant problems. Nobody really understands, what it means, to have p.ex. 100 ng/l of an individual substance in a given water, but one hundred looks like a large number and therefore 100 ng/l seem to be much more than 0.0001 mg/l although both values means the same. On the other side we have to consider the following thesis:

> All dissolved organic substances present in sources of drinking water are micropollutants regarding their concentration. But only a very small percentage can be analysed individually. Therefore the unknown micropollutants may be most important.

Fig. 2

There can be no doubt, that all defined organic substances occuring in drinking water sources have concentrations below 1 mg/l, that means they are in the µg/l range and should be called micropollutants regarding their concentration. But are we allowed to say, that only these substances can be referred to as pollutants, which we can analyse individually? The following examples of this problem might help for a better understanding.

The next table gives some analytical data for micropollutants in the Rhine river and for some group and sum parameters on three different places along the river.

The data given in this table have been analysed in 1975 and show that we could determine in those years most of the nonpolar organic chlorine compounds as single substances. But these defined substances are only about 10 % of the overall chlorinated organics and only about 1 % of the total organics measured as DOC. While the surrogate parameters are increasing from Basel to Duisburg along the Rhine river, we very often find a decrease resulting from removal processes for some of the individual substances, especially for the volatile organics.

How can we conclude from these data, that only the well defined micropollutants are of real importance with respect to drinking water quality? What sort of data have to be used to determine for example, treatment efficiencies. We have got another important problem resulting from the large fluctuation in the river flow water. The following table gives some information about this question.

compound	Basel	Köln	Duisburg
Chloroform	1.1	0.8	1.1
Carbon Tetrachloride	0.3	1.4	3.3
Chloroethanes	2.5	0.2	0.6
Trichloroethylene	0.9	0.8	0.6
Tetrachloroethylene	0.3	0.8	1.5
Chlorobutadienes	4.4	0.4	0.3
Chlorobenzenes	0.9	0.6	5.3
Chlorotoluenes	–	0.5	1.3
Benzene	0.2	0.3	0.8
Toluene	0.8	0.7	1.9
Dimethylbenzenes	0.2	0.7	1.4
Trimethylbenzenes	0.1	0.4	1.0
Tetramethylbenzenes	–	0.1	0.5
$DOCl_{total}$	55	78	100
$DOCl_{nonpolar}$	10	5	15
DOC (mg/l)	2.3	4.6	5.0

concentration in µg/l

Fig. 3 Organic compounds in the Rhine river (1975)

Wasserchemie Karlsruhe   1985/3804

compound	$\bar{x}_g$	σ_g	$c_{95\%}$
Chlorobenzene	0.06	3.26	0.42
1,2–Dichlorobenzene	0.32	2.03	1.0
1,3–Dichlorobenzene	0.05	2.27	0.19
1,4–Dichlorobenzene	0.19	1.92	0.56
Trichlorobenzenes	0.04	2.46	0.18
Tetrachlorobenzenes	n.n.		
Chlorotoluenes (3)	0.06	2.50	0.27
Nitrobenzene	0.42	2.26	1.6
Nitrotoluenes (3)	0.25	2.19	0.91
Nitroxylenes	0.07	3.03	0.44
Methoxynitrobenzenes	0.09	4.45	1.1
Chloronitrobenzenes	0.32	3.24	2.2
Dimethylanilines	<0.05		
Ethyltoluidine	<0.05		
Nitrophenols	<0.05		
Dichlorophenols	<0.05		
Dichloroisopropylether	0.06	3.34	.44
Dichloroisobutylether	<0.05		
Diethylphthalat	0.23	2.77	1.2
Dibutylphthalates	0.81	1.66	1.9

concentration in µg/l

Fig. 4 Geometric mean values of organic micropollutants from composite samples over 2 weeks (Rhine river near Düsseldorf – 1984)

Wasserchemie Karlsruhe   1985/3806

These data don't result from samples taken immediately but from composite samples over 2 weeks. Nevertheless for these mixed samples we observe fairly big fluctuations. This can be seen from the results for the geometric standard deviation. If one multiplies or divides the geometric mean values with the standard deviation you can calculate the limits where 66 % of all data lie within. The $c_{95\%}$ value can be seen as a maximum concentration. Only 5 % of all data are higher than this value. Although none of the mean values are higher than 1 µg/l, the table shows that the maximum concentrations for some individual compounds will be higher than this general goal.

It should be mentioned here, that these fluctuations are fairly typical for all micropollutants in river waters. This can be seen from the next diagram too.

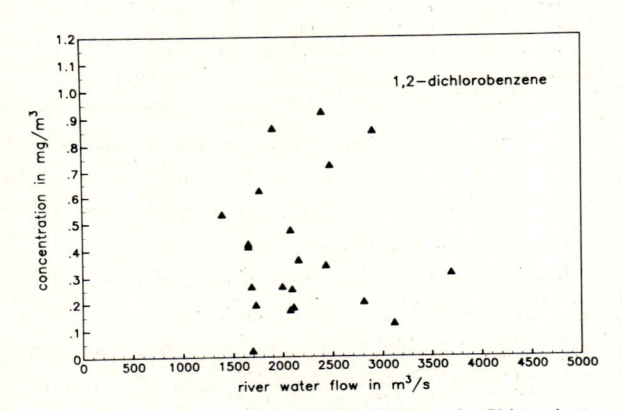

Fig.5 concentration of 1,2-dichlorobenzene in Rhine river water in dependance of water flow (Düsseldorf-1984)
Wasserchemie Karlsruhe   1985/3815

Here the concentrations of 1,2-dichlorobenzene are given as a function of the river water flow. One can see that the concentration fluctuations are so large here, that there seems to be no worthwhile correlation between the two parameters. The same has been found true for 1,4-dichlorobenzene and here we also don't see any tendency regarding seasonal changes etc. The fluctuations will even be greater, if we take separate samples.

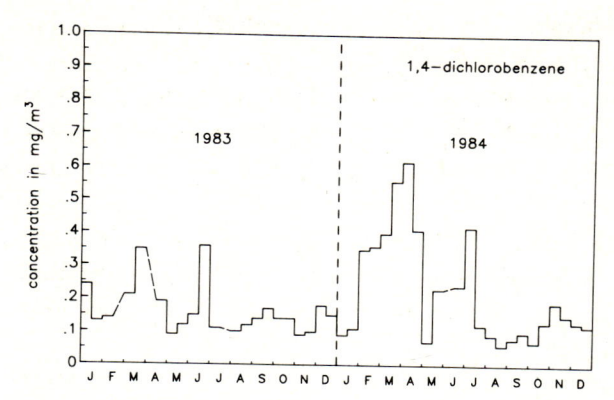

Fig.6 concentration of 1,4-dichlorobenzene in Rhine river water vs. time (near Düsseldorf)
Wasserchemie Karlsruhe   1985/3814

The data given here result from two-weekly samples. Therefore all sudden changes in the water quality cannot been seen from these data. This means that the sampling method is very important for the results and for the conclusion drawn out of those data. Usually we need the mean values for the evaluation of health impacts, especially in those cases where water treatment doesn't lead to a high removal efficiency. On the other side we need the maximum concentrations for treatment considerations, as we can have a much higher breakthrough for peak concentrations when using certain treatment steps like powdered activated carbon with a constant dose. It is for this reason that peak pollution may determine the medium concentrations in the treated water. Here it has been proven worthwhile to calculate the maximum values as the mean values for one day at normal water flow conditions. This proposal has first been made in the IWAR-memorandum (IAWR-memorandum 1973) and methods have been proposed to calculate these maximum concentrations for river waters (Tagungsbericht 1979).

All the considerations discussed till now may be summarized with the following conclusion:

>   Analytical control of single organic micropollutants may be insufficient to maintain a safe drinking water quality. Additional summary and group parameter measurements will always be required.

Fig. 7

If we want to get enough information of all organic micropollutants occuring in a river water used for drinking water purposes we always need summary parameters like DOC or COD on one side and group parameters like TOX or TOS etc. on the other side. Without these data we cannot be sure about the water quality.

These data allow furthermore decisions on the types of micropollutants we have to expect in a water source, as it usually will be impossible to determine all single substances which can occur as micropollutants.

This is especially true for river waters like the Rhine river water, where we are not able to analyse more than about 10 % of all micropollutants. In order to underline this general observation and to present some more information about it, some results of gel chromatography studies are given with the next figures.

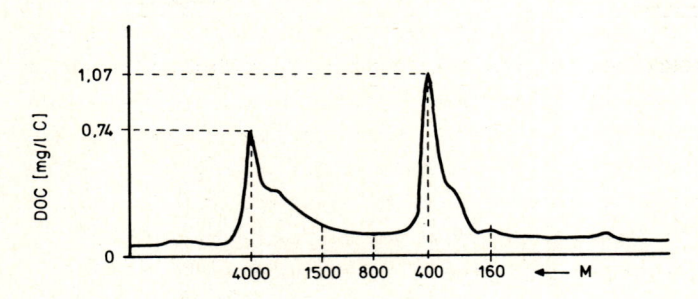

Fig. 8   **Molekülmassen - Verteilung Rhein bei Basel**
Wasserchemie Karlsruhe   1985 / 3801

The measurements presented in the figure have been made by Fuchs in our institute using TSK gels and DOC measurement for the effluent control after the TSK-gel-column. The procedure developed by Fuchs allows a fairly good separation, where not only the molecular weight is of importance but adsorbability to some extend too. For this the molecular weight data given within figure 8 cannot be used as correct figures in all cases but they give a rough estimate of the type of organics and their molecular weight distribution. One can see that we have roughly two fractions if we make measurements with the Rhine river water above Basel with about 50 - 60 % of humics or better fulvic substances. But there occur some lower molecular weight substances too, but here a part of this peak may result from better adsorbable substances.

If we do the measurement with the Rhine river in Karlsruhe one can see from the following figure 9, that here we find many more peaks. This increase in the number of peaks continues as we go north to Wiesbaden and then to the lower parts of the Rhine river at Düsseldorf and Orsay.

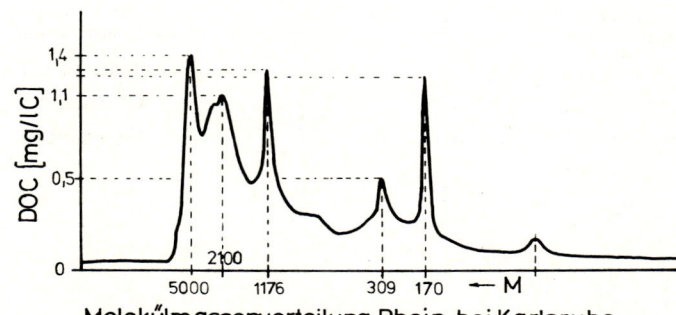

Fig.9 Molekülmassenverteilung Rhein bei Karlsruhe.

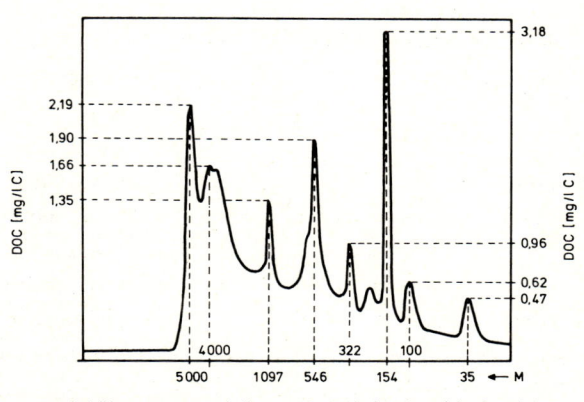

Fig.10 Molekülmassen-Verteilung Rhein bei Düsseldorf-Flehe

Wasserchemie Karlsruhe   1985 / 3803

Here we have many more groups of different substances, especially in the lower molecular weight fractions. It should be mentioned that the water samples analysed here didn't contain volatile organics. This means, that most of the usually analysed organic micropollutants couldn't be found in these waters as the enrichment has been partially done by stripping.

These waters still contain many different micropollutants as it can be seen from the next example. Here the gel separation has been done in a larger column so that samples for group-parameter measurements could be prepared. In addition DOC-, TOX- and TOS-values have been measured for 10 different fractions. The results are given in the following table.

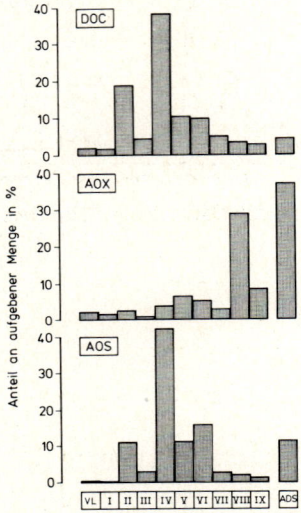

Fig.11 Molmassentrennung der trinkwasserrelevanten Stoffe der BASF Ludwigshafen: Verteilung von DOC, AOX, AOS auf die Fraktionen I bis IX (VL = Vorlauf, ADS = adsorbierter Anteil)

The data given for the composition of the different groups are oriented in the following way. Beginning on the left hand side are the high molecular weight substances, while on the right hand side are the low molecular weight organics and those which remain adsorbed onto the column material. All values are given in % of the total organics in the water. This water has been a biologically treated waste water after removal of the slowly biodegraded and the easily adsorbed organics and i.e. only the "drinking water relevant" substances remained. These substances pass all the usually applied treatment barriers in the water treatment plant and will therefore occur in the drinking water.

So it is important whether or not these substances are micropollutants or not.

One can see from this diagram, that the medium molecular weight substances comprise the largest fraction of the DOC. A very similar distribution has been found for the organic sulfur compounds. In contrast the organic chlorine compounds belong to the low molecular weight fraction and to the adsorbable substances. The figure also indicate that we don't achieve a full separation into individual substances using this method. So we can't say, that the different groups are micropollutants. They are however man made pollutants and they can be dangerous for human health.

We have to recognize, that organic micropollutants in river waters can never be completely defined. We have in most cases many other pollutants in micro-amounts, which have to be removed during drinking water treatment. It is for this reason that group parameter measurements like TOX and TOS are very important for water quality control. This very often is totally different, when we have organic micropollutants in ground water sources. Here we usually have only to consider a few substances, present in the water. An example for this can be seen out of the following figure.

parameter	
Trichloroethylene	60
Tetrachloroethylene	0.9
cis–1,2–Dichloroethylene	188
Σ single compound as chlorine	184
POX	180
AOX	13
TOX	193

concentration in µg/l

Fig.12 Single and surrogate parameters of polluted groundwater
Wasserchemie Karlsruhe   1985/3819

If we compare here the sum of all micropollutants with the TOX or also the POX values, we can conclude that there will be no other organic chlorine compounds in this water. We know all micropollutants in this water. In this case we can control treatment efficiency with either group parameter measurements or individual compound analyses. But this may not be true for ground waters with many different pollutants e.g. with waters coming from a solid waste deposition area. Here as with river waters, we cannot determine all the micropollutants. We have to recognize this and choose the right analytical control methods for each special case.

ORGANIC MICROPOLLUTANTS AND DRINKING WATER TREATMENT

If we have to consider the impact of different organic micropollutants present in raw water sources on the final drinking water quality, we also have to know something about the efficiency of different treatment processes in removing the organic micropollutants. This problem will be discussed within many different papers given at this conference, so that I can stay with some more general remarks and some examples. Generally speaking the following has to be considered in this respect:

> Removal of organic micropollutants through
> different treatment processes depends not only
> on their chemical structure but on all other
> water constituents as well.

Fig. 13

Micropollutants don't determine the overall effect of a water treatment process used in drinking water plants. For this reason, we first have to know more about the water in general than about the chemical structure of the micropollutants, if we want to predict the removal efficiency for each pollutant and a given treatment process. In order to get a better understanding of this, a few examples of micropollutant reduction in practical plants will be presented-especially in activated carbon treatment plants at the Rhine river.

The first example gives data for trichloroethylene in the Rhine river as well as in the river bank filtrate and after different treatment steps. The results are given as mean concentrations over one week and show primarily that there is no biodegradation during river bank filtration with a retention time of about 6 weeks. In this special case there has been observed a sudden increase of the concentration in the river in January, and one can see that there is some increase about 2 months later in the bank filtrate too. The fluctuations in the bank filtrate are lower, but we find about the same mean concentration as observed in the river. The concentration equalization depends on adsorption processes within the

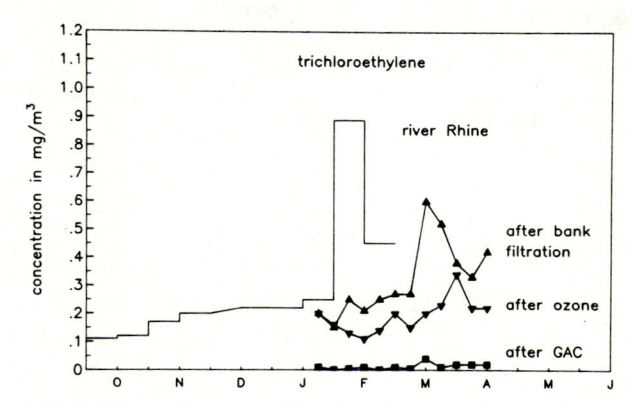

Fig.14 concentration of trichloroethylene after different treatment steps for the period Oct. 82 – April 83
Wasserchemie Karlsruhe   1985/3816

ground. This is one of the reasons, why water works using bank filtration don't have many problems with sudden concentrations changes within the river. The equalization effect is even more pronounced for aromatic substances as they are better adsorbed onto the organic material within the river banks.

One also can see a small decrease through ozonisation and this is due to some stripping and some chemical degradation. The reduction is not very impressive however compared with the result from the carbon filter, where practically all trichloroethylene can be removed, since regeneration always occurs before breakthrough of the discussed micropollutants. We have to acknowledge further, that the concentrations found in the Rhine river are usually very low if we take composite samples as it has been done in this case.

The situation regarding the concentrations of these sorts of micropollutants are very different if we consinder some ground waters. An example for this will be given with the next figure.

Here we have nearly about 200 µg/l, but we have three different volatile chlorinated hydrocarbons. And one can observe a very different adsorption behavior for these three substances. We have to consider further that besides the micropollutants, we also have about 1 - 2 ppm of other organics (humics) competing with the micropollutants. It is for this reason, that we have a lot of difficulties if we want to predict the breakthrough behavior in such a case. The reason for this lies in the fact, that these substances show a non ideal behavior regarding adsorption competition. This effect can be seen from the following figure.

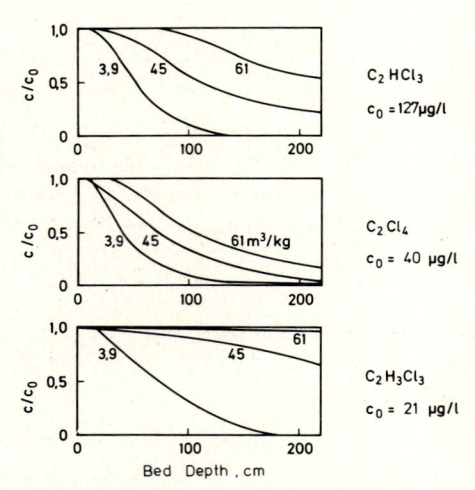

Fig.15 Concentration Profiles of Volatile Organic Contaminants in a GAC-Adsorber

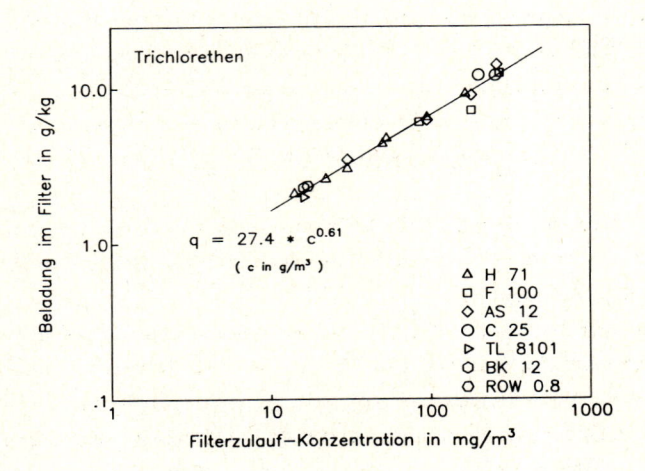

Fig.16 Maximum filter loading vs influent concentration for trichlorethylene

The data summarized within this figure are taken from many different treatment plants and they give the maximal loading found in equilibrium with the inflowing water for seven different carbons. These equilibrium loadings depend on the influent concentration of the substance only and are independent of all other micropollutants and the total organics concentration as well as from the type of activated carbon used. This is a very astonishing result and can be explained only with different assumptions on the type of pores available for the different substances within this water. These data prove the non ideal adsorption behavior of the three chlorinated hydrocarbons and we have to take this into account, if we want a nearly complete removal of these micropollutants, especially of trichloroethane.

Regarding the figure, we have to consider that it only gives the maximum loading which can be reached in practical plants. But usually this loading cannot be reached as we have an earlier breakthrough. But here technical solutions can really help and are therefore very important for micropollutant removal also. One example for this can be seen in the next figure.

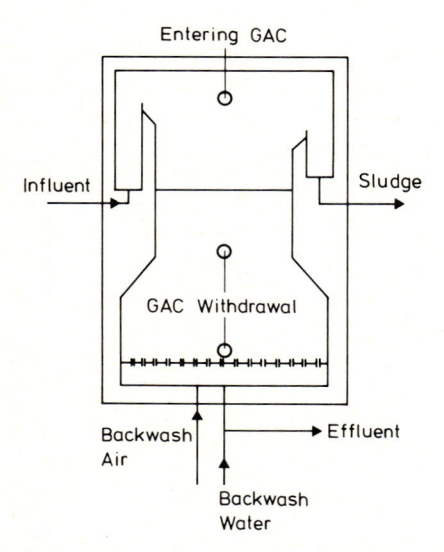

Fig. 17   **GAC-Filter with Changing Cross-Section**

Wasserchemie Karlsruhe   1985 / 3593

This is a schematic drawing of a GAC-filter, as it is used in Mannheim (Kretschmar, 1985) and this filter has two different cross-sections and these have a difference in the area for the water flow of about 25 %. This allows a separate backwashing of the upper part of the filter and the carbon therein has a higher loading at total filter breakthrough than the lower part. So this upper part can be fluidized and be pumped to reactivation or an intermediate tank. After this one can fluidize the lower part and pump this carbon which is not fully loaded in a separate tank. Then new or reactivated carbon is placed in the lower part of the filter and on top of this the carbon previously coming from the lower part can be put on top. Thus only one filter is necessary to reach the same effect as when having two separate filters in series. This type of filter operation doesn't only reduce the carbon costs by about 25 %, but it also gives more safety for the removal of organic micropollutants and this may be still more important. The reason for the greater safety lies in the fact, that we have half of the filter volume always as new and un- loaded carbon and for this competition is not so important in the first part of the filter run. This example shows that processes considerations play an important role, if we look on the importance of micropollutants on drinking water quality.

In order to present some data on the behavior of micropollutants when using different treatment steps in series, the next two figures present data from two different water works at the Rhine river, which show prac- tical results typical for these plants. The data are given as geometric mean values over a period of about 3 months and include, besides the river water quality, the bank filtered raw water data as well as the results after ozonisation and after the activated carbon filters but before the final desinfection which, in most cases, is done with 0.1 - 0.15 mg/l chlorine dioxide.

Regarding the effect of bank filtration one can see, as has been shown before, that there is some biodegradation of substances like benzenes, chlorobenzenes and chlorotoluenes as well as for naphthalene. Ozonisation doesn't have some real impact on the concentrations but GAC, as expected, has a large influence if we forget about the small concentrations of ben- zene and toluene.

The last example gives very similar data but contains in addition to the single substance data, the values for TOC as well as for TOX and TOS. Here the effect of ground filtration is fairly high showing that some of the more polar micropollutants, which can't be analysed individually, are biodegradable. Here, the effect of carbon filtration is worthwhile

	Rhine water	after bank filtration	after ozone	after GAC
Chloroform	0.40	0.29	0.32	0.24
Carbon Tetrachloride	0.19	0.08	0.09	0.02
Trichloroethylene	0.23	0.30	0.17	0.01
Tetrachloroethylene	0.37	0.40	0.32	0.01
Tetrachloroethane	0.03	0.01	0.01	<0.005
Hexachloroethane	n.n.	n.n.	n.n.	n.n.
Hexachlorobutadiene	n.n.	0.05	0.04	n.n.
Benzene	0.13	0.05	0.05	0.05
Toluene	0.06	0.03	0.03	0.03
Dimethylbenzenes	0.29	0.04	0.04	0.01
Styrene	0.01	<0.005	<0.005	n.n.
Naphthalene	0.09	0.01	0.01	<0.005
Chlorobenzene	0.09	<0.005	<0.005	n.n.
1,2–Dichlorobenzene	0.36	0.02	0.01	n.n.
1,3–Dichlorobenzene	0.09	0.03	0.01	n.n.
1,4–Dichlorobenzene	0.17	0.09	0.06	n.n.
1,2,4–Trichlorobenzene	0.05	0.09	0.04	n.n.
Chlorotoluenes	0.08	<0.005	<0.005	n.n.

concentration in µg/l

Fig.18   Removal of organic micropollutants in a water works on the lower Rhine (1983)

Wasserchemie Karlsruhe   1985/3807

parameter	Rhine water	after bank filtration	after ozone	after GAC
Chloroform	1.00	0.61	0.35	0.14
Carbon Tetrachloride	0.44	0.29	0.18	0.01
Trichloroethylene	0.56	0.60	0.21	0.02
Benzene	0.60	0.05	0.05	n.n.
Toluene	0.74	0.04	0.03	0.02
Xylenes	0.75	0.07	0.07	0.02
Dichlorobenzenes	0.81	0.45	0.14	0.03
Chlorotoluenes	0.20	<0.01	<0.01	0.01
AOX	51	22	19	10
TOS	93	38	25	22
TOC (mg/l)	3.4	1.6	1.3	0.8
UV(254) (m^{-1})	8.2	4.4	1.9	0.8

concentration in µg/l

Fig.19   Efficiency of drinking water treatment steps in a water works on the lower Rhine

Wasserchemie Karlsruhe   1985/3805

for the removal of the organic chlorine concentration, but not as good for the removal of sulfonic acids. We have to ask in this case, if the removal efficiency of GAC, which is very good for the defined compounds can be different for the unknown micropollutants. But, the combined treatment with the three different processes is highly efficient and, if we would only control UV-absorbance or DOC we would get similar results, compared to controlling all individual micropollutants.

SUMMARY AND CONCLUSIONS

If we consider the possible impact of micropollutants present in the water source on drinking water quality, we usually come to the following points:

> Possible impact of organic micropollutants
> - Direct or indirect health effects.
> - Necessity of additional treatment steps.
> - Corrosion activation and inhibition.
> - Impact on water uses.

Fig. 20

Besides the always most important health effects, we have to consider corrosion effects and also special water uses. Very often micropollutants lead to the necessity of other treatment steps. But the possible health aspects usually are most important. There is no doubt, that we can't decide without knowing the chemical structure, if an organic micropollutant has some adverse health effects or not. Also when we have this knowledge, we very often don't have the necessary toxicological data. So we have difficulties to fix an acceptable maximum concentration. And if we don't have the necessary toxicological information, we don't need the exact chemical structure either. This is one of our big problems in connection with organic micropollutants.

We can overcome this, partially measuring group parameter data like TOX and by trying to reduce this concentration as far as possible. This very often is a more worthwhile control method compared to analysing the individual substances in the ng/l range.

The second problem we have to deal with is safety. Here, besides analytical control, which usually is very limited, we can use a combination of different treatment processes.

> Effective removal of organic micropollutants can only be achieved for most pollutants using a combination of different treatment steps including biological and chemical oxidation as well as granular activated carbon.

Fig. 21

Most of the problems we have in practice with organic micropollutants present in the water source occur in water works without sufficient treatment steps. Especially biological oxidation processes are of large importance, here especially if they are done through ground filtration. This has been found true for two reasons. First biological oxidation within the ground needs time and second longer retentions times lead to a better equalization and this reduces peak concentrations. Furthermore experience with chemical wastewaters has shown that the more toxic substances are very often slowly biodegradable and are therefore removed during ground filtration. Besides this after biological oxidation the remaining substances are most easily adsorbed and this helps for a better overall treatment efficiency too. In addition to biological oxidation within the ground, activated carbon filters are very important for a safe drinking water. Here, we can now predict the breakthrough behavior of micropollutants too.

As a final conclusion one can say the following. We usually call only these substances as micropollutants, which can be analysed as individual substances. But, if we have solved this analytical problem, most of our questions are answered too. We can decide then more easily, if we have to remove the identified substances and to what final concentration. We also can study treatment steps to remove the pollutant within the water work or, last but not least, at the place where it comes from.

But after this we still stay with the question if there are any unknown micropollutants within our water which we can't determine individually and these unknown substances are in most cases more problematic and possibly more dangerous for human health than all our well known organic pollutants. Here we have to find the right control method and strategy and here we have to spend more work and activity in the future if we want a safe drinking water.

LITERATURE

/1/ Rook, J.J.:
Formation of Haloforms during Chlorination of Natural Waters.
Water Treatment and Exam. 23 (1974), 2, 234

/2/ Drinking Water and Health
National Academy Press, Washington, D.C. Vol. 3 (1980)
u. Vol. 4 (1982)

/3/ Memorandum der IAWR (Internationale Arbeitsgemeinschaft der Wasserwerke im Rheineinzugsgebiet)
Rheinwasserverschmutzung und Trinkwassergewinnung, Mai 1973

/4/ Tagungsbericht 7. ARW-Arbeitstagung, Basel 1979
(Arbeitsgemeinschaft Rhein-Wasserwerke e.V.)

/5/ Kretschmar, W.:
Chlorinated Hydrocarbons in Drinking Water
Water Supply, Vol. 3, Berlin "B"-Pergamon Press Ltd.
pp 197 - 202 (1985)

THE CONTROL OF ORGANICS IN DRINKING WATER IN CANADA AND THE UNITED STATES (STANDARDS, LEGISLATION AND PRACTICE)

P. TOFT, DEPARTMENT OF NATIONAL HEALTH AND WELFARE, OTTAWA, CANADA

ABSTRACT

Both the United States and Canada have a federal form of government, but approaches used in the two countries to ensure the safety of drinking water supplies differ. The Environmental Protection Agency currently enforces regulations for 10 organic chemicals (including 6 pesticides) under the Safe Drinking Water Act and provides advice on others through its health advisory program. Canada, however, does not have similar legislation, but rather provides health-related guidelines for 21 organic chemicals (including 16 pesticides) which are used by the provincial agencies responsible for drinking water supplies. Both countries are in the process of revising their standards and will include a variety of additional synthetic organic chemicals.

Where possible, standards are set using a calculated acceptable daily intake usually derived from animal feeding experiments. Procedures for setting standards for carcinogens involve a blend of risk estimation coupled with consideration of the feasibility of reducing the risk in light of socio-economic factors. Most drinking water treatment plans in North America utilize 'conventional' treatment. Some now employ modifications in order to minimize trihalomethane formation. A few use aeration or granular activated carbon to remove synthetic organic chemicals.

INTRODUCTION

An adequate supply of clean, wholesome drinking water is one of the principal requirements for good health. Traditionally the primary aim of the water supply industry has been to ensure that the consumer is provided with water which is free from pathogenic micro-organisms. Only secondarily was attention focused on its chemical content, and then mainly on the dissolved inorganic substances, usually from natural sources. With the exception of pesticides, the organic content of drinking water was thought to be of concern only in so far as it contributed to taste and odour. It is only in the past decade that interest has been directed to any extent towards the vast array of trace organic substances which are present in many drinking water supplies.

Not surprisingly drinking water standards and regulations in both Canada and the United States have mirrored this. The first drinking water standards in North America were promulgated in the early part of the twentieth century and prescribed the bacteriological quality of drinking water supplied on common carriers (Stacha and Pontius, 1984, and Health and Welfare Canada, 1969). Even today, the single most important cause of waterborne disease in the United States and Canada is the presence of pathogenic micro-organisms due to a failure to maintain an adequate concentration of disinfectant (Craun, 1979, and Health and Welfare Canada, 1981).

It was the demonstration by Rook (Rook, 1974) and Bellar (Bellar, 1974), that trihalomethanes could be formed during the treatment process, that awakened interest in the presence and possible health significance of organic chemicals in drinking water supplies. Advances in analytical methods, such as the development of gas chromatography/mass spectrometry, permitted scientists to uncover a large number of organic chemicals in drinking water supplies that had previously gone undetected. In 1981 a NATO committee report listed 744 organic compounds that had been identified in the drinking water of 14 countries (Borzelleca, 1981). More recently it has been estimated that more than 2000 such chemicals are known (Packham, 1984). This growing number has led regulatory agencies in North America and elsewhere to question the health significance of ingesting these chemicals daily in drinking water and the need for standards.

LEGISLATION

Both the United States and Canada have a federal form of government, but the division of powers between the levels of government differs in the two countries. These differences are reflected in the approaches used to ensure safe drinking water supplies.

United States

The first drinking water standards in the United States were promulgated in 1914 under the authority of the Interstate Quarantine Act of 1893, and were limited to bacteriological quality (Stacha and Pontius, 1984). They were administered by the United States Public Health Service (USPHS) and regulated drinking water supplied on interstate carriers, but were widely adopted as guidelines by state and local governments. They were revised four times and by 1962 set mandatory limits for health related chemical and microbiological contaminants (U.S. Government Printing Office, 1962). The 1962 USPHS standards were used in all fifty states either as regulations or guidelines, and were legally binding at the federal level for about 700 water systems which supplied interstate common carriers.

The United States Safe Drinking Water Act was promulgated in 1974 and required the Environmental Protection Agency (EPA) to establish a framework for action to improve public drinking water supplies (U.S. Government, 1974). Regulations were to be developed by the Federal Government and supplemented by states which accept primary enforcement responsibility. In particular, EPA had to:

a) identify substances in potable water that may have adverse impacts on health;

b) specify for each contaminant either (i) a maximum contaminant level (MCL) for each substance, where monitoring is economically and technologically feasible, or (ii) specify treatment techniques that would reduce contaminant concentrations to the extent feasible.

The Act called for the establishment of a regulatory framework with the following features:

1) promulgation of National Interim Primary Drinking Water Regulations (NIPDWR) and National Secondary Drinking Water Regulations (NSDWR);

2) the National Academy of Sciences to conduct a study to assess the health effects of contaminants in drinking water and to provide recommended contaminant levels;

3) EPA to publish a list of Recommended Maximum Contaminant Levels (RMCL), that in the administrator's judgement, assures that "no known or anticipated adverse effects on the health of persons occur and which allows an adequate margin of safety";

4) EPA to propose and promulgate National Revised Primary Drinking Water Regulations, which will include RMCLs, MCLs, as well as monitoring and reporting requirements for those contaminants that may have adverse health impacts.

National Interim Primary Drinking Water Regulations were promulgated in December, 1975, and became effective on June 24, 1977 (Cotruvo, 1984). They contained standards, monitoring requirements and analytical methods for six organic chemicals (all pesticides). An amendment was issued in 1979 to regulate trihalomethanes (chloroform, bromoform, bromodichloromethane and dibromochloromethane). The values are listed in Table I.

TABLE I

Current MACs and MCLs for Organic Chemicals (mg/L)

	Canada	U.S.A.
2,4-D	0.1	0.1
2,4,5-TP	0.01	0.01
Aldrin + dieldrin	0.0007	
Carbaryl	0.07	
Chlordane	0.007	
DDT	0.03	
Diazinon	0.014	
Endrin	0.0002	0.0002
Heptachlor + heptachlor epoxide	0.003	
Lindane	0.004	0.004
Methoxychlor	0.1	0.1
Methyl parathion	0.007	
Nitrilotriacetic acid	0.05	
Parathion	0.035	
Pesticides, total	0.1	
Trihalomethanes	0.35	0.1
Toxaphene	0.005	0.005

Canada

In Canada the legislative base derives from the constitutional distribution of powers between the federal and provincial governments, as set out in the British North America Act of 1867. This act does not deal specifically with water resources (and therefore with drinking water supply), but judicial interpretation over the years has resulted in a situation where drinking water is a shared federal-provincial jurisdiction.

Federal authority stems primarily from the Department of National Health and Welfare Act under which responsibility is assigned to investigate and conduct programs related to public health. The Minister of National Health and Welfare also has authority to prescribe regulation standards for food (and consequently for water). Although this has been done for bottled water, standards have not been prescribed for tap water because of the traditional role assumed by the provinces. The Federal Government is solely responsible for drinking water supplied on common carriers which cross interprovincial and international boundaries and for drinking water supplies in the Northern territories, defence bases, and national parks. The first drinking water standards in Canada provided a standard for the bacteriological quality of water used for drinking purposes on ships on the Great Lakes and Inland Waters, and were promulgated through an Order-in-Council in 1923 (Health and Welfare, 1969). They were extended in 1930 and 1937, and adopted as regulations under the Department of National Health and Welfare Act in 1954 (Government of Canada, 1954).

The provinces have the primary authority to legislate for municipal water supplies owing to their ownership of natural resources, including water. The municipal or equivalent acts in each of the provinces empower municipalities to construct, operate and maintain potable water supplies. Generally, the provinces assume the lead role in ensuring an adequate and safe supply of drinking water, whereas the Federal Government provides leadership in developing adequate guidelines for drinking water quality, especially for the parameters which affect human health. These guidelines are developed through a joint federal-provincial mechanism and are not legally enforceable unless promulgated as regulations by the appropriate provincial agency. The first comprehensive Canadian drinking water guidelines were published by the Department of National Health and Welfare in 1968. They were completely revised in 1978 and specify health-related limits for 21 organic chemicals, including 16 pesticides. Quebec and Alberta are the only provinces that have enacted drinking water quality standards by legislation (Government of Quebec, 1984, and Government of Alberta, 1982). Other provinces use the Federal-Provincial guidelines as the basis for controlling municipal drinking water supplies.

STANDARDS

The primary impetus for setting standards or guidelines for drinking water quality is to safeguard human health. According to Whyte and Burton (Whyte and Burton, 1980):

> Standards are prescribed levels, quantities or values which are regarded as authoritative measures of what is a safe enough, or acceptable, amount of. . .contamination or exposure to risk.

A wholesome or healthful drinking water is one which not only does not cause disease but is also aesthetically satisfying. Standards are therefore determined for both health-related parameters and factors affecting taste and odour. Two types of limits are recognized in both the United States (Environmental Protection Agency, 1975 and 1977) and Canada (Health and Welfare Canada, 1980):

- health-related or primary standards (legally enforceable in the United States)
- aesthetic or secondary standards

In addition both countries identify maximum exposure values for chemical contaminants, called <u>maximum contaminant levels (MCL)</u> in the United States National Interim Primary Drinking Water Regulations and <u>maximum acceptable concentrations (MAC)</u> in the Guidelines for Canadian Drinking Water Quality. The United States MCL for a contaminant is determined such that it "shall protect health to the extent feasible". The Canadian MAC is defined as the limit beyond which the contaminant "is either capable of producing deleterious health effects or is aesthetically objectionable".

Because of their relatively low concentrations, the chemical constituents of drinking water are, with very few exceptions, unlikely to cause adverse health effects in the short term. The effects that might occur would be manifested over a long time period. In deriving limits, therefore, comprehensive information on chronic toxicity, and especially carcinogenicity, is more relevant than acute toxicity data. The current standards for organic chemicals in drinking water were developed in similar ways in both the United States and Canada. Substances to be considered were selected from the large number known to occur in water supplies based on factors such as:

- relatively high frequency of detection in drinking water supplies;
- found in relatively high concentrations;
- evidence that the substance can cause adverse effects;
- structure similar to a known toxic chemical.

Chemicals which appear to be of local concern only are not usually considered for inclusion in national standards or guidelines.

Comprehensive scientific reviews were then prepared for each substance, which include information such as:

- major sources and concentrations in drinking water;

- information on human exposure from all sources (e.g. drinking water, food and air);
- information on health effects, including information on the relationships between exposure and magnitude of effects;
- identification of the most sensitive groups at risk.

In the United States, criteria reviews are assembled by the National Academy of Sciences on behalf of EPA (United States National Academy of Sciences, 1977). Some 128 reviews were prepared by the NAS on organic substances, 55 of them pesticides. Since then a further 48 organic chemicals have been reviewed. In Canada the Federal-Provincial Working Group on Drinking Water prepared 67 background documents used to derive the 1978 guidelines, of which 28 were on organic chemicals or groups of chemicals (Health and Welfare Canada, 1980).

For those substances for which a threshold can be demonstrated in the dose-response relationship, an acceptable daily intake (ADI) can be calculated by applying appropriate safety factors to the no-observed-adverse-effect dose. The ADI approach is used in both Canada and the United States when developing standards for non-carcinogens. The safety factors used vary according to the quantity and quality of the available data; similar factors are used in both countries. It should be recognized, however, that the ADI is a value that represents a certain degree of confidence about the relative safety of a chemical. It is the maximum dose of a single chemical taken daily for a lifetime which is not anticipated to produce an observable toxic response in man. It does not, however, imply absolute safety.

Drinking water is not, for most substances, the major source of exposure to chemicals. Due account must be taken of exposures that result from food, air, occupation, and lifestyle factors (such as smoking) when deriving a drinking water standard by this method. It is therefore necessary to apportion the calculated ADI for each parameter between the major exposure routes. For example, 20% of the ADI for pesticides was allocated to drinking water and 80% to food in both Canada and the United States.

For "non-threshold toxicants" it is not possible to calculate an ADI by the method described above. The procedures followed in setting standards in both Canada and the United States include a blend of estimating the magnitude of risk of the particular health effect, usually cancer, with consideration of the feasibility of reducing the risk, in light of socio-economic factors. Except for radionuclides the only substances regulated as possible carcinogens under the U.S. NIPDWR are the group of four trihalomethanes (THMs - chloroform, bromoform, bromodichloromethane, and chlorodibromomethane). In an amendment promulgated in 1979 the U.S. EPA established a MCL for total trihalomethanes of 0.1 mg/L computed as a running annual average of their sum (Environmental Protection Agency, 1979). In contrast, the Guidelines for Canadian Drinking Water Quality 1978 established a MAC for the same group of chemicals of 0.35 mg/L.

The reasons for establishing drinking water standards for these chemicals are similar in both countries, and include the following factors:
- the potential human health risks;
- the major source of human intake is drinking water;
- THMs are the most common synthetic organic chemicals found in drinking water, and the most abundant;
- they are introduced during the treatment process and can be readily controlled.

In addition EPA felt, most importantly, that THMs could be used as indicators of concurrent formation of other undesirable by-products.

Exposure to chloroform at high doses can elicit a variety of adverse health effects, but carcinogenesis is the hazard of concern at the lower concentrations found in drinking water. A number of epidemiology studies have been conducted on the possible relationships between cancer mortality and morbidity and the concentrations of THMs in drinking water supplies. Although some positive correlations were reported, such retrospective epidemiological studies cannot establish causality because of the many unknown or uncontrollable factors that cannot be taken into account. In establishing limits for THMs both countries therefore relied on data obtained from studies of chloroform toxicity in laboratory animals and assumed that the other trihalomethanes would exhibit similar toxic effects.

Both Canada and the United States based their conclusions on studies conducted by the National Cancer Institute (National Cancer Institute, 1976) using rats and mice dosed repeatedly with chloroform. Hepatocellular carcinomas were observed in male and female mice, and kidney epithelial tumours were found in male rats but not in females. The U.S. EPA included both studies in its assessment, but Canadian officials rejected the mouse data on the grounds that the doses used were too high to be useful in predicting the cancer potential at the low doses found in drinking water.

Health and Welfare Canada recommended its guideline value for THMs of 0.35 mg/L based primarily on health considerations and the acceptability of a calculated maximum risk of incurring cancer of about 1 in 2.5 million per year. In reality such calculations are fraught with uncertainty and may differ by several orders of magnitude depending on the models used. In contrast, the selection by EPA of an interim MCL for THMs of 0.1 mg/L was based largely on the technical feasibility of achieving a reduction of THM levels with the concomitant benefits to public health. In practice the difference between these two standards is less than it would at first appear. The lower United States number is a running annual average whereas the higher Canadian figure is a maximum value which should not be exceeded, and THM concentrations in drinking water can vary considerably from winter to summer, commonly by more than an order of magnitude.

REVISIONS TO CURRENT STANDARDS

The United States and Canada are both in the process of revising their drinking water standards and guidelines. In addition to re-evaluating its guidelines for phenols, pesticides, and THMs, Canada is considering the need for guidelines for a variety of other organic substances. Substances under evaluation include benzene and alkylated derivatives, chlorobenzenes, chlorinated ethylenes, dichloromethane, chlorinated ethanes, carbon tetrachloride, dioxins, and benzo(a)pyrene. It is also intended that the list of pesticides for which MACs were developed should be updated to reflect current usage. As in the past this process of revision is being undertaken as a joint Federal-Provincial activity.

In the United States the Safe Drinking Water Act requires that EPA propose and promulgate 'National Revised Primary Drinking Water Regulations', which will include 'Recommended Maximum Contaminant Levels' (RMCLs) for those contaminants that may have an adverse effect on human health. Legally enforceable MCLs will then be determined which are as close to the RMCLs as technically feasible. EPA is now in the process of developing revised regulations through a step-wise program consisting of five phases (see Table II). Each phase involves publication of an Advance Notice of Proposed Rule Making (ANPRM), proposal of RMCLs, promulgation of RMCLs and proposal of MCLs, and finally promulgation of MCLs. Each stage is followed by a period of time for public comment. Phase I began in March, 1982, with the issuing of an ANPRM for

TABLE II

Regulatory Framework for Revisions to the
National Interim Primary Drinking Water Regulations

Phase I	Volatile organic chemicals
Phase II	Synthetic organic chemicals, inorganic chemicals, microbiological contaminants
Phase IIA	Fluoride
Phase III	Radionuclides
Phase IV	Disinfectant by-products including trihalomethanes
Phase V	Other chemicals not considered above

TABLE III

Proposed RMCLs for Volatile Organic Chemicals

Tetrachloroethylene	0
Trichloroethylene	0
1,2-Dichloroethane	0
Carbon tetrachloride	0
Vinyl chloride	0
Benzene	0
1,1-Dichloroethylene	0
1,1,1-Trichloroethane	200 ug/L
p-Dichlorobenzene	750 ug/L

volatile organic chemicals. This was followed by a proposal in June 1984 for RMCLs for nine chemicals (shown in Table III), seven of which were treated as potential carcinogens. 1,1,1-Trichloroethane and para-dichlorobenzene were not considered to be carcinogenic, and RMCLs for these substances were developed by the ADI method, allowing a 20% contribution from drinking water. EPA considered the following approaches in setting RMCLs for carcinogens:

1. Set the RMCL at zero;
2. Set the RMCL at the analytical detection limit;
3. Set the RMCL at a concentration based upon a calculated negligible lifetime risk.

Establishing RMCLs at zero could be considered appropriate for chemicals for which no effect threshold exists. If necessary, the approach could be limited, for example, to geno-toxic substances or known human carcinogens rather than the larger number of presumed carcinogens. It also embodies the concept that drinking water should be free from avoidable risk. Using the detection limit of a carcinogen as its RMCL may be a more practical way of stating the same objective, since it is theoretically not possible to demonstrate the complete absence of a chemical. Detection limits, however, are constantly changing as analytical techniques improve. The third option was to set the RMCL at a level considered to present negligible risk based on suitable risk models. Just as analytical detection limits change so do risk calculation methods (and possibly also what is thought to be negligible risk). Cothern also has recently pointed out that the range of uncertainties inherent in some risk calculations due to the choice of model may exceed six orders or magnitude! (Cothern, 1984). EPA has proposed that the RMCLs for carcinogens be set at zero, but a final decision has not been made. It is interesting to note that the concept of the RMCL is similar to that of the "objective" limit defined in the Canadian guidelines as "the ultimate quality goal for both health and aesthetic considerations" (Health and Welfare Canada, 1980). The objective limit was generally set at the detection limit obtainable by a laboratory of good standing using conventional analytical techniques.

TABLE IV

Synthetic Organic Chemicals Being Considered for Inclusion in the U.S. Revised National Primary Drinking Water Regulations

Endrin	Methoxychlor
2,4-D	
Toxaphene	2,4,5,-TP
1,2-Dichloroethylene	Dichlorobenzenes
Aldicarb	Chlordane
Endothall	Carbofuran
Heptachlor	Styrene
PCBs	Dibromochloropropane
1,2-Dichloropropane	Pentachlorophenol
Alachlor	Ethylene dibromide
Epichlorohydrin	Xylene
Toluene	2,3,7,8-TCDD (dioxin)
Chlorobenzene	Hexachlorobenzene
Lindane	Ethyl benzene

Phases 2 and 3 of the development of the revised standards were initiated with the publication by EPA of an ANPRM in October, 1983 (Cotruvo, 1984). Included for consideration are a number of synthetic organic chemicals and pesticides (see Table IV). Determination of which of these substances will be included in the revised primary drinking water regulations will be based on a detailed analysis of their potential health significance.

As an adjunct to its regulatory program, EPA has developed an advisory program which provides advice on the health effects, analytical methods and treatment techniques of chemicals in drinking water (Lappenbusch and Moskowitz, 1984). Health advisories are intended to provide information to deal with short-term exposures such as those resulting from spills. They are frequently developed in response to specific requests. Draft health advisories have been calculated for 22 chemicals (Table V).

TABLE V

Chemicals for Which EPA Has Drafted Health Advisories

Trichloroethylene	n-Hexane
Tetrachloroethylene	Methyl ethyl ketone
1,1,1-Trichloroethane	Toluene
Dichloromethane	1,4-Dioxane
Carbon tetrachloride	1,2-Dichloroethane
1,1-Dichloroethylene	Fuel oil #2/kerosene
Benzene	Carbofuran
Xylene	Uranium
Ethylene gylcol	PCBs
Cis & trans 1,2-Dichloroethylene	Formaldehyde
Chlordane	

DRINKING WATER TREATMENT

In the United States there are more than 220,000 public water supply systems serving over 200 million persons (Stacha and Pontius, 1984). Of these, 60,000 (26%) are community systems which serve consumers in residential areas who use the water almost every day over a long period of time. (The remainder provide water to intermittent users, such as travellers, and provide service primarily to campgrounds, motels, restaurants, etc.). Approximately two-thirds of the population served by community systems receive surface water. In Canada there are about 2500 drinking water treatment systems serving about 20.2 million persons, approximately 87% of the population. The remaining 13% rely on private sources of supply using mainly groundwater.

The majority of plants in North America which treat surface waters utilize so-called 'conventional treatment'. Complete conventional treatment is primarily aimed at removing turbidity and microbiological contaminants. Typically it consists of prechlorination, coagulation, flocculation, sedimentation, filtration, and final disinfection (post chlorination).

The bulk of the organic matter in natural waters consists of humic materials, and it is well established that chlorination of such materials results in formation of trihalomethanes. The rate and extent of THM formation is dependent upon a number of factors including pH, temperature, chlorine dose, contact time with chlorine, nature of organic precursor material and its concentration. In light of the recent regulations and guidelines for maximum THM concentrations, disinfection practices have been altered in many systems. THM formation can be effectively reduced in some instances by relatively simple changes in the treatment process. Several alternatives that have found use are as follows:

1. Reducing the quantity of chlorine used, especially in the prechlorination step.
2. Changing the point of application of chlorine to after the removal of most of the humic precursors in the coagulation-flocculation-filtration steps.
3. Improving pH control to minimize THM formation and to optimize removal of precursors. (Since the haloform reaction is base catalyzed, decreasing the pH will decrease the rate of formation of THM).
4. Using an alternative disinfectant, either alone or in conjunction with chlorine.
5. Removing THMs after they have been formed, either by aeration or by

treatment with activated carbon, or by a combination of both.

The other organic chemicals regulated in the United States NIPDWR are 6 pesticides. The current guidelines in Canada have health-related MACs for 17 additional organic chemicals - 16 pesticides and nitrilotriacetic acid (NTA). The limits for these substances have rarely, if ever, been exceeded in all public water supply systems in the United States and Canada for which data are available. However, although inorganic coagulants can remove up to 80% of the natural organic matter in water, and hence considerably reduce the formation of THMs, they have little effect on the concentrations of dissolved synthetic organic chemicals. Recent advance in coagulation chemistry have led to the development of new types of coagulants containing prepolymerized gels. Health and Welfare Canada funded studies to determine the organic removal performance of these new coagulants and coagulant aids (Benedek, 1983). While they were found to be equally efficient as alum in removing humic and fulvic acid and other trihalomethane precursors, they were not effective for the removal of dissolved organic chemicals. 'Conventional' treatment practices have therefore not played any significant role in providing drinking water which contains acceptable concentrations of synthetic organic chemicals.

Although chlorine is by far the most common disinfectant used in North America, alternative oxidants have found use in some systems - in many instances to solve particular taste and odour problems unaffected or sometimes exacerbated by the use of chlorine. Ozone is the most powerful commonly available oxidant and disinfectant and is now used in about 25 treatment plants in the United States and in about 50 plants in Canada, mainly in Quebec (Rice, 1985). Chlorine is usually added as a secondary disinfectant to maintain a residual in the distribution system. Chloramines were popular in the United States in the 1930s, but their use decreased with the introduction of breakpoint chlorination. They are finding favour again in some systems because they produce only insignificant quantities of THMs and can sometimes control taste and odour problems. Chlorine dioxide has limited application in North America; its principal use is for taste and odour control.

Granular activated carbon is now used in a few systems where organic chemical contamination is a problem. Several communities in the United States have installed aeration facilities to remove volatile organic chemicals, often coupled with carbon absorption. These have found particular use in treating contaminated groundwaters (McKinnon and Dyksen, 1984).

A proposed amendment was made in 1978 to the U.S. National Interim Primary Drinking Water Regulations to control synthetic organic contaminants in drinking water derived from the raw water (Symons, 1984). It proposed the use of granular activated carbon treatment to reduce the concentrations of chemicals for which it was not thought feasible to set MCLs. However, in March 1981 the proposal was withdrawn, primarily because of the resistance to it by the water supply industry.

In conclusion both the United States and Canada have standards for a limited number of organic chemicals in drinking water supplies. The range of substances considered is being extended as the need is foreseen. Conventional water treatment only marginally affects the content of organic chemicals although it can be adapted to reduce the formation of trihalomethanes. The removal of other organic chemicals requires the use of different techniques such as aeration and granular activated carbon.

REFERENCES

Bellar, T.A., Lichtenberg, J.J. and Kroner, R.C., 1974. The occurrence of organohalides in chlorinated drinking water. J. Am. Water Works Assoc., 66: 703-706.

Benedek, A., 1983. The removal of trihalomethane precursors and synthetic organic chemicals from potable water supplies by coagulation. Unpublished report. Health and Welfare Canada. Ottawa, Canada.

Borzelleca, J., 1981. Report of the NATO/CCMS Drinking Water Pilot Study on Health Aspects of Drinking Water Contaminants. In: H. Van Lelyveld and B.C.J. Zoeteman (Editors), Water Supply and Health. Elsevier, Amsterdam, pp. 205-217.

Cothern, C.R., Coniglio, W.A. and Marcus, W.L., 1984. Uncertainty in population risk estimates for environmental contaminants. Presented at the Annual Meeting of the Society for Risk Analysis. Knoxville, U.S.A.

Cotruvo, J.A. and Vogt, C., 1984. Development of revised primary drinking water regulations. J. Am. Water Works Assoc., 76(11): 34-38.

Craun, G.F., 1979. Waterborne disease outbreaks in the United States. J. Environ. Health, 41: 259-265.

Environmental Protection Agency, 1975. National Interim Primary Drinking Water Regulations. Federal Register, 40: 59566. Washington, U.S.A.

Environmental Protection Agency, 1977. National Secondary Drinking Water Regulations. Federal Register, 42: 17144. Washington, U.S.A.

Environmental Protection Agency, 1979. Amendments to the National Interim Primary Drinking Water Regulations: Control of Trihalomethanes in Drinking Water. Federal Register, 44: 68624. Washington, U.S.A.

Government of Alberta, 1982. The Clean Water Act, Clean Water (Municipal Plants) Regulations; Alberta Regulation 37/73 with amendments to Alberta Regulations 83/82. Alberta, Canada.

Government of Canada, 1954. Department of National Health and Welfare Act. Potable Water Regulations for Common Carriers, P.C., 1954 - 1213.

Government of Quebec, 1984. Loi sur la qualite de l'environnement. Reglement sur l'eau pour la consommation humaine. Quebec, Canada.

Health and Welfare Canada, 1969. Canadian Drinking Water Standards and Objectives 1968. Queen's Printer. Ottawa, Canada.

Health and Welfare Canada, 1981. Food-borne & Water-borne Diseases in Canada. Annual Summary (1977). Ottawa, Canada.

Health and Welfare Canada, 1980. Guidelines for Canadian Drinking Water Quality, 1978 - Supporting Documentation. Supply and Services Canada. Ottawa, Canada.

Lappenbusch, W.L. and Moskowitz, S.B., 1984. EPA Health Advisory Program in Drinking Water and Human Health. American Medical Association. Chicago, U.S.A., pp. 183-196.

McKinnon, R.J. and Dyksen, J.E., 1984. Removing organics from groundwater through aeration plus GAC. J. Am. Water Works Assoc., 76(5): 42-47.

National Cancer Institute, 1976. Report on carcinogenesis bioassay of chloroform. Washington, U.S.A.

Packham, R.F., 1984. Evaluation of human health hazards from drinking water - a global problem. In: Evaluation of methods for assessing human health hazards from drinking water. International Agency for Research on Cancer Workshop. Lyon, France.

Rice, R.G., 1985. Ozone for drinking water treatment - evolution and current status. In: Safe Drinking Water - the impact of chemicals on a limited resource. Drinking Water Research Foundation. Alexandria, U.S.A. pp. 123-160.

Rook, J.J., 1974. Formation of haloforms during chlorination of natural waters. Water Treat. Exam., 23: 234-243.

Stacha, J.H. and Pontius, F.W., 1984. An overview of water treatment practices in the United States. J. Am. Water Works Assoc., 76(10): 73-85.

Symons, J.M., 1984. A history of the attempted federal regulation requiring GAC adsorption for water treatment. J. Am. Water Works Assoc., 76(8): 34-43.

United States Government Printing Office, 1962. Public Health Service Drinking Water Standards - 1962. USPHS Publication 956. Washington, U.S.A.

United States Government, 1974. The Safe Drinking Water Act of 1974. PL 523, 93rd Congress.

United States National Academy of Sciences, 1977. Drinking Water and Health, 1-5. National Academy Press. Washington, U.S.A.

Whyte, A.V. and Burton, I., 1980. Environmental Risk Assessment. John Wiley and Sons. New York, U.S.A.

LEGISLATION AND POLICY FOR THE PROTECTION OF THE DRINKING WATER SUPPLY IN THE NETHERLANDS

Dr.Ir. A.M. van Dijk-Looijaard and Dr. H.A.M. de Kruijf
National Institute of Public Health and Environmental Hygiene, P.O.Box 150, 2260 AD Leidschendam (The Netherlands)

The drinking water supply in The Netherlands is particularly influenced by the pollution of surface water with organic micropollutants as the country is located at the delta of the polluted rivers Rhine and Meuse. Also ground water pollution, resulting from intensive industrial and agricultural activities in this densely populated country, is becoming increasingly important.

Consequently the Dutch Government has great interest in international research, discussions and agreements concerning the protection of raw water sources.

This paper summarizes the drinking water quality regulations together with the present legislation and activities carried out for the protection of both surface water and ground water. Most measures are now taken in the international frameworks of the EC (European Community) or IRC (International Rhine Commission), but in the Dutch legislation and sanitation policy additional activities are being carried out to safeguard the quality of drinking water in The Netherlands.

Finally the policy of the Dutch government to continue the safe and durable provision of drinking water in the future is discussed.

THE WATER SUPPLY SITUATION IN THE NETHERLANDS

The first public water supply systems in the Netherlands were founded over 100 years ago - Amsterdam in 1853, Den Helder in 1856 - as a reaction to the cholera outbreak of 1848. Later in the 19th century, other cities also obtained a public water supply system. The country side was connected during the 20th century, increasing the overall percentage of people connected to water supply to over 99% in 1970. The present drinking water consumption in the Netherlands is 1050 million m^3/year, of which 66% originates from ground water and 34% from surface water, mostly taken from the rivers Rhine and Meuse. Ground water abstraction is limited to those areas of the country (figure 1) where fresh ground water is available and where sufficient recharge occurs to prevent damage to agriculture and the environment.

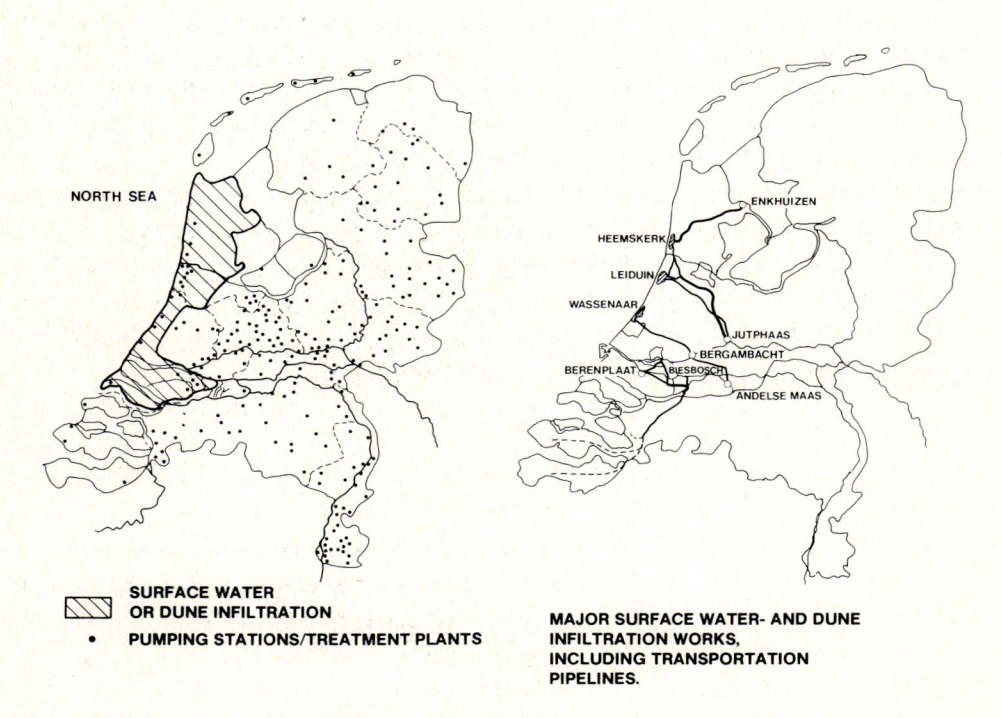

Figure 1 Location of surface and ground water treatment plants in the Netherlands

In the western part of the country fresh ground water is scarcely available hence surface water is used, either after infiltration in the dunes or after storage in reservoirs to bridge periods of low flow and/or bad quality. Surface water is rarely used directly from the river. Locally bankfiltration is applied as purification treatment.

It is beyond doubt that the energetic provision of reliable drinking water has contributed significantly to public health with respect to waterborne diseases, experienced in The Netherlands in the last 100 years.
However the problems associated with pollution of raw water sources - which have attracted broad attention only in the last decades - are also particularly relevant here. Some surface waters are heavily polluted as ·The Netherlands is located at the delta of the rivers Rhine and Meuse, downstream of many (foreign) municipal and industrial discharges. Pollution of soil and ground water is wide spread, as a result of the high demands of population, industry and agriculture on the small area of the country and formerly inadequate legislation and control. In addition soil pollution might have a direct impact on the quality of drinking water by permeation of compounds through drinking water pipelines.
Consequently it is concluded that the watersupply situation in The Netherlands should provide ample data on research, sanitation and control of organic micropollutants.

In this paper, the experiences in The Netherlands with micropollutants and the policy for the protection of raw water sources will be discussed. For clarity distinction will be made between surface water and ground water in view of their specific characteristics (refer to table I). Furthermore, the present legislation with respect to drinking water quality will be elucidated also in view of the policy and guidelines in the EC. Finally, the policy of the Dutch government to reduce pollution of raw water sources and to safeguard the quality of drinking water in the future, will be discussed.

PROTECTION OF SURFACE WATER

Located at the delta of the polluted rivers Rhine and Meuse (De Kruijf, 1982), The Netherlands obviously have a vital interest in international agreements for the protection of the quality of these rivers. The need for continuous efforts against pollution of these major water resources can be illustrated by looking back at the well-known endosulfan spill of 1969 (Greve, 1971), when mass fish mortality revealed the pollution of the river Rhine and the Dutch waterworks had to interrupt the intake of water for 14 days. In 1983 the situation with respect to spills of chemicals still was not satisfactorily (Table II).

Table I
General characteristics of raw water sources with respect to micropollutants

Characteristic	Surface water	Ground water
Number of micropollutants	can be high (hundreds)	generally low
Concentration of micro-pollutants:	generally low (1 µg/l)	generally low, in specific cases high (100 µg/l)
Wash-out time of micro-pollutants:	days	many years
Origin of micropollutants:	diffuse: agricultural and urban run-off, point source: waste (communities, industries, accidents) cannot always be traced	diffuse: deposition point source: concentrated waste (dump, spill) can usually be traced
Sanitation measures: i.e. measures to prevent pollution of drinking water	interrupt intake, extra treatment step	isolate pollution by selectively trapping or pumping activities or purification

Although the quality of the rivers is improving there is still a diffuse pollution of the river Rhine with low concentrations of organo-chlorine micropollutants (figure 2). It is noted that the indicated parameters represent only a fraction of the total amount of organic compounds. From these examples it will be clear that The Netherlands' waterworks face a difficult task to safeguard the provision of drinking water from such a polluted source. The waterworks have to apply the best technical means to treat the water, including emergency storage in reservoirs or by dune infiltration, coagulation/filtration, ozonization, activated carbon treatment, slow sand filtration and sometimes safety chlorination. Figure 3 as an example shows the treatment schemes for the Amsterdam and Rotterdam water works. The results with respect to the removal of organic matter are given in table III.

Table II

MICROPOLLUTANTS WITH A CONCENTRATION > 2 UG/L IN THE RIVER RHINE AT LOBITH

	Classe
Dimethylaniline	B
N.N. diethylaniline	B
Sugarderivative	C
Nitrobenzene	C
Nitroaniline	B
Bischloroisobutylether	B
Dichlorobenzene	B
Ethylidenebis(oxy)bispropane	C
Nitrophenetol	C
Trimethylazidocyclohexene	C
T.butylcresol	C
Bischloropentylether	B
Xylene	C
C-3 benzenes	C
Chinoline	B

1983

0 1 2

o = fase 0 > ug/l
1 = fase 1 > 3 ug/l class B, > 6 ug/l class C
2 = fase 2 > 5 ug/l class B, > 10 ug/l class C

FROM ANNUAL REPORT WRK, 1983

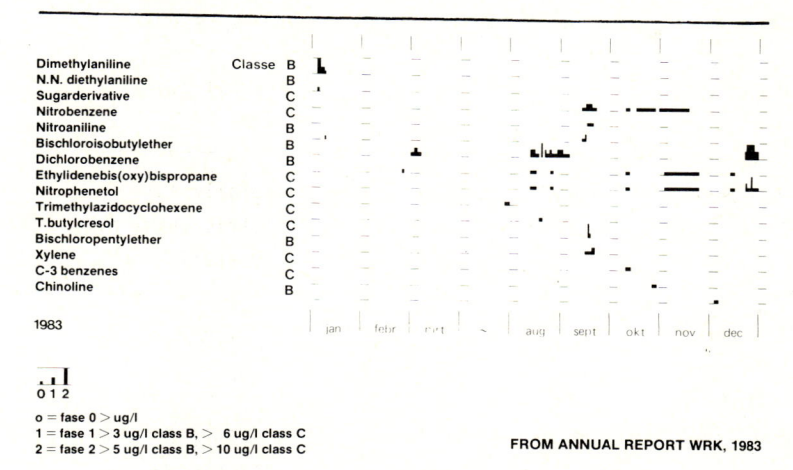

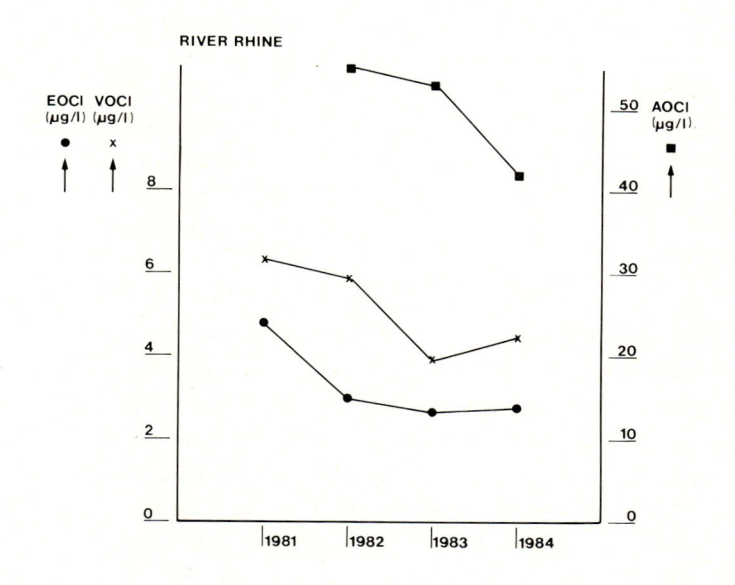

Figure 2. Mean year value of EOCl, VOCl and AOCl in the River Rhine (Data form RIZA and RIWA)

TABLE III Removal of organic compounds by the treatment process.

Component	Unit	River Meuse* at Keizersveer	Tap water* at Kralingen	River Rhine** at Ochten	Tap water** Leiduin
COD	mg/l	14	1.8	13	7
KMnO$_4$-use	mg/l	16	3	13	5.5
UV-extinction (254 nm)	1/m	11	1.6	9.9	
color	Pt-mg/l	13	1	15	5
fenols	µg/l	4	0.5	4	1
mineral oil	"	60	10	110	16
hexachloro-benzene	"	0.01	0.01	0.2	0.01
α HCH	"	0.01	0.01	0.01	0.01
γ HCH	"	0.02	0.01	0.01	0.01
choline-esterase inhibitors	"	0.03	0.01	0.5	0.1
anionic detergents	"	30	10	60	20
total trihalomethanes	"			1.4	3.2
fluoranthene	"			0.08	0.01

* from annual report Rotterdam Waterworks 1983
**   "   "   "   Amsterdam Waterworks 1981

AMSTERDAM DUNE WATERWORKS

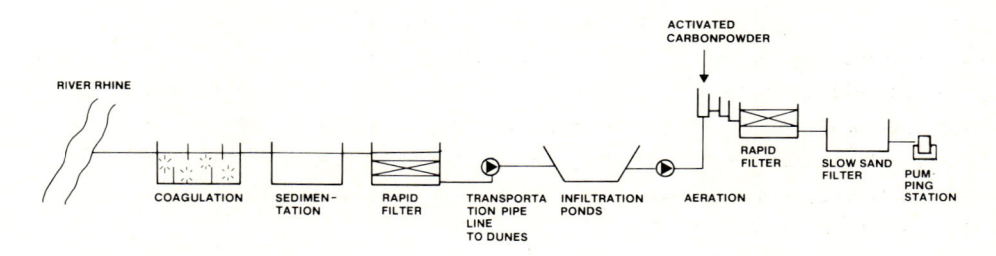

ROTTERDAM WATERWORKS

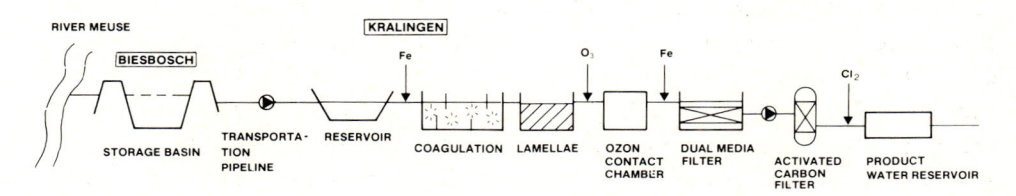

Figure 3

In spite of the high standards of workmanship that the waterworks impose on themselves, there is a growing concern of the long-term effects of compounds that are not removed completely or will be formed by the treatment processes. This applies in particular to the hundreds of organic micropollutants that are present in the Rhine at the $\mu g/l$ level or below. For compounds that incidentally occur in higher concentrations - mostly because of spills - the waterworks have established a continuous monitoring system that will interrupt the water intake if the contamination level exceeds a certain level. Obviously this system cannot be used for the many low level pollutants.

It was shown (Kool et al, 1982) by experiments using the Ames-test that concentrates (200-fold) of the river Rhine water containing these pollutants show an increased mutagenic activity, 10 times as much as river Meuse water. The mutagenic activity of unchlorinated drinking water prepared from Rhine water is low, so it appears that at least those compounds contributing to the Ames-test are removed to some degree by the treatment processes.

Nevertheless, reducing pollution of the surface water is of vital importance to The Netherlands' waterworks. Consequently, The Netherlands have stimulated international discussion and agreements to combat pollution of surface water. These activities have been carried out in 2 main frameworks, the European Community (EC) and the International Rhine Commission (IRC).

INTERNATIONAL RHINE COMMISSION

In 1963, the International Commission for the protection against the pollution of the River Rhine (International Rhine Commission, IRC) was officially founded by the Bern Convention. Member states are Switzerland, France, West-Germany, Luxembourg and The Netherlands (in 1972 the European Community also joined the Commission). At first, the position of the IRC was rather weak, but after the endosulfan-incident in 1969 The Netherlands' government took the initiative to come to regular Conferences of the Ministers of the member states. In 1972 at the first Minister Conference it was decided that the IRC would draft regulations against a.o. the chemical pollution of the river Rhine. After some delay, caused by the desire to harmonize the regulations for the river Rhine with the EC-regulations, the Convention against the pollution of the river Rhine was signed in 1976. The content of the Convention is essentially similar to the EC-directive on surface water pollution (76/464/EC), with respect to the definitions of the so-called black list and grey list substances. The former list includes substances which are both toxic, persistent and lead to bioaccumulation. The discharge of these substances should be minimized as far as possible ("best technical means"). The greylist substances are less harmfull and the criteria for discharge of these substances should be based on economic criteria ("best feasible means") and on the intended use of the surface water also.

Differences between the work of the EC and the IRC appeared in the selection of substances for the black and grey lists. Naturally the IRC wanted to focus primarily on substances relevant to the river Rhine. The original EC list of 1500 black-list substances was reduced to 83 by the IRC by applying the following criteria (Jansen and Dekker, 1984):

1. the substance should be produced in the Rhine area at an amount of 500 ton/year, equivalent to an average concentration of 10 µg/l;
2. the substance should have been detected in the river Rhine;
3. the substance should be of great importance to the environment.

These criteria also lead to a selection of 34 compounds out of the list of 83 for priority study.

The formal results i.e. actual guidelines are limited to mercury from the chlorine-alkaline industry (1979, in force 1983), and cadmium (in force 1983). Regarding the drins (aldrin, dieldrin and eldrin) it has been decided to await the EC measures, as the only producer of drins is located downstream of Rotterdam (Jansen and Dekker, 1984). For chlordane and heptachlore no measures will be taken as these pesticides are no longer produced in the Rhine basin. Regarding DDT, γHCH, HCB and endosulfan the EC measures are awaited as no important discharges occur in the Rhine basin. PCB's are present in the Rhine, but mainly as a result of diffuse discharges from PCB containing objects. Consequently the IRC has adopted the EC directives concerning the use, collection and disposal of PCB-containing objects.

Finally the IRC has decided not to prepare discharge criteria for arsenic, benzidine and polynuclear aromatic hydrocarbons (PAH's) as insufficient evidence is available to establish a negative health effect at the concentrations found in the Rhine (benzidine) or the emission is too diffuse to enable efficient control through the Rhine Convention. A review of the status of the black list substances of the IRC is given in table IV.

For the grey-list substances so far only for chromium a quality goal and discharge criteria have been formulated.

Reviewing the official results of the IRC, the conclusion must be drawn that the actual results in terms of official directives for black listed substances are really modest. However, the publicity resulting from the IRC-activities together with the political impact of IRC activities in the EC have already led to a number of sanitation measures, especially with respect to industrial discharges, and thus to a decrease of the pollution of the river Rhine.

EUROPEAN COMMUNITY

The European Community's first Environmental Action Programme was adopted by the Council of Ministers in 1973. It ran for four years and has been followed by two further Action Programmes - the second from 1977 - 1981 and the third from 1982- 1986.

In all 3 programmes, considerable emphasis has been placed on the importance of water quality and of action to protect it. The first Action Programme set up the basic structure of legislation, while the second Action Programme developed environmental quality objects, standards and guidelines. The third Programme seeks to plan for an overall improvement in environmental quality based on the principles that each action must be applied at the most appropriate level and that prevention rather than cure should be the rule.

TABLE IV Present status of black-list substances in the EC and the IRC

IRC nr.	EG nr	Compound	In study IRC	In study EC	Proposal of the commission IRC	Proposal of the commission EC	Status 1984 IRC	Status 1984 EC
62	92	mercury					x	x
7	12	cadmium					x	x
1		aldrin				x		
51	71	dieldrin				x		
54	77	endrin				x		
9	15	chlordane			x	x		
56	82	heptachlor (and epoxide)			x	x		
33	46	DDT and metabolites				x		
59	85	HCH					x	x
57	83	hexachlorobenzene		x				
68	101	PCBs and PCTs		x			x	x
53	76	endosulfan		x				
58	84	hexachlorobutadiene	x	x				
69	102	pentachlorofenol	x			x		
2	4	arsenic				x	x*	
6	8	benzidine		x			x*	
66	99	PAH		x			x*	
83	128	vinylchloride			x			
41	59	1,2-dichloroethane	x	x				
55	78	epichlorohydrine	x					
34	48	1,2-dibromoethane	x	x				
	122	2,4,6-trichlorofenol	x	x				
5	7	benzene	x	x				
8	13	tetrachloromethane	x			x		
15	13	chloroform	x			x		
78	121	trichloroethene	x	x				
73	111	tetrachloroethene	x	x				
18	28	o chloronitrobenzene						
19	29	m " "						
20	30	p " "						
75	117/118	trichlorobenzenes	x	x				
10	17	o-chloroaniline		x				
11	18	m- "		x				
12	19	p- "		x				
42	60	1,1-dichloroethene		x				
43	61	1,2-dichloroethene		x				
	119	1,1,1-trichloroethane		x				
		1,1,2- "		x				
		1,1,2,2-tetrachloroethane						
	20	monochlorobenzene		x				
	62	dichloromethane		x				
	65	1,2-dichloropropane		x				
	89	malathion		x				
	64	2,4-dichlorofenol						
	22	2-chloroethanol						
	66	1,2-dichloropropanol						
	100	parathion		x				

* no discharge criteria have been given
** 18 other chemicals have not been selected yet; some investigations have started

A considerable number of directives dealing with water quality has been prepared in the past 12 years, which can be divided in 2 main categories:
1. Directives that lay down quality objectives for water for particular uses (refer to table V).
    These include the directives for drinking water (80/778/EC) and for surface water destined for drinking water production (surface water directive, 75/440/EC).
2. Directives concerning the discharge of dangerous substances (refer to table VI).
    These include two "parent" directives for the discharge of dangerous substances into the aquatic environment (76/464/EC) and into ground water (80/68/EC).

Table V   EC Directives

Nr.	Short title
70/659/EEC	Freshwater fish directive
75/440/EEC	Surface water directive
76/160/EEC	Bathing water "
77/795/EEC	Exchange of information on surface water (Decision)
79/869/EEC	Frequency of sampling and analysis for surface water
79/923/EEC	Fish and shell fish waters directive
80/777/EEC	Marketing of natural mineral water
80/778/EEC	Drinking water directive

Table VI   EC Directives

Nr.	Short title
76/403/EEC	Discharge of PCBs and PCTs
76/464/EEC	Dangerous substances into aquatic environment
78/176/EEC	Titanium dioxide industry
80/ 68/EEC	Ground water directive
82/176/EEC	Mercury from chlorine-alkaline electrolyses industry
83/513/EEC	Cadmium discharge directive
84/156/EEC	Mercury from other sectors
84/491/EEC	HCH discharge directive

The directive related to the discharge of dangerous substances into the aquatic environment follows the same pattern as the Chemical Convention of the Rhine. It also defines black-list and grey-list substances. The present status of the selection of black-list substances in this directive is also given in table IV (Dekker, 1985). After a substance has been studied, a proposal is made which should be approved of by the memberstates before an official directive comes into force. Compared to the IRC, additional discharge criteria have been prepared by the EC for Hg-emissions from other sectors than the chlorine--alkaline industry as well as for Cd and HCH (lindane).

The directive relating to quality objectives for water for particular uses and functions had a major impact on the sanitation of surface water in the member states. These have the obligation to incorporate the quality objectives in their national legislation. The quality criteria for organic compounds of the surface water directive, together with the recently adjusted Dutch legislation, are presented in tabel VII. In the Dutch surface water legislation (WVO, 1983), the EC quality class A2 has been taken as the objective; this code refers to surface water destined for drinking-water production that can be treated by means of conventional physical-chemical treatment. Two organic parameters have been replaced by comparable parameters (EOCl and VOCl).

It should be stipulated that the definition of quality objectives is only the first step in the sanitation process. At present not all surface waters in the Netherlands comply with all criteria. The central Government outlines its general policy and priorities for improvement of water quality in the so called IMP- 5 year plan (Water Action Program). The regional agencies that control the surface water quality (Waterboards) now face the task to reduce the discharges by modifying discharge licenses and by influencing other governmental authorities to impose limits on diffuse discharges (e.g. pesticides) (de Ruijter, 1984). Next to the WVO, the EC- surface water directive has also been included in the new Drinking water supply by Law (1984). Depending on the purification system which is available category I, II or III must be used. The legal obligation in the Drinking water supply by Law includes that surface water which does not meet the standards in column B of the specified category of purification may not be substracted for drinking water production. The standards in column A are triggering values to inform the Waterboards and the Regional Inspector of Public Health that the surface water does not meet the EC and WVO-criteria.

TABLE VII   Surface water quality regulations

Parameter	Unit	EC Surface water directive 1975							WVO 1983	Drinking water supply by Law 1984						IMP-5 year plan water 1981-1984
		Class A1		Class A2		Class A3			Class I		Class II		Class III		
		G	I	G	I	G	I		A	B	A	B	A	B	
Color	mg/l Pt/Co	10	20	50	100	50	200	50	-	20	50	100	-	200	
Suspended solids	mg/l	25	-	-	-	-	-	50	25	-	-	50	-	-	
Smell	dilution factor	3	-	10	-	20	-	16	3	-	16	-	20	-	
Taste	"														
With chloroform extractable compounds	residue mg/l	0.1	-	0.2	-	0.5	-								
Mineral oils or emulgated or dissolved carbo-hydrates	µg/l	-	50	-	200	500	1000	200	-	50	200	200	-	1000	200
Fenols	µg/l	-	1	1	5	10	100	5	-	1	-	5	10	10	10
Detergents which react with methylene blue	µg/l	200	-	200	-	500	-	200	200	-	200	-	500	-	200
Pesticides - individual - total - choline-esterase inhibitors	µg/l	-	-	-	2.5	-	5	0.05 0.1 1.0	- - -	- 0.05 0.5	0.05 0.1 1	- 0.5 2	- - -	- 0.5 5	0.05 0.1 1
PAH's COD EOCl VOCl BOD	µg/l µg/l O$_2$ µg/l µg/l mg/l O$_2$	- - 3	0.2 - -	- - 5	0.2 - -	- 30 7	1 - -	0.2 30 10 20 7	- - 3	0.2 - -	0.2 30 6	0.2 - -	- 40 7	1 - -	0.2

PROTECTION OF GROUND WATER

The Netherlands are amongst the most densily populated countries of the world, with 14.4 million inhabitants on a total area of 36900 km^2 (population density 426/km^2). Moreover, the country is highly industrialized and in agriculture, horticulture and cattle breeding intensive production techniques are widespread. Consequently, a high stress is placed on the soil and its capacity to adsorb and degrade any pollution and waste products resulting from human activities. In the past decade it became quite clear that soil pollution in many countries presents a serious threat to the ground water resources. This danger particularly exists in The Netherlands as ground water is the major water source for the Dutch Waterworks (700 million m^3/year or 66% of the total use) and in many areas ground water is almost exclusively used for water supply purposes.

Ground water pollution can be caused by two major groups of activities (Zeilmaker, 1984), i.e.
1. Activities that can be located clearly
    - disposal of domestic and industrial waste
    - storage and transport of substances
    - spills and leakages of substances
    - infiltration of polluted surface waters
2. Diffuse activities
    - agricultural use of manure, fertilizers and pesticides
    - precipitation of airborne pollutants.

The first group of activities have created a number of well-known soil pollution incidents, since the case of "Lekkerkerk" in 1980 (Brinkman, 1981). At present some 4000 cases of soil pollution have been registered in The Netherlands, of which some 1000 are situated in areas destined for housing or water supply. Pollution of soil and ground water is many times discovered only many years after the original polluting activity. This is due to low velocity of the transport of the pollution as a result of the low flow velocities of the ground water and effects of adsorption and degradation. Emergency measures against pollution usually consist of either removal of the polluted soil (and treatment of the polluted ground water) and/or isolation of the pollution, by means of impervious materials (plastic sheets, steel walls, clay layers and the like) or of the ground water by means of discharge wells. These sanitation measures are carried out under the Emergency Soil Purification Act, which acts as a temporary Act until the new Act on Soil Protection will come into force in 1985. At present in sanitation projects a list of A, B and C values, for concentrations of 51 compounds or classes of compounds in soil/sludge and in ground water/surface water, is being used as a reference framework (table VIII).

Table VIII   Guidelines for soil sanitation

	Ground and surface water (ug/l)		
Component Level	A	B	C
- benzene	0.2	1	5
- ethylbenzene	0.5	20	60
- toluene	0.5	15	50
- xylenes	0.5	20	60
- fenols	0.5	15	50
- total aromatic hydrocarbons	1	30	100
- naphtalene	0.2	7	30
- anthracene	0.1	2	10
- fenanthrene	0.1	2	10
- fluoranthene	0.02	1	5
- pyrene	0.02	1	5
- benzpyrene	0.01	0.2	1
- PAH's total	0.2	10	20
- individual halogenated hydrocarbons	1	10	50
- total halogenated hydrocarbons	1	15	70
- chlorobenzene individual	0.02	0.5	2
- chlorobenzene total	0.02	1	5
- chlorofenols individual	0.01	0.3	1.5
- chlorofenols total	0.01	0.5	2
- chloro PAH's total	0.01	0.2	1
- PCB's total	0.01	0.2	1
- EOCl	1	15	70
- pesticides total	0.05	0.2	1
- org.chloro pesticides individual	0.1	0.5	2
- org.chloro pesticides total	0.1	1	5
- tetrahydrofuran	0.5	20	60
- pyridine	0.5	10	30
- tetrahydrothiophene	0.5	20	60
- cyclohexanone	0.5	15	50
- styrene	0.5	20	60
- gasoline	10	40	150
- mineral oil	20	200	600

These values are primary meant for watercatchment-, housing- and preservation areas (Guidelines for soil sanitation) and indicate respectively the "natural" reference value, the situation where additional research may be required and the situation where sanitation might be required eventually. The diffuse polluting activities have only recently become known, mainly because of the concern around "acid rain" and nitrate-pollution of ground water, caused by animal waste disposal and festilizer use. Because of the diffuse nature of this pollution, measures resulting in an immediate effect are hard to find. Efforts are being concentrated on prevention rather than cure.

The EC-directive on ground water pollution (80/68/EC) basically deals with the first group of activities only (Zeilmaker, 1984). It defines black-list and grey-list substances (table IX).

Table IX   EC Groundwater Directive (80/68/EC)
___

Black-list substances
- organic halogenated-, P- and Sh-compounds
- carcinogenic, mutagenic or teratogenic compounds
- Hg and Hg-compounds
- Cd and Cd-compounds
- cyanides
- mineral oil and hydrocarbons

Grey-list substances
- metalloïds and metals with metal-compounds:
  Zn, Cu, Ni, Cr, Pb, Se, As, Sb, Mo, Fe, Sn, Ba, Be, B, Co, V, U, Th, Te, Ag
- biocides and derivatives not on list I
- some organic Si-compounds
- taste and smell causing compounds
- inorganic P-compounds and P
- fluorides, ammonia and nitrites
___

It is more strict than the directive on discharge of dangerous substances in the aquatic environment because the definition includes groups and families of substances, and one has to prove that a compound does not belong to one of these lists before it can be discarded. Discharge of black-listed substances is to be reduced by the "best technical means", whereas discharge of grey-listed substances should be limited to such an extent that pollution of ground water is prevented. This EC-directive will be implemented in the new Dutch Act on Soil Protection, that also details specific soil protection areas and ground water recharge areas.

These areas will receive additional protection, to be detailed in regional plans and provincial ground water plans.

Regarding the ground water recharge areas, a separate committee has already proposed several protection measures, including the definition of special zones based on the residence time of the ground water (respectively 60 days, 10 years and 25 years, refer to figure 4 (From report on guidelines and recommendations for the protection of water catchment areas).

CLASSIFICATION OF PROTECTION AREAS

SAND- AND GRAVEL SEDIMENTS

- 60D RESIDENCE TIME
- 10YR RESIDENCE TIME
- 25YR RESIDENCE TIME
- BORDER RECHARGE AREA

- 25 YEAR ZONE
- 10 YEAR ZONE } PROTECTION AREA
- WATER CATCHMENT AREA

LIMESTONE AND SANDSTONE SEDIMENTS

BORDER RECHARGE AREA

- REST RECHARGE AREA
- PROTECTION AREA R = 2 KM
- WATER CATCHMENT AREA

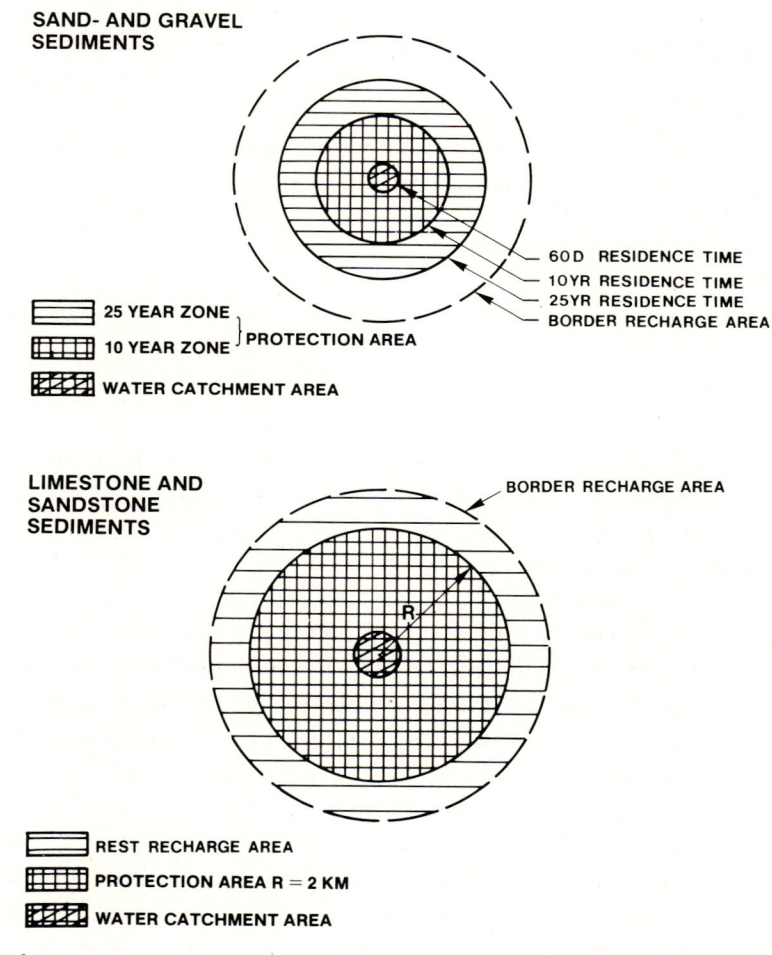

Figure 4.

In these zones different activities will be prohibited (refer to table X). This will provide a partial base for protection against some of the more diffuse activities causing pollution, such as pesticides.

The effects of airborne pollutants are more difficult to combat. In the summer of 1981 the VOCl in rain water at The Bilt (annual report 1981 KNMI/RIV)) was 0,7 µg/l e.g. and in the same period in 1983 the sum of several volatile aromatic compounds was 0,24 µg/l (Van Noort, 1985).

Recent measurements indicate an increasing concentration of several contaminants in ground water. Table XI gives an example of diffuse pollutants that appear to be widespread in the environment (Van Duyvenboden et al, 1985).

Although concentrations in ground water are generally still low, in some pumping stations high concentrations of organic micropollutants have already been found. Good control of pumping stations based on adequate parameters need still more attention as numerous organic micropollutants may occur in ground water for which specific parameters not yet exist or for which existing parameters are not sufficient. Development of bioassay parameters suitable for toxicological monitoring are also important in this context.

TABLE X

Activities which are usually prohibited in a protection area:
___

- transport, storage, handling, desposit, production and processing of compounds which could influence the quality of ground water
- disturbance of ground and surface layers and exposure of aquifers
- traffic infrastructure
- infiltration of surface water
- military activities and installations, air traffic infrastructure
- use of fertilizer and pesticides
- intensive farming, fertilizer storage, fertilizer disposal, use of sewage treatment sludge
- use of building materials which could lead to pollution, e.g. tar, bitumen and
- scoria
- infiltration of cooling water
- washing and maintenance of cars etc.
___

TABLE XI

Example from ground water quality monitoring network

Soil use/ soil type	8-10 m below surface level			23-25 m below surface level		
	TOC (mgC/l)	EOCL (μgCl/l)	VOCl (μgCl/l)	TOC (mgC/l)	EOCL (μgCl/l)	VOCl (μgCl)
Farmland	7.7	0.8	1.1	6.2	0.9	0.2
Grassland	11	0.5	0.1	7.6	0.5	0.1
Preservation areas	3.9	0.5	0.2	3.8	0.6	0.2
Sand	7.3	0.6	0.3	5.4	0.5	0.2
River clay	6.0	0.4	0.2	5.9	0.5	0.3
Sea clay	11	0.5	0.2	8.1	1.8	0.3
More-peat	13	0.6	3.7	15	0.4	0.1
Peat	16	0.6	0.1	12	0.7	0.1
Loam	8.4	0.9	0.1	11	1.5	0.3

number of samples: 297   number of samples 279
From RIVM report nr. 840382001 1985

DRINKING WATER QUALITY REGULATIONS

Mainly as a result of the increasing pollution of raw water sources, the quality of drinking water requires permanent attention. The EC-directive on drinking water quality (80/778/EC) specifies a list of 62 parameters for which maximum allowable concentrations (MAC) or guidance levels (GL) are given. In the EC-directive a distinction was made between 5 groups of parameters.
1 Organoleptic parameters
2 Physical-chemical parameters
3 Parameters concerning undesirable substances
4 Parameters concerning toxic substances
5 Microbiological parameters

Table XII gives the values for parameters which are related to organic compounds.

The MAC values may only be exceeded as a result of natural causes (e.g. color) or exceptional weather conditions, whereas the GL values serve as a quality goal.

The EC-directive has been the base for the new Drinking water supply by Law of 1984. In the Dutch Act, 4 groups of parameters have been distinguished.
A. Parameters which may never exceed the given levels (toxic substances and microbiological parameters)

B. Parameters that may only exceed the indicated levels as a result of natural causes or exceptional weather conditions.
C. Parameters which may be exceeded, considering the water source, the preparation and distribution. Exceedence should be reported.
D. Parameters of which a minimal concentration is required.

The parameters ad B. consist of those organoleptic, fysical-chemical and undesirable parameters for which the EC has given MAC values. For the parameters ad C. the EC only gave guideline values.

Next to an analysing and sampling scheme, the frequency of measurement of all parameters has been given in the Drinking water supply by Law.

Simultaneously with the new Drinking water supply by Law, the VEWIN (The Netherlands Waterworks Association) has prepared new "Recommendations for water supply companies". These Recommendations contain some additional quality goals for group B- and C-parameters

For organic substances a quality goal is given for color, TOC and trihalomethanes

For trihalomethanes a value of 70 μg chloroform/l has been given, which is rather high compared to the values measured by the waterworks at the moment. The new drinking water regulations (EC, Drinking water supply by Law and VEWIN Recommandations) give a more sound base for the supervision and control of drinking water quality in the light of the environmental pollution occurring in present-day society. Nevertheless, it should be remembered that these regulations only function as a primary check. Still continuous attention with respect to the occurrence of organic micropollutants is required to safeguard the quality of drinking water and to prevent pollution. This does not solely apply to the water sources, but also to water treatment as e.g. formation of haloformes during chlorination may occur and reactions with yet unknown pollutants during treatment. Similarly during water distribution migration of organic micropollutants (such as methylbromide) through plastic pipes may occur or the leaching from coatings. In view of the vital importance of water supply, one should strive for proper management of the whole water system from source to tap with continuous attention for the quality of the product. It should be mentioned in this respect, that the KIWA (Netherlands Waterworks Testing and Research Institute and the Public Institute of Health) have developed a system of testing products applied in the water supply sector, such as chemicals, pipeline materials, coatings etc. This system aims to arrive at a positive list of products that are allowed to come into contact with drinking water from a toxicological point of view (Noordam, 1985).

At this moment the series of known organic micropollutants and its hazards for men is steadily growing due to new analytical techniques.

TABLE XII Drinking water quality regulations

Parameters	Units	EC Drinking-water Directive 1980 GL	MAC	Drinking water supply by Law
color	mg/l Pt/Co	1	20	20
turbidity	mg/l SiO	1	10	10
smell	dilutionfactor	0	2 at 12°C 3 at 25°C	2 at 12°C 3 at 15°C
taste	dilutionfactor	0	2 at 12°C 3 at 25°C	2 at 12°C 3 at 15°C
$KMnO_4$	mg/l O_2	2	5	5
with cloroform extractable compounds	mg residue/l	0.1	-	1
mineral oil, emulgated and dissolved hydrocarbons	µg/l	-	10	10
fenols	µg/l C_6H_5OH	-	0.5	0.5
detergents which react with methylene blue	µg/l laurylsulfate	-	200	200
chlorinated hydrocarbons (no pesticides)	µg/l	1	-	1
Pesticides individual	µg/l	-	0.1	0.1*
Pesticides total		-	0.5	0.5
PAH's	µg/l	-	0.2	0.2

* including choline-esterase inhibitors

Present day series of parameters do not at all cover the list of organic micropollutants. The question remains whether for each new hazardous low concentration compound a guideline value should be indroduced or that other adequate control measures should be created.

POLICY OF THE DUTCH GOVERNMENT

An effective policy for the protection of drinking water quality should follow several tracks to cope with the many threats of the industrialized society. This is particularly true for The Netherlands. The policy of the Dutch Government thus has to focus on 5 areas:
1. International discussions and co-operation;
2. Prevention of pollution;
3. Reduction of emissions
4. Best technical treatment techniques;
5. Inspection and supervision.

Ad. 1.
International co-operation is of vital importance in view of the large imports of pollutants through the rivers Rhine and Meuse and through the air. Moreover, international measures have the advantage that the economic consequences are borne by all European partners. Consequently the Dutch Government strongly promotes the environmental protection activities of IRC and EC (and Meuse commission).
Ad. 2.
Prevention of pollution is the most logical measure in environmental protection policy (IMP-M). Although it is in many cases impossible to achieve, the Dutch Government pursues this by promoting the development of clean technologies, the efficient use of chemicals in agriculture (e.g. fertilizer) and stimulation of the use of clean raw materials (e.g. natural gas).
Ad. 3.
The reduction of emissions in the field of water supply are pursued by:
- reduction of discharges to surface water following IMP-5 year plan Water;
- reduction of discharges to groundwater following IMP-5 year plan Soil;
- reduction of airborne pollutants following IMP-5 year plan Air;
  (These strategies per compartment are integrated in the IMP-M which attempts to present a total view on the environment and its protection.)
- reduction of fertilizer discharge by collection and central treatment (through Fertilizer Act).

- Guideline for the use of chemicals during the drinking-water preparation;
- Guideline for the prevention of permeation of pollutants through pipeline materials.

Ad. 4.

The best technical treatment techniques in the field of water supply include:
- storage facilities to interrupt intake during spills (either in reservoirs or infiltration in soil);
- optimization of treatment processes for removal of organic micropollutants (coagulation, ozonation, activated carbon filtration);
- use of alternative disinfectants to chlorine.

Ad. 5

Inspection and supervision includes not only routine analyses but also the research into and the evaluation of existing parameters. Basically control always must be based on sanitary survey and insight on threats of the quality of the source.

Although one always should be alert to individual toxic substances, it is impossible to screen 600 - 1500 organic micropollutants or more on a routine base. Additional research therefore should be performed on groupparameters such as VOX, EOX AOX, TOC and AOC. Until now only a small fraction (5 - 15%) of the total amount of organic micropollutants in water can be traced with the parameters presently in use. It is important to assess the toxicological significance of the unknown organic substances. The development of biological group parameters can possibly be helpful in indicating the actual hazards of these compounds. The development of sensitive (i.e. covering most pollutants), economic attractive parameters, which are suitable for routine analysis is an excellent present-day challenge.

The safeguarding of drinking water with special respect to hazardous organic compounds could in our view be reached best by such parameters.

REFERENCES

Brinkmann, F.J.J..Proc. Second European Symp., Killarney, 1981. Analysis of Organic Micropollutants in water p.51-53.
Dekker, R.J. (RIZA), Personal communication, 1985.
Duijvenboden, W., van, Taat, J., Gast, L.F.L. Landelijk Meetnet Grondwaterkwaliteit RIVM Report. nr. 840382001, 1985.
Greve, P.A., Wit, G.L. J. Water Poll. Control Fed., 43 (12), 2338-2348 (1971).
Jansen, J.H., Dekker, R.H. H_2O (17) nr. 10, 213-218, 1984.
Kool, H.J., Kreijl, C.F., van, Greef, E., de, Kranen, H.J., van, Environ. Health. Perspect. 46, 207-214, 1982.
Kruijf, H.A.M., de, Toxicol. Environ. Chem., 6, pp. 41-63, 1982.
Guidelines for Soil Sanitation, VROM 1983, The Hague ISBN 9012044030.
Noordam, P.C., Graveland, A., H_2O (18), nr. 8, 158-159, 1985
Noort, P. van (RIVM), Personal Communication, 1985.
Report on the water quality of the Rhine in The Netherlands. 1970-1983 and Suppl. 1983, nr. 84-097 RIZA (State Institute for Waste Water Treatment).
Guidelines and recommendations for the protection of water catchment areas

(1980 KIWA/RID Report.

Ruiter, M.A., de, H_2O (17), nr. 22, 522-526, 1984.

Chemical Composition of Precipitation over the Netherlands, Annual Report KNMI/RIV (nr. 217 810 006), 1981.

Trouwborst, T. H_2O 814), nr.1, p.4-10, 1981.

Zeilmaker, D.A. H_2O (17) nr.22, p.500-507, 1983.

WORLD HEALTH ORGANIZATION, GUIDELINES ON ORGANIC MICROPOLLUTANTS

W.M. Lewis
Consultant, Promotion of Environmental Health, World Health Organization, Regional Office for Europe, Copenhagen, Denmark

INTRODUCTION

A vast amount of information has accumulated over the years associating disease and ill-health in the population with drinking water of unsatisfactory quality. In an endeavour to meet the need for some form of international agreement on requirements for a safe, potable water supply, the World Health Organization in 1958 published its first International Standards for Drinking Water. These were revised in 1963 and again in 1971, but in each, quality emphasis was primarily directed to microbiological aspects although some inorganic contaminants also received consideration.

It was generally considered, and until recent years accepted, that special considerations applied to drinking water supplies in the European Region and this was reflected in the 1970 European Standards for Drinking Water, which updated the 1961 publication, by the inclusion of the first ever quality standard for an organic contaminant in drinking water.

Over the years the W.H.O. Standards have been beneficial in providing soundly based information on quality criteria for a safe and acceptable drinking water supply. However, useful though such publications have been, a fair amount of justifiable criticism has been levelled against them. Basically, such criticism emphasized the intransigent, authoritative nature of the standards and the lack of helpful advice when quoted criteria were breached and, secondly, the distinction between acceptable water quality for the European Region and the rest of the world.

CRITICISM OF W.H.O. STANDARDS

At the time of their formulation it was the general concensus that the European problems associated with source water quality, were different from those elsewhere, due essentially to the problems created in Europe by the discharge of industrial waste to the aquatic environment. If this premise was indeed a true reflection of conditions existing at that period, it has become obvious during the course of the past decade that the gap, if such there was, has gradually been reduced and many of the problems relating to water are today universal. Moreover, the modern philosophy is to consider judgement of water quality essentially on toxicological data acknowledging that the human system will, in general, react similarly when subjected to a given concentration of toxic agent. It was, however, recognized that under differing socioeconomic conditions, varying climates and other specific circumstances a common universal water quality standard would not be applicable.

W.H.O. GUIDELINES FOR DRINKING WATER QUALITY

The World Health Organization decided therefore that guidelines for drinking water quality, with values based upon health-related criteria, would enable countries to develop their own national standards using their own risk/benefit criteria and taking into account their own indigenous circumstances to interpret the data supplied by W.H.O. The need for some universal guidance was urgent in view of the impending pressure as the International Drinking-Water Supply and Sanitation Decade progressed, and thus in 1979 the Regional Office for Europe, Copenhagen, and World Health Organization, Geneva, with generous support from the Danish International Development Agency (DANIDA), commenced the preparation of the "W.H.O. Guidelines for Drinking Water Quality". The work in preparing Guidelines covered a period of two and a half years from the first meeting in December 1979 to the plenary finalizing the whole in March 1982. The publication is intended to comprise three volumes: Volume 1, "Recommendations", which appeared October 1984; Volume 2 which will consist of criteria monographs for each substance or aspect considered in Volume 1; and Volume 3 concentrating basically on microbiological recommendations particularly related to rural areas and small communities, especially in the developing countries of the world. The particular aspect of Guidelines which concerns this International Symposium, the organic micropollutants, was the specific responsibility of the Regional Office for Europe of the World Health Organization

Modern society depends to an ever-increasing degree upon the supply of chemicals, whether for consumer products, agriculture or health care, etc. 70 000 chemicals are available on the commercial market with perhaps an additional 1000 new chemical entering into commerce each year. The production of synthetic organic chemicals has increased exponentially since about 1940 and shows little sign of diminishing. Problems associated with industrial effluent disposal have resulted in chemical pollution of water resources and, consequently, drinking water supplies to such an extent that in excess of 700 chemical compounds have been identified in drinking water, over 600 of which are organic compounds many of which are pharmacologically active. The benefits of chemicals are often clear but the risks to health less so and it is therefore to be expected that there is increasing public concern about the possibility of harmful effects on health associated with the chemical contaminants of water. The biological effects of a number of organic chemicals can range from acute to the long-term mutagenic, carcinogenic, neurotoxic or behavioural changes. Thus, in considering the problem of micro-organic pollution in drinking water supplies, W.H.O., attempting to offer guidance on safe levels of intake, was faced with a quite formidable task. It is beyond comprehension that an attempt should be made to define safe limits for all such organic compounds at present identified as contaminants in drinking water. To resolve the dilemma a consultation decided upon a selection of five criteria to evaluate the choice from the known range of organic contaminants. Those included

(a) existence of toxicological evidence of possible harmful effects;
(b) frequency of occurrence in various drinking waters;
(c) concentration levels recorded in drinking water;
(d) availability of suitable analytical techniques;
(e) remedial measures available for removal.

Using these criteria a selection of just under 50 organic compounds was then examined in detail before finally allocating firm guideline values to fifteen organic compounds, and tentative guideline values to a further three. Before continuing further, it would be helpful to define what is meant by a "guideline value" as it is important that the quoted guideline values are not just abstracted by those people who are disciples of the syllogism of Descartes and translated into maximum acceptable concentrations.

"A guideline value represents a concentration or a number that ensures aesthetically pleasing water and does not result in any significant risk to the health of the consumer. A guideline value when exceeded is to be used as a signal

(i) to investigate the cause with a view to taking remedial action;

(ii) to seek advice from authorities responsible for public health."

The guideline values specified have been established to safeguard health on the basis of lifelong consumption. Short-term exposures to higher levels, such as might occur following an accidental spill, may be tolerated. The amount by which, and the duration for which, any guideline value can be exceeded without affecting public health depends upon the specific substance involved.

Throughout Volume 1 many pages have been devoted to offering guidance relating to potential practical remedial measures in the event of divergence from a recommended guideline value in a drinking water supply. As the main objective of surveillance activities is the detection of deviations in a drinking water supply as quickly as possible, it is also axiomatic that such deviations from acceptable quality be corrected with the least possible delay.

The actual guideline values set tend to err on the side of caution because of insufficient qualitative and quantitative toxicological evidence and also the uncertainties associated with interpretation. The most important influence on the quality of drinking water of some organic compounds which were examined is their influence on the aesthetic or organoleptic aspects rather than their direct health effects.

PRINCIPLES UPON WHICH GUIDELINE VALUES WERE SET FOR ORGANIC CONTAMINANTS

It was assumed that the average consumption of water is 2 litres per person per day and that the average weight of a person is 70 kg. For those organic compounds whose toxic effect only becomes apparent after a dose threshold has been exceeded, an acceptable daily intake (ADI) - previously established by the joint F.A.O./W.H.O. Expert Committee on Food - was used as the basis for calculating a guideline value. When a published ADI was not available, a value was calculated using data available from the scientific literature by applying a "safety factor" to the "no-observable-adverse-effect dose" (i.e., the maximum ineffective dose). It was then decided what proportion of the ADI should be allocated to drinking water and what proportion to other sources of exposure (food, air, etc.).

The magnitude of the "safety factor" is determined by the strength of the toxicological evidence and the nature and magnitude of the health risk. A safety factor as low as 10, for example, is used only when evidence is available of human health effects plus supportive data from other species. When the health data are limited or incomplete either when relating to acute or chronic effects, the safety factor may be 1000. Intermediate values are also used where appropriate. The ADI allocated to drinking water was determined as an inverse function of the tendency for the chemical to accumulate in food chains. Thus, for chemicals like the chlorinated pesticides, which readily accumulate, as little as 1% of the ADI was allocated to drinking water, whereas for those organics that accumulate to a lesser extent a greater proportion was allocated, but not more than 10%.

For very many organic compounds there are few data on actual or potential sources of human exposure but, where known, consideration was given to the proportion of normal intake derived from those alternative sources. However, when the guideline value calculated was such that at this level a detrimental effect on taste and odour or the aesthetic quality of the drinking water would result, a level at or beneath the taste threshold value was chosen as the recommended guideline value. With all the compounds classified above, the intensity of the effect or response decreases with a reduction in dose, and the biological reaction often reaches zero before the dose becomes equal to zero.

The assumption was made that below a certain limiting exposure level, or dose, i.e., below the threshold, a toxic effect was not observable. Thus, the threshold for an adverse effect is the dose that gives rise to biological changes beyond the limits of homeostatic adaptation.

SUBSTANCES POSSESSING CARCINOGENIC PROPERTIES

A view is held that carcinogenic or mutagenic chemicals are subject to the same physicochemical and biological interactions that are considered to result in a threshold dose for other chemicals. The alternate view is that for chemicals whose toxic effects give rise to neoplastic disease or mutations of genetic material, a single molecule of a chemical is sufficient to initiate a process that may ultimately lead to an observed harmful effect, and that it may not be possible to demonstrate that a threshold dose exists.

A World Health Organization Scientific Group concluded in 1974 that "the existence of a threshold may be envisaged, nevertheless the difficulties of determining a threshold for a population are great, therefore a mathematically derived conclusion that it is impossible to demonstrate a "no-effect" level experimentally cannot be ignored".

In arriving at guideline values for substances within this category, values were computed from a very conservative, hypothetical, mathematical model and a realistic application of these values could therefore include uncertainties representing an order of magnitude either way. The "multistage" model employed assumes that there is a finite risk from any exposure to these chemicals, however small, and that the risk is proportional to the dose. The model is designed to estimate the highest possible upper limit of incremental (excess over background) risk from a lifetime of exposure to a particular daily amount of a substance. An "acceptable risk" of 1 in 100 000 per lifetime was arbitrarily selected by W.H.O. as the criterion. However, it will be appreciated that the actual risk could also be "zero" if the "no threshold" assumption in the model is invalid. Values recommended in this case are therefore conservative and err on the side of safety.

The limitations of animal experimental models are appreciated in that the accuracy and reliability of a quantitative prediction of toxicity in man depends upon a number of conditions which are recognized. For this reason it was decided that guideline values would only be set where reliable data were available from two species of animals, preferably with supporting evidence such as the mutagenicity tests, in addition to relevant, direct evidence from population studies.

For the reasoning set forth above, it is recommended that because the guideline values for these substances were developed differently from those having a dose response, they should not therefore be interpreted in the same way when judging water quality.

Logical though the above rationale may be, the World Health Organization is conscious of the fundamental question concerning the validity of applying animal data to man. Granted that the basic biological processes of molecular, cellular and organ functions are similar from one mammalian species to another, there are marked differences between the standard relatively homogeneous rodent strain in a laboratory environment and the cosmopolitan human in his complex environment. However, as supportive evidence to the choice of carcinogenic chemicals, almost all of the chemicals or industrial processes that have been positively associated with human cancer, through the programme of the International Agency for Research on Cancer, are known to be carcinogens to animals. Thus it may be reasonably safely concluded that chemicals carcinogenic to laboratory animals are likely also to be carcinogenic in man. This is the basis upon which the selection of organic chemicals in drinking water was made when compiling Guidelines.

TENTATIVE GUIDELINE VALUES

Consideration of the available animal data relating to some selected chemicals did not reach the criteria selected and used as the basis for guideline consideration. Nevertheless, for such compounds there was a body of toxicological evidence available to justify consideration of such compounds, (present in several drinking waters at levels which could not be ignored), as a potential hazard to health. Evidence of carcinogenicity did not justify a full guideline value and to such compounds were ascribed a tentative value which, when more positive toxicological data become available, may be substantiated or be withdrawn. Such tentative values were derived using the "multi-stage" model even though the selected chemicals did not reveal significant carcinogenic properties. Consequently, a greater degree of uncertainty is applicable to those tentative values than to the remainder of the guideline values.

From the foregoing it is obvious that considerable uncertainties are attached to the guideline values set for many of the toxic constituents. In setting limits for such substances, acute toxicity data are generally an irrelevancy, and reliance must be placed upon the (limited) available chronic toxicity data which are frequently less than comprehensive. Compounding the problem attempting to judge the acceptable daily intake of a particular chemical and its apportionment between the various routes of human exposure, and in particular that due to water, was a constant reminder of the inadequacy of factual data. It was discovered at an early stage that reliable information on the health effects and the routes of exposure of almost all the organic micropollutants in water was less than adequate. Additionally, and perhaps not surprisingly, the almost total lack of factual data relating to health effects of these substances at the concentrations recorded in drinking water was characteristic of the problems facing W.H.O. in formulating guidelines.

Generally, compared with other routes of human exposure the organic micropollutants' contribution via drinking water is minute, in fact it would be difficult to identify many substances for which water represents a significant proportion of the dietary intake.

Because the various difficulties which are presented when evaluating safe drinking water quality - especially in relation to the organic micropollutants, it is emphasized that the numerically quoted guideline values should not be used directly from the tabulated values, but such values must be interpreted in conjunction with the information contained under the appropriate monographs of Volume 1 which are a synopsis of the more detailed health-related criteria to appear in Volume 2.

ANALYTICAL CONTROL

Various "recommended" methods of water analysis are published by both national and international organizations. Experience shows, however, that adequate analytical accuracy, even using standard methods, is not always achievable due to a variety of factors. In compiling Guidelines it was not considered essential to specify standard analytical methods except in the case of those entities the result of which is dependent upon the method employed, colour, turbidity, etc.

Many of the micro-organic contaminants set forth in Guidelines may be present in water at very low concentrations, and therefore the limit of detection is often likely to be the most important criterion in selecting a method of analysis. It was also decided that any analytical method could be selected which was capable of meeting the required accuracy and for this reason Guidelines expressed in clear terms the accuracy which should be achieved. To set an accuracy target which is too stringent increases the time, effort and often equipment needed to perform the task, and consequently also the cost which is an important item. The setting of needlessly stringent targets for accuracy is therefore to be avoided. What is stressed is that whatever analytical method is ultimately chosen as meeting the target, appropriate analytical quality control procedures must be implemented to check that adequate accuracy is being achieved on a continuous basis.

INTERPRETATION OF GUIDELINES

It is intended that Guidelines should be used by nations as the basis for developing national standards - not only for community piped-water supplies, but also for all water used for drinking purposes, including standpipes, wells, tanker-distributed or bottled water, including that for serving transient populations. Furthermore, it is accepted that water is essential to sustain life and must be available even if the quality is not entirely satisfactory. Adoption of too stringent a drinking water standard could limit the availability of water supplies to meet those standards, which could be a significant consideration in regions of water shortage. This fact W.H.O. has recognized.

In compiling the guideline values a risk/benefit approach has been adopted as illustrated in the assumption of the potential additional risk of one carcinoma per 100 000 population per lifetime when deciding recommendations for carcinogenic micropollutants. The process of risk assessment comprises usually four phases: risk identification, risk estimation, risk evaluation and risk management. The latter two are frequently a unified operation but the sequence assists in separating the scientific from the socioeconomic aspect. Governments with the responsibility for protecting their citizens have to rely upon the available scientific evidence (guideline) prior to a thorough evaluation of risk.

Risk embodies hazard and the probability of its occurrence, but once exposure has taken place, the probability of an inadventitious event occurring still depends on the susceptibility of the host. For this reason, Guidelines emphasizes the necessity to identify population groups particularly at risk, such as pregnant women, infants (nitrate) and other consumers with specific physiological defects.

Nevertheless, nations in developing their own national drinking water standards will not only be conscious of the prevailing environmental, social, economic and cultural conditions but will also be the judge of the benefits and the risks in adopting values which differ from the Guideline recommendations.

CONCLUSIONS

There is no clear unequivocal evidence of health effects associated with organic micropollutants at the levels recorded in drinking water.

The presence of an increasing number of such compounds in water is reported almost annually and, no doubt, as the analytical expertise becomes more sophisticated, the identified compounds will increase accordingly.

Limits for the presence of such material in drinking water is helpful in the management of drinking water quality, but an understanding of the basis for the choice of such limits is essential if they are to be correctly interpreted by those involved in water quality management.

There is an embarrassing absence of reliable toxicological information on a large number of organic micropollutants in drinking water, especially the long-term health effects, and especially at the levels normally found in potable water.

Compared with other dietary sources, drinking water contributes a very small proportion of the daily intake of such chemicals. Although it is recognized that social, cultural and economic considerations should be taken into account when setting national water quality standards, never should these considerations be to the detriment of health protection which should be of primary concern.

THE EFFECTS OF A HAZARDOUS AND A DOMESTIC WASTE LANDFILL ON THE TRACE ORGANIC QUALITY OF CHALK GROUNDWATER AT A SITE IN EAST ANGLIA

K.M. BAXTER

WRc Environmental Protection, P.O.Box 16, Marlow, Bucks, SL7 2HD (UK)

ABSTRACT

As part of a research project to assess the effects of landfill leachates on groundwater quality thirteen boreholes within, around and down groundwater gradient from two adjacent domestic and hazardous wastes sites were sampled for trace organic quality.

Results so far show that significant inorganic and gross organic (TOC) groundwater contamination does not occur more than a few metres from the landfill boundaries. Trace organic analyses indicate that the mineral oils, phenolic wastes and chlorinated solvents, known to have been tipped into the hazardous waste landfill, have now reached the water table, although only at low concentrations.

INTRODUCTION

In view of the increasing concern about the concentration and distribution of trace organics in groundwater used for public supply the Water Research Centre (WRc) has recently begun a study to determine the total organic loading to groundwater from a number of point and dispersed sources. Included in this study will be the identification of the volatile organic microcontaminants together with an assessment of their persistance in the groundwater system. One of the major point sources of concern are landfills which are usually located near centres of population, often within the same groundwater catchment as boreholes used for potable supply. Over the last ten years the WRc has conducted a considerable amount of research into the effects of both domestic and hazardous waste landfilling on the inorganic and gross organic (as TOC or DOC) quality of groundwater (ref.1). Only very recently, however, has there been any consideration as to the effects of landfilling on the trace organic quality of groundwater.

As part of a detailed study into the effects of landfill leachate on the trace organic quality of groundwater by WRc, a site at Ingham in East Anglia was chosen where both a hazardous waste and a domestic waste landfill overlie the Chalk, a major water supply aquifer in the UK (Fig. 1). There is a considerable history of research at this site with over 50 boreholes having been drilled for a variety of reasons over the last 10 years (ref.2).

SITE DETAILS

The solid geology of the Ingham area is within the lower part of the Cretaceous Upper Chalk (a micrite composed almost entirely of coccolith fragments approximately 2 µm in size). Solution of the chalk has produced an irregular surface over which Pleistocene Ingham sands and gravels lie unconformably. These are in turn overlain by a chalk-rich till. The sands and gravel reach a maximum local thickness of 16 m beneath the site.

At Ingham, the two landfills have been created at sites that were initially quarried for gravel extraction. The hazardous waste (Folly) site is situated about 300 m west of the domestic (Culford Road) site.

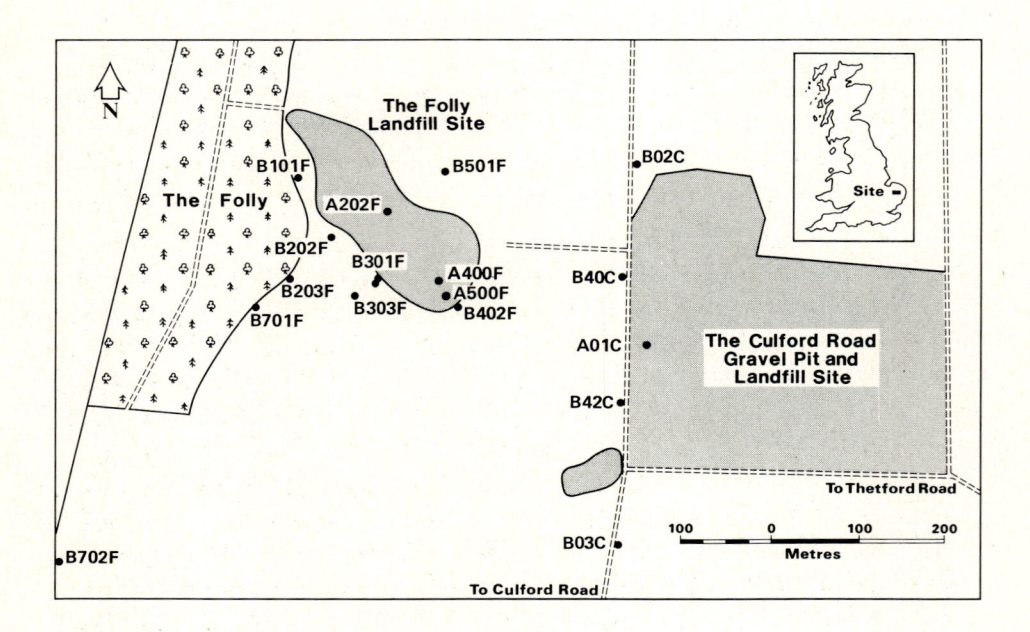

Fig. 1. Borehole locations.

The age of the hazardous waste site is unknown but landfilling began in 1968 and continued until 1973, initially with builders' rubble and paper but shortly after this by sawdust, synthetic latex and egg packing waste. Between 1970 and 1972 approximately 470 m³ of aqueous industrial wastes, including oil and oily sludges, were discharged into three lagoons excavated into the fill material. The site was completed by a clay cap in 1974.

The landfilling commenced at the northern end of the domestic waste site in the late 1970's and followed progressive gravel extraction south-eastwards across the site. Landfilling has continued through the late 1970's and early 1980's and it is expected that the site will be full by the mid 1980's. Only domestic and commercial wastes are allowed to be filled at this site. However, over the last five years mixed organic and inorganic liquid wastes and solvents have been discharged at an annual rate of approximately 900 m^3. These liquids are poured into small hollows made in the current tipping face.

In addition to landfilling, approximately 1660 m^3/a of mixed industrial and abbatoir wastes are discharged into two lagoons (one just to the north and the other 1.5 km to the north of the landfills) while sludge mixed with industrial and abbatoir wastes are spread onto nearby fields at a rate of around 10 m^3/ha/a.

The groundwater table is between 7 m and 23 m below surface and has a gradient of 0.002 towards the west south-west. Two groundwater level surveys carried out in August 1984 and January 1985 show there to be little annual change in level. The gradient is uniform across the site indicating that the landfills have a negligible effect on the rate or distribution of infiltration.

A survey of groundwater quality carried out in 1984/5 gave similar results to that seen in a survey of the site carried out 11 years earlier. Total organic carbon (TOC) and chloride levels (Figs. 2 and 3) in the groundwater are elevated above background only in those boreholes adjacent to, or within, the landfill area. Significant groundwater contamination does not extend more than a few metres from the landfill boundaries. The form of the chemical quality depth profile through the unsaturated zone at borehole A202(F) in the centre of the landfill is again similar to results from 1974/5. In both cases the unsaturated zone was shown to be highly polluted to just above the water table; there being little evidence, therefore, for any downward movement of leachate since 1974. The reason for this is uncertain and is the subject of additional research.

ORGANICS SURVEY

The aim of the trace organic survey was to determine the areal distribution of trace organics in the groundwater below, around and down gradient from the two landfills. Altogether thirteen boreholes were chosen as suitable for sampling using an electric submersible pump.

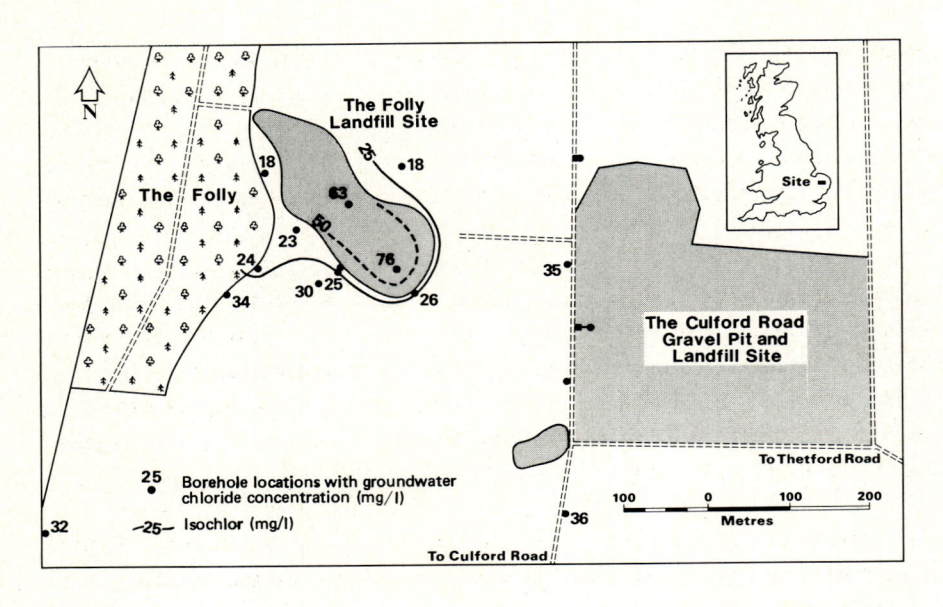

Fig. 2. Groundwater chloride concentrations 1984/85.

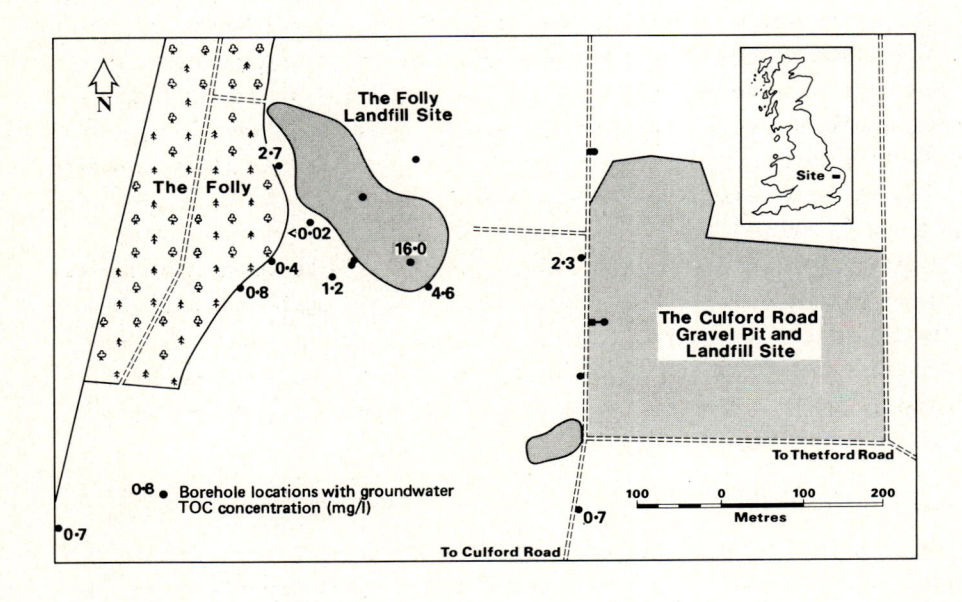

Fig. 3. TOC concentrations in groundwater 1984/85.

Before the field sampling programme began the CG/MS analytical technique was validated to determine the recovery effeciencies of the XAD resin extraction and concentration stages and to assess the reproducibility of the instrumentation. A blank run was also carried out to assess any interference that might arise from the use of the submersible pump and textile-reinforced elastomer flexible rising main.

On site the submersible pump was lowered to about 5 m below the water table in each borehole and the borehole pumped to waste for at least 6 hours before sampling. Fifteen litres of water were collected from each borehole in organically-clean winchester flasks. Samples were taken to the laboratory for spiking with deuterated standards within 24 hours. Extraction was undertaken within 48 hours of collection.

Two private pumped boreholes located up groundwater gradient from the site and one public supply borehole that was well away from the two landfills, were also sampled to establish the 'background' levels of trace organics in Chalk groundwater.

Results so far available are given in Table 1. These are only qualitative, as yet, but do show boreholes located within or close by the landfills to be highly polluted while the others show only slight traces of contamination. These initial results and those of earlier findings (ref.3) show that mineral oils, phenolic wastes and chlorinated solvents, which are known to have been discharged into the hazardous waste landfill, have now reached the groundwater, but only at relatively low concentrations.

Generally the three background samples were found to be clean with only low levels of phthalates (<0.1 µg/l), carboxylic acids (<1 µg/l) and toluene (<1 µg/l).

TABLE 1

Initial results of trace organic survey.

Borehole No	Compound (see key below)	
B101F	TL, TCE, EB, TMB/MEB, A, PT	(2 UNKNOWNS)
B202F	A, NT, TRE, TL, TCE, MEB, PB, TMB, DCB, AB, TMC, NP, DEOB, MNP, BP, OB, TMNP, PT	(8 UNKNOWNS)
B203F	TCE, P	(2 UNKNOWNS)
B303F	TRE, TL, TCE, A, P	(2 UNKNOWNS)
A400F	TRE, MCH, TL, TCE, AR, DMB, MEB, PB, TMB, DCB, BB, AB, NP, DEOB, MNP, BP, OB, DMPN, S	(10 UNKNOWNS)
A701F	TRE, TL, TCE, P	(4 UNKNOWNS)
B03C	TL	(2 UNKNOWNS)
B40C	TRE, TL, TCE, MPH, NP, A, S	(2 UNKNOWNS)

KEY:- A - alkane, AB - alkyl benzene, AR - aromatic, BB - butyl benzene,
BP - biphenyl, DCB - dichlorobenzene, DEOB - diethoxybenzene,
DMB - dimethylbenzene, DMPN - dimethylnaphthalene, EB - ethylbenzene,
MCH - methylcyclohexane, MEB - methylethylbenzene,
MNP - methylnaphthalene, MPH - methylphenol, NP - naphthalene,
NT - nitrile, OB - 1,1'-oxybis-benzene, PB - propylbenzene,
PT - phthalate, S - sulphur, TCE - tetrachloroethane,
TL - toluene, TMB - trimethylbenzene, TMC - trimethylcyclohexanone,
TMNP - trimethylnaphthalene, TRE - trichloroethene.

REFERENCES

1   Department of the Environment (DoE). Co-operative Programme on the Behaviour of Hazardous Wastes in Landfill Sites. Final Report of the Policy Review Committee, 1978, HMSO, London.
2   P.A. Towler, N.C. Blakey, T.E. Irving, L. Clark, P.J. Maris, K.M. Baxter and R.M. Macdonald. A Study of the Bacteria of the Chalk Aquifer and the Effects of Landfill Contamination at a Site in Eastern England. In: Hydrology in the Service of Man, 18th Congress IAH, 1985, Cambridge, UK.
3   P.J. Maris, D.L. Readhead, B.L. Brown and C. Barber. Ingham Landfill, Suffolk: Preliminary Survey, August - September 1984. Preliminary Report for the Department of the Environment on Analyses of Organic Wastes. WRc Internal Report, 1975.

BIOLOGICAL PROCESSES AND ORGANIC MICROPOLLUTANTS IN TREATMENT PROCESSES

B. E. RITTMANN

Environmental Engineering and Science, University of Illinois at Urbana-Champaign, 208 North Romine, Urbana, IL 61801 (U.S.A.)

ABSTRACT

This paper characterizes the factors that control biodegradation of organic micropollutants in biologically active drinking water processes. Particularly important are the dominance of attached microorganisms, the aerobic potential, the low concentration of organic matter, the very low concentrations of specific micropollutants, and the presence of inorganic electron donors. Biodegradation of the specific compounds is feasible if sufficient microbial growth occurs through utilization of natural organic material and if the secondary-utilization kinetics of the micropollutants are sufficiently rapid.

INTRODUCTION

With respect to microbiological processes that can occur during water treatment, organic micropollutants are comprised of a wide range of compounds and materials. Foremost are the anthropogenic compounds of industrial origin: common examples include the chlorinated aliphatic solvents, the chlorinated benzenes and phenolics, and the pesticides. Also of great concern, especially with groundwaters, are the petroleum derivatives associated with leaked or spilled gasoline and oils. All of these compounds are acknowledged to be health hazards and often create serious taste and odor problems when present in very low concentrations.

Numerous other organic compounds and materials are important to microbiological processes, even though these micropollutants do not necessarily constitute a health or aesthetic problem. The naturally occurring humic and fulvic acids are present in all waters. Polymeric carbohydrates and proteins also comprise a large fraction of organic matter in waters. Carboxylic acids are significant in most natural waters. The natural organic materials, typically present in low concentrations, are important to biological processes for two reasons: they can serve as growth-supporting substrates for microbial growth in a treatment process or in the distribution system, and they can be products of microbial activity. Additionally, the natural organic materials are important determinants for the fate of the more hazardous compounds because they serve as precursors to trihalomethane (THM) formation during chlorination and compete for adsorption sites in adsorbing processes, such as granular activated carbon (GAC).

The objective of this paper is to develop several of the key concepts that control the microbiological fate of organic micropollutants during drinking water treatment. Anthropogenic and natural organic materials, as well as their interactions, are addressed.

CHARACTERISTICS OF MICROBIOLOGICAL ACTIVITY IN DRINKING WATER TREATMENT

Table 1 lists five key characteristics of microbiological activity as it occurs in drinking water treatment. Although exceptions probably can be found for most of the characteristics, they are the most common determinants of what kinds of microbiological reactions can occur during drinking water treatment. The following sections describe the implications of these characteristics for organic micropollutant removal.

TABLE 1
Microbiologically relevant characteristics found in drinking water treatment

Characteristic	Comments
1. High specific surface area	-Allows sufficiently large accumulation of attached (biofilm) biomass.
2. Oxidized environment	-Dissolved oxygen normally is present. -Strongly reducing conditions never encountered.
3. Low concentration of total biodegradable organic material	-Low DOC in source (mg/L) -Most DOC is polymeric -Only a fraction of DOC may be readily degradable. -Minimal capacity to synthesize new biomass.
4. Very low concentrations of individual micropollutants	-Often in mg/L levels or less -Little selective pressure by one compound.
5. Inorganic electron donors often relatively prevalent	-E.g., NH_4^+, Fe^{2+}, Mn^{2+} -Can constitute major source of biomass growth.

Biological processes

All the drinking-water-treatment processes which exhibit microbiological activity have high specific surface area for the attachment, growth, and accumulation of biofilm microorganisms. Rittmann and Snoeyink (1984) reviewed the use of a variety of biofilm processes employed primarily in Europe for biological water treatment: they include fixed-bed gravel media filters, fixed-bed pozzolana filters, upflow fluidized beds, and slow sand filters. In each case, the specific surface area is at least 120 m^{-1}, but it can be much higher for the processes using very small attachment media, such as the 0.2-mm sand in an English fluidized bed (Short, 1975), which had a specific surface area of about 6000 m^{-1}.

Other engineered treatment processes show biodegradation, even though biological activity is not the main process objective. Rapid sand filters (Richard, 1979; Baliga, 1969; Bourbigot et al., 1982; Jekel, 1977) and GAC filters (Jekel, 1977; Rice et al., 1980; Gomella and Versanne, 1980) showed biological reactions, especially when ozonation was applied prior to the process. Again, these processes have very high specific surface areas.

The pretreatment steps of bank filtration (Sontheimer, 1980) and dune filtration (Piet and Zoetaman, 1980) are further examples of biologically active processes that have very high specific surface areas. Although abiotic reactions also occur during bank and dune filtration, biodegradation of organic and inorganic constituents is brought about by the attached microorganisms.

That all the biological processes which occur in drinking water treatment are of the biofilm type should be of no surprise to someone who understands biological systems. Biofilm processes are most advantageous when substrate concentrations are low, because attachment and accumulation as a biofilm allow retention of a large biomass, even though the growth potential is low (i.e., low substrate concentration) and liquid washout rates are rapid (i.e., very short liquid detention times). Attachment confers several advantages to biofilm microorganisms (Rittmann and McCarty, 1980a; Stratton et al., 1983), but the most important seems to be cell retention and accumulation far superior to that feasible in any suspended-growth process. Thus, the biofilm microorganisms attached to a very high surface area make biodegradation feasible in drinking water treatment.

Biofilm kinetics

The kinetics of biodegradation in biofilm processes are controlled by at least four factors. First are the intrinsic biodegradation kinetics of the individual compounds, a subject which is discussed more in the following sections and which is not particular to biofilms. The second item is the mass transfer resistance, which occurs because the biofilms are examples of heterogeneous catalysts. In order to contact metabolizing bacteria within the biofilm, substrate must be transported from the bulk liquid to the biofilm surface and then from the surface into the biofilm. Each transfer step offers resistance to mass transport and brings about reductions in substrate concentration in the biofilm, compared to the bulk liquid.

The third factor determining the biofilm reaction rate is the amount of active biomass accumulated per unit surface area. In general, a large biomass accumulation per unit area gives a faster substrate removal rate per unit area, or flux. The amount of biomass that accumulates depends on a balance of biofilm growth and loss rates. Biofilm grows in proportion to the amount of substrate which is metabolized; the true yield coefficient, Y (typically in kg cell mass per kg substrate removed), describes the growth proportionality. The main biofilm losses are cell maintenance and shear losses to the bulk liquid. Although the specific maintenance-loss rate depends on the microorganism type and temperature, the shear loss rate depends on reactor conditions which affect hydrodynamic shear. Rittmann (1982b) illustrated how shear stress controls the shear-loss rate for relatively smooth surfaces, but the relationships for irregular surfaces still are unknown, although losses probably are less than for smooth surfaces.

Shear loss rates can be controlled in practice. If too much biofilm is accumulating and causes clogging, the shear loss rate can be increased by increasing flow velocity or by employing various means of scour and backwash. When too little biomass accumulates, shear losses can be reduced by slowing flow velocities, by eliminating backwash, or by providing a more irregular surface that offers protection from shear stress.

The fourth factor affecting reaction rates is the substrate concentration. Substrate concentration plays two inter-related roles in a biofilm process. First and most obviously, the intrinsic reaction by a microorganism increases as that organism contacts a higher substrate concentration. The classic Monod relationship is most often used to express the intrinsic rate,

$$r_{ut} = -\frac{kS_f X_a}{K_s + S_f} \qquad (1)$$

in which r_{ut} = rate of substrate accumulation due to substrate utilization (kg/m^3-day); k = maximum specific rate of substrate utilization (kg/kg cells-day); S_f = substrate concentration at a point in the film (kg/m^3); X_a = concentration of active cells in the film at that point (kg cells/m^3); and K_s = half-maximum rate concentration (kg/m^3). As long as S_f is not much greater than K_s, the absolute value of the removal rate increases as S_f increases. For a given amount of biomass per unit surface area, the overall biofilm reaction rate also increases as the bulk liquid concentration increases, because a higher bulk concentration gives higher S_f values. The relationship of biofilm reaction rate to bulk concentration is somewhat more complicated than Equation 1, because of the mass transfer resistance; however, solutions for flux as a function of bulk concentration are available (Rittmann and McCarty, 1981; Rittmann, 1982a).

The second role played by substrate concentration is to allow biomass growth. Because cell growth is related to substrate utilization through the true yield (i.e., $r_{gr} = -Yr_{ut}$, in which r_{gr} = rate of new biomass growth (kg cells/m^3-day)), a higher substrate concentration allows a higher cell growth rate. When the cell growth rate increases, more biofilm mass accumulates. Thus, higher substrate concentrations also give greater accumulations of biomass, which usually increase the surface reaction rate.

Steady-state biofilm kinetics for growth-limiting substrate

Rittmann and McCarty (1980a,b) and Rittmann (1982a) combined together and evaluated the four factors that control biofilm kinetics. They defined a steady-state biofilm as one in which growth and loss rates just balance each other; thus, the amount of attached biomass remains at a unique steady-state value for a given substrate concentration. Although the reader should refer to the original papers for details of the model and evaluations, one critical observation is necessary here. As the concentration of growth-limiting substrate decreases, it reaches a concentration sufficiently low that the biofilm growth rate never can be as great as the loss rate. Hence, the biofilm always is in net loss and either disappears or never exists. The concentration below which a steady-state biofilm cannot exist is called S_{min} and is defined by

$$S_{min} = K_s \frac{b'}{Yk-b'} \qquad (2)$$

in which b' = the overall biofilm loss rate coefficient (days^{-1}). For simple organic compounds in aerobic systems, S_{min} values typically range from 0.1 to 1.0 mg/L as Chemical Oxygen Demand (COD) (Stratton et al., 1983; Rittmann and McCarty, 1980b).

The main conclusion from the S_{min} concept is that there exists a lower limit on the concentration of growth-limiting substrate. As the raw water concentration approaches S_{min}, biodegradation ceases because biofilm cannot be sustained. The low concentrations of total organic matter and of individual organic compounds found in raw waters suggest that biodegradation in drinking water treatment can be limited by a lack of growth-supporting organic substrate.

Removal of organic materials

As explained previously by Rittmann and Snoeyink (1984), measurements of the removals of and residual concentrations of biodegradable organic material--the likely growth-limiting substrates in a raw water--are compounded by lack of an accurate means to analyze for the low concentrations typically encountered. However, results for removals of dissolved organic carbon (DOC) or total organic carbon are available for several biologically active treatment processes.

Milliner et al. (1972) and Short (1975) reported on the performance of a fluidized bed treating water from the River Thames. Whereas they obtained essentially complete NH_4^+ oxidation, they found that TOC decreased only 6% (3.7 to 3.5 mg/L), while 5-day Biochemical Oxygen Demand (BOD_5) was reduced from 2.8 to 2.0 mg/L. Jekel (1977) observed removal of 1.0 mg/L of DOC from Ruhr River water in a fluidized bed that received 4.5 mg/L of DOC and that achieved 92% nitrification.

Organic removals across rapid sand filters are quite variable. Eberhardt, et al. (1977) found an average DOC removal of 2.6 mg/L, or 43%, across flocculation, sedimentation, and filtration processes at Bremen, FRG. Sontheimer (1978) saw DOC removals of less than 0.2 mg/L across rapid sand filters at Mulheim, FRG, but about 1.0 mg/L of the input 4.2 mg/L was removed previously by flocculation and sedimentation. At Shreveport, Louisiana, Glaze et al. (1982) obtained TOC removals and total inorganic carbon (TIC) increases of 9%, or roughly 0.5 mg/L, without preozonation and 11% with preozonation.

Slow sand filters are known to contain biological activity. Eberhardt et al. (1977) obtained an additional 0.9 mg DOC/L removal using slow sand filters at Bremen. In The Netherlands (KIWA, 1983), a removal of 3.1 mg TOC/L was observed across a slow sand filter having a detention time of 2.6 hours.

The process which has been studied most extensively for organic matter removal is GAC. Analyzing biodegradation is difficult, because adsorptive mechanisms remove TOC or DOC abiotically. Therefore, increases in TIC are better measures of biodegradation. Jekel (1977) showed TIC increases of 1.2-1.4 mg/L at Mulehim, where the input DOC was 2.6-2.8 mg/L. Since the total DOC removal was 1.6-1.8 mg/L, the large majority of organic-compound removal was attributed to biodegradation. KIWA (1983) reported biological depletion of 1.6-2.1 mg/L of dissolved oxygen across GAC columns. Glaze et al. (1982) obtained 0.9-1.5 mg/L increases in TIC with no preozonation and about 1.0 mg/L with preozonation. In general, TIC increases were about twice as high for GAC columns as for parallel sand columns.

Sontheimer (1980) reported significant DOC removal through bank filtration. At Duisburg-Hamborn, approximately 3.5 mg DOC/L was removed from an input of 5-7 mg/L. At Duisburg-Wittlaer, bank filtration produced a relatively constant

DOC of 2 mg/L when the raw water varied from 4-7 mg/L. Although adsorption and dilution had some impact on the results, significant removals of organic material were achieved by bank filtration of a relatively polluted source.

In all cases, biodegradable organic matter was removed in biofilm processes. The degree of removal varied with water quality and process type. For example, the greatest mass was removed by bank filtration, in which the raw DOC was highest and the liquid retention time was by far the greatest. In another report (van der Kooij, 1982) in which Assimilable Organic Carbon (AOC) was measured, the fractional removal by GAC was more than 80%, but mass removals were small (up to 0.16 mg AOC/L). Clearly, biodegradability and process conditions affect the rate and extent of organic matter removal. The degree of which the S_{min} concentration controls these removals cannot be well defined by the field results alone.

Effects of oligotrophy

Microorganisms that survive and function metabolically when their substrate concentrations are very low are called oligotrophs (Poindexter, 1981). Reported characteristics of oligotrophs include efficient utilization of multiple and varied compounds, relatively low maximum growth rates, and low K_s values (Poindexter, 1981; Morita, 1980; Hirsch et al., 1979; Novitski and Morita, 1978; Matin and Veldkamp, 1978). Numerous soil, freshwater, and marine bacteria are reported to have oligotrophic characteristics, although the definitions used to classify oligotrophs are not constant.

Low concentration environments favor attached growth and oligotrophy. Thus, it is no surprise to find that oligotrophs often occur in biofilms and that drinking-water-treatment processes would select for oligotrophs. However, virtually no information is available to characterize biofilm oligotrophs.

Recent work in the author's laboratory was undertaken to isolate and kinetically characterize oligotrophic bacteria grown as biofilms. Acetate or salicylate was the growth-limiting substrate, present in the feed at only 1 mg/L. Completely mixed biofilm reactors (Tuck, 1984) were constructed by putting a 20-times effluent recycle around 2.5-cm columns filled with 3-mm glass beads (Stratton et al., 1983; Namkung et al., 1983). A special data-analysis technique was developed to calculate the kinetic parameters K_s and k from in situ biofilm bacteria (Rittmann, B. E. et al., "In situ determination of kinetic parameters for biofilm bacteria: oligotroph isolation and characterization," in preparation). Table 2 lists the results. Compared to "conventional" values, K_s and k are low and in accord with the hypothesis that oligotrophs have low K_s and maximum rate values. Calculations of S_{min} (from Equation 2) show that the biofilm oligotrophs should be able to achieve relatively low minimum concentrations, especially culture A, which was taken

TABLE 2

In situ kinetic parameters for biofilm oligotrophs

Substrate	K_s, mg COD/L	k, mg COD/mg cells-day	S_{min}, mg COD/L[1]
A1 Salicylate	0.0005	0.15	0.00063
B Salicylate	0.11	0.17	0.11
C Salicylate	0.08	0.15	0.1
A2 Salicylate	0.02	1.74	0.001

[1]Assumes that Y = 0.6 mg cells/mg COD and b' = 0.05/day^{-1}.

from an oligotrophic lake. Exactly how these results for salicylate and acetate correlate to capabilities for natural organic matter is not yet known, but they indicate potential for achieving very low effluent concentrations of biodegradable organic matter.

Other effects on biofilm kinetics for natural materials

Even if biofilm processes are capable of achieving effluent concentrations that approach very low S_{min} values, other kinetic effects can prevent this from occurring in practical processes. One factor that likely plays an important role in slowing biofilm kinetics is the relatively large molecular size of the natural materials. Most natural organic compounds in freshwater are found in the molecular-weight range of 1000-10,000 daltons (Bruchet et al., 1984). The compounds in the 1000-5000-dalton range mainly are fulvic acids, while compounds in the 5000-10,000-dalton range contain polymeric carbohydrates and proteins (Bruchet et al., 1984). Small solutes, such as acetate, salicylate, glucose, and amino acids, have molecular weights in the order of 100 daltons.

Two phenomena can act to retard degradation kinetics for large molecules. The first is slow diffusion and mass transport caused by the molecules being large. The Wilke-Chang relation (Perry and Chilton, 1973) is useful for estimating the diffusion coefficient in water.

$$D = 1.48 \times 10^{-8} (V_b)^{-0.6} \tag{3}$$

in which D = molecular diffusion coefficient (m^2/day)

V_b = molar volume of the solute at its boiling point (mL/mole).

If molecular size is approximately proportional to the molar volume, then D is proportional to $(MW)^{-0.6}$. Hence when the molecular weight increases from about 100 to about 1000 daltons, D decreases by a factor of $(1000/100)^{0.6} = 4.0$. For a molecular weight of 10,000 daltons, the reduction is by a factor of about 16. The impact of such reductions in diffusivity is to slow the kinetics by an amount up to the factor that D decreases (Williamson & McCarty, 1976a,b).

In addition, as mass transfer becomes more important, the region of first-order kinetics is extended to higher bulk concentrations.

Figure 1 illustrates the reaction rate for a monomeric compound (galactose) and two polymers, peat fulvic acid (PFA) and commercial humic acid (CHA). PFA and CHA have approximate average molecular weights of 10,000 daltons and 5000 daltons, respectively. Results were taken in completely mixed biofilm reactors having constant biofilm accumulation for all tests. Differences in reaction rate (i.e., flux into the biofilm) are caused only by changes in intrinsic kinetics, including diffusivity. The two expected trends are observed. First, fluxes of the polymers are lower in their first-order (straight-line) regions, which indicates slower mass transport for the polymers. Second, the first-order regions extend to higher bulk concentrations for the larger molecules. The largest molecules, PFA, has the slowest first-order kinetics and the farthest extent of first-order kinetics. Another interesting aspect of the results in Figure 1 is that the maximum rate for the CHA apparently is greater than for galactose, a simple sugar-molecule.

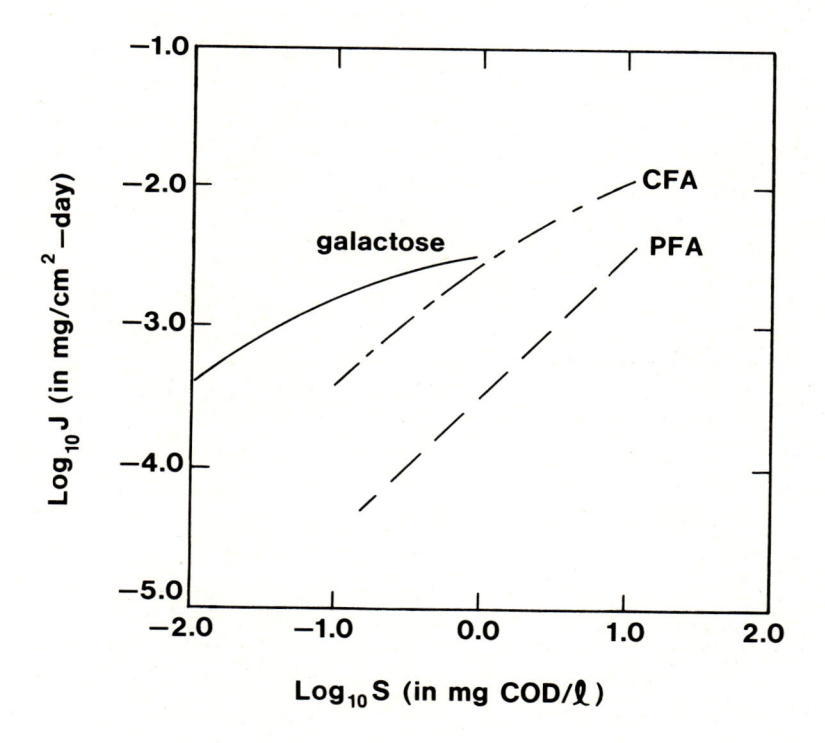

Fig. 1. Effect of increasing molecular weight on substrate flux and extent of first-order flux region for completely mixed biofilm reactors. Fulvic and humic acid CODs are calculated as 2 g COD/g TOC.

The results shown here demonstrate that the polymeric materials are biodegradable, but that their kinetics are limited by mass transfer moreso than for the simple monomers. Thus, S_{min} values for polymers may be equally low as for simple compounds tested. However, the reaction times needed for similar fractional removals probably are longer for the polymers. For example, in similar biofilm reactors, having detention times around 12 minutes, the polymeric materials showed 9-40% removals, whereas most monomers showed removals in excess of 90%. The field data that show greater fractional removals for bank filtration, which has long detention times, tend to support these findings, although different input concentrations and qualities confound direct comparison of field results.

The second phenomenon that can slow the biodegradation kinetics of polymers is the need for an initial hydrolysis reaction carried out by extracellular enzymes. Such a first step is not required for monomers, which can be transported across the cell membrane directly. Adding an extra step could slow the overall reaction rate. Too little information is available to know how much effect hydrolysis has on the overall kinetics.

BIODEGRADATION OF SPECIFIC ANTHROPOGENIC COMPOUNDS

Of ultimate concern is the fate of the hazardous organic compounds of industrial origin. The two factors of paramount interest for determining the fate of anthropogenic compounds in biologically active drinking water processes are the intrinsic biodegradability of the compounds by aerobic microorganisms and the interactions of the anthropogenic compounds with natural organic material. Each topic is discussed below.

Biodegradability of compound class

Recent reviews of biodegradability (Kobayashi and Rittmann, 1982; Rittmann et al., 1985) indicate that aerobic biodegradation is possible for only a few chlorinated aliphatics, for most aromatics and phenols, for some chlorinated biphenyls, and for numerous pesticides. In addition, the light fractions from petroleum (e.g., gasoline and light-oil constituents) are known to be degradable under aerobic conditions (Atlas, 1981). Notable about aerobic degradation are the compounds not known to be degraded. They include unsubstituted aromatic hydrocarbons and most one- to three-carbon chlorinated aliphatics. Much work remains to be performed before the biodegradabilities of all relevant compounds are known for environments pertinent to drinking-water treatment.

Degradability is inextricably tied to removal kinetics. Unfortunately, elucidation of the intrinsic removal kinetics for anthropogenic compounds is still in an infant state. Current research in the author's laboratory and elsewhere is making progress towards obtaining degradation parameters, but progress is slow because of the many compounds and many different environmental conditions that can occur in reactors.

One convenient simplification that often is realistic for organic micropollutants is to reduce the Monod relation (Eqn. 1) to its first-order approximation,

$$r_{ut} = -k_1 X_a S_f \qquad (4)$$

in which k_1 = an apparent first-order rate coefficient (m^3/kg-cells-day) that is equal to k/K_s from the Monod relation. The intrinsic reaction kinetics can be classified as rapid or slow by the value of k_1 (Rittmann et al., 1985). For example, k_1 = 10,000 m^3/kg cells-day represents quite rapid biodegradation, while a value of 100 m^3/kg cells-day represents slow kinetics.

Secondary utilization

When micropollutant concentrations are very low, such as at the µg/L level or lower, support of biofilm biomass by utilization of only one compound is problematical and probably impossible. Although selection of oligotrophs could mitigate some of the limitation of S_{min}, no current evidence indicates that K_s and S_{min} values are very low for the anthropogenic compounds of health concern.

The mechanism by which micropollutants can be removed, even though they are present at concentrations less than S_{min}, is secondary utilization (Kobayashi and Rittmann, 1982; Stratton et al., 1983; Namkung et al., 1983). Secondary utilization occurs when the micropollutant is utilized by microorganisms which are grown and sustained through utilization of a more plentiful substrate. The growth-supporting substrate, present at a concentration greater than S_{min}, is called primary substrate; the micropollutant, present at a sub-S_{min} concentration, is called a secondary substrate.

Biodegradation of secondary substrates has been demonstrated numerous times (Stratton et al., 1983; Namkung et al., 1983; McCarty et al., 1981; Bouwer and McCarty, 1983). The necessary features for secondary utilization are that sufficient primary substrate be available to support the biomass and that the secondary substrate be degraded by the microorganisms present. Primary substrate can be one compound or the aggregate of many individual compounds; individual secondary substrate can be part of an aggregate primary substrate.

Because secondary utilization of micropollutants requires a primary substrate, the presence and biological fate of the natural organic matter, which constitutes the large majority of DOC in waters, are paramount to assessing the fate of micropollutants. Work in the author's laboratory indicates that humic and fulvic materials in waters are from 19 to 30% biodegraded as part of an aggregate primary substrate (Namkung, 1985) in lab-scale biofilm reactors similar to those described previously. As primary substrate alone, PFA was only 9% removed, but the biofilm grown supported removals of 8 to 95% for a range of monomeric compounds. Thus, natural organic polymers can support biofilm growth that is capable of removing other compounds present in very low concentrations.

The current weakness is our lack of knowledge about biodegradation rates for specific compounds when biofilm is grown on natural materials. Although data on intrinsic kinetics are sparse, the secondary utilization concept provides a good structure for designing experiments and interpreting results.

INTERACTIONS WITH AUTOTROPHS

Often present in water supplies are inorganic compounds that can serve as electron donors for certain aerobic bacteria. The most common inorganic species are NH_4^+, Fe^{2+}, and Mn^{2+}. The amount of biodegradable COD in the inorganic electron donors often can be greater than that present in biodegradable organic matter. For instance, 1 mg/L of NH_4^+-N contains 4.5 mg COD/L, while 1 mg/L of Fe^{2+} and Mn^{2+} contain 0.14 and 0.29 mg COD/L, respectively.

Oxidation of the inorganic electron donors and growth of autotrophic biofilm is well documented (Rittmann and Snoeyink, 1984). The growth of autotrophic biofilm could interact in two ways with the heterotrophic bacteria that oxidize organic material. First, autotrophs can compete with heterotrophs for biofilm space and for dissolved oxygen. Such competition would be deleterious to organics removal. The second possible interaction is positive and of interest in more situations: the growth and metabolism of the autotrophs can form organic products which are utilized by the heterotrophs, thus augmenting the amount of primary substrate for heterotrophs and the accumulation of heterotrophic biomass. Since primary substrate usually is limiting in most water-treatment situations, the autotrophic activity can improve removal of organic compounds by allowing accumulation of more heterotrophic biomass. A possible example of such enhancement was reported by Rittmann and Brunner (1984).

SUMMARY

Biodegradation of organic micropollutants is controlled by the environmental characteristics of biological drinking water treatment. Key among those characteristics are high specific surface area, oxidized potential, low concentration of biodegradable organic matter, very low concentrations of particular organic micropollutants, and the presence of inorganic electron donors.

The high surface area and low concentrations cause biofilm processes to be the only feasible type of biological treatment. Documented examples of biologically active fixed-film processes include rapid and slow sand filters, fluidized beds, pozzolana and gravel filters, GAC columns, and ground filtration.

The biodegradation of natural organic polymers, including humic and fulvic acids, is important in drinking water treatment for two reasons. One, these polymers constitute the majority of DOC and biodegradable organic material in waters. Second, biofilm growth and accumulation are sustained mainly through the utilization of the polymers. This biofilm then is able to effect removal of the organic micropollutants through secondary utilization, even though the micropollutants are present at concentrations too low to allow growth of microorganisms through their utilization alone.

Oxidation of inorganic electron donors--NH_4^+, Fe^{2+}, and Mn^{2+}--can provide additional primary substrate for growth of heterotrophic bacteria, thus enhancing organic micropollutant removal.

Many organic micropollutants are known to be biodegradable under aerobic conditions, although a few notable exceptions exist. The greatest weakness for predicting the biological fate of specific micropollutants is a lack of knowledge of the intrinsic reaction rates of the specific compounds when natural polymers support microbial growth.

ACKNOWLEDGMENT

The author acknowledges the valuable discussions and input from Dr. Eun Namkung and Dr. Vernon Snoeyink.

REFERENCES

Atlas, R. M., 1981. Microbial degradation of petroleum hydrocarbons: an environmental perspective. Microb. Rev., 45: 180-209.
Baliga, K, Y., 1969. Biologically mediated chemical changes in the filtration of aerated ground waters. Ph.D. dissertation, Department of Civil Engineering, University of Illinois, Urbana, Illinois.
Bourbigot, M. M., Dodin, A., and Lheritier, R., 1982. Limiting bacterial aftergrowth in distribution systems by removing biodegradable organics. Presented at the Annual Conference of the Amer. Water Works Assn., Miami, Florida.
Bouwer, E. J. and McCarty, P. L., 1983. Transformations of 1- and 2-carbon halogenated aliphatic organic compounds under methanogenic conditions, Appl. Environ. Microb., 45: 1286-1294.
Bruchet, A., Tsutsumi, Y., Dugvet, J. P., and Mallevialle, J., 1984. Use of gel permeation chromatography to study water treatment processes. Presented at the Amer. Chem. Soc. Conf., Philadelphia.
Eberhardt, M., Madsen, S., and Sontheimer, H., 1977. Investigations of the use of biologically effective activated carbon filters in the processing of drinking water. U.S. Environ. Prot. Agency, EPA-TR-77-503.
Glaze, W. H., Wallace, J. L., Dickson, K. L., Wilcox, D. P. Johansson, K. R., Chang, E., Basch, A. W., Scalf, B, G., Noack, R. K., and Smith, D. P., Jr., 1982. Evaluation of biological activated carbon for removal of trihalomethane precursors. Municipal Environ. Res. Lab., U.S. Environ. Prot. Agency, Cincinnati, Ohio.

Gomella, C. and Versanne, D., 1980. Nitrification biologique et affinage d'une eau de forage. Tech. et Sc. Municipales, 211.

Hirsch, P., Bernhard, M., Cohen, S., Ensign, J., Jannasch, H., Koch, A., Marshall, K., Matin, A., Poindexter, J., Rittenberg, S., Smith, D., and Veldkamp, H., 1979. Life under conditions of low nutrient concentration, group report. In: M. Shilo, ed., Strategies of Microbial Life in Extreme Environments, Dahlem Konferenzen, Verlag Chemie, New York, pp. 357-372.

Jekel, M., 1977. Biological treatment of surface waters in activated carbon filters. Presented at the Water Resources Center, KIWA, and Engler Bunte Institute meeting, Karlsruhe, F.R.G.

KIWA, 1983. Activated carbon in drinking water technology. In: J. C. Krvithof and R. C. van der Leer, eds., Report of the Study Group on Activated Carbon, Nievwegein, The Netherlands, in press.

Kobayashi, H. and Rittmann, B. E., 1982. Microbial removal of hazardous organic compounds. Environ. Sci. Technol., 16: 170A-181A.

Malcomb, R. L., 1985. The geochemistry of stream fulvic and humic substances. In: Humic Substances, I - Geochemistry, Isolation, and Characterization, International Humic Substances Society, in press.

Matin, A. and Veldkamp, H., 1978. Physiological basis of the selective advantage of _Spririllum_ sp. in a carbon-limited environment. J. Gen. Microb., 105: 187-197.

McCarty, P. L., Rittmann, B. E., and Reinhard, M., 1981. Trace organics in groundwater. Environ. Sci. Technol., 15: 40-52.

Milliner, R., Bowles, D. A., and Brett, R. W., 1972. Biological pretreatment at Tewkesbury. Proc. Soc. Water Trmt. Exam., 21: 318.

Morita, R. Y., 1980. Microbial life in the deep sea. Can. J. Microb., 26: 1375-1285.

Namkung, E., 1985. Kinetics and mechanisms of low-concentration-multisubstrate utilization by biofilms. Ph.D. dissertation, Department of Civil Engineering, University of Illinois, Urbana, Illinois.

Namkung, E., Stratton, R. G., and Rittmann, B. E., 1983. Predicting removal of trace-organic compounds by biofilms. J. Water Poll. Control Fedn., 55: 1366-1372.

Novitski, J. A. and Morita, R. Y., 1978. Possible strategies for the survival of marine bacteria under starvation conditions. Marine Biol., 48: 289-295.

Perry, R. H. and Chilton, C. H., 1973. Chemical Engineer's Handbook, fifth ed., McGraw-Hill Book Co., New York.

Piet, G. J. and Zoeteman, B.C.J., 1980. Organic water quality changes during sand bank and dune filtration of surface waters in The Netherlands. Wastewater Reuse for Groundwater Recharge, California State Water Resources Control Board, Sacramento, California, pp. 195-215.

Poindexter, J. S., 1981. Oligotrophy. Feast and famine existence. Adv. Microb. Ecol., 5: 63-89.

Rice, R. G., Gomella, L., and Miller, G. W., 1978. Rouen, France water treatment plant. Civil Engr., May, 1978, pp. 76-82.

Richard, Y., 1979. Biological methods for the treatment of groundwater. In: H. Sontheimer and W. Kuhn, eds., Oxidation Techniques in Drinking Water Treatment, U.S. Environ. Prot. Agency, EPA-570/9-79-020.

Rittmann, B. E., 1982a. Comparative performance of biofilm reactor types. Biotech. Bioengr., 24: 1341-1370.

Rittmann, B. E., 1982b. The effect of shear stress on biofilm loss rate. Biotech. Bioengr., 24: 501-506.

Rittmann, B. E. and Brunner, C. W., 1984. The nonsteady-state process for advanced organics removal. J. Water Poll. Control Fedn., 56: 874-880.

Rittmann, B. E., Jackson, D., and Storck, S. L., 1985. Potential for treatment of hazardous organic chemicals with biological processes. In: D. L. Wise, ed., Biotechnology Applied to Environmental Problems, CRC Press, Inc., Boca Raton, Florida, in press.

Rittmann, B. E. and McCarty, P. L., 1980a. Model of steady-state-biofilm kinetics. Biotech. Bioengr., 22: 2343-2357.

Rittmann, B. E. and McCarty, P. L., 1980b. Evaluation of steady-state-biofilm kinetics. Biotech. Bioengr., 22: 2359-2373.

Rittmann, B. E. and McCarty, P. L., 1981. Substrate flux into biofilms of any thickness. J. Environ. Engr. Div., Amer. Soc. Civil Engr., 107(EE4): 831-849.

Rittmann, B. E., and Snoeyink, V. L., 1984. Achieving biologically stable drinking water. J. Amer. Water Works Assn., 76: 106-114.

Short, C. S., 1975. Removal of ammonia from river water. Tech. Report TR3, Treatment Div., Water Research Centre, Medmenham Laboratory, Medmenham, England.

Sontheimer, H., 1978. The Mülheim process. J. Amer. Water Works Assn., 70: 393.

Sontheimer, H., 1980. Experience with river bank filtration along the Rhine river. Wastewater Reuse for Groundwater Recharge. California State Water Resources Control Board, Sacramento, California, pp. 195-214.

Stratton, R., Namkung, E., and Rittmann, B. E., 1983. Biodegradation of trace-organic compounds by biofilms on porous medium. J. Amer. Water Works Assn., 75: 463-469.

Tuck, C., 1984. The enrichment and kinetic characterization of oligotrophic bacteria by biofilm techniques. M. S. thesis, Dept. Civil Engr., University of Illinois, Urbana, Illinois.

van der Kooij, D., Visser, A., and Hijnen, W.A.M., 1982. Determining the concentration of easily assimilable organic carbon in drinking water. J. Amer. Water Works Assn., 74: 540.

Williamson, K. and McCarty, P. L., 1976a. A model of substrate utilization by bacterial films. J. Water Poll. Control Fedn., 48: 9.

Williamson, K. and McCarty, P. L., 1976b. Verification studies of the biofilm model for bacterial substrate utilization. J. Water Poll. Control Fed., 48: 281.

INFLUENCE OF MICROBIOLOGICAL ACTIVITY IN GRANULAR ACTIVATED CARBON FILTERS ON THE REMOVAL OF ORGANIC COMPOUNDS

J. DE LAAT, F. BOUANGA and M. DORE

Laboratoire de Chimie de l'Eau et des Nuisances, Université de Poitiers, 40, Avenue du Recteur Pineau, 86022 Poitiers cedex (France)

ABSTRACT

In order to show the influence of bacterial activity in GAC filters on the removal of organic matter, experiments were carried out with diluted aqueous solutions of biodegradable and of non-biodegradable compounds under sterile and non sterile conditions. Productions of CO_2 show clearly that the bacterial activity can lead to a complete elimination of the biodegradable solute and to a partial microbiological regeneration of the adsorbent. By its influence on the competitive adsorption, the biological removal of biodegradable compounds leads to an increase of GAC adsorptive capacities for the non-biodegradable organic matter.

INTRODUCTION

The use of granular activated carbon filters in drinking water treatment plants leads to a removal of organic contaminants and of organic substances, such as humic and fulvic acids which are produced naturally. This removal is obtained mainly by a physical adsorption of the solutes on the adsorbent. Bacterial activity on granular activated carbon, which can be shown by different analytical methods (Bacterial counts, ATP determinations, photography by electron-microscope) (Bourbigot et al., 1981 ; Den Blanken, 1982 ; Rizet and Coute, 1981 ; Weber et al., 1978) also contributes to the removal of the biodegradable part of the dissolved organic compounds (10-20 %) and can be increased after ozone oxidation (Brunet, 1981 ; Van Der Kooij, 1982).

EXPERIMENTAL

Experiments were carried out with aqueous synthetic solutions of pure organic compounds, prepared daily as follows :

- Phosphate buffer pH 8
  - Na_2HPO_4 : 335 or 670 mg l^{-1}
  - KH_2PO_4 : 16 or 32 mg l^{-1}
- Magnesium sulfate : 50 mg l^{-1}

---

*This work was carried out within the framework of the scientific programme of the G.S. "Traitements chimiques des eaux", C.N.R.S./Société Lyonnaise des Eaux.

- Manganese sulfate : 0.14 mg l^{-1}
- Ammonium sulfate : 4.7 mg l^{-1}
- Silver sulfate : 0.5 mg l^{-1} for the sterile solutions

Salicilyc acid and 4- nitrophenol were used as biodegradable solutes, 4- chlorobenzoic acid and 2- methyl 4,6- dinitrophenol were used as non-biode-gradable ones. Several granulometric fractions of a commercial activated carbon (CHEMVIRON F400) were sieved, washed, and dried at 105°C for 3 days.

Batch experiments

Batch experiments were carried out with powdered activated carbon (∅ < 80 µm) at 20°C so as to obtain the equilibrium adsorption capacities of the adsorbent for the different organic compounds used.

Micro-column filtrations

Filtrations were carried out with microcolumns filled with 20 grams of GAC (0.8 < ∅ < 1.0 mm). The experimental conditions were the following :
- Section : 1 cm2
- Bed height : 70 cm
- Velocity : 4 m h^{-1}
- Empty bed contact time : 10 mn

The temperature of the columns was fixed at 20°C by means of water flowing around the exterior of the GAC bed.

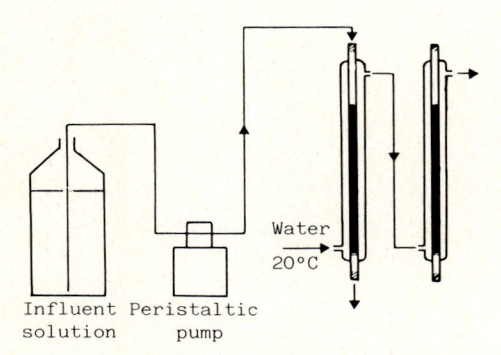

Fig. 1. Apparatus for GAC filtrations

The influence of the bacterial activity on the organic substance removal was shown by comparing the breakthrough curves obtained from sterile and non-sterile filtrations. Sterile conditions were obtained by adding a silver salt to the influent solutions. Bacteria present in distilled water (100-1 000 bacteria ml^{-1}) provided a continuous inoculation of the non-sterile column.

Analytical methods

The organic compound concentrations in the solutions were determined by UV spectrophotometry or by High Performance Liquid Chromatography analysis.

During the filtration experiments, the change in the organic matter and the bacterial activity were measured every 24 hours : Dissolved Organic Carbon determinations ; Inorganic Carbon (CO_2) production measurements ; Dissolved oxygen consumptions measurements ; Bacterial counts (sterilization control).

RESULTS

Adsorption capacities (sterile conditions)

The equilibrium adsorption capacities obtained from batch tests and from one-organic compound solutions agree with Freundlich's law : $x/m = kCe^{1/n}$, where x/m represents the amount of adsorbed solute ($\mu mol\ g^{-1}$) for a given liquid phase concentration Ce ($\mu mol\ l^{-1}$). The constants k and n are listed in Table 1 :

TABLE 1

k and n contant values

Organic compound	k	n
Salicylic acid	211.52	6.707
4- chloro benzoic acid	218.52	5.169
4- nitrophenol	527.93	4.692
2- methyl 4,6- dinitrophenol	536.85	7.72

Moreover, the dynamic capacities obtained from filtration experiments agree with those calculated from the Freundlich isotherm.

Influence of the bacterial activity

Microbiological regeneration. Figure 2 gives typical results obtained for a easily biodegradable compound, salicylic acid. In non-sterile conditions, oxygen consumption and inorganic carbon production curves show clearly that the biodegradation of salicylic acid begins after about 3 days filtration. After 6 days filtration, the column reaches an equilibrium which corresponds to a mineralization yield of nearly 50 % : an influent T.O.C. concentration of 4.2 mg l^{-1} is biologically oxidized into 2.1 mg l^{-1} of inorganic carbon. High carbon dioxide productions formed between the third and the sixth day of filtration indicate that the bacterial activity leads to a complete biodegradation both of the solute present in the influent and of the solute adsorbed during the time that the bacteria acclimatize to the substrate.

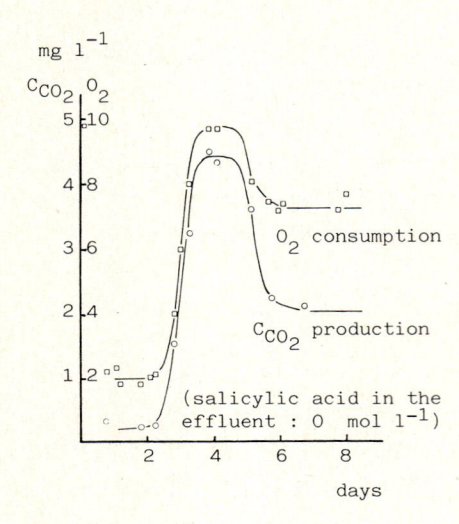

Fig. 2. GAC filtration of a salicylic acid solution (Influent concentration : 50 µmol l^{-1} ; TOC = 4.2 mg l^{-1})

This microbiological regeneration of the granular activated carbon can also be observed during the filtration of a 4- nitrophenol solution on a saturated adsorbent (Fig. 3) and explains why there is an inorganic carbon production of a value higher than the T.O.C. value of the influent.

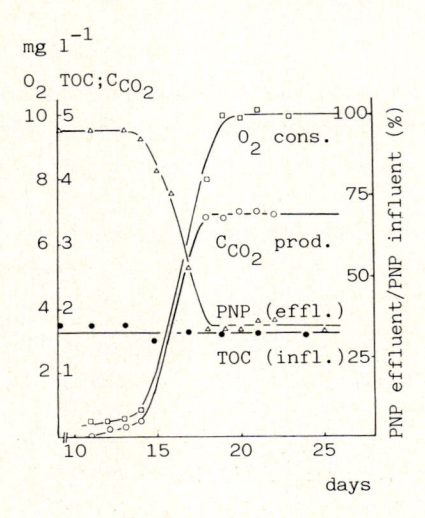

Fig. 3. GAC filtration of a 4-nitrophenol solution (PNP : 16.66 µmol l^{-1} ; TOC = 1.2 mg l^{-1}) on a saturated adsorbent.

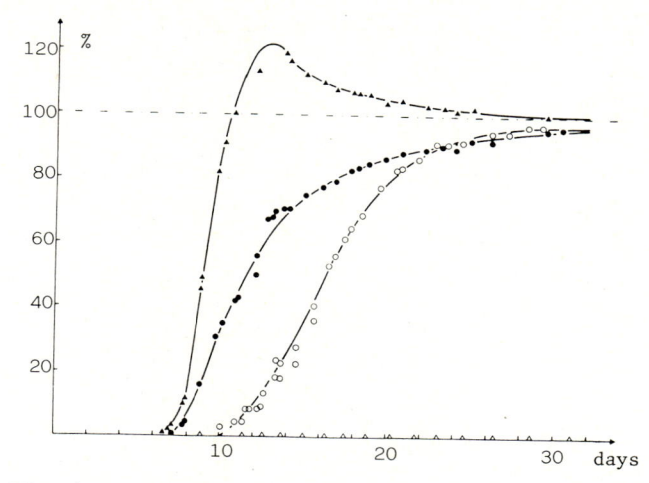

Fig. 4. Experimental breakthrough curves of salicilyc acid (C_o = 25 µmol l^{-1}, ▲, Δ) and of 4- chloro benzoïc acid (C_o = 25 µmol l^{-1}, ●, o) in mixture obtained in sterile (▲, ●) and non-sterile conditions (Δ, o)

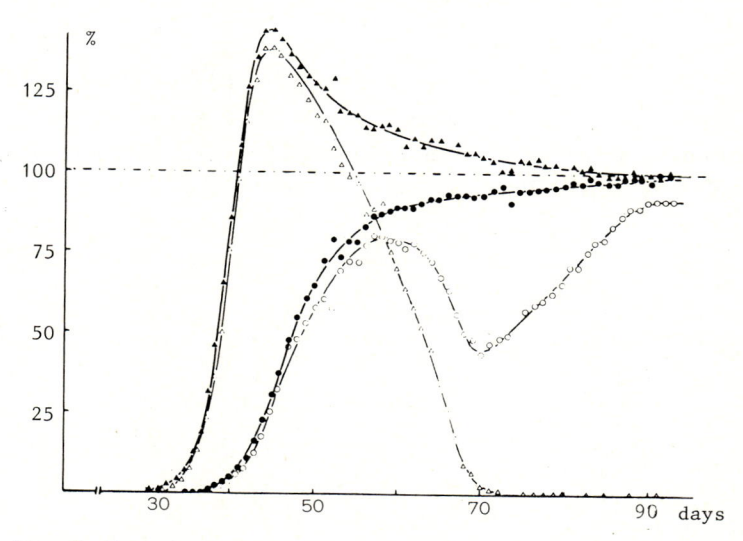

Fig. 5. Experimental breakthrough curves of 4- nitrophenol (C_o = 20 µmol l^{-1}, ▲, Δ) and of 2- methyl 4,6- dinitrophenol (C_o = 20 µmol l^{-1}, ●, o) in mixture in sterile (▲, ●) and in non-sterile conditions (Δ, o).

<u>Influence of bacterial activity on the organic compound removal.</u> As shown by figures 4 and 5, the biodegradations of salicylic acid and of 4- nitrophenol, which begin respectively after 3 and 40 days of filtration (O_2 consumption, CO_2 production) lead to an increase in the removal of the non- biodegradable compounds (4- chlorobenzoic acid, 2- methyl 4,6- dinitrophenol). For these compounds the calculated adsorptive capacities are very much the same as those

obtained from pure solution filtrations (Table 2).

TABLE 2

Comparison between experimental adsorption capacities obtained from one-solute and two-component solution filtrations.

	One-solute solution	Two-component solution	
	Sterile conditions $\mu mol\ g^{-1}$	Sterile conditions $\mu mol\ g^{-1}$	Non sterile conditions $\mu mol\ g^{-1}$
Salicylic acid ($C = 25\ \mu mol\ l^{-1}$)	366.5	192.0	Biodegradation
4- chlorobenzoic acid ($C = 25\ \mu mol\ l^{-1}$)	445.4	340.5	430.7
4- nitrophenol ($C = 20\ \mu mol\ l^{-1}$)	999	425.7	Biodegradation
2- methyl 4,6- dinitrophenol ($C = 20\ \mu mol\ l^{-1}$)	791	670	798

By its effects on the microbiological regeneration of the adsorbent and on the competitive phenomena, the bacterial activity contributes to an increase of the GAC filter time life.

REFERENCES

Bourbigot, M.M., Lhéritier, R., et Benezet-Toulze, M., 1981. Traitement biologique de l'eau potable. Mesure de l'activité bactérienne dans les milieux filtrants. Techn. Sci. Munic., 12: 639-648.
Brunet, R., 1981. Evolution chimique de la micropollution organique en cours d'ozonation des eaux de surface filtrées. Incidence du taux d'ozonation sur la biodégradabilité des composés organiques résiduels. Thèse de 3e cycle, Univ. Poitiers.
Den Blanken, J.G., 1982. Microbial activity in activated carbon filters. J. Env. Eng. Div., 108: 405-425.
Rizet, M. and Coute, A., 1981. Evaluation des micro-organismes des milieux granulaires utilisés pour la filtration des eaux de surface. Techn. Sci. Munic., 6: 371-379.
Van Der Kooij, I.D., Visser, A. and Hijnen, W.A.N., 1982. Determining the concentration of easily assimilable organic compounds in drinking water. J. Am. Wat. Wks Ass., pp 540-545.
Weber, W.J., Pirbazari, M. and Melson, G.L., 1978. Biological growth on activated carbon : an investigation by scanning electron microscopy. Envir. Sci. Technol., 12: 817-820.

DEVELOPMENTS IN BIOTECHNOLOGY OF RELEVANCE TO DRINKING WATER PREPARATION

DICK B. JANSSEN and BERNARD WITHOLT

Groningen Biotechnology Center, University of Groningen, Nijenborgh 16,
9747 AG Groningen (The Netherlands)

ABSTRACT

This paper discusses strategies to increase the feasibility of microorganisms for the removal of toxic xenobiotics from waste water and drinking water. Based on the principles of adaptational mutations and genetic exchange of catabolic activities, it becomes possible to select and engineer microorganisms that are suitable for the degradation of recalcitrant compounds. The detailed biochemical knowledge that is required for this is now rapidly evolving, and especially for the degradation of chlorinated organics several detoxifying dehalogenation mechanisms have been studied in detail. The feasibility of specialized bacteria for waste and water treatment will be dependent on the possibility to obtain stable performance and maintenance in treatment systems.

INTRODUCTION: BIOTECHNOLOGY AND THE ENVIRONMENT

Applications of microorganisms for waste-treatment processes are well established. Especially in the area of aerobic and anaerobic waste water treatment, very successful microbiological processes are used for the removal of biodegradable organic matter. Thus, biological treatment can largely prevent damage to ecosystems by excessive loadings of organic compounds that are present in municipal and industrial waste water. Nevertheless, the amount of research into the microbiological aspects of these waste treatment systems is limited, certainly when compared to the efforts put into medically and industrially important organisms. During the last several years, there is an increasing demand for a better understanding of microbiological processes of relevance to waste treatment. The major reason is shock-effects caused by the discovery of numerous cases of soil and water pollution by synthetic chemical compounds. Increased analytical power (Keith, 1976) has led to the worldwide identification of toxic chemicals in various environmental samples, drinking water, and animal tissues.

Treatment techniques for the efficient removal of toxic compounds are required, both to prevent point-source related pollution and for the preparation of drinking water. Current treatment processes are extremely expensive or technically unsuitable. Thus, there is a need for improvement of treatment techniques. Regarding biological treatment, improvements should be directed towards the biological side of the systems and towards technological aspects.

This paper will discuss a number of microbiological principles that are of relevance to the improvement of biological treatment techniques for the removal of toxic or recalcitrant xenobiotic compounds. Possibilities for enlarging the capacity of microorganisms will be presented, and these are based on a better understanding of the molecular basis of recalcitrance and the biochemistry and genetics of biodegradation.

DEGRADATION OF XENOBIOTICS BY MICROORGANISMS

Numerous man-made organic chemicals that enter the environment are characterized by a low susceptibility to biodegradation in various ecosystems. Many of these compounds may persist in the environment for up to decades, and phenomena such as the accumulation of organochlorine pesticides and polychlorinated biphenyls in animals and the occurence of several chlorinated hydrocarbons in drinking water must be considered to be the result of very slow biodegradation rates (Alexander, 1973; Lal and Saxena, 1982; Furukawa, 1982).

In general, environmental conditions and physico-chemical characteristics of the compound itself may inhibit biodegradation (Alexander, 1965, 1973; Atlas, 1977) (Table 1). For compounds that are of special interest to drinking water technologists, however, the main cause of poor biodegradation is related to the aberrant chemical structure of many xenobiotics. Water that is used for the preparation of drinking water usually has passed several stages that allow efficient removal of rapidly degradable organics, and the compounds that remain are present at trace levels and have a recalcitrant chemical structure (Keith, 1976). These structures are so different from the natural substrates of microorganisms that microbial proteins do not recognize the molecules. Thus, enzymes that can perform the required degradative reactions are not present or have extremely

TABLE 1
Some causes of poor biodegradation of organic compounds.

Unfavourable environmental conditions	Lack of molecular oxygen Lack of an electron acceptor Lack of nutrients (phosphorus, nitrogen) Lack of water Unfavourable pH High salinity Extreme temperature Presence of toxins
Unfavourable physico-chemical properties of the compound	Recalcitrant structure Insolubility Adsorption and binding to soil components Low concentration

low conversion rates. Compounds like carbon tetrachloride, perchloroethylene, hexachlorobutadiene and several polychlorinated biphenyls and pesticides do not have natural analogs and are extremely recalcitrant to microbial conversion.

CRITICAL ENZYMES

Easily degradable xenobiotics show enough similarity to natural substrates of microorganisms to allow at least partial conversions by enzymes involved in the metabolism of natural compounds. Other conversions must be carried out by catabolic enzymes with elevated activities towards xenobiotic substrates. 3-Chlorobenzoate, for example, can be efficiently degraded by some *Pseudomonas* sp. provided that the bacteria can produce enzymes that catalyze the cleavage of the aromatic ring (catechol-1,2-dioxygenase and cycloisomerase) with some 20- to 200-fold higher activities towards chlorinated substrates, when compared to normal ring-cleavage enzymes (Knackmuss, 1981).

Many degradative routes involve a sequence of steps, some of which are catalyzed by 'specialized' enzymes, and some by enzymes that play a role in the degradation of natural substrates. This may be illustrated by the pathway for 1,2-dichloroethane degradation in *Xanthobacter autotrophicus* GJ10 (Janssen et al., 1985) (Fig. 1). Steps I and IV are catalyzed by dehalogenases, which may be regarded as enzymes with extraordinary catabolic activities towards

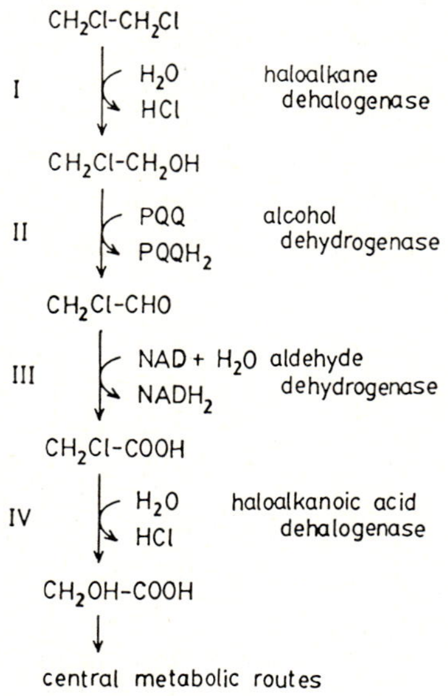

$$
\begin{array}{c}
CH_2Cl\text{-}CH_2Cl \\
\text{I} \quad \downarrow \begin{array}{l} \nwarrow H_2O \\ \swarrow HCl \end{array} \quad \text{haloalkane dehalogenase} \\
CH_2Cl\text{-}CH_2OH \\
\text{II} \quad \downarrow \begin{array}{l} \nwarrow PQQ \\ \swarrow PQQH_2 \end{array} \quad \text{alcohol dehydrogenase} \\
CH_2Cl\text{-}CHO \\
\text{III} \quad \downarrow \begin{array}{l} \nwarrow NAD + H_2O \\ \swarrow NADH_2 \end{array} \quad \text{aldehyde dehydrogenase} \\
CH_2Cl\text{-}COOH \\
\text{IV} \quad \downarrow \begin{array}{l} \nwarrow H_2O \\ \swarrow HCl \end{array} \quad \text{haloalkanoic acid dehalogenase} \\
CH_2OH\text{-}COOH \\
\downarrow \\
\text{central metabolic routes}
\end{array}
$$

Fig. 1. Degradation of 1,2-dichloroethane in *Xanthobacter autotrophicus* GJ10.

xenobiotic compounds (see also Table 6). The alcohol dehydrogenase (step II), in contrast, is a very common enzyme. In fact, it is a methanol dehydrogenase that as a result of its broad substrate specificity is also able to convert other alcohols, fortuitously including 2-chloroethanol, and the enzyme is also present in several organisms that cannot degrade chlorinated compounds. The aldehyde dehydrogenase (step III) probably too is a normal enzyme of *Xanthobacter*. Thus, the metabolism of 1,2-dichloroethane in this bacterium is possible due to capability of the organism to produce only two enzymes with extraordinary activities towards chlorinated substrates.

ADAPTATION OF MICROORGANISMS

Microorganisms, especially bacteria, are known to have an enormous potential to adapt their properties to changing environmental conditions, including the introduction of large amounts of chemicals in their environment. A number of genetic adaptive responses that enable bacteria to degrade xenobiotics and utilize them for energy generation and cell mass synthesis have been identified (Clarke, 1981) (Table 2).

Important responses seem to be mutational events that lead to changes in substrate specificity of catabolic enzymes and that enable new substrates to be degraded. Mutations of catabolic genes may lead to enzymes that have an altered three dimensional structure. This is believed to alter the catalytic site of enzymes and to allow binding and conversion of compounds that are not converted by the original protein. Such mutations have been identified in the gene encoding amidase activity in *Pseudomonas aeruginosa* (Clarke, 1981), and have led to the extension of the catabolic activities with the result that butyramide, valeramide and phenylacetamide can be degraded by the evolved proteins. Phenotypically, the effect of this class of mutations is the evolution of microorganisms that degrade xenobiotic compounds with structural similarities to compounds that

TABLE 2
Genetic alterations that enlarge the catabolic range of bacteria.

Intracellular
1. Structural mutation in a gene encoding a catabolic enzyme
2. Regulatory mutations (promotor- or regulatory gene-located)
3. Loss of unproductive enzymes by mutation
4. Mutation causing loss or activation of transport protein
5. Activation of silent genes by a combination of 1 and 2

Intercellular
6. Acquisition of new catabolic genes by plasmid transfer
7. Introduction of cloned catabolic genes
8. Future: introduction of genes encoding engineered proteins

are degraded by the parent organism.

Another class of genetic adaptive responses involves regulatory mutations. Since catabolic proteins are inducible only by a restricted number of substrates, poor biodegradation may be caused by the inability of xenobiotics to induce the synthesis of enzymes that are required for essential degradation steps. Mutations that lead to constitutive high level expression or elevated induction of catabolic proteins have been described (Clarke, 1981; Slater et al., 1979).

A combination of the above events can lead to the activation of so called 'silent genes'. These are considered to be nonfunctional segments of DNA that can be activated by mutations and then lead to the production of proteins involved in the degradation of xenobiotics. Direct proof for the conversion of silent genes to active genes, encoding enzymes involved in the degradation of synthetic chemicals, however, is still lacking (Clarke, 1981; Hall, 1978).

Besides genetic alterations that lead to the acquisition of new catabolic activities, it may also be necessary that certain enzyme activities are absent. A critical step in the degradation of chloroaromatics (e.g. 3-chlorobenzoic acid or chlorobenzene) is ring-cleavage of chlorocatechols (Knackmuss, 1981). The presence of catechol-2,3-dioxygenase, which is a normal enzyme for the metabolism of non-halogenated aromatics is unfavourable, since in the case of chlorinated catechols the enzyme produces acylhalides, which are toxic for the enzyme and the cell, or it produces a chlorinated hydroxymuconic acid semialdehyde that cannot be further degraded and accumulates. Thus, for efficient degradation, mutational events that lead to the loss of catechol-2,3-dioxygenase are required and the bacteria must instead exclusively produce catechol-1,2-dioxygenase. The latter enzyme produces chlorinated muconic acids that can be further degraded and channelled into central metabolism after dehalogenation (Knackmuss, 1981; Motosugi and Soda, 1983).

The adaptive responses desribed above show that mutations may lead to the evolution of organisms with increased catabolic activities. Furthermore, genetic exchange between bacteria must be considered as a main mechanism for the evolution of strains with new degradative abilities. The conjugational introduction of genes encoding proteins that can catalyze critical steps in a catabolic route may enable a compound to be (faster) degraded. The importance of this process is strongly supported by the observation that degradative genes are so often localized on transmissible plasmids (Haas, 1983; Ghosal et al., 1985). Probably, these exchanges of genetic information between microorganisms spread newly acquired degradative abilities among different microorganisms, analogous to the well-known worldwide spreading of the capability of bacteria to resist antibiotics.

ISOLATION AND GENETIC CONSTRUCTION OF NEW STRAINS

Classical enrichment techniques, i.e. batch culture enrichment or enrichment in continuous cultures, have been used with great success for the isolation of bacteria that have the potential to degrade xenobiotics (Harder, 1981; Cook et al., 1983). Of main importance seems to be a good inoculum, which is a sample from a site that has been exposed to the chemicals under study for long periods of time and under favourable environmental conditions. This approach leaves evolution and development of the required catabolic activities to nature. On the other hand, long-term operation of continuous cultures (several months) may allow adaptational processes to take place and allow the isolation of bacteria that are not readily obtained from natural samples (Kellogg et al., 1981).

An interesting development is the genetic construction of organisms that can degrade specific recalcitrant problem compounds. This approach is certainly

TABLE 3
Strategies for the isolation of organisms with degradative abilities.

1. Enrichment in batch culture or continuous culture
2. Breeding in the presence of plasmid bearing strains
3. Transfer of catabolic pathways by conjugation
4. Transfer of catabolic activities by cloning techniques
5. Protein engineering and the introduction of engineered genes

promising and may be used for the combination of essential genes for a given catabolic route in a single organism (Table 3). Up to now, recombinant DNA techniques have yielded organisms that can degrade 4-chlorobenzoic acid, 3-, 4- and 5-chlorosalicylic acid, and 2-chloronaphthalene (Weightman et al., 1984). A drawback is that the construction of a catabolic route in a bacterium requires extensive biochemical knowledge about the degradation pathway and it may be laborious to obtain this information. So far, genetic construction has been restricted to minor increases of the substrate range of bacteria, and it is possible that strains with the same activity could also have been obtained by direct selection from environmental samples. More drastic increases in catabolic potential will require the introduction and expression of genes that are involved in cometabolism of xenobiotics in physiologically unrelated organisms, or even of genes ecoding detoxifying liver enzymes.

Rearrangement of genetic information among different microorganisms does not produce new enzymes with new activities. To achieve this in the laboratory, it would be necessary to engineer catabolic enzymes to accept new substrates or to increase turnover rates of inefficient enzymes. Protein engineering, however, is still in its infancy and will require considerably more biochemical information about the relationship between protein structure and catalytic activity than is available now.

So far, we must conclude that our polluted environment will be, at least in the near future, the best supplier of microorganisms with extraordinary degradative abilities.

ADAPTATION TO EXTREME ENVIRONMENTS

Efficient conversion of xenobiotics in treatment systems may require the use of organisms that are well adapted to specific environmental conditions. Some organisms have this property, as may be illustrated by the degradation of n-alkanes by *Pseudomonas*, which has extensively been studied (Shapiro et al., 1984; de Smet, 1982). The organisms show extreme tolerance to apolar compounds and can actively grow in an environment containing more than 99% of an n-alkane. The substrate is degraded to an aldehyde by the action of a three component ω-hydroxylase complex and an alcohol dehydrogenase. Further metabolism involves conversion to a carboxylic acid and β-oxidation. The genes of the first steps of the catabolic route, i.e. the conversion of n-alkanes to alcohols, are plasmid localized and they have now been sequenced. This allows the development of more insight into the structure-activity relation of their protein products. After cloning on a small plasmid, it has also been possible to transfer the genes of alkane oxidation to *Escherichia coli* and this has yielded an *E. coli* derivative that is able to degrade n-alkanes (G. Eggink, R. Lageveen and B. Witholt, manuscript in preparation).

Such transfer of catabolic activities to other bacteria opens the possibility to obtain degradative routes in microorganisms that are more suitable for application in the extreme environments of certain treatment systems (Kobayashi, 1984). Extreme environments include acid or alkaline pH, the presence of polar or toxic compounds, high salinity and low or high temperature. Genetic recombination has been used for the construction of a strain that degrades chlorophenols and shows increased tolerance to high concentrations of phenol, which is toxic for most bacteria (Schwien and Schmidt, 1982). Furthermore, it will possibly become important to express catabolic activities in organisms that do better fit into a certain biological system, which could be related to such diverse properties as good adherence in a biofilm reactor or the ability to proliferate in the soil environment.

AEROBIC DEGRADATION OF CHLORINATED HYDROCARBONS

Chlorinated hydrocarbons that are recognized as important environmental pollutants include several solvents and degreasing agents (mainly chlorinated C1- and C2-hydrocarbons), pesticides, polychlorinated biphenyls, several intermediates in chemical synthesis, and a number of waste- and byproducts of industrial synthesis. For drinking water technologists, chlorinated organics that are produced during chlorination, among which trihalomethanes (Rook, 1976), must be added to this list.

Table 4 lists chlorinated hydrocarbons that are known to be biodegradable under aerobic conditions and for which pure cultures have been isolated (Leisinger, 1983; Motosugi and Soda, 1983, Janssen et al., 1985). Although these data illustrate the potential of microorganisms to degrade chlorinated organics, the compounds listed only represent a fraction of the total xenobiotics that have been identified in surface water or ground water that is used for the preparation of drinking water (Keith, 1976). So far, no unequivocal positive evidence for biodegradation has been obtained for trihalomethanes, carbon tetrachloride,

TABLE 4
Chlorinated solvents that can be degraded by pure bacterial cultures.

Compounds		
Dichloromethane	*Hyphomicrobium, Pseudomonas*	Stucki et al., 1981
2-Chloroethanol	*Pseudomonas*	Stucki et al., 1981
Methylchloride, ethylchloride, propylchloride, butylchloride, 1,2-dichloroethane, 1,3-dichloropropane, allylchloride	*Xanthobacter*	Janssen et al., 1985 Keuning et al., 1985
Chlorobenzene	unidentified bacterium	Reineke and Knackmuss, 1984

1,1,1-trichloroethane, 1,2-dichloropropane and chlorinated ethylenes, including such frequent contaminants of resource water as trichloroethylene and perchloroethylene. Although several chlorinated biphenyls are biodegradable (Furukawa, 1982), the higher chlorinated components may also be regarded as recalcitrant.

Pure culture studies with bacteria that degrade chlorinated organics have enabled the establishment of several mechanisms of chloride release from organic substrates. Halogenated compounds must be dehalogenated before they enter the central metabolic routes that lead to carbon dioxide or cell mass. Thus, dechlorination reactions play a crucial role in de detoxification and metabolism of halogenated compounds. A summary of dehalogenation mechanisms that have been demonstrated to be operative in aerobic bacteria is given in Table 5.

An attractive mechanism for dehalogenation of chlorinated aliphatics is provided by the recently discovered hydrolytic haloalkane dehalogenase, which does not require cofactors, oxygen or energy (Janssen et al., 1985; Keuning et al., 1985) (Table 6). A severe limitation, however, comes from the limited range of compounds that is actively converted and used for growth. The haloalkane dehalogenase does not hydrolyze trihalomethanes, 2-chloropropane, 1,2-dichloropropane and 1,1,1-trichloroethane, for example. Similar limitations have the dichloromethane dehalogenase and haloaromatic dehalogenase (Table 5). Further enzyme evolution, in nature or in the laboratory, will possibly produce new enzymes that do have the ability to degrade more compounds. Dehalogenases that split off chlorine from a secondary carbon atom (as in 2-chloropropionic acid) or from a doubly or triply substituted carbon atom (2,2-dichloropropionic acid and trichloroacetic acid) have already been described. Thus, there is no reason a priori to exclude the possibility of hydrolytic dehalogenation of compounds like 1,2-dichloropropane and 1,1,1-trichloroethane.

ANAEROBIC DEGRADATION OF CHLORINATED HYDROCARBONS

Recent experiments with chlorinated aliphatics under methanogenic conditions have provided evidence for the occurence of degradative reactions that do not take place in aerobic cultures. At low redox potential, very recalcitrant compounds such as chloroform, tetrachloroethylene and 1,1,2,2-tetrachloroethane may be slowly dehalogenated by reductive mechanisms (Bouwer et al., 1981; Bouwer and McCarty, 1983). Often, these reactions yield only partial conversions and they appear to be extremely slow. It is beyond doubt that anaerobic conversions in sediments, subsurface soil, and groundwater are of enormous value for the ultimate disappearance of many synthetic halocarbons. However, because of the low reaction rates, considerable effort will be required to develop waste treatment processes on the basis of anaerobic degradation. Unless the process can be fastened, impractical reactor volumes and residence times will be necessary.

The mechanisms of reductive dehalogenations remain obscure. Probably, reduced sulfur compounds or heavy metal compounds act as a mediator of electron transfer from organic compounds to chlorinated substrates, thus releasing chloride (Kobayashi and Rittmann, 1982). Such reactions are believed to take place only at low redox potential and in the absence of oxygen. In the soil environment, abiotic chemical processes are possibly also involved in the slow degradation of halogenated organics.

Up to now, no anaerobic degradation has been observed for chlorinated benzenes and chlorinated biphenyls. It is assumed that all biological degradation of aromatic hydrocarbons is dependent on molecular oxygen (Atlas, 1981).

TABLE 5

Dehalogenation mechanisms of halogenated xenobiotics in aerobic bacteria.

---

1. Nucleophilic substitution by dichloromethane dehalogenase (glutathione-S-transferase).

    Reaction: $CH_2Cl_2 + GSH \longrightarrow GS-CH_2Cl + HCl$
    $GS-CH_2Cl + H_2O \longrightarrow GS-CH_2OH + HCl$
    $GS-CH_2OH \longrightarrow GSH + HCHO$

    Substrates: CH_2Cl_2, CH_2BrCl, CH_2Br_2, CH_2I_2

    Organisms: *Hyphomicrobium*, *Pseudomonas* (Stucki et al, 1981; Leisinger, 1983)

2. Hydrolytic dehalogenation of carboxylic acids (haloalkanoic acid dehalogenase).

    Reaction: $CH_2Cl-COOH + H_2O \longrightarrow CH_2OH-COOH + HCl$

    Substrates: dichloroacetate, trichloroacetate, bromoacetate, dibromoacetate, fluoroacetate, chloroacetate, iodoacetate, 2-chloropropionate, 2,2-dichloropropionate, 2-bromopropionate, 2-chlorobutyrate, 2-bromobutyrate

    Organisms: *Pseudomonas*, *Moraxella* (Motosugi and Soda, 1982)

3. Hydrolytic dehalogenation of haloalkanes (haloalkane dehalogenase).

    Reaction: $CH_2Cl-CH_2Cl + H_2O \longrightarrow CH_2OH-CH_2Cl + HCl$

    Substrates: some n-haloalkanes (see Table 6)

    Organism: *Xanthobacter autotrophicus* (Keuning et al., 1985)

4. Hydrolytic dehalogenation of aromatics (4-chlorobenzoate dehalogenase).

    Reaction: 4-chlorobenzoate + $H_2O \longrightarrow$ 4-hydroxybenzoate + HCl

    Substrates: 4-chlorobenzoate, 4-fluorobenzoate, 4-bromobenzoate

    Organism: *Arthrobacter* (Marks et al., 1984)

5. Oxidative dehalogenation of aliphatics (haloalkane hydroxylase).

    Reaction: $CH_2Cl-(CH_2)_7-CH_2Cl + O_2 + NADPH_2 \longrightarrow$
    $CHO-(CH_2)_7-CH_2Cl + H_2O + NADP + HCl$

    Substrates: 1,9-dichlorononane, 1,6-dichlorohexane, 1,5-dichloropentane, 1-bromoheptane, 1-iodoheptane, 1-chloroheptane

    Organism: *Pseudomonas* (Omori and Alexander, 1978)

6. Oxidative dehalogenation of aromatics by monooxygenase (4-chlorobenzoate-4-hydroxylase).

    Reaction: 4-chlorobenzoate + O_2 + $NADPH_2 \longrightarrow$
    4-hydroxybenzoate + HCl + $NADP$ + H_2O

    Substrates: 4-chlorobenzoate, 3-chlorobenzoate

    Organism: *Pseudomonas* (Klages and Lingens, 1980)

7. Oxidative dehalogenation of aromatics by dioxygenase (2-fluorobenzoate-dioxygenase).

    Reaction: 2-fluorobenzoate + $O_2 \xrightarrow{NADH_2 \quad NAD}$ catechol + HF

    Substrates: 2-fluorobenzoate, 4-chlorophenylacetate

    Organisms: *Pseudomonas* B13 (Knackmuss, 1981), *Pseudomonas* CBS3 (Markus et al., 1984)

TABLE 5 - Continued

---

8. Anti-elimination after cycloisomerization of ring-fission products (non-enzymatic after cycloisomerase action).

   Reaction: 2-chlorohydroxymuconic acid ⟶ 4-carboxymethyl-Δ^2butenolide + HCl

   Substrates: halomuconic acids derived from 3-chlorobenzoate, 4-chlorobenzoate, 4-fluorobenzoate, chlorobenzene, 3-chlorophenol, 3,5-dichlorocatechol, 4-fluorophenylacetic acid, 3,5-dichlorobenzoate, 2,4-dichlorophenoxyacetic acid

   Organisms: *Pseudomonas*, *Alcaligenes* (Knackmuss, 1981; Motosugi and Soda, 1983)

9. Reductive dechlorination after ring-fission (chloromaleoylacetate reductase, chlorosuccinate reductase).

   Reaction: chloromaleoylacetate + $NADH_2$ ⟶ 3-oxoadipate + NAD + HCl

   Substrates: ring-cleavage products from 3,5-dichlorocatechol

   Organism: *Pseudomonas* (Chapman, 1979; Knackmuss, 1981)

APPLICATION OF SELECTED CULTURES

The feasibility of selected microorganisms for waste treatment will normally require that organisms can grow and degrade the unwanted compounds under the prevailing environmental conditions. The maintenance of organisms can be based on 1) selective pressure caused by the limited availability of easily degradable substrates, or 2) selective pressure caused by environmental conditions that favour the desired population of organisms. It is much easier to meet these prerequisites in treatment systems for industrial effluents than for the removal of low levels of xenobiotics during the preparation of potable water. The availability of easily degradable substrates and other interfering compounds may be limited by performing waste treatment close to the source of production (Finn, 1983). Selective environmental conditions may include the limited

TABLE 6

Substrate specificity of haloalkane dehalogenase from *Xanthobacter autotrophicus*.

Substrates degraded	Not degraded	Degraded by other dehalogenases
1,2-dichloroethane	tr-1,2-dichloroethene	dichloromethane
ethylchloride	1,1-dichloroethane	chloroacetic acid
1-chloropropane	1,1,1-trichloroethane	dichloroacetic acid
1-chlorobutane	1,1,2-trichloroethane	trichloroacetic acid
1,3-dichloropropane	2-chloropropane	2-chloropropionic acid
allylchloride	1,2-dichloropropane	2,2-dichloropropionic acid
ethylbromide	1,3-dichloropropene	
1-bromopropane	1-chloropentane	
1-iodopropane	chlorinated carboxylic acids	
1,2-dibromoethane		

availability of nitrogen (Finn, 1983) or the presence of inhibitory compounds.

BIOLOGICAL TREATMENT OF WATER BY SAND INFILTRATION

Infiltration techniques are used by drinking water plants that are dependent on polluted surface water as a resource. During sand or soil infiltration, organic micropollutants are removed as a result of sorption to organic material and clay particles, and by biodegradation (Bouwer et al., 1984; McCarty et al., 1981). In the long term, only biodegradation can yield efficient removal of organics. Some notoriously recalcitrant compounds may not be removed, especially when their octanol-water partition coefficient is low. Although the effectivity of dune and bank infiltration should increase in time by adaptation, this may be hindered by extreme low nutrient levels (McCarty et al., 1981).

A future development could be to inoculate such treatment systems with cultures of microbes that have extraordinary catabolic activities towards specific unwanted chemicals. Furthermore, on the basis of known characteristics of pollutant components, it may be necessary to obtain aerobic conditions at some stage, since, as stated above, some compounds are only subject to biodegradation in the presence of molecular oxygen.

BIOLOGICAL DEGRADATION IN GRANULAR ACTIVATED CARBON

Microbial activity on activated carbon filters is involved in the long-term stable removal of organics from drinking water (van der Kooij and Hijnen, 1982; Wilcox et al., 1983). The filters can remove compounds at trace levels by absorption and by the presence of organisms that are capable of growth at extremely low nutrient levels and which show very high affinities for their substrates. These so-called oligotrophs considerably extend the periods during which the filters are active, but lifetime is still limited by the absorption capacity for non-degradable compounds, e.g. chlorinated aliphatics and trihalomethanes.

An attractive way to increase the performance of activated carbon filters would be inoculation with specialized organisms for the removal of problem compounds. It is improbable that selection and construction techniques (Table 3) can yield organisms that can degrade the whole range of micropollutants encountered during water treatment, but it is conceivable that certain compounds that are present at high levels in resource water or are produced during chlorination can be efficiently removed.

The main halogenated products from chlorination procedures are trihalomethanes and di- and trichloroacetic acid. Microorganisms that can degrade and grow at the expense of the latter two compounds have already been described in the literature (Motosugi and Soda, 1983). Other frequently observed problem halogenated compounds are trichloroethylene, 1,1,1-trichloroethane, 1,2-dichloropropane (in The Netherlands) and perchloroethylene. Bacteria that degrade these latter recalcitrant chemicals have not yet been described.

In principle, specialized cultures immobilized on granular activated carbon should be capable of degrading xenobiotic compounds to extremely low final levels and without leaving toxic intermediates. Although biodegradation of xenobiotics in batch samples of polluted water is often characterized by first-order kinetics, i.e. degradation rates are proportional to the concentration of the compound (Alexander, 1985), this does not exclude extremely low levels in the effluents of activated carbon filters. Efficient and long-term removal of chlorobenzene, the three dichlorobenzenes, and 1,2,4-trichlorobenzene has been observed in activated carbon columns inoculated with bacteria from activated sludge (Bouwer and McCarty, 1982). Specialized organisms thus may reduce regeneration costs and increase the application potential of the filters. The prospects of such techniques will be dependent on the future availability of microorganisms that contain enzyme activities for the degradation of recalcitrant problem compounds.

REFERENCES

Alexander, M., 1965. Biodegradation: Problems of molecular recalcitrance and microbial fallibility. Adv. Appl. Microbiol. 7: 35-80.
Alexander, M., 1973. Nonbiodegradable and other recalcitrant molecules. Biotechn. Bioeng. 15: 611-647.
Alexander, M., 1985. Biodegradation of organic chemicals. Environ. Sci. Technol. 18: 106-111.
Atlas, R.M., 1977. Stimulated petroleum biodegradation. C.R.C. Crit. Rev. Microbiol. 5: 371-386.
Atlas, R.M., 1981. Microbial degradation of petroleum hydrocarbons: an environmental perspective. Microbiol. Rev. 45: 180-209.
Bouwer, E.J., and McCarty, P.L., 1982. Removal of trace chlorinated organic compounds by activated carbon and fixed-film bacteria. Environ. Sci. Technol. 16: 836-843.
Bouwer, E.J. and McCarty, P.L., 1983. Transformations of 1- and 2-carbon halogenated aliphatic organic compounds under methanogenic conditions. Appl. Environ. Microbiol. 45: 1286-1294.
Bouwer, E.J., Rittmann, B.E. and McCarty, P.L., 1981. Anaerobic degradation of halogenated 1- and 2-carbon organic compounds. Environ. Sci. Technol. 15: 596-599.
Bouwer, E.J., McCarty, P.L., Bouwer, H. and Rice, R.C., 1984. Organic contaminant behavior during rapid infiltration of secondary wastewater at the Phoenix 23rd Avenue project. Water Res. 18: 463-472.
Chapman, P.J., 1979. Degradation mechanisms. In: A.W. Bourquin and P.H. Pritchard, Editors), Proceedings of the Workshop: Microbial Degradation of Pollutants in marine environments. Env. Prot. Agency, Gulf Breeze, pp. 28-66.
Clarke, P.H., 1981. Adaptation. The fifteenth Marjory Stephenson memorial lecture. J. Gen. Microbiol. 126: 5-20.
Cook, A.M., Grossenbacher, H. and Hütter, R., 1983. Isolation and cultivation of microbes with biodegradative potential. Experientia 39: 1191-1198.
de Smet, M.-J., 1982. A biotechnological approach to the synthesis of epoxides. PhD Thesis, University of Groningen, Groningen.

Finn, R.K., 1983. Use of specialized microbial strains in the treatment of industrial waste and in soil decontamination. Experientia 39: 1231-1236.

Furukawa, K., 1982. Microbial degradation of polychlorinated biphenyls. In: A.M. Chakrabarty (Editor), Biodegradation and Detoxification of Environmantal Pollutants. CRC Press, Boca Raton, pp. 33-57.

Ghosal, D., You, I.-S., Chatterjee, D.K. and Chakrabarty, A.M., 1985. Microbial degradation of halogenated compounds. Science 228: 135-142.

Haas, D., 1983. Genetic aspects of biodegradation by pseudomonads. Experientia 39: 1199-1213.

Hall, B.G., 1978. Experimental evolution of a new enzymatic function. II. Evolution of multiple functions for *ebg* enzyme in *E. coli*. Genetics 89: 453-465.

Harder, W., 1981. Enrichment and characterization of degrading organisms. In: T. Leisinger, A.M. Cook, R. Hütter and J. Nüesch (Editors), Microbial Degradation of Xenobiotics and Recalcitrant Compounds. Academic Press, London, pp. 77-96.

Janssen, D.B., Scheper, A., Dijkhuizen, L. and Witholt, B., 1985. Degradation of halogenated aliphatic compounds by *Xanthobacter autotrophicus* GJ10. Appl. Environ. Microbiol. 49: 673-677.

Keith, L.H. (Editor), 1978. Identification and Analysis of Organic Pollutants in Water. Ann Arbor Science, Ann Arbor, Mich.

Kellogg, S.T., Chatterjee, D.K. and Chakrabarty, A.M., 1981. Plasmid-assisted molecular breeding: new technique for enhanced biodegradation of persistent toxic chemicals. Science 214: 1133-1135.

Keuning, S., Janssen, D.B. and Witholt, B., 1985. Purification and characterization of hydrolytic haloalkane dehalogenase from *Xanthobacter autotrophicus* GJ10. J. Bacteriol., in press.

Klages, U. and Lingens, F., 1980. Degradation of 4-chlorobenzoic acid by a *Pseudomonas* sp. Zbl. Bakt. Hyg., I. Abt. Orig. C 1: 215-223.

Knackmuss, H.-J., 1981. Degradation of halogenated and sulfonated hydrocarbons. In: T. Leisinger, A.M. Cook, R. Hütter and J. Nüesch (Editors), Microbial Degradation of Xenobiotics and Recalcitrant Compounds. Academic Press, London, pp. 189-212.

Kobayashi, H.A., 1984. Application of genetic engineering to industrial waste/wastewater treatment. In: G.S. Omenn and A. Hollaender (Editors), Genetic Control of Environmental Pollutants. Plenum Press, New York, pp. 195-214.

Kobayashi, H. and Rittmann, B.E., 1982. Microbial removal of hazardous organic compounds. Environ. Sci. Technol. 16: 170A-183A.

Lal, R. and Saxena, D.M., 1982. Accumulation, metabolism, and effects of organochlorine insecticides on microorganisms. Microbiol. Rev. 46: 95-127.

Leisinger, T., 1983. Microorganisms and xenobiotic compounds. Experientia 39: 1183-1191.

Marks, T., Smith, A.R.W. and Quirk, A.V., 1984. Degradation of 4-chlorobenzoic acid by *Arthrobacter* sp. Appl. Environ. Microbiol. 48: 1020-1025.

Markus, A., Klages, U., Krauss, S. and Lingens, F., 1984. Oxidation and dehalogenation of 4-chlorophenylacetate by a two-component enzyme system from *Pseudomonas* sp. strain CBS3. J. Bacteriol. 160: 618-621.

McCarty, P.L., Reinhard, M. and Rittmann, B.E., 1981. Trace organics in groundwater. Environ. Sci. Technol. 15: 40-51.

Motosugi, K. and Soda, K., 1983. Microbial degradation of synthetic organochlorine compounds. Experientia 39: 1214-1220.

Omori, T. and Alexander, M., 1978. Bacterial dehalogenation of halogenated alkanes and fatty acids. Appl. Environ. Microbiol. 35: 867-871.

Reineke, W. and Knackmuss, H.-J. 1984. Microbial metabolism of haloaromatics: isolation and properties of a chlorobenzene degrading bacterium. Appl. Environ. Microbiol. 47: 395-402.

Rook, J.J., 1976. Haloforms in drinking water. J. Am. Water Works Assoc. 68: 168-172.

Schwien, U. and Schmidt, E., 1982. Improved degradation of monochlorophenols by a constructed strain. Appl. Environ. Microbiol. 44: 33-39.

Shapiro, J.A., Owen, D.J., Kok, M. and Eggink, G., 1984. *Pseudomonas* hydrocarbon oxidation. In: G.S. Omenn and A. Hollaender (Editors), Genetic Control of Environmental Pollutants. Plenum Press, New York, pp. 229-240.

Slater, J.H., Lovatt, D., Weightman, A.J., Senior, E. and Bull, A.T., 1979. The growth of *Pseudomonas putida* on chlorinated aliphatic acids and its dehalogenase activity. J. Gen. Microbiol. 114: 125-136.

Stucki, G., Brunner, W., Staub, D. and Leisinger, T., 1981. Microbial degradation of chlorinated C1 and C2 hydrocarbons. In: T. Leisinger, A.M. Cook, R. Hütter and J. Nüesch (Editors), Microbial Degradation of Xenobiotics and Recalcitrant Compounds. Academic Press, London, pp. 131-137.

van der Kooij, D. and Hijnen, W.A.M., 1983. Verwijdering van organische stoffen door microorganismen bij filtratieprocessen. H_2O 16 (13): 306-311.

Weightman, A.J., Don, R.H., Lehrbach, P.R. and Timmis, K.N., 1984. The identification and cloning of genes encoding haloaromatic catabolic enzymes and the construction of hybrid pathways for substrate mineralization. In: G.S. Omenn and A. Hollaender (Editors), Genetic Control of Environmental Pollutants. Plenum Press, New York, pp. 47-80.

Wilcox, D.P., Chang, E., Dickson, K.L. and Johanssen, K.R., 1983. Microbial growth associated with granular activated carbon in a pilot water treatment facility. Appl. Environ. Microbiol. 46: 406-416.

THE INFLUENCE OF WATER TREATMENT PROCESSES ON THE PRESENCE OF ORGANIC SURROGATES AND MUTAGENIC COMPOUNDS IN WATER

M.A. VAN DER GAAG, J.C. KRUITHOF and L.M. PUIJKER
The Netherlands Waterworks' testing and research Institute, KIWA Ltd.
Research Department, P.O. Box 1072, 3430 BB Nieuwegein (the Netherlands)

ABSTRACT

The effects of granular activated carbon filtration and of the combination of ozonation and GAC filtration on the quality of Rhine water were studied in a pilot plant. The scope of the study was to compare both systems in relation to the removal of organic contaminants in water, and to the reduction of the side effects of chlorination. The water quality was measured with organic surrogate parameters (organohalogen, -nitrogen, -phosphorus and -sulphur) and in bacterial mutagenicity assays.

In this particular setting, the combination of ozonation and GAC filtration was superior in all points to GAC filtration alone. The effects of ozonation are sometimes quite different, depending on the type of water treated. Its positive influence should be confirmed in a local situation.

As GAC treatment causes a shift towards formation of more brominated THM after chlorination, special attention was given to this item. A higher inorganic bromide/DOC ratio resulted in higher brominated THM concentrations after chlorination. However, the mutagens formed during chlorination in presence of more inorganic bromide could be inactivated more easily by rat liver homogenate than in the normal setting. The results of this study confirmed earlier findings stating a negative influence of chlorination on water quality.

INTRODUCTION

Many new parameters have been introduced quite recently for the monitoring of water quality, with emphasis on possible implications for health. The major problems in the field of drinking water are related to pollution of river water with toxic substances, and to the side effects of water treatment processes (Rook, 1974; Bellar et al., 1974; Poels et al., 1980; van Kreijl et al. 1980; Cheh et al., 1980; Loper, 1980; Bull, 1982). In 1974 both Rook (1974) and Bellar and coworkers (1974) demonstrated the formation of trihalomethanes (THM) during chlorination. Later, many investigators showed that halogenated compounds with a higher molecular weight are also created. Efforts made to remove precursors for haloforms resulted in a shift towards formation of more brominated haloforms, because of the relatively high bromide/organic carbon ratio (Luong et al., 1980; Kruithof et al., 1982; Merlet et al., 1982; Kruithof et al., 1985).

The emphasis for the methodology developed by our institute was laid upon the integration of analytical-chemical measurements and short therm assays for toxicity (Noordsij et al., 1983). The same concentrates of organic substances from water can be used for GC/MS identification, fractionation with HPLC, measurement of organic surrogate parameters and short term assays for mutagenicity, starting with the Amestest. Pilot studies showed that the method could be used for the analysis of water treatment processes (Van der Gaag et al., 1982) and to determine side effects of chlorination (Kruithof et al., 1982).

Based on these results, a pilot plant experiment was set up with two objectives. The usefullness of the newly developed parameters had to be confirmed, and the effectivity of two parallel water treatment processes was studied. The treatment processes included rapid sand filtration (RF), followed by either granular activated carbon (GAC) filtration or by a combination of ozonation and GAC filtration (O3-GAC). The first topics of this study, about halfway the project, have been presented by Kruithof et al. (1985). The project has now come to an end.

In this paper we present a few highlights. The presence of extractable and XAD adsorbable organonitrogen, -phosphorus and -sulphur in river water, and their behaviour in purification processes will be discussed. Most attention will be given to the removal of mutagenic contaminants, and to the side effects of chlorination.

EXPERIMENTAL SET-UP

Pilot plant

The pilot plant was located in Nieuwegein, and received water from the Rhine, which had been settled for about 2 hours (Poels et al., 1980). The pilot plant is shown in figure 1. The operation conditions are given in table 1.

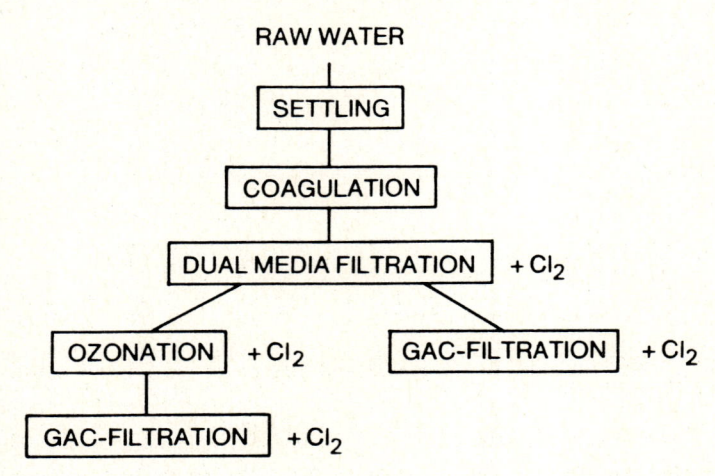

Fig. 1. Schematic design of the pilot plant.

TABLE 1

Pilot plant conditions

Process	Conditions	
* Coagulation with 7 mg/l Fe(III)	pH	: 7.8 - 8.0
	flocculation time	: 20 minutes
* Sedimentation in a plate separator	surface load	: 1.0 m/h
	residence time	: 3 minutes
* Dual media filtration over 0.45 m anthrite and 0.55 m sand	filtration rate	: 5.2 m/h
* Ozonation with 2.7 mg/l O3	initial contact time	: 7 minutes
	reaction time	: 80 minutes
* GAC filtration over 1.0 m Norit ROW 0,8S	EBCT (Empty Bed Contact Time)	: 20 minutes
* Post chlorination with a criterion of 0.2 mg/l after 20 minutes	contact time	: 120 minutes

Quality parameters

The measurement of quality parameters included the determination of THM, and of purgeable-, petroleum ether extractable- and carbon-adsorbable organohalides (POX, EOX and AOX). The other surrogates and the Ames test were determined in the organic material which was isolated on XAD at pH7 and pH2, and concentrated in ethanol (Noordsij et al., 1983). The THM were measured with a headspace/GC/ECD technique. The organohalide content of purged volatiles desorbed from Tenax, of petroleum ether extracts (EOX), of XAD samples (X7OX and X2OX) and of the AOX samples were measured with a microcoulometric titration system, after pyrolysis at 850-1000°C. The organosulphur and -phosphorus contents were measured after conversion into hydrogen sulphide respectively phosphine. The details of these techniques have been described by Kruithof et al. (1985).

The Amestest was performed with strains TA98 and TA100 (Maron and Ames, 1983) with some slight modifications. The S9 (liver homogenate) was prepared from Aroclor induced Sprague-Dawley rats by CIVO-TNO (Zeist, the Netherlands). In the assays with S9, 0.5ml of S9-mix containing 0.075ml of S9 was added to 3ml soft agar. The XAD samples were concentrated to 25 µl ethanol per liter equivalent. They were assayed in both strain without and with addition of S9-mix at six dose levels, ranging from 10 to 140 ul of sample per plate.

EFFECTIVITY OF GAC FILTERS FOR THE REMOVAL OF ORGANIC COMPOUNDS

Precence and removal of organitrogen (-ON), -phosphorus (-OP) and -sulphur (-OS) compounds

Extractable -ON, -OP and -OS. The EOP and EOS contents of the rapid filtrate (RF) remained below the detection limits (0.2 µg/l resp. 1 µg/l). The EON concentration varied between 1 and 2 µg/l. Both GAC filtration and the O3-GAC combination removed the EON for longer than 15.000 ebv (empty bed volumes).

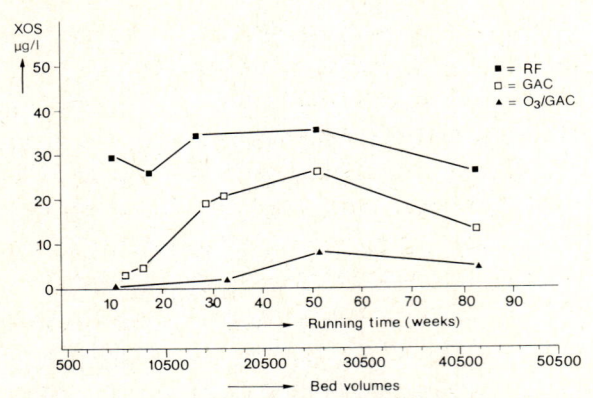

Fig. 2. $X_{7+2}OS$ (as µg S/l) in unchlorinated effluents.

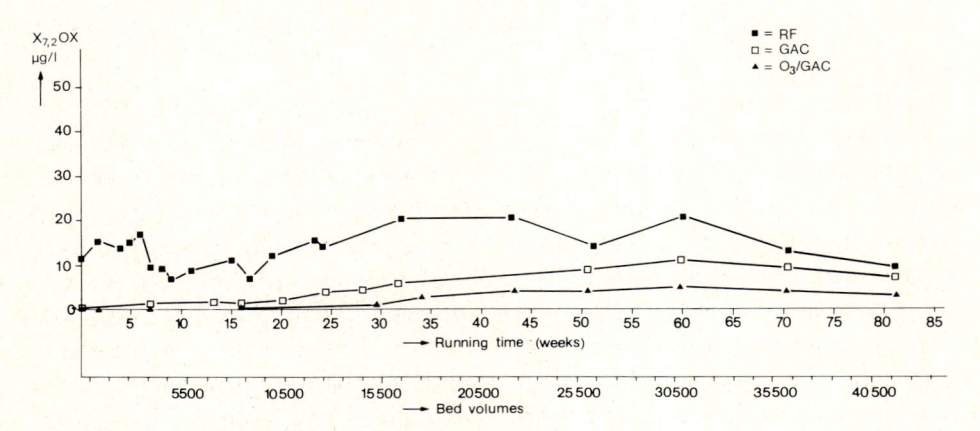

Fig. 3. $X_{7+2}OX$ (as µg Cl/l) in unchlorinated effluents.

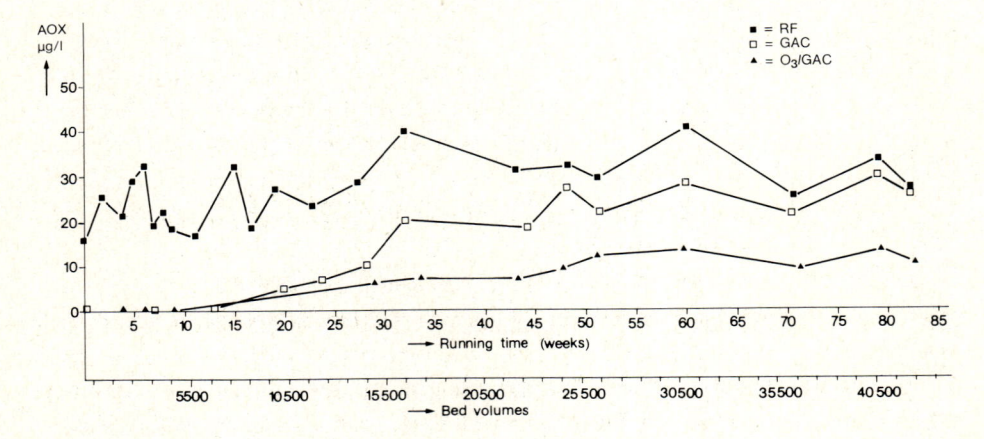

Fig. 4. AOX (as µg Cl/l) in unchlorinated effluents

XAD adsorbable -ON, -OP and -OS. The XOP concentrations were low in the rapid filtrate (0.4 - 1 µg/l). From the beginning of the GAC filter run, an XOP breakthrough was observed, especially in the pH2 fraction.

The X7ON concentration of rapid filtrate ranged from 19 to 25 µg/l. At first, it was completely removed by GAC filtration and by O3-GAC treatment, but it gradually rose to 10 µg/l after 17,000 ebv. The behaviour of XOS was quite similar (figure 2). The XOS content of the rapid filtrate varied between 25 and 35 µg/l. After a gradual increase of the XOS content in the GAC effluent, a steady state was reached after 25,000 ebv. The O3-GAC combination showed better results.

Use of -ON, -OP and -OS surrogates for monitoring of carbon life

As no EOS and EOP was detected in coagulated and rapid filtered Rhine water, the use of these parameters is limited to monitoring of incidental (industrial) spills in river water.

The XOP concentrations were low but the X2OP in particular was poorly adsorbed by carbon. The identity of this type of compounds is not known at this time, and there are no indications on possible health implications of this group. The breakthrough characteristics of XOS and XON are very similar to those of the DOC. As the DOC, these surrogates are indicative for bulk organics, and can therefore be omitted under normal circumstances.

Removal of organohalogens (-OX)

Volatile and extractable organohalogens are poorly removed by both treatment processes. THM concentrations in the rapid filtrate were low (0.1-1.0 µg/l), and consisted only of CHCl3. The purgeable organohalogens (POX) were always found in higher concentrations (0.5-5 µg/l) than CHCl3, indicating that other volatile organohalogens are present in the surface water. The breakthrough of POX also started earlier, after some 2,500 evb., while it dit not occur before 6,000 ebv. for chloroform. The concentration range of EOX was similar to that of POX, but it was first detected in the GAC effluent after 8,000 ebv (Kruithof et al. 1985).

Adsorbable organohalogens are present in much higher concentrations in river water than the volatile and extractable OX. The X7+2OX (the summed concentrations of the XAD pH7 and pH2 fractions) fluctuated between 7 and 20 µgCl/l (figure 3). The presence of XOX was detected in the GAC effluents almost from the start of the filter run, but a gradual rise only started after some 10 to 15,000 ebv.. It reached a maximum in RF-GAC of about 11 µg/l after 30,000 ebv. The steady state level in the O3-GAC effluent was lower (about 5 µg/l after 30,000 ebv.), mainly because of a 40-50% reduction caused by the ozonation of the rapid filtrate (Kruithof et al., 1985). The situation is quite the same for the AOX, but the concentrations were higher (figure 4).

Mutagenic activity

Mutagenicity of the GAC influents. The highest mutagenic effect was recorded in the rapid filtrate (table 2). Ozonation reduced the mutagenic activity of the rapid filtrate, in particular in the assay with S9 mix (table 2). This reduction of the mutagenicity by ozone is not uncommon (van der Gaag et al., 1982, Kool and van Kreijl, 1984). Ozonation however, has very different effects on mutagenicity depending on the ozone dose and the type of water treated. Van Hoof (1985) showed a sharp increase of the mutagenicity of Meuse water due to preozonation. A marked increase of mutagenic activity was seen in bank filtered water after ozonation (van der Gaag, unpublished results), and marginal enhancement of mutagenicity by ozonation has been mentioned for other water types (van der Gaag et al., 1982; van Hoof, 1982). Therefore, it seems necessary to analyse the effect of ozone at each location seperately in order to establish the optimal treatment.

TABLE 2

Mean mutagenicity of rapid filtrate and ozonate as induced revertants/1.6 l.equ*

	TA98 -S9 mix	TA98 +S9 mix	TA100 -S9 mix	TA100 +S9 mix
Rapid Filtrate				
pH7	290	900	170	270
pH2	100	130	90	80
Ozonate				
pH7	160	50	90	105
pH2	30	15	85	70

* from Kruithof et al., 1985.

Breakthrough of mutagenic activity. No breakthrough of mutagenic activity was measured in the GAC filter effluents for a very long period. Marginal dose related increases of the induced revertants were found in the RF-GAC effluent in the period between 22 and 32 weeks. The highest increase over the spontaneous revertants never went far above 40 induced revertants. After 44 weeks, significant mutagenicity was found for strain TA98. The steady state of the mutagenic activity was reached after 60 weeks. The level of induced revertants remained low, at 40-50/1.6 l.equ. for both fractions without S9-mix, and 50-60/1.6 l.equ. for both fractions with S9-mix. The first mutagenicity for TA100 was measured after 51 weeks in the pH2 fraction. This activity then remained stable at a level of about 55 rev./1.6 l.equ. Mutagenicity of pH7 and pH2 fraction for TA100 with S9-mix was only recorded incidentally during the last 30 weeks of the experiment.

No systematic occurrence of mutagenicity was found in the effluent of the RF-03-GAC effluent during the 82 weeks of this study. Incidentally, a low mutagenic effect (<40 rev./1.6 l.equ.) was recorded with strain TA100 in the pH2 fraction. It must be noted that in this study, the ozonization caused a strong reduction of the mutagenicity, compared to that of the rapid filtrate. The mean number of induced revertants in ozonate was much lower than in the rapid filtrate. This varied from 2-3 times for TA98 -S9 up to 10-20 times for TA98 +S9 (Kruithof et al., 1985).

Surrogates were not related to mutagenicity

The concentrations of organohalogen surrogates in unchlorinated water did not show any relation with the results of the Amestest. Correlation coefficients for a linear regression analysis in samples of the rapid filtrate ranged from r=0.08 (X20X vs TA98 -S9 mix) to r=0.33 (X70X vs TA98 +S9 mix), each analysis with 20 paired data. There was no indication for non-linear regressions. Although these results do not exclude a major influence of organohalogenated compounds on the mutagenicity, it is evident that the organohalogens in the river water are not the only compounds contributing to the mutagenic activity.

GAC FILTRATION FOR THE CONTROL OF SIDE EFFECTS OF CHLORINATION
Formation of brominated THM

The investigation of the formation of brominated THM after chlorination of treated waters was hindered by an exceptionally dry summer. The low flow of the Rhine in the summer of 1983 led to a higher Br^-/DOC ratio. Chlorination of the rapid filtrate caused the formation of more bromoform during this period, lasting from week 20 to week 45 (figure 5). Normally, $CHCl_3$ and $CHBrCl_2$ are predominantly present in this water, which is evident from the figures in the other periods. The total molar THM production in the rapid filtrate is quite constant, and is determined by the chlorine dose.

THM formation in the GAC effluent started to increase after some 3000 ebv. (figure 6). Based on earlier experiments, a bromoform peak was expected to occur in the chlorinated effluent between 3,000 and 10,000 ebv., followed by a return to higher chloroform contents in later periods (Kruithof et al., 1982). The THM concentrations in the chlorinated GAC effluent rose gradually during the first 10,000 ebv. of the filter run (figure 6). In this period, the precursor content of the water was the limiting factor for the THM formation. The higher bromoform contents were detected a bit later than in the previous experiments. The second shift, back to a predominance of $CHCl_3$ after 10,000 ebv. failed to happen because of the high bromide concentrations in Rhine water. The bromoform content starts

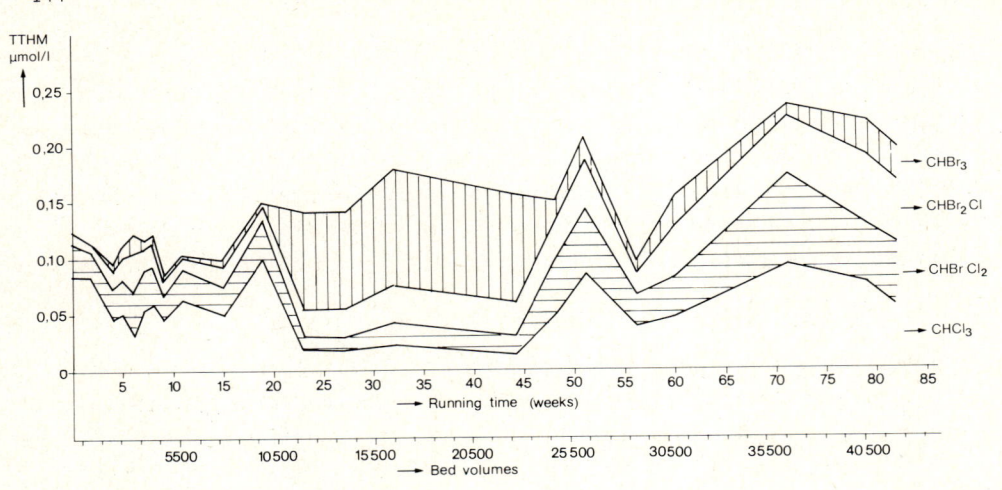

Fig. 5. Formation of THM after chlorination of the rapid filtrate.

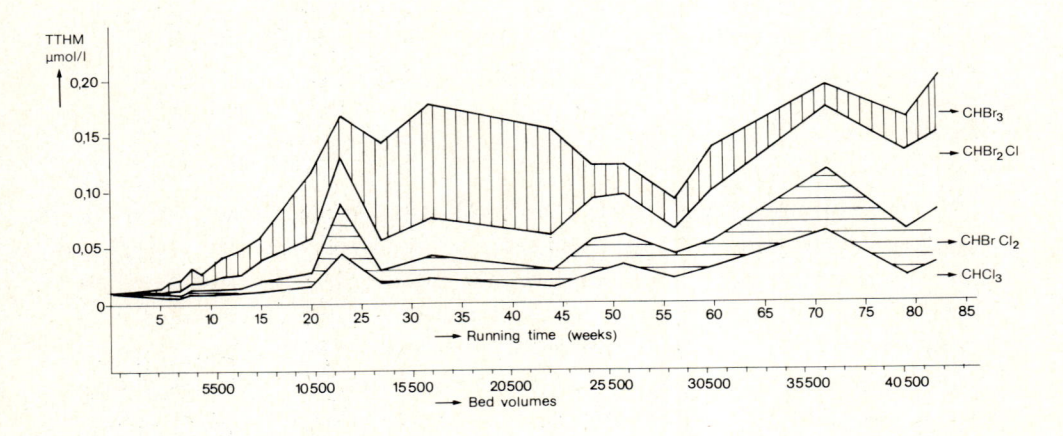

Fig. 6. Formation of THM after chlorination of the GAC-effluent.

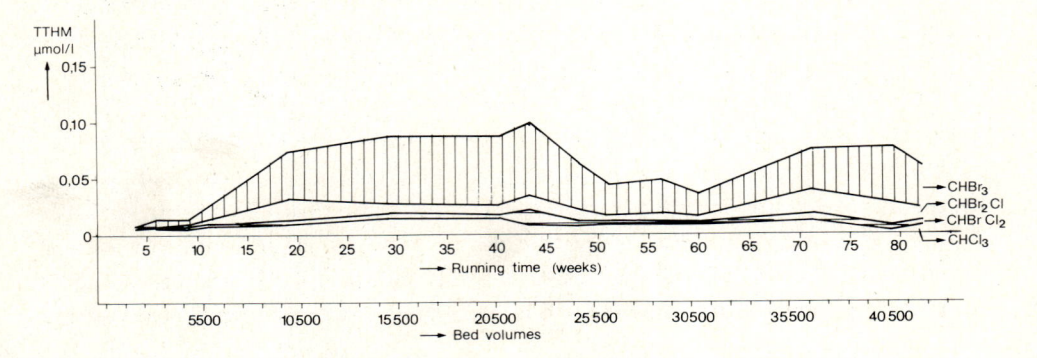

Fig. 7. Formation of THM after chlorination of the O_3-GAC effluent.

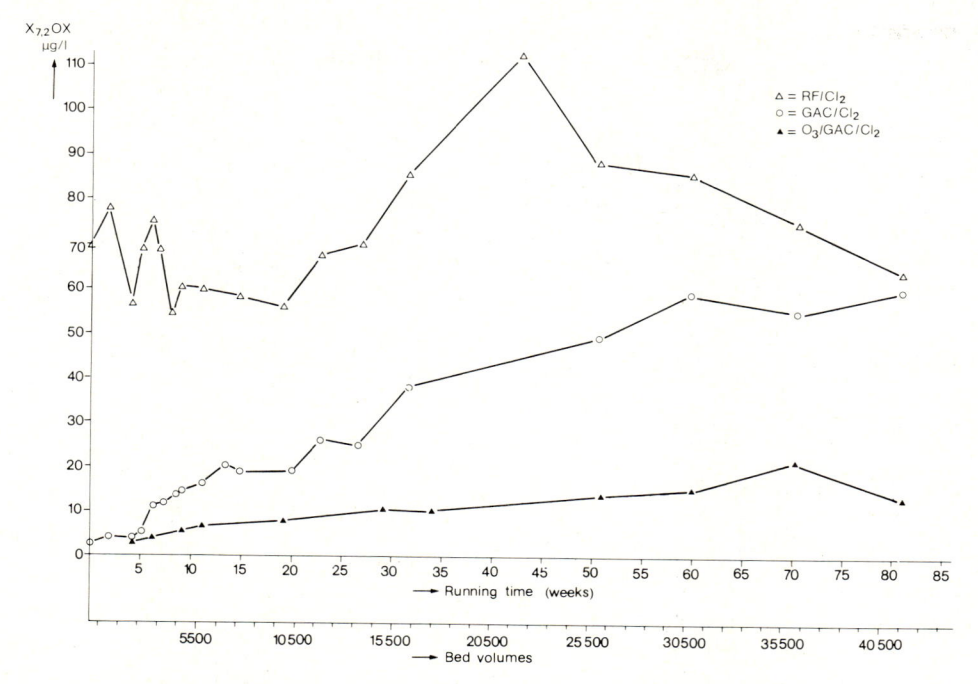

Fig. 8. $X_{7+2}OX$ concentrations (as µg Cl/l) in chlorinated effluents.

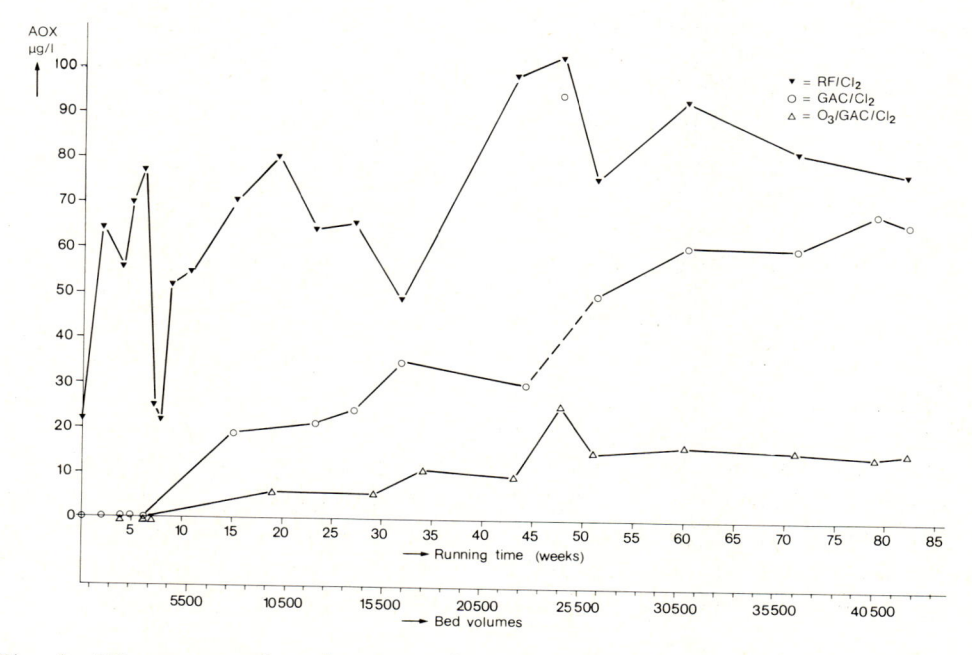

Fig. 9. AOX concentrations (as µg Cl/l) in chlorinated effluents.

to diminish after 23,000 ebv., at the same time as the change in the rapid filtrate. After 15,000 ebv., the formation of THM remained stable (0.18 µmol/l). From that time on, the THM distribution is about the same as in chlorinated rapid filtrate, and the chlorine dose determines the final THM production.

No THM were detected in the effluent of the O3-GAC filter during the first 2,300 ebv. (figure 7). The highest THM concentrations were found after 10,000 ebv.. In the following time, THM concentrations of the chlorinated O3-GAC remained at a level of 0.8 µmol/l until the end of the period of low Rhine water flow in week 45. This also was the period in which the highest concentrations of brominated THM were observed. This indicates that most precursors for the THM formation have been removed by ozonation (Kruithof et al., 1985). Only the more hydrophilic precursors are left after ozonation, and they are poorly adsorbed by GAC. The combination of ozonation and carbon filtration gave the better results in reducing the THM formation due to chlorination. Throughout the experiment, the THM production in the RF-O3-GAC system was 50% lower than in the RF-GAC system.

Non-volatile organohalogens

High concentrations of non-volatile organohalogens measured as X7-, X2- and AOX were formed upon chlorination of rapid filtrate (figures 8 and 9). A gradual rise of the concentration of XAD adsorbable organohalogens is observed almost from the start in the chlorinated GAC effluent (figure 8). From 5,500 ebv. on, the X2OX is constantly higher than the X7OX, both reaching a steady state after some 25,000 ebv.. This happened at an earlier stage in the chlorinated O3-GAC effluent. After some 15,000 ebv., the highest XOX levels were noted, remaining at the same level during the whole experiment (figure 8). The same pattern is observed, with some small differences, for AOX values (figure 9).

Other organic surrogates

Post chlorination had no effect on the concentrations of -ON, -OP and -OS surrogates.

Mutagenicity

Mutagenicity of the GAC influents after chlorination was most of the time higher than in the unchlorinated water. The major effect of chlorination is seen with strain TA100 -S9 mix (table 3). The mutagenicity is about ten times higher than before chlorination. Part of it is inactivated by liverhomogenate. Ozonation reduces the side effects of chlorination in Rhine water where it comes to mutagenicity (table 3).

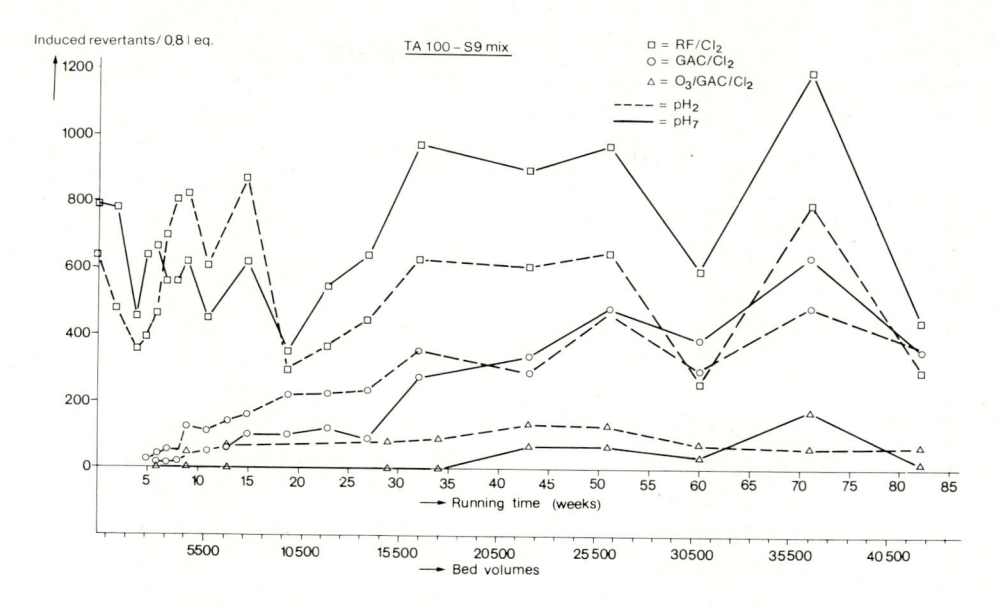

Fig. 10. Mutagenicity of chlorinated effluents for strain TA100-S9.

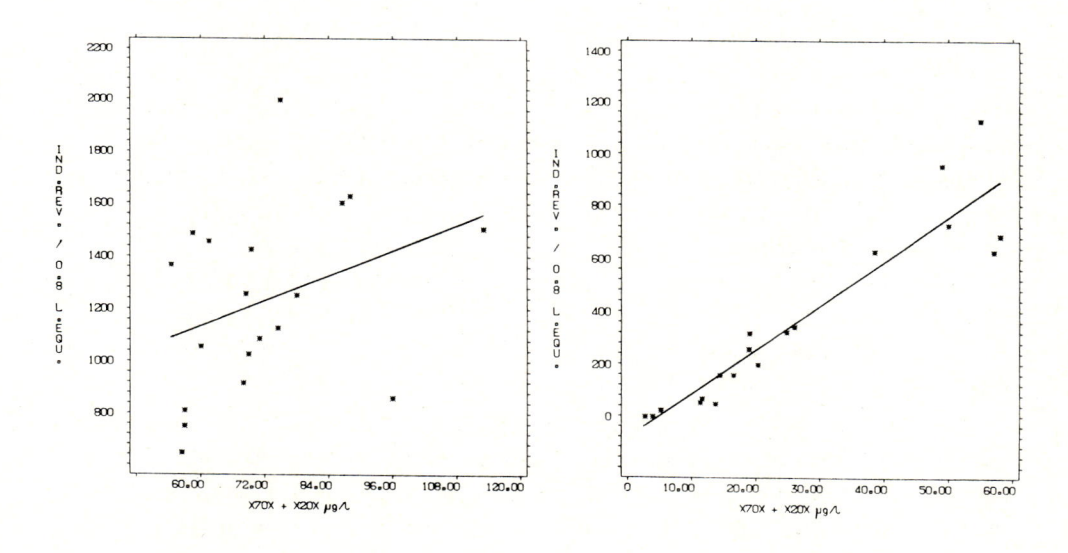

Fig. 11. Summed mutagenicity of XAD fractions for TA100-S9 versus $X_{7+2}OX$.
Left: chlorinated rapid filtrate. Right: Chlorinated GAC-effluent.

TABLE 3

Mean mutagenicity of chlorinated rapid filtrate and ozonate (rev./1.6 l.equ)*

	TA98 -S9 mix	TA98 +S9 mix	TA100 -S9 mix	TA100 + S9 mix
Chlorinated Rapid Filtrate				
pH7	550	790	1240	460
pH2	190	100	880	210
Chlorinated Ozonate				
pH7	160	60	190	110
pH2	35	10	170	70

* from Kruithof et al., 1985.

<u>Chlorination of GAC effluents</u> caused a gradual increase of the mutagenicity with the running time (figure 10). The mutagenicity of the chlorinated GAC effluent for TA100 -S9 mix was first observed after about 2,000 ebv.. The breakthrough was initially higher in the pH2 fraction, but after some 16,000 ebv. the mutagenic activity of the pH7 fraction reached a same level. Mutagenic effects following chlorination of the O3-GAC effluent only started appearing after 5,000 ebv. in the pH2 fraction, and after 22,000 ebv. in the pH7 fraction. In both fractions, this mutagenic activity settled almost directly at a level of 150-200 induced revertants/ 1.6 liter equivalents, where it remained for the whole filter run. This level was the same as in the chlorinated ozonate, which means that the effect of the carbon filter in the O3-GAC system on reducing the precursors for mutagenic side effects of chlorination does not exist after that time.

<u>Can surrogates predict the breakthrough of mutagenicity?</u>

Surrogates are a cheaper and more rapid parameter for water quality than the Amestest. It could therefore be interesting if measurements of surrogates can replace an Amestest in some parts of the evaluation of the carbon life. This would assume that a specific category of organic compounds is responsible for the mutagenic activity. In the case of chlorinated water, it is known that organohalogens are formed, and that these substances could also be responsible for the mutagenic effect.

The results from the chlorinated rapid filtrate did not show any relation between organohalogen surrogates and the mutagenicity in strain TA100 -S9 mix, which is most representative for chlorination (table 4 and figure 11). The regression analysis in the chlorinated GAC effluent was in sharp contrast with the results of the rapid filtrate (table 4 and figure 11). A very significant relation between mutagenicity and XAD organohalogen surrogates was found for the chlorinated GAC-filtrate. The relation found for the chlorinated O3-GAC were not quite so good, but is was based on fewer results.

It appears that most organohalogens formed upon chlorination of the rapid filtrate are not mutagenic. GAC filtration removes a part of the DOC, and in this way many precursors which can react with chlorine. After GAC filtration, a smaller amount of organohalogens is formed, but a significant part of these organohalogens seem to be mutagenic. The relation between organohalogen surrogates and mutagenicity should therefore have to be established for each type of water. Global relations which were suggested by Kool and coworkers (1984) are the most appropriate approach of this problem in our views.

TABLE 4

Linear correlation coefficients between mutagenicity in TA100 -S9 mix as ind. rev./0.8 l.eq. and organohalogen surrogates in µg Cl/l (n = 19 pairs of data)

	X7OX	X2OX	X7+X2OX	AOX
RF+Cl	0.31	0.11	0.15	0.00
GAC+Cl	0.83	0.86	0.87	0.63 *
O3-GAC+Cl **	0.43 ***	0.61	0.75	0.47

*   n = 13 pairs of data;
**   n = 12;
*** the possibility of non-linear regression is not excluded but was difficult to check because of the small number of data

INORGANIC BROMIDE AND SIDE EFFECTS OF CHLORINATION

The inorganic bromide/DOC ratio has a marked influence on the THM content and especially on the THM composition. We started a closer investigation of this phenomenon, focussing on the effect of bromide on adsorbable organohalogen formation and on mutagenicity in the Amestest. The experiments were carried out in GAC filtrate after a carbon life of 15 months (30,000 ebv.). The GAC effluent was chlorinated under the normal post chlorination conditions (table 1) without addition of bromide (a Br$^-$concentration of 250 µg/l) and after adding 500 or 1750 µg Br$^-$/l (as NaBr).

Bromide and THM

Addition of bromide to the water caused a slight decrease of the formation of CHCl3 and CHBrCl2 upon chlorination (figure 12). The concentration of CHBr2Cl was hardly affected, but a substantial rise of the CHBr3 content was observed. Besides a shift towards the formation of more brominated THM, there is also a rise of the total molar THM production.

Bromide and non volatile organohalogens

Addition of bromide had no influence on the formation of adsorbable organohalogens after chlorination (figure 13). This phenomenon may be explained

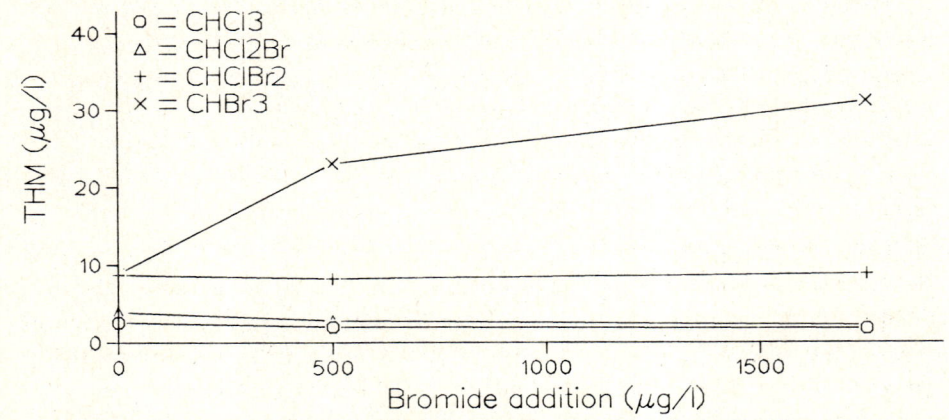

Fig. 12. THM formation at different inorganic bromide concentrations.

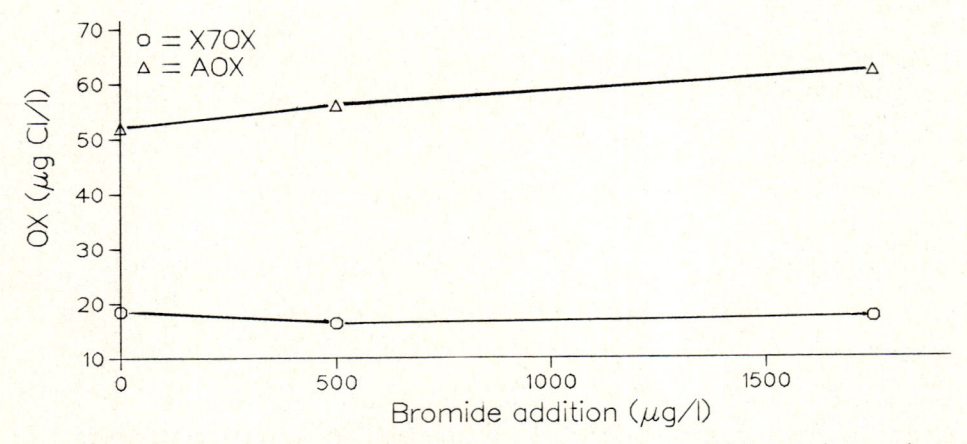

Fig. 13. Formation of X_7OX and AOX after chlorination at different inorganic bromide concentrations.

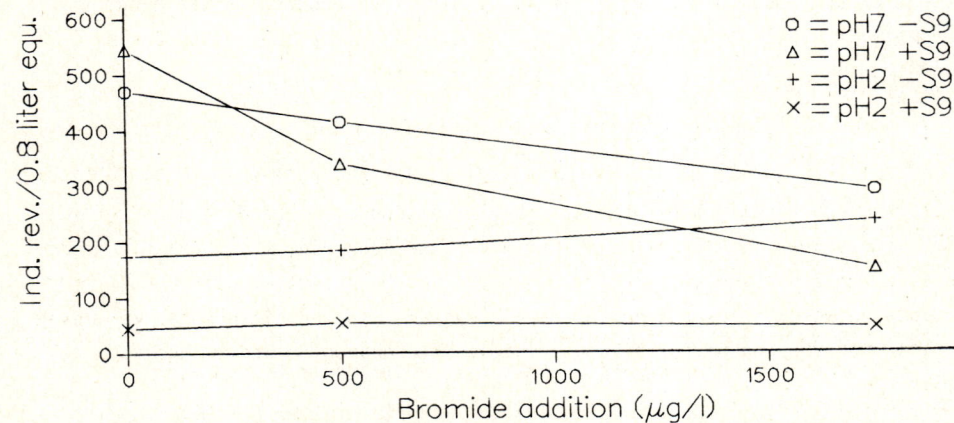

Fig. 14. Mutagenicity of XAD fractions for strain TA100 at different inorganic bromide concentrations.

by a lower adsorbable organochlorine content, occurring together with an increased organobromine formation. The expected increase due to the higher molecular weight of bromide is not seen because the OX values are expressed in µg Cl/l, and because the recovery of OBr could be lower than that of OCl. We expect to obtain more details on this topic, in a study based on separation of OCl and OBr (Puijker et al., in preparation).

Bromide and mutagenicity

Only the mutagenic activity in the XAD pH7 fraction was affected by the addition of bromide (figure 14). At the high bromide dose level, a reduction of the number of revertants was noticed in the assay without S9-mix. Increasing bromide concentrations in the water enhanced the inactivation of the mutagenicity by S9-mix.

Many questions remain

The results of this preliminary study confirm that the bromide content of the chlorinated water has an effect on the quality parameters. Further analysis of this influence will only become possible after a reliable method has been developed to discern OCl from OBr.

CONCLUSIONS

Removal of organic micropollutants and mutagenic activity from Rhine water

For all parameters examined in this study, treatment of Rhine water with a combination of coagulation, rapid filtration, ozonation and GAC filtration showed better results than without the ozonation step. Volatile and extractable organohalogens were poorly removed by both purification systems. Moderately hydrophilic organics showed a gradual breakthrough, which settled at a much lower steady state level in the RF-O3-GAC system than in the RF-GAC system. A slight breakthrough of mutagenic activity was first observed after about 22 weeks in the RF-GAC system. No mutagenicity was detected in the RF-O3- GAC system during the 82 weeks of the experimental period.

Control of the side effects of chlorination

As for the removal of organic micropollutants, the performance of the O3-GAC system in controlling side effects of chlorination of treated Rhine water was superior to that of GAC filtration only. On the long run, the total difference between the two systems could be attributed to the effect of ozonation. The THM production was 50% lower than in the GAC system, and the differences were even greater for the production of non volatile organohalogens and mutagenic activity.

A correlation was found between the XAD-adsorbable organohalogen concentrations in the chlorinated GAC-effluents and the mutagenicity induced by chlorination. Such a relation was not found when filtered Rhine water was chlorinated, indicating that a selective breakthrough happens of compounds which react with chlorine to form mutagens.

The bromide content of the water has an influence on the qualitative and quantitative aspects of the side effects of chlorination. Improvements of techniques which can discriminate OCl from OBr is needed to evaluate the role of bromide.

RECOMMENDATIONS

The results of this study show that micropollutants cannot be completely removed from water by treatment with coagulation, rapid sand filtration, ozonation and GAC filtration, followed by chlorination. As long as the evaluation of the health effects of these micropollutants is not completed, the concentrations of these compounds in water should be kept as low as possible. For the treatment of Rhine water, this can be achieved by a combination of ozonation and GAC filtration.

The side effects of chlorination deserve special attention. Post chlorination should be omitted, when the biological quality (hygienic aspects and aftergrowth) is sufficient, and when the distribution system does not need protection.

Based on these principles, the dutch waterworks investigate the following philosophy on disinfection
- no chemical disinfection is applied when sufficient physical, mechanical and biological barriers (slow sand filtration, UV disinfection) are present.
- when chemical disinfection is needed, the use of alternative disinfectants (chloramine, chlorine dioxide) can be considered, if their side effects have been evaluated and compared to those of chlorine.
- in the meantime, when disinfection is needed, a limited dose of chlorine may be applied.

REFERENCES

Bellar, T.A., Lichtenberg, J.J. and Kroner, R.C., 1974. The occurrence of organohalides in chlorinated drinking water. JAWWA, 66, 703-706.
Bull, R.J., 1982. Health effects of drinking water desinfectants and disinfectant by-products. Environ. Sci. Technol. 16, 554A-559A.
Cheh, A.M., Skochdopole, P. and Cole, L., 1980. Nonvolatile mutagens in drinking water: production by chlorination and destruction by sulfite. Science 207, 90-93.
Janssens, J.G., van Hoof, F. and Dirickx, J., 1984. Ozonation and activated carbon filtration: a critical evaluation. Aqua 2, 102-107.
Kool, H.J., van Kreijl, C.F. and van Oers, H., 1984. Mutagenic activity in drinking water in the Netherlands. A survey and a correlation study. Toxic. Envir. Chem. 7, 111-129.

Kool, H.J. and van Kreijl, C.F., 1984. Formation and removal of mutagenic activity during drinking water preparation. Water Res. 18,.

Kruithof J.C., Nuhn P.A.N.M., van Paassen J.A.M., 1982. Verwijdering van trihalomethanen en precursors voor trihalomethanen door actieve koolfiltratie (Removal of trihalomethanes and precursors for trihalomethanes by granular activated carbon filtration) H2O 15, 277-284.

Kruithof, J.C., Noordsij, A., Puijker, L.M. and van der Gaag, M.A., 1985. The influence of water treatment processes on the formation of organic halogens and mutagenic activity by post chlorination. In: "Water chlorination: Chemistry, Environmental impact and Health effects, vol. 5", (Jolley, R.L., Bull, R.J., Davis, W.P., Katy, F., Roberts jr, M.H. and Jacobs, V.A. eds.), Chelsea, MI: Lewis Publishers (in press).

Loper, J.C., 1980. Mutagenic effects of organic compounds in drinking water. Mutation Res. 76, 241-268.

Luong T., Peters C.J., Young R.J. and Perry R., 1980. Bromide and Trihalomethanes in Water Supplies. Environ. Technol. Lett. 1, 299-310.

Maron, D.M. and Ames, B.N., 1983. Revised methods for the Salmonella mutagenicity test. Mutation Res. 113, 173-215.

Merlet N., de Laat, J. Dore, M., 1982. Oxydation des bromures au cours de la chloration des eaux de surface: incidence sur la production de composés organohalogénés (Bromide oxidation during surface water chlorination: influence upon organohalogenated formation). Revue francaise des sciences de l'eau 1, 215-231.

Noordsij, A., van Beveren, J. and Brandt, A., 1983. Isolation of organic compounds from water for chemical analysis and toxicological testing. Intern. J. Environ. Anal. Chem. 13, 205-217.

Poels, C.L.M., van der Gaag, M.A. and van de Kerkhoff, J.F.J., 1980. An investigation into the long-term effects of Rhine water on Rainbow trout. Water Res. 14, 1029-1035.

Rook, J.J., 1974. Formation of haloforms during chlorination of natural water. Water Treatment Exam. 23, 234-245.

Van der Gaag, M.A., Noordsij, A. and Oranje, J.P., 1982. Presence of mutagens in dutch surface water and effects of water treatment processes for drinking water preparation. In: "Mutagens in our environment" (Sorsa and Vainio eds), p. 277-286. Alan R. Liss Inc., New York.

Van Hoof, F., 1982. Formation and removal of mutagenic activity in drinking water by ozonisation. Aqua 5, 475-478.

Van Hoof, F., Janssens, J.G. and van Dijck, H., 1985. Formation of mutagenic activity during surface water preozonization and its removal in drinking water treatment. Chemosphere (in press).

Van Kreijl, C.F., Kool, H.J., de Vries, M., van Kranen, C.F. and de Greef, E., 1980. Mutagenic activity in the rivers Rhine and Meuse in the Netherlands. Sci. Tot. Environ. 15, 137-147.

REMOVAL OF ORGANIC MICROPOLLUTANTS BY COAGULATION AND ADSORPTION

V. L. SNOEYINK and A.S.C. CHEN
Department of Civil Engineering, University of Illinois, 208 North Romine, Urbana, Illinois 61801

ABSTRACT

The factors which affect removal of organic micropollutants by coagulation, sedimentation, filtration and activated carbon adsorption will be reviewed. Removal of specific compounds by coagulation, sedimentation and filtration is often slight, unless the pollutants adsorb on particles or associate with humic substances which are then coagulated. By comparison, removal of humic substances by these processes can be substantial, depending upon the water chemistry and the process conditions. Activated carbon may be applied in both the powdered (PAC) and granular (GAC) form. PAC and GAC have been used successfully throughout the world to remove odorous compounds. PAC has been used to a much smaller extent for removal of other micropollutants, but there is much potential for improvement of the application procedure so that good results can be achieved. GAC is widely used to remove micropollutants other than odor in Europe but has not been extensively used for this purpose in North America. The compounds which can be removed by GAC are presented and process monitoring procedures are discussed. Factors which limit its use include incomplete knowledge about which compounds must be removed and what effluent concentrations are acceptable.

INTRODUCTION

Humic substances along with trace quantities of organic micropollutants constitute the bulk total organic carbon (TOC) pool of natural waters. Humic substances are naturally occurring products of biological activity in the aquatic and terrestrial environments. These substances, although of little health concern by themselves, can form trihalomethanes (THMs) and other halogenated organic compounds during chlorination at concentration levels required for water disinfection. They also can transport toxic materials such as pesticides and heavy metals from the environment into the finished water. The organic micropollutants, derived primarily from man's activities, find their ways to water supplies, usually in trace quantities, through a variety of sources such as domestic and industrial discharges and agricultural runoff. These organic pollutants are often toxic or carcinogenic and may be harmful even at very low concentration levels. Further, these pollutants are usually small molecules that are, in general, poorly removed by conventional water treatment processes.

The growing concern over the presence of different types of organic compounds in drinking water supplies and its public health significance has prompted many studies during the past decade on the effectiveness of conventional water treatment processes in removing potentially toxic and health-hazardous organic pollutants from water supplies. In this paper, we review the role of the physico-chemical processes, including coagulation, sedimentation, filtration and activated carbon adsorption, on the behavior of humic substances and organic micropollutants during drinking water treatment. The factors that affect removal of these compounds from water will be discussed.

COAGULATION, SEDIMENTATION AND FILTRATION
Coagulation of naturally occurrring organic matter

Naturally occuring organic matter can be removed effectively by coagulation with alum or ferric salts and subsequent sedimentation and filtration. The NORS survey (see Table 1) showed that removal of nonvolatile TOC averaged 30% in 63 water treatment plants, but there was no apparent attempt to optimize the removal. Higher removals from river water (40-60%) were reported, presumably when an attempt was made to get good removal, and up to 90% removal of selected natural organic fractions has been observed in laboratory tests. The factors that affect natural organic removal by coagulation are summarized as follows:

TABLE 1

Summary of reported removals organic constituents in water by coagulation.

Water Source/ Type of Organics	Test Conditions	Constituents	% Removals
National Organics Reconnaissance Survey (63 plants)	Treatment plant studies, coag/sed/filt	NVTOC	~30
EPA - Ohio River	Pilot plant studies, coag/sed/filt, Al(III)	TOC	60
Rhine River, Alpine lakes	Al(III), pH > 7	DOC	25-40
Humic acids	Lab tests, Al(III) and Fe(III)		60-90
Fulvic Acids	Lab tests, Al(III) and Fe(III)		10-60

NOTE: These data are from a literature review by Kavanaugh (1978).

(1) pH. The removal is much better from slightly acidic water; the optimum pH is 5-6 for alum and 4-5 for ferric salts (Semmens et al., 1978). At these values of pH, anionic humates can interact with polymeric cations of coagulant salts by formation of chemical bonds or by simple neutralizing reactions.

(2) Coagulant dose. The coagulant dose required is usually proportional to the concentration of humic substances in water and decreases with decreasing pH, since more carboxylic groups become protonated and do not interact with the polymeric cations of coagulant salts (Semmens et al., 1978).

(3) Coagulant type. Both alum and ferric salts are effective in humic substances removal. Cationic polyelectrolytes do not readily form a settleable floc, but may produce a filterable floc and thus may be suitable for removal of humics by coagulation and direct filtration (Amy and Chadik, 1984). Further research is needed on this topic, however. The use of high molecular weight polymers with alum does give satisfactory results for humic substances removal. However, the polymer should be added after the alum to bridge destabilized floc particles.

(4) Preozonation. Ozonation of water before coagulation has been reported to increase the removal of color, particles, organic compounds, and THM precursors by coagulation with aluminum salts (Saunier et al., 1983), but although it works in certain waters, further research is needed to show why this effect occurs. The increased removal of THM precursors after ozonation suggests that polar compounds produced by ozonation can be more easily removed by aluminum coagulants. A study of water supplies in Florida (Elefritz et al., 1984) also reported a significant reduction of THM formation potential with a very slight dose of ozone followed by coagulation with lime. The effect of preozonation on the removal of organic micropollutants by coagulation has not been studied in detail, but the effect is probably small.

Removal of humic substances by softening

Removal of humic substances by lime-soda softening process also can be substantial. For example, up to 40-50% fulvic acid or THM precursors were removed from groundwater (Singley et al., 1977) and water at an English water treatment plant (Wilson, 1960) by the softening process. Randtke and co-workers (1982) and Liao (1984) also reported effective removal (up 40-80%) of several humic substances and attributed this removal to mechanisms involving direct precipitation of calcium-humate aggregates, or through coprecipitation, in which humic substances adsorb onto calcium carbonate solid surfaces and are then removed. Favorable conditions for the adsorption and coprecipitation of humic substances include high pH (>11.0), a high calcium but a low carbonate concentration in water, presence of some inorganics such as magnesium and phosphate and poor crystallinity of calcium carbonate solids.

Coagulation of specific organic compounds

Specific organic compounds may be removed from water by coagulation by different means. They may associate with humic substances or adsorb on colloidal material in water. They then can be removed from water by coagulation of the humic substances and the particles, sedimentation, and filtration. The degree of association will depend on the nature of the organic compounds, the characteristics of the colloidal materials in suspension and the water quality (Semmens et al., 1978). Organic compounds may also adsorb directly onto the floc which results from coagulant addition, or they may be precipitated with coagulant metal ions and then removed.

Very few studies have been done on the role of conventional water treatment processes on the removal of organic micropollutants. A few reports concerning removal by coagulation are summarized in Table 2. In most of these studies, the solutions examined were very complex (e.g., sea water or lake water) and the mechanism of removal could not be determined. The removal was most frequently attributed to coprecipitation with ferric ions, and factors including solution pH, nature of the organic molecules and the quality of the water studied were considered to have an important effect. Both Semmens and Ocanas (1976) and Albert (1975) investigated the removal of specific compounds from distilled and/or deionized water by alum or ferric sulfate and found only 0-30% removal of dihydroxybenzoic acid (DHBA) and resorcinol, and no removal of vanillic acid.

Pesticides removal by coagulation

Only a few pesticides can be effectively removed by the coagulation process. As indicated in Table 3, coagulation with alum or ferric salts, followed by sedimentation and filtration, removed DDT and methoxychlor up to 98% at the parts per billion concentration level. By comparison, the coagulation process is much less effective in removing aldrin, endrin, dieldrin, lindane, parathion or 2,4,5-T at parts per billion level. There is essentially no removal of 2,4-D, rotenone, toxaphene by either alum or ferric salts.

Removal of specific organic compounds by softening

Lime-soda softening does not effectively remove specific organic compounds from water. Liao (1984) and Randtke and co-workers (1982) reported that simple monomeric molecules, except those containing phosphoryl/phosphonyl groups, are not removed by softening to any significant extent. The softening process,

TABLE 2

Removal of specific organic compounds by coagulation.

Compound	Concentration Level	Coagulant Used	% Removal	Source of Water	Reference
phenol	10 µg/L	$FeCl_3$	60-87	Lake water	Sridharan & Lee
citric acid	125 µg/L	$FeCl_3$	20-30	Lake water	Sridharan & Lee
citric acid	0.1-1.0 mg/L	$FeCl_3$/NaOH	90	Sea water	Chapman & Rae
resorcinol	2-10 mg/L	$Fe_2(SO_4)_3$	0-8	DDW[1]	Semmens & Ocanas
vanillic acid	10 mg/L	Alum	0	DDW	Albert
DHBA[2]	2-10 mg/L	$Fe_2(SO_4)_3$	1-30	DDW	Semmens & Ocanas
glucose	0.1-1 mg/L	$FeCl_3$/NaOH	16	Sea water	Chapman & Rae
glycine	0.1-1 mg/L	$CuCl_2$/NaOH	48	Sea water	Chapman & Rae
glycine	8 µg/L	$FeCl_3$	25-50	Lake water	Sridharan & Lee
phenylalanine	-	$Fe(OH)_3$	53	Sea water	Tatsumoto et al.
glutamic acid	0.1-1 mg/L	$FeCl_3$/NaOH	90	Sea water	Chapman & Rae
aspartic acid	0.1-1 mg/L	$FeCl_3$/NaOH	77	Sea water	Chapman & Rae
succinic acid	0.1-1 mg/L	$FeCl_3$/NaOH	60	Sea water	Chapman & Rae
glycollate	0.1-1 mg/L	$FeCl_3$/NaOH	45	Sea water	Chapman & Rae
lysine	0.1-1 mg/L	$FeCl_3$/NaOH	73	Sea water	Chapman & Rae

[1] DDW-distilled-deionized water
[2] DHBA-dihydroxybenzoic acid

TABLE 3

Removal of pesticides by coagulation.

Pesticide	Concentration Level	Coagulant Used	% Removal	Reference
DDT	0.1-10 mg/L	Alum, $Fe_2(SO_4)_3$, or $FeCl_3$	40-80	Carollo
DDT	10-25 µg/L	Alum	98	Robeck et al.
DDT	-	Alum	30-40	Whitehouse
methoxychlor	1-10 mg/L	Alum or $Fe_2(SO_4)_3$	85-95	Steiner & Singley
lindane	1-10 µg/L	Alum	<10	Robeck et al.
dieldrin	1-10 µg/L	Alum	55	Robeck et al.
endrin	1-10 µg/L	Alum	35	Robeck et al.
2,4,5-T	1-10 µg/L	Alum	63	Robeck et al.
parathion	1-10 µg/L	Alum	20	Robeck et al.
aldrin	-	Alum	10	Whitehouse
2,4-D	-	Alum or $Fe_2(SO_4)_3$	0-3	Aly & Faust
rotenone	<170 µg/L	Alum	0	Cohen et al.
toxaphene	<400 µg/L	Alum	0	Cohen et al.
toxaphene	<400 µg/L	Alum	0	Nicholson et al.

however, is effective in removing polymeric electrolytes possessing oxygen-containing functional groups. The removal increases with increasing degree of polymerization. The removal is also affected by other molecular characteristics including degree of hydrophilicity, molecular charge, and molecular geometry.

The removal of volatile acids (Randtke et al., 1982), THMs (Singley et al., 1977), and other purgeable halogenated organic compounds (Wood and Demarco, 1979) have been occasionally reported, but it is not apparent in these studies if the removal of these compounds was a result of loss to the atmosphere or incorporation into the softening sludge. Therefore, no conclusions can be drawn about the efficiency of softening in removing these compounds.

ACTIVATED CARBON ADSORPTION

Activated carbon adsorption is a common drinking water treatment process. Activated carbon may be applied in both the powdered (PAC) and granular (GAC) form. PAC is added primarily for control of taste and odor and its dose can be varied as the problem demands. The predominant reason for the use of GAC initially was to control taste and odor, but now is also used to remove other organic compounds.

The widespread occurrence of groundwater supply contamination with volatile organic compounds, such as trichloroethylene and tetrachloroethylene, and the increasing recognition of the vulnerability of many of our surface water supplies or organic contamination have highlighted the need for improved purification procedures, including the use of GAC adsorption processes for removal of organic compounds. In fact, GAC has been widely used in Europe to remove organic micropollutants other than taste and odor although it has not been extensively used for this purpose in North America. The major problem confronting utility managers and regulatory officials who consider the possible use of adsorption for surface water treatment is quantifying the benefit that would be obtained for the expenditure. The majority of the organic compounds that in surface waters have not been identified and the long-term health effects of those that have been found are largely unknown. The frequency of occurrence of spills which might cause periodic high concentrations of harmful compounds is largely unknown. Thus, if adsorption is to be used, its health benefits will be largely unknown and design parameters and regeneration or replacement frequency will have to be set arbitrarily.

Removal of specific organic compounds

The ability of GAC to remove a broad spectrum of organic compounds from water has been extensively reported in the literature during the last 20 years. For example, Dobbs and Cohen (1980) reported single-solute isotherm data for 128 organic compounds of different structures, sizes, functionalities, etc., that exhibit different adsorption characteristics. Data compiled by McGuire and Suffet (1980) show some are strongly adsorbed, whereas others are weakly adsorbed (see Figure 1). Many of the compounds that have been detected in surface and groundwaters at low concentrations, e.g., THMs, volatile halogenated compounds, polynuclear aromatic hydrocarbons (PAH), nitrosamines and organic pesticides, have been shown to be toxic or carcinogenic at higher concentrations. Effective removal of most of these compounds can be achieved by GAC treatment, and the removal efficiency, especially for PAH, nitrosamines and pesticides, often exceeds 95% or more (see Faust and Aly (1983) for a review).

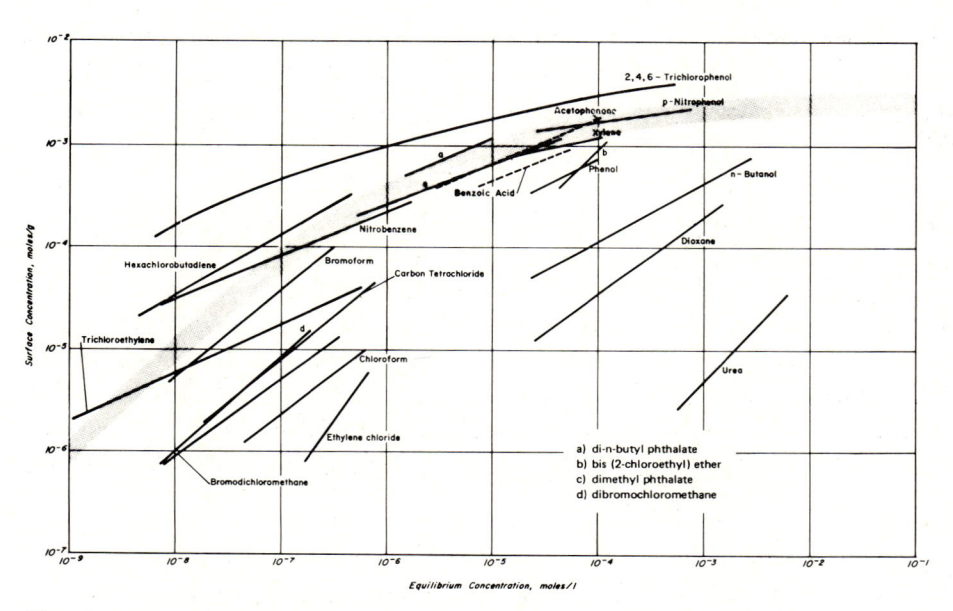

Fig. 1. General and specific adsorption isotherms (after McGuire and Suffet, 1980).

Removal of THMs and other volatile organic compounds (VOCs)

The discovery of THMs resulting from the chlorination of naturally occurring organic matter, e.g., humic substances, in water supplies has led the EPA to amend the National Interim Primary Drinking Water Regulations to include a maximum contaminant level (MCL) of 0.1 mg/L for total THMs (TTHMs). Consequently, removal of these compounds by GAC adsorption was studied extensively in both pure systems and pilot- and full-scale studies. Review of the data from most of the pilot- and full-scale studies indicate that the carbon bed-life or time to breakthrough of the THMs is much shorter, ranging from 4-26 weeks, than that of normal operation for taste and odor control (see Figure 2 and the review by Faust and Aly (1983). Therefore, the use of GAC for THMs

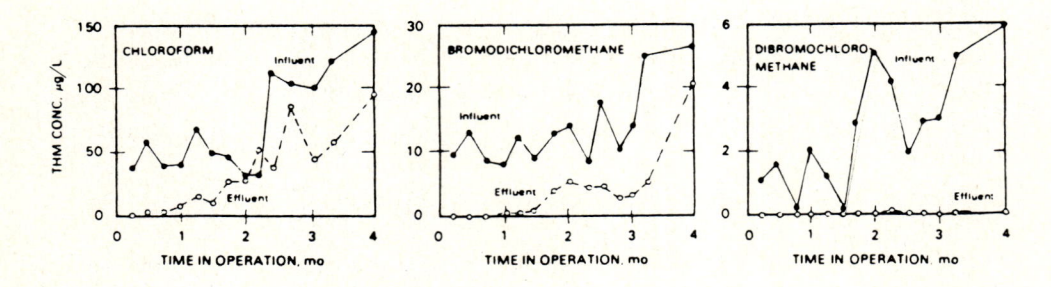

Fig. 2. Removal of THM from Cincinnati, Ohio tapwater (after Symons et al., 1981).

removal requires frequent replacement or regeneration of the exhausted carbon, and other alternatives often are more economical.

GAC is highly effective for VOC removal (see Snoeyink in AWWA Research Foundation-KIWA Report for a review). It has the advantage over air stripping that both volatile and nonvolatile compounds can be removed and ultimately disposed of by incineration. The gaseous effluent from air stripping is discharged to the atmosphere and this may not be permitted in some locations. However, GAC is more costly to use than air stripping if no gas stream cleanup is required for air stripping.

Removal of THM precursors

The adsorptive capacity of GAC for humic substances, or THM precursors, is generally low and variable, depending on the source and nature of the humic compounds being adsorbed. As indicated by several pilot- and full-scale studies (see Faust and Aly (1983) for a review), the removal by virgin GAC is initially effective but a rapid breakthrough occurs after a short operating period (see Figure 3). A steady state develops during which a rather constant percentage of precursor material continues to be removed (see Figure 3), perhaps because of biological activity and the slow rate of adsorption of large molecules. Since THM precursors will compete for adsorption sites with other micropollutant molecules, and GAC is an expensive process, it is desirable to optimize coagulation, sedimentation and filtration for organics removal to reduce the load to the carbon column. The GAC bed-life can thus be extended and it should be possible to reduce the cost of treatment.

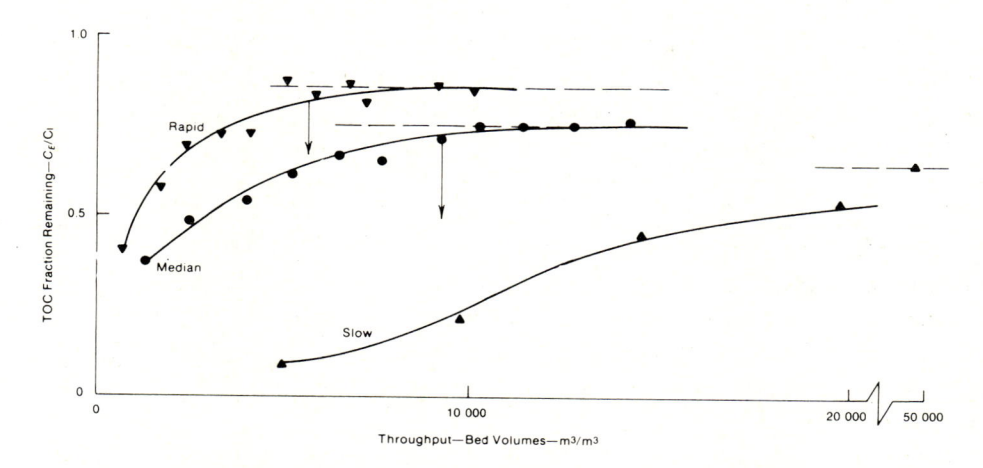

Fig. 3. Representative TOC breakthrough curves (after Roberts and Summers, 1982).

Factors affecting GAC adsorption

Quality of GAC. Specific surface area and pore size distribution are the principal characteristics affecting the adsorptive capacity of GAC. Micropores constitute a major portion of the specific surface area, and many possess molecular dimensions. Thus certain adsorbate molecules can penetrate some pores but are excluded from other smaller pores. For example, pore volumes in pores with a radius of less than 70 Å were found to correlate well with the adsorption capacity of GAC for commercial humic acid and peat fulvic acid with a molecular weight of less than 1000; however, pore volume in pores with a radius of less than 400 Å correlated with GAC capacity for peat fulvic acid with molecular weight of more than 50,000 (Lee et al., 1981).

Characteristics of adsorbate molecules. Solubility, molecular weight and polarity of organic solutes are the most significant properties affecting the adsorptive capacity. In general, adsorption of organic molecules from aqueous solutions increases as an homologous series is ascended (known as Traube's rule). For example, the adsorption of a series of aliphatic acids on GAC increases as the chain length of the molecule increases. The increase in adsorption is due to the decrease in polarity and the decrease in solubility as the molecule chain length (or molecular weight) increases. Factors related to solubility and polarity, such as functionality of the adsorbate molecule and substitution in the ring structure, also affect the adsorptive capacity.

pH. The adsorption of electrolytes, such as acids and bases, by GAC from aqueous systems is generally affected by the solution pH. In general, both the undissociated and ionized forms of an adsorbate can be adsorbed on GAC with the undissociated form being more strongly adsorbed than the ionized form. At alkaline pH values greater than the pK_a of a weak organic acid, the adsorptive capacity is greatly reduced. The adsorptive capacity increases as pH decreases and a maximum occurs in the range where pH is numerically equal to pK_a. At pH values more acidic than the pK_a value, the adsorptive capacity may decrease due to changes in the surface properties of the carbon.

Competitive adsorption. Organic pollutants do not exist as single solutes in natural waters, but rather as mixtures of many compounds. Since the adsorbability of different organic compounds can vary widely, displacement of weakly adsorbed by strongly adsorbed compounds is highly possible. Competitive effects on adsorption should therefore be considered when evaluating the overall efficiency of the GAC process.

Because of the competitive adsorption, it is possible that concentrations of pollutants may appear in the effluent of a GAC filter at concentrations higher than are found in the influent. This event can only occur when 1) the column is saturated, or nearly so, with a pollutant and a more strongly adsorbing compound appears in the influent, and 2) the column is near saturation with a pollutant and the influent concentration decreases so that desorption occurs (Thacker et al., 1984). Operation of the adsorber so that complete saturation at the influent concentration does not occur will prevent this problem.

Pretreatments. Water treatment prior to GAC adsorption often affects the GAC performance during water treatment. Coagulation in conjunction with sedimentation and filtration can reduce the organic content of water supplies, thereby reducing the amount of organic material to be removed by the GAC column and improving the product water quality. If the objective of the GAC treatment becomes one of removal of specific organic micropollutants, rather than the removal of TOC, pretreatment with synthetic resin or activated alumina before GAC should reduce the amount of natural organic matter in the water and thus the competitive interactions on the GAC; the result may be a more cost-effective process for removal of the micropollutant (Snoeyink, 1985).

Reactions of chlorine, or other oxidative pretreatment chemicals, such as ozone, chlorine dioxide, and permanganate, with GAC or organic compounds in aqueous systems or on the carbon surface can alter the GAC performance to different degrees. For example, ozone can react with humic substances to produce more polar intermediates that are less adsorbable on GAC but usually more biodegradable. Microbial activity in a GAC contactor often leads to improved TOC removal by GAC.

Current research at the University of Illinois also indicates that, when a chlorine-containing disinfectant (HOCl, ClO$_2$ or NH$_2$Cl) is applied to GAC with adsorbed organic compounds, it will react both with the carbon and the adsorbed compounds. Unusual products not characteristics of solution reactions are formed when carbon is present. For example, the HOCl-GAC-adsorbed 2,4-dichlorophenol (DCP) reaction will give unusual products including a series of hydroxylated PCBs at HOCl concentrations normally encountered in drinking water treatment practice (see Table 4). A similar product mixture was also obtained when the GAC was first treated with HOCl and then 2,4-DCP was adsorbed. Further, some of these products may desorb from the carbon column. Additional data on this effect are given by Voudrias et al. (1985). Therefore, the application of chlorine-containing disinfectants to GAC adsorbers, as is very common in the USA, needs to be reevaluated and eliminated where possible.

TABLE 4

Reaction products from HOCl-2,4-dichlorophenol-GAC reactors

Compound	Reaction
(OH)$_x$ — benzoquinone — Cl (x = 0-1)	I
OH — C$_6$H(COOMe)(Cl)$_y$ (y = 2-3)	I-IV (y = 2) I (y = 3)
(OH)$_2$ — biphenyl — (Cl)$_z$ (z = 3-4)	I-IV
(OH)$_v$ — dibenzodioxin — (Cl)$_u$ [u = 2-3, v = 1-2] ?	I (u = 2 & v = 2 or u = 3 & v = 1) I-IV (u = 2 & v = 1)

I. Preadsorbed 2,4-DCP reacted with 10 mg/L HOCl as Cl$_2$.
II. Preadsorbed 2,4-DCP reacted with 1.5 mg/L HOCl as Cl$_2$.
III. Preadsorbed 2,4-DCP reacted with 1.5 mg/L HOCl as Cl$_2$ in the presence of peat fulvic acid.
IV. 2,4-DCP reacted with chlorine-preoxidized F-400 GAC.

CONCLUSIONS

Good removal of natural organic matter can be achieved by coagulation, sedimentation and filtration. This removal results in the production of fewer chlorinated compounds when the water is post-chlorinated, and in reduced quantities of organics which must be removed by adsorption if GAC adsorbers are employed. Specific micropollutants are generally not well-removed, except for highly surface active compounds, such as DDT or polynuclear aromatic hydrocarbons, which adsorb to particles in the water. These particles can then be removed by coagulation.

GAC gives excellent removals of many micropollutants, but its performance depends upon the constituents of the water. The presence of substances which compete with the micropollutant for adsorption sites on the GAC significantly affects the replacement frequency of the GAC. Pretreatment of the water before adsorption is also important. Prechlorination, for example, can result in the formation of many compounds which otherwise would not be present.

REFERENCES

Albert, G., 1975. Influence of dissolved organic compounds on flocculation. Heft 9, Engler-Bunte Inst., University of Karlsruhe, FRG.

Aly, O. M. and Faust, S. D., 1965. Removal of 2,4-dichlorophenoxyacetic acid derivatives from natural waters. J. Amer. Water Works Assoc., 57: 221.

Amy, G. L. and Chadik, P. A., 1984. Cationic polyelectrolytes as primary coagulants for removing trihalomethane precursors. J. Amer. Water Works Assoc., 76: 527.

AWWA Research Foundation-KIWA, 1983. Occurrence and removal of volatile organic chemicals from drinking water. Cooperative Research Report published by the AWWA Research Foundation, Denver, CO.

Carolla, J. A., 1945. Removal of DDT from water supplies. J. Amer. Water Works Assoc., 37: 1310.

Chapman, G. and Rae, A. C., 1967. Isolation of organic solutes from sea water by co-precipitation. Nature, 214: 627.

Cohen, J. M., Kamphake, L. J., Lemke, A. E., Henderson, C. and Woodward, R. L., 1960. Effect of fish poisons on water supplies, Part I. Removal of toxic materials. J. Amer. Water Works Assoc., 52: 1551.

Dobbs, R. A. and Cohen, J. M., 1980. Carbon adsorption isotherms for toxic organics. EPA-600/8-80-023, U.S.E.P.A., Cincinnati, OH.

Elefritz, R. A., Porter, D. W., and Morris, S. F., 1984. The application of ozone in softening processes for cost-effective THM control: Two case histories. Presented at Seminar on Strategies for the Control of Trihalomethanes at the AWWA Southeast Annual Conference, Jekyll Island, GA.

Faust, S. D. and Aly, O. M., 1983. Chemistry of Water Treatment. Ann Arbor Science, Woburn, MA.

Kavanaugh, M. C., 1978. Modified coagulation for improved removal of trihalomethane precursors. J. Amer. Water Works Assoc., 70: 613.

Lee, M. C., Snoeyink, V. L., and Crittenden, J. C., 1981. Activated carbon adsorption of humic substances. J. Amer. Water Works Assoc., 73: 440-446.

Liao, M. Y., 1984. Removing soluble organic contaminants from water supplies by softening. Ph.D. Thesis, University of Illinois, Urbana, IL.

McGuire, M. J. and Suffet, I. H., 1980. The calculated net adsorption energy concept. In: I. H. Suffet and M. J. McGuire, eds., Activated Carbon Adsorption of Organics from the Aqueous Phase. Proc. of the 1978 ACS Symp. in Miami Beach, FL, Ann Arbor Science Publishers, Inc., Ann Arbor, MI.

Nicholson, H. P., Grizenda, A. R. and Teasley, J. I., 1966. Water pollution by insecticides, a six and one-half year study of a water shed. In Proc. Symp. on Agri. Waste Water. U.S.E.P.A., Atlanta, GA.

Randtke, S. J., Thiel, C. E., Liao, M. Y. and Yamaya, C. N., 1982. Removing soluble organic contaminants by lime-softening. J. Amer. Water Works Assoc., 74: 192-202.

Robeck, G. G., Dostal, K. A., Cohen, J. M. and Kreissl, J. F., 1965. Effectiveness of water treatment processes in pesticide removal. J. Amer. Water Works Assoc., 57: 181-200.

Roberts, P. V. and Summers, R. S., 1982. Performance of granular activated carbon for total organic carbon removal. J. Amer. Water Works Assoc., 74: 113-118.

Saunier, B. M., Selleck, R. E. and Trussell, R. R., 1983. Preozonation as a coagulant aid in drinking water treatment. J. Amer. Water Works Assoc., 75: 239.

Semmens, M., Edzwald, J. K., Taylor, M., and Sanks, R., 1978. Organics removal by coagulation - a review and research needs. Presented at the 98th Annual AWWA Conference, Atlantic City, NJ.

Singley, J. E. et al., 1977. Minimizing trihalomethane formation in a softening plant. U.S.E.P.A., Water Supply Res. Div., Municipal Environ. Res. Lab., Cincinnati, OH.

Snoeyink, V. L., 1985. Trends in water treatment technology: disinfection, oxidation and adsorption. Presented at Cambridge Meeting on Environmental Technology Assessment, Cambridge, UK.

Sridharan, N. and Lee, G. F., 1972. Coprecipitation of organic compounds from lake water by iron salts. Environ. Sci. and Technol., 6: 1031.

Symons, J. M. et al., 1981. Treatment techniques for controlling trihalomethanes in drinking water. EPA-600/2-81-156, U.S.E.P.A., Cincinnati, OH.

Tatsumoto, M., Williams, W. T., Prescott, J. M. and Hood, D. W., 1961. Amino acids in samples of surface sea water. J. Marine Res., 19: 89.

Thacker, W. E., Crittenden, J. C. and Snoeyink, V. L., 1984. Modeling of adsorber performance: variable influent concentration and comparison of adsorbents. J. Water Poll. Cont. Fed., 56: 243.

Voudrias, E. A., Larson, R. A. and Snoeyink, V. L., 1985. Effects of activated carbon on the reactions of free chlorine with phenols. Environ. Sci. and Technol., 19: 441.

Whitehouse, J. D., 1967. A study of the removal of pesticides from water. Kentucky Water Resources Institute.

Wilson, A. L., 1960. The removal of fulvic acids by water-treatment plants. J. Appl. Chem., 10: 377.

Wood, P. R. and Demarco, J., 1979. Treatment of groundwater with granular activated carbon. J. Amer. Water Works Assoc., 71: 674.

ORGANIC MICROPOLLUTANTS AND TREATMENT PROCESSES: KINETICS AND FINAL EFFECTS OF OZONE AND CHLORINE DIOXIDE

J. Hoigné

Swiss Federal Institute of Water Resources and Water Pollution Control (EAWAG), 8600 Dübendorf (Switzerland)

ABSTRACT

Both, ozone and chlorine dioxide are highly selective oxidants. Based on reaction kinetics predictions of their reactions with micropollutants are possible. The availability of reaction-rate constants is discussed. Of further interest is the role of secondary oxidants (•OH radicals) which are produced during decomposition of ozone. Also their effects on micropollutants can be quantified when the •OH radical scavenging-efficiency of the water and the rate of ozone decomposition are calibrated.

INTRODUCTION

In water treatment, both ozone and chlorine dioxide are used as disinfectants and as oxidants. Although their primary application was to improve colour, odor and taste, or to oxidize manganese and iron compounds (Masschelein, 1980; Miller et al., 1978), recent interest has been focussed on the use of ozone or chlorine dioxide to remove micropollutants. Corresponding with the Symposium's aim I will summarize on this last topic and describe the reaction kinetics involved in order to illustrate which reaction parameters must be defined to make results generalizable.

The scheduled title "Removal by Oxidation Processes" has been changed. In general neither ozonation nor a treatment with chlorine dioxide leads to a final "removal", i.e. a full mineralisation of micropollutants. Rather these processes transform primary micropollutants into secondary solutes. Only the kinetics with which these are formed and further oxidized determine the chemical transformations achieved. Since ozonation is used in combination with other processes, the succeeding treatment steps and the formulated quality criteria will determine if the oxidation process can be considered beneficial to the overall water treatment scheme.

Many useful practical results in improving water quality by using oxidative processes have been published. However, most of these holistic studies are based on operations which in respect to kinetics either do not directly correspond with the real water treatment processes or they have been performed on particular treatment lines and could therefore not be generalized.

In order to compile the literature and to plan effective future studies, we must find the fundamental variables and rules which will allow for reduction of the masses of phenomenological information to be reduced to sets of minimal data needed for generalizing the results. Therefore we aim to break up complex sequences of reactions into subsets which allow for (kinetic) quantification and predictions of product formations. If these subsets of reactions correspond to those which are encountered in other research fields, the use of the wealth of information already accumulated in such related fields becomes possible.

KINETIC ASPECTS OF OZONATION

We discussed the most important pathways of ozonation reactions such as listed in Fig. 1 in earlier reviews (see e.g. Hoigné, 1982a). In general to difficulties have arisen thus far as the concepts are applied in further studies or to include results obtained by other researchers. The concepts have shown to

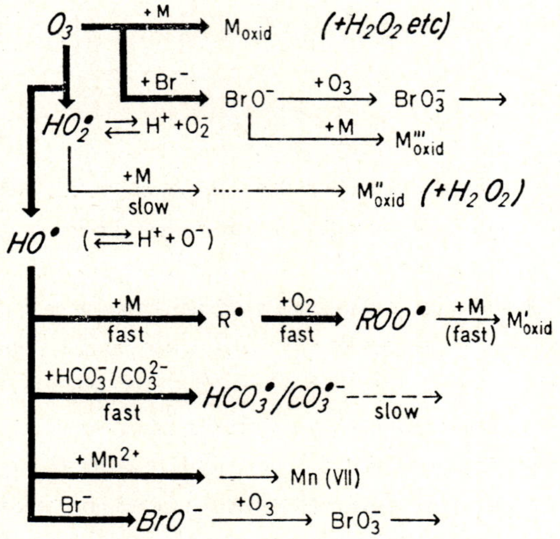

Fig. 1. Different pathways of the reactions of ozone and the formation of secondary oxidants (Hoigné, 1982a). For details also compare Fig. 9.

be widely applicable whenever sufficient effort is invested allowing for correct interpretations. Therefore, I will re-summarize the points from that review which are of main interest for our Symposium's topic and show where further progress has been made and where further progress can be expected in the near future.

Direct ozonation reactions

The primary interest is to predict which micropollutants M become oxidized directly by molecular O_3. Experience shows that these reactions are always first order with respect to the concentrations of ozone and the micropollutant. If we consider a plug-flow or a batch-type reactor, the rate-law for the elemination of M can therefore be written as:

$$\ln \frac{[M]}{[M]_o} = -k_M[O_3] \cdot t \quad (1)$$

Thus the logarithm of the relative residual concentration of M declines linearly with the time during which a given concentration of ozone is present. The slope increases with the concentration of ozone and the substrate-specific reaction-rate constant, k_M. In Fig. 2 and 3 a few exemplifying organic micropollutants are arranged on a scale based upon their second-order reaction-rate constants. On the right hand a corresponding scale indicates the time t_{37} within which the micropollutants are eliminated by a factor e (i.e. to 37 %);

$$\tau = t_{37} = \frac{1}{k_M[O_3]} \quad (2)$$

This scale is calibrated for a concentration of O_3 of 10^{-5}M (about 0.5 mg/ℓ), e.g. a typical order of magnitude applied for drinking water treatment. (For the conversion of k_{O_3} to k_M values the stoichiometric yield factor was approximated by 1.0 although real values would range between 0.2 to 1.0 $\Delta M/\Delta O_3$. Appropriate corrections would not show up in this compressed scale which covers 10 orders of magnitudes.)

From the few examples presented in these figures we learn:
- Saturated alkyl-groups do not react.
- Olefinic compounds react within seconds. However, when α-standing H atoms are substituted by chlorine, these substances become inert. (Perchlorethylene or even trichlorethylene cannot be oxidized by this direct reaction within a reasonable ozonation time.)
- Phenols react within seconds. The phenolate anion reacts 10^6 times faster

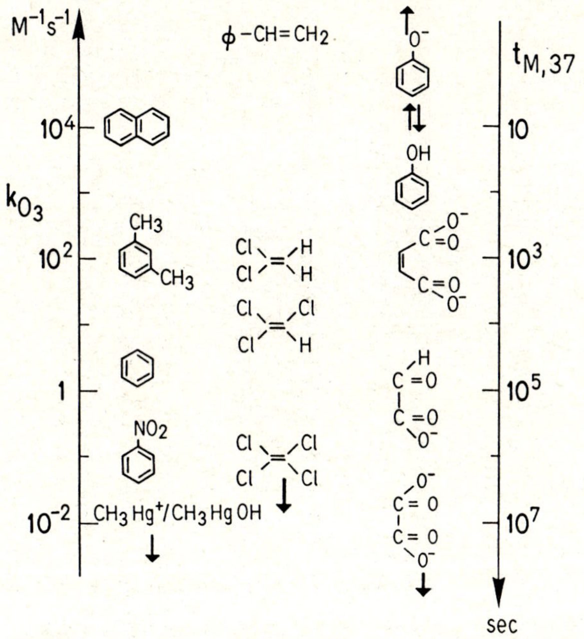

172

Fig. 2. Rate constants for direct reactions of ozone with organic solutes. *)

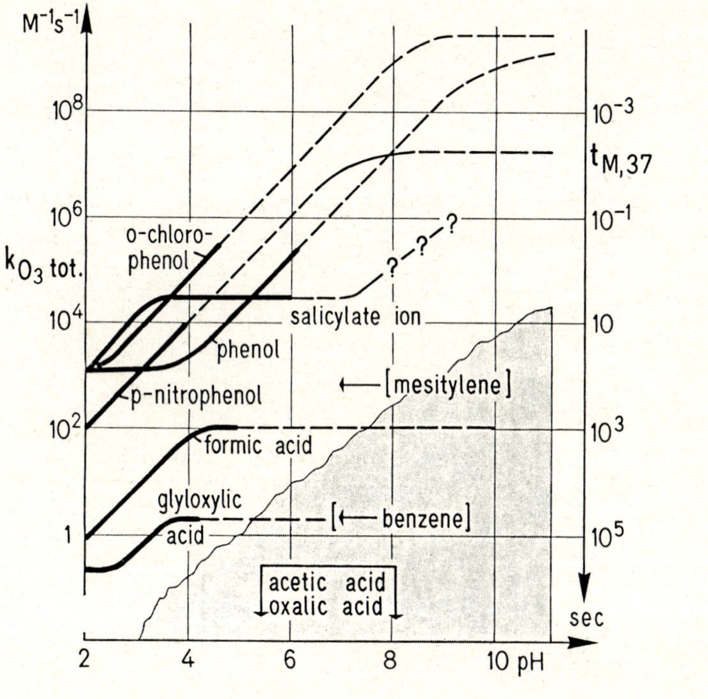

Fig. 3. Rate constants for direct reactions of ozone with organic solutes vs. pH. *) ▽ zone where "Radical-Type Mechanisme" generally dominates

*) Right-hand scale: $t_{M,37}$ is the reaction time required to reduce the concentration of the solute by a factor e (37 %) if the ozone is present in a concentration of 10^{-5} M (0.5 mg/ℓ); batch-type or plug-flow reactor. (Hoigné and Bader, 1983a,b).

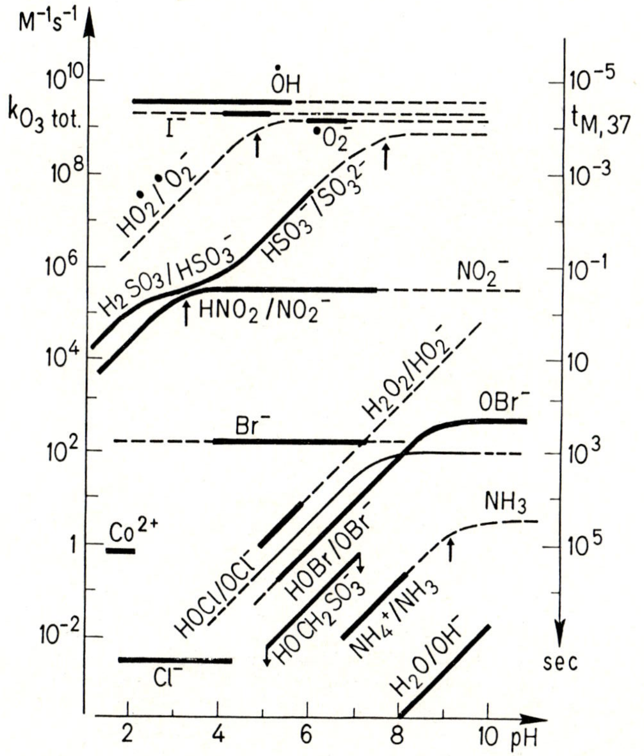

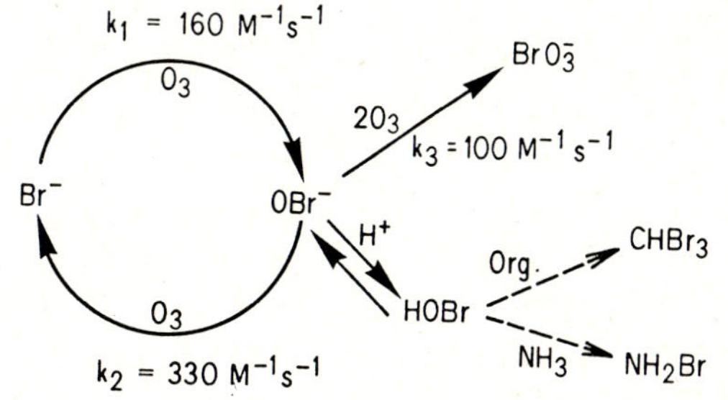

Fig. 5. Reactions of ozone with bromide and hypobromite (Haag and Hoigné, 1983).

Fig. 4. Rate constants for direct reaction of ozone with inorganic solutes vs. pH (data selected from Hoigné et al., 1985).
*) See footnote Fig. 3.

than the non-dissociated phenols. Thus even above pH 4, where only 10^{-6} parts of the phenol are dissociated, the rate of reaction becomes controlled by the small amount of this phenolate anion and the apparent rate constant therefore increases with the degree of dissociation, i.e. by a factor 10 per pH unit. Such pH dependencies must be considered when the relative rates with which two substrates become eliminated during an ozonation process are compared at different pH's (Hoigné and Bader, 1983b). They also become important when mass transfer of ozone becomes accelerated by chemical reactions (Masschelein and Goossens, 1984).

The oxidation products produced by ozonolytic cleavage of aromatic ring systems are glyoxylate-, malate-, oxalate-, acetate-, or formate- ions. Of these products only the formate ion reacts slightly during further ozonation. All others will accumulate as apparent final products when no alternative pathways of oxidations are operative (see entries in Fig. 2 and 3).

These examples show that direct ozonation reactions are highly selective. Only those compounds become oxidized which contain functional groups which are easily attacked by the electrophilic ozone.

Such lists of rate constants indicate which compounds will react with ozone during a given ozonation time. However, for organic solutes in water the rate constants of only about 110 representative substances have been published (Hoigné and Bader, 1983a,b). Few additional compounds have been added to these lists from compatible studies performed also by other authors. However, a small research group (1 1/2 persons) could determine about 20 (pH-dependent) further rate constants per project year. But generalisations will become easier rather when further critically evaluated structure-reactivity relationships become established, even though such relationships will hold only within series of comparable types of compounds which react by the same rate determining reaction step.

The many rate constants published for non-aqueous media can only be converted to the water phase, using the relevant conversion factors which depend on the specific classes of compounds (Hoigné and Bader, 1983a). In addition, in water many compounds of toxicological interest dissociate into ions or become complexed by cations, thus forming new species with very different rate constants (Hoigné and Bader, 1983b).

However, experimental methods for the determination of reaction-rate constants of aqueous organics are developed and well tested. If some fundamental rules are obeyed, they can mostly be easily applied in any laboratory for determining the constants of additional compounds of interest. Therefore, we conclude that is is most appropriate to determine further rate constants within the case studies of direct interest. If such measurements are properly done, the results can be periodically reviewed and integrated into existing comprehensive lists of rate data.

A rather comprehensive review listing reaction-rate constants for about 60 inorganic species has recently been compiled (Hoigné et al., 1985a). Such data are required whenever inorganic reactions compete with the organic micropollutants or when ozone is applied in combination with other inorganic disinfectants such as chlorine (Hoigné, 1985b).

Many research groups have identified the succession of products formed when particular aquatic substrates were ozonated. Whenever the direct action of ozone predominates, these results correspond with those expected based upon the known rate constants. (Also compare e.g. the list of product eliminations included in the conference paper by Sontheimer with the rate-constants given in Fig. 2.)

The efficiencies of these direct reactions of ozone with micropollutants are only defined when the relative decrease of the concentration of the micropollutant is based on the mean concentration of the ozone acting during the stated process time (also see Table 1).

Interactions of bromide

Many European drinking waters contain bromide concentrations in the 0.1 to 1 mg/ℓ range. (Examples: locations below alkali mines such as those situated along the Rhine below Seltz or around Bercelona; a few groundwaters in Northern Germany and Belgium; the main water supply in Israel; some mineral waters.) In such waters ozone reacts with bromide within minutes, thus producing bromine (HOBr) which can be further oxidized to bromate. Alternatively, bromine can react with either ammonia or with organic substrates. The latter results in the formation of brominated organic compounds such as bromoform (Haag and Hoigné, 1983). Details of these complex schemes of reactions could be kinetically analysed in detail (see Fig. 5). They allow conclusions for water treatment. In this respect the contradictuous observations from van der Gaag et al. shown at this Conference (also see Kruithof et al., 1985) merit special interest. They report that the level of brominated organics in Rhine water ([Br]$\sim$0.2 mg/ℓ) was much lower when pre-chlorination treatment was replaced with pre-ozonation. However, based on our kinetic data, about half of the total bromide should be transferred into HOBr/OBr$^-$ during the ozonation treatment they applied (compare Fig. 10 in Haag and Hoigné, 1983). Does this bromine produce less organo halogens when present together with oxidants such as ozone? A comparable situation is known to occur when chlorine is acting in presence of chlorine dioxide. Also in this mixture no haloforms were produced.

The possible interactions of these different pathways of reactions show that brominated organics will depend on many process parameters which must well be observed and noted whenever an experiment is quoted (compare Table 1).

TABLE 1
Minimal key parameters which control the oxidation reactions in drinking water and which must be considered and quoted for any experiment.

Ozonation process[a]	Chlorine dioxide process[a]
For direct reaction of O_3:	
$[\overline{O_3}]$; t; pH[b]	$[\overline{ClO_2}]$; t; pH[b]
For HOBr/BrO$^-$ intermediates:	
$[\overline{O_3}]$; t; pH; $[Br^-]$; DOC[c]	----------
For reactions of secondarily produced $^\bullet OH$:	
$(\Delta O_3 \rightarrow {^\bullet OH})$; HCO_3^-/CO_3^{2-}; DOC[c]; [other radical scavengers].	----------

[a] Also the type of the reactor and the temperature must be stated (compare Hoigné, 1982a).
[b] The pH is only of importance when the speciation of the substrate changes is pH dependent. (In cases where the substrates form metal complexes, also the concentrations of the relevant metal ions must be considered.
[c] A qualification of the type of the DOC gives a further help for later comparison.

Oxidations by $^\bullet OH$ radicals

Aqueous ozone decomposes. For example, half of the ozone introduced into rawwater from Lake Zürich (pH ∼ 7.8, $[HCO_3^-]$ ∼ 1.2 mM) decomposes within the ozonation time of 10 minutes, even after the spontaneous ozone demand of less than 0.1 mg/ℓ ozone is met. In a groundwater of low DOC the lifetime can be about twice this value but in many other types of waters decomposition is faster (Hoigné and Bader, 1979). About half of the decomposed ozone becomes converted into the $^\bullet OH$ radical which is the most reactive aqueous oxidant. $^\bullet OH$ can act on organic micropollutants either by $^\bullet H$-abstraction, by $^\bullet OH$ addition to a double bond, or by an electron transfer reaction (see Fig. 6). The radicals thereby formed from organic reactants easily add to the oxygen molecule, a biradical, which in aerated water is present in relatively high concentrations. The resulting peroxy radicals disproportionate or combine with each other, forming

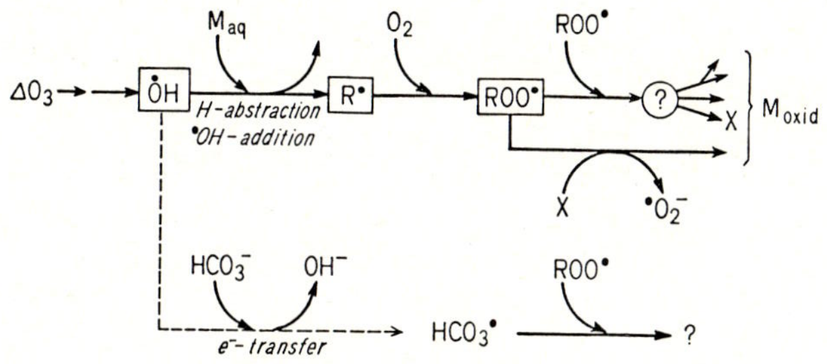

Fig. 6. Reactions of •OH radicals with solutes and successive formations of secondary radicals.

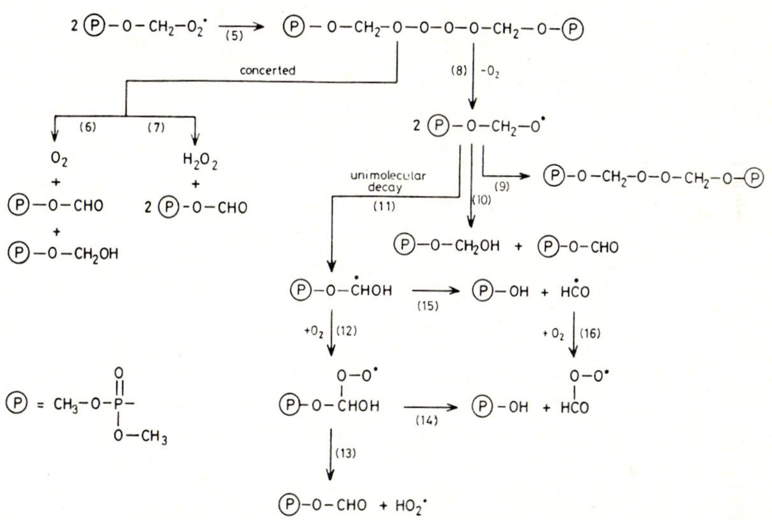

Fig. 7. Examples of the formation of secondary oxidation products from peroxy radicals which result from the reaction of •OH with the substrate. (Schuchmann and Sonntag, 1984). (Only the bimolecular pathway has been considered here.)

many types of mostly labile intermediates which undergo further reactions to produce peroxides, aldehydes, acids, hydrogen peroxide etc. Many types of organic peroxy radicals also release superoxide anions (O_2^-) or hydrogen peroxide which then selectively react with further ozone (see later).

Details on the formation of final products from ·OH radical reactions are only known for a few types of exemplifying substrates. So far these have mostly been studied by radiation chemists, who have been concerned with the byproducts formed when ·OH radicals are initiated by ionizing radiation such as that applied for cancer treatment or for food irradiation. An example of product formations from a specified substrate is shown in Fig. 7. Evidently, the product formations depend on so many parameters that our comprehension will remain restricted to a few selected examples on which detailed chemical analyses have been performed. Even such exemplifications can be of general value only if the amount of the occurring ·OH radical processes can be calibrated or predicted.

A kinetic analysis of the ·OH reactions can be reduced to the following conclusions: ·OH radicals are so reactive that they are rapidly consumed (within microseconds) by the sum of bicarbonate and all the organic compounds present in drinking water. Only a few will survive long enough to react with a specified micropollutant. In such a situation the kinetics with which a specified micropollutant M becomes eliminated in a batch-type reactor can be described by

$$\ln \frac{[M]}{[M]_o} = - \eta(\Delta O_3) \frac{k'_M}{\Sigma k'_i [S_i]} \qquad (3)$$

Thus, the logarithm of the relative residual concentration of M declines linearly with $\eta(\Delta O_3)$, the amount of ozone decomposed to OH· radicals. The elimination of M increases with the rate constant with which °OH reacts with M, (k'_M), but it decreases with the rate with which the sum of all other substrates, S_i, scavenge ·OH, i.e. with $\Sigma(k'_i[S_i])$.

Fig. 8 shows the depletion of ozone-resistant micropollutants by this pathway in water from Lake Zürich. Here toluene was spiked into an additional sample as a reference solute. Its relative rate of depletion just corresponds with that of micropollutants present in 10^4 times lower concentrations. Based on equation 3 and much experiences, such as quoted above, the effect of ·OH radicals on further micropollutants can easily be calibrated by analysing the depletion of any probe substance which is not oxidized by ozone in the direct reaction. Conversion factors to be applied can be determined from the many lists of rate constants for ·OH radical reactions published in the literature (for a review see Farhatazis and Ross, 1972) and quoted in Fig. 9.

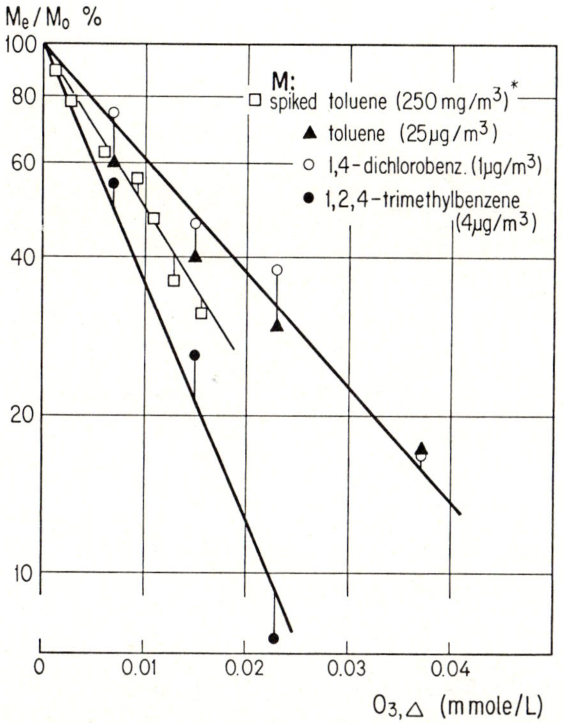

Fig. 8. Elimination of ozone-resistant organic micropollutants in water of Lake Zürich, following the decomposition of ozone to the more reactive ·OH radical. (Only the trimethylbenzene was also directly degraded by ozone itself.) The elimination of the micropollutants are compared with that of spiked toluene which was used as a reference at a 10,000 times higher concentration. (Zürcher et al., 1982).

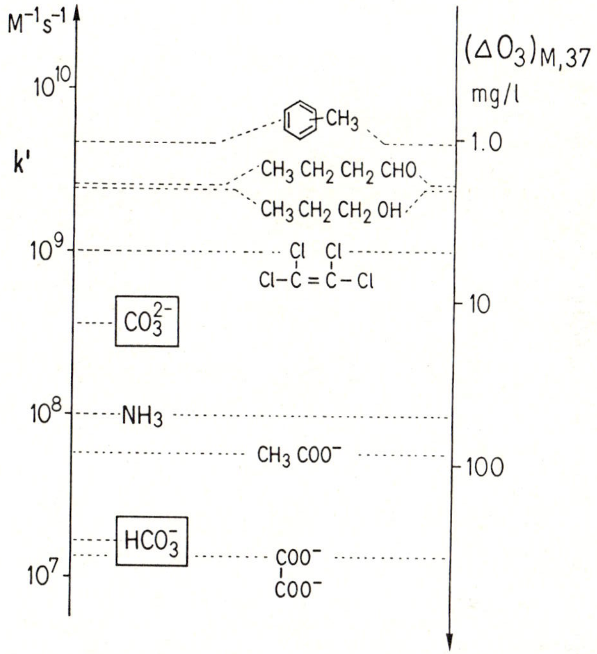

Fig. 9. Rate constants for reactions of ·OH radicals with different solutes. $(\Delta O_3)_{37}$ is the required amount of decomposed ozone which results in an elimination of the quoted substrate to 37 % of the initial value. (Batch-type or plug-flow reactor.) This scale is calibrated for an eutrophic type of lakewater (Lac de Bret, DOC = 4 mg/ℓ, $[HCO_3^-]$ = 1.6 mM, pH = 8.3. (It changes prop. to the DOC of a water.)

They generally range from 10 to 10^{10} M⁻¹s⁻¹. On the right-hand scale of this figure the kinetic values have been converted to values which indicate the amount of ozone which must decompose to achieve an elimination by a factor of e (to 37 %):

$$(\Delta O_3)_{37} = \frac{\Sigma(k'_i [S_i])}{\eta \cdot k'_M} \tag{4}$$

(In our earlier publications we called (ΔO_{37}) the "Ω-value".

The conversion is based on a scavenging rate for ˙OH radicals which is typical for an eutrophic type of lakewater. This scale increases proportional to the DOC which here controls the lifetime of ˙OH radicals.

The rate of decomposition of ozone and the formation of ˙OH radicals depend on many water parameters due to chain reactions which accelerate the decomposition of ozone (see Fig. 10). Such chain reactions can be initiated by OH⁻ and therefore often proceed more rapidly with higher pH. In addition, many other solutes which are present in a drinking water can act as initiators. The ozonide

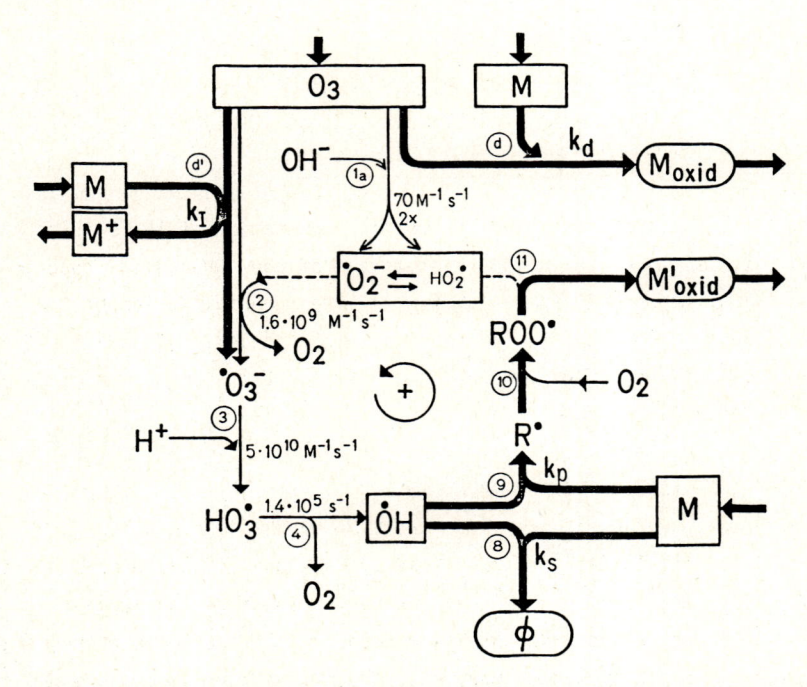

Fig. 10. Reactions of aqueous ozone in presence of solutes M which react with O_3, or which react with ˚OH radicals by scavenging ˙OH and by converting ˙OH to HO_2˙ (Staehelin and Hoigné, 1983).

anions, $\cdot O_3^-$, formed by the reactions of the initiators decompose immediately to ·OH radicals (Bühler et al., 1984), which are consumed by organic compounds. Some intermediates formed release O_2^- (typical reactions for alcohols, or sugars, or formic acid etc.), which transfers an electron to additional ozone in a selective reaction. Therefore, solutes which transform non-selective ·OH radicals into O_2^- act as promoters of the chain reaction. In contrast, other solutes can scavenge ·OH, producing intermediates which cannot react with further ozone. These types of solutes inhibit the chain reaction and thus stabilize the ozone (example: bicarbonate). The situation is complicated because the effects of different solutes do not additively contribute to the rate of the chain reaction; rather it is the relative amount of promoter to inhibitor which controls the rate of decomposition of ozone to ·OH by this chain reaction (Staehelin and Hoigné, 1983).

The direct reaction of ozone and the reactions succeeding the formation of ·OH radicals produce other products. Therefore, the importance of the two pathways have to be considered and distinguished whenever a product formation is to be discussed. But in most drinking waters all sorts of compounds are present which act as initiators, or promoters, or inhibitors of the chain reaction. Therefore, a quantitative prediction of the fate of ozone in a real water is not possible without measurements.

Because of these complexities we recommend that the rate of decomposition of ozone is determined experimentally and that the efficiency of the °OH radical reactions is calibrated by observing an ozone resistant reference solute. For the planning of such experiments and for the interpretation of the results the kinetic characteristics of the chain reaction and of the °OH radical reactions must be considered (compare Hoigné, 1982). A further generalisation of the results in respect to the oxidations of micropollutants is only possible if the tested water has been well characterized and the amount of decomposed ozone determined (see Table 1).

KINETIC ASPECTS OF CHLORINE DIOXIDE REACTIONS

Recently an extended review paper on reactions of chlorine dioxide with aquatic organic materials has been published by Rav-Acha (1984). Some kinetic laws and data have also been summarized in a paper by Hoigné and Bader (1982). From these two papers we conclude: At low concentrations and temperature, such as that typical in drinking water treatment, the rate controlling steps proceed according to rate laws which are first order with respect to the concentration of both, chlorine dioxide and the solute. Chlorine dioxide is a much

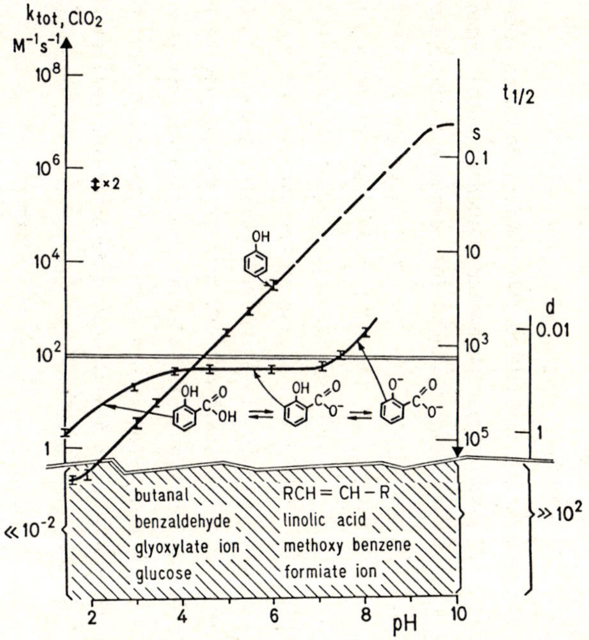

Fig. 11. Rate constants of reactions of chlorine dioxide with organic solutes vs. pH. Right-hand scale: t_{37} is the reaction time required to reduce the concentration of the solute to 50 % if the chlorine dioxide is present in a concentration of 10^{-5}M (1 mg/ℓ); batch-type or plug-flow reactor. Selected from Hoigné and Bader (1982b).

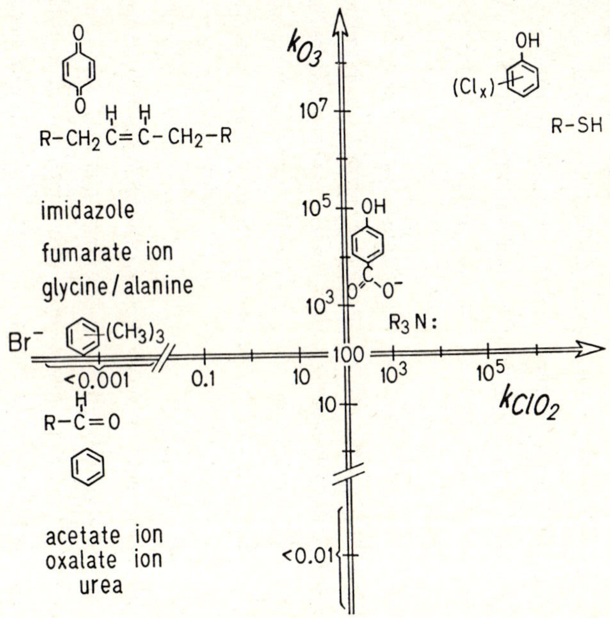

Fig. 12. Rate constants of different solutes for reactions with ozone vs. the rate constants with which they react with chlorine dioxide. (Hoigné and Bader, 1982b).

more selective oxidant than ozone. Therefore, also the reactivity of chlorine dioxide can best be characterized by quoting second-order reaction-rate constants. As shown by the exemplifications in Fig. 11, these are high in case of phenolic compounds (which only react according to their degree of dissociation to phenolate anions), non-protonated tertiary amines, thio compounds and polycyclic aromatic hydrocarbons (Rav-Acha et al., 1983, not shown in Fig. 11). Unreactive are olefinic C = C double bonds, non-condensed aromatic hydrocarbons, amines (except tertiary), or the amino groups present in amino acids, or aldehydes, ketones, quinone, carbohydrates, and ammonia and bromide ions. The differences in selectivity between ozone and chlorine dioxide is illustrated in Fig. 12 where the rate-constants for reactions with ozone are plotted vs. those for reactions with chlorine dioxide. Most striking are the differences for bromide, ammonia and quinones which react with ozone but not with chlorine dioxide. Already from these relations many differences considering product formations in drinking water can be understood.

Further rate constants will be published, soon. Rate constants and product formation arising in non aquatic systems or in cellulose or food bleaching are non comparable because other reactions occur in these systems and when high ClO_2 concentrations are applied.

LIMITATIONS OF KINETIC CONSIDERATIONS
Role of specified reference substances

The kinetics of an elimination process can only be quantified if specified substances of known speciations are considered. (Exception: In cases where the kinetic order of reaction is zero.) Group parameters, such as total organic carbon (TOC) or total organic halogen compounds (TOX) contain a mixture of individual substances of different reactivity. The kinetics of their transformations therefore depend on their actual composition. Because this composition changes as the reaction proceeds, a non-defined shift in the apparent rate constants is observed. In other words, an "aging of the kinetics" occurs. Thus, no general kinetic laws and constants can be formulated when group parameters are observed. Therefore, kinetic systems or oxidation processes must be calibrated by the observation of specified reference substances. (In this respect the chemical kinetics are not different from those encountered in disinfection studies which also must be calibrated by observing well specified organisms.)

Formation of new micropollutants

This paper has focussed on the elimination kinetics of specified micropollutants. However, new types of micropollutants may be produced during reactions of oxidants or during processes using a combination of oxidants. For example: Lakewater ozonation performed before chlorination generally reduces the haloform precursors but the same ozonation process enhances the formation potential for chloropicrin.

This example shows, while the kinetic considerations are important for the interpretation of the elimination of specific micropollutants, they generally cannot give a direct answer for processes occurring with non-specified solutes, such as represented by dissolved humics.

ACKNOWLEDGMENT

I thank Mrs. Susanne Masten for discussions and correcting the manuscript.

REVIEWS ON KINETICS AND ADDITIONAL REFERENCES

Bühler, R., Staehelin, J. and Hoigné, J., 1984. Ozone decomposition in water studied by pulse radiolysis. I. HO_2/O_3^- as intermediates. J. Phys. Chem. 88: 2560-64; II. OH and HO_4 as chain intermediates, ibid. 5999-6004.

Haag, W.R. and Hoigné, J., 1983. Ozonation of bromide-containing waters: Kinetics of formation of hypobromous acid and bromate. Environ. Sci. & Technol. 17: 261-267.

Hoigné, J. and Bader, H., 1979. Ozonation of water: Oxidation competition values of ·OH radical reactions of different types of waters used in Switzerland. Ozone. Sci. & Eng. 1: 357-372.

Hoigné, J., 1982a. Mechanisms, rates and selectivities of oxidations of organic compounds initiated by ozonation of water. In: Handbook of Ozone Technology and Application, R.G. Rice & A. Netzer, Editors, Ann Arbor Science, Vol. 1: 341-379.

Hoigné, J. and Bader, H., 1982b. Kinetik typischer Reaktionen von Chlorodioxid mit Wasserinhaltsstoffen. Vom Wasser 59: 254-267.

Hoigné, J. and Bader, H., 1983a. Rate constants of reactions of ozone with organic and inorganic compounds in water: I. Non-dissociating organic compounds. Water Res. 17: 173-183.

Hoigné, J. and Bader, H., 1983b. Rate constants of reactions of ozone with organic and inorganic compounds in water: II. Dissociating organic compounds. Ibid: 185-194.

Hoigné, J., Bader, H., Haag, W.R., Staehelin, J., 1985a. Rate constants of reactions of ozone with organic and inorganic compounds in water: III. Inorganic compounds and radicals. Water Res. 19: (in press).

Hoigné, J., 1985b. Verhalten anorganischer Ionen und Desinfektionsmittel bei Ozonungsprozessen. Intern. Ozon Symposium, Wasser Berlin, IOA, (in press).

Kruithof, J.C., Noordsij, A., Puiker, L.M., and van der Gaag, M.A., 1985. The influence of water treatment processes on the formation of organic halogens and mutagenic activity by post chlorination. In: Water Chlorination, Environmental Impact and Health Effects, Vol. 5, R.L. Jolley, Editor (in press).

Masschelein, W.J. (Editor), 1982. L'ozonation des eaux: Manuel pratique. Technique et Documentation, Paris, 224 pp.; in Engl.: Ozonation Manual for Water and Wastewater Treatment, John Wiley, New York, (1982), 324 pp.

Masschelein, W.J. and Goossens, R., 1984. Nitrophenols as model compounds in the design of ozone contacting and reacting systems. Ozone: Science and Engin., 6: 143-162.

Miller, G., Rice, R.G., Robson, C.M., Scullin, R.L., Kühn, W. and Wolf, H., 1978. An assessment of ozone and chlorine dioxide technologies for treatment of municipal water supplies. EPA-600/2-78-147 Municipal Environmental Res. Lab., Cincinnati, USA, 571 pp.

Rav-Acha, Ch., 1984. Review Paper: The reactions of chlorine dioxide with aquatic organic materials and their health effects. Water Res. 18: 1329-1341.

Schuchmann, M.N. and von Sonntag, C., 1984. Radiolysis of di- and tri-methyl phosphates in oxygenated aqueous solution: A model system for DNA strand breakage. J. Chem. Soc., Perkin Trans II: 699-704.

Staehelin, J. and Hoigné, J., 1983. Reaktionsmechanismus und Kinetik des Ozonzerfalls in Wasser in Gegenwart organischer Stoffe. Vom Wasser 61: 337-348.

Zürcher, F., Bader, H. and Hoigné, J., 1982. Verhalten organischer Spurenstoffe bei der Ozonung von Trinkwasser. In: Concerted Action Analysis of Organic Micropollutants in Water (Cost Project 64 bis), Vol. 2, Brussels, Commission of the European Communities 1982, 198-213.

FORMATION OF LINEAR ALDEHYDES DURING SURFACE WATER PREOZONIZATION AND THEIR REMOVAL IN WATER TREATMENT IN RELATION TO MUTAGENIC ACTIVITY AND SUM PARAMETERS

François VAN HOOF, Jan JANSSENS and Hanja VAN DIJCK
Study Syndicate for Water Research, c/o Antwerp Waterworks
Mechelsesteenweg 64, 2018 - ANTWERP - BELGIUM.

ABSTRACT

Low molecular weight aldehydes were formed during surface water preozonization, their levels showing a positive correlation with increasing ozone dose applied and with increasing water temperature. A strong negative correlation was observed between aldehyde levels and U.V. absorbance at 254 nm.

Coagulation had no influence on the aldehydes present and the influence of rapid double layer filtration varied strongly with temperature : significant removals were only observed above 10°C.

Mutagenic activity generated by preozonization in Salmonella typhimurium TA98 shows an ozone dose depending relationship different from the formation of linear aldehydes. Its removal by coagulation is not effective but rapid double layer filtration reduces mutagenic activity to marginal levels.

In this respect too no clear parallel can be drawn between the presence of low molecular weight aldehydes and mutagenic activity.

INTRODUCTION

Ozone has long been known to have a profound impact on organic matter in water. This is clearly evidenced by the decrease in UV_{254} absorption of ozonated water. The effects of ozone on organic matter are thought to be related to different effects which ozone produces : micro-flocculation, biological aftergrowth, mutagenic effects. (1-4).

Linear aldehydes (C_6-C_{16}) were previously detected in ozonated waters by different authors. (5,6). Recently we showed that ozonization of different types of water produces mainly low molecular weight (C_1-C_3) aldehydes, increasing levels of aldehydes being formed after application of increasing ozone doses. We also demonstrated that humic substances are the precursors for the formation of these aldehydes. (7).

In this paper we shall discuss the formation of linear aldehydes during surface water preozonization, their removal in further water

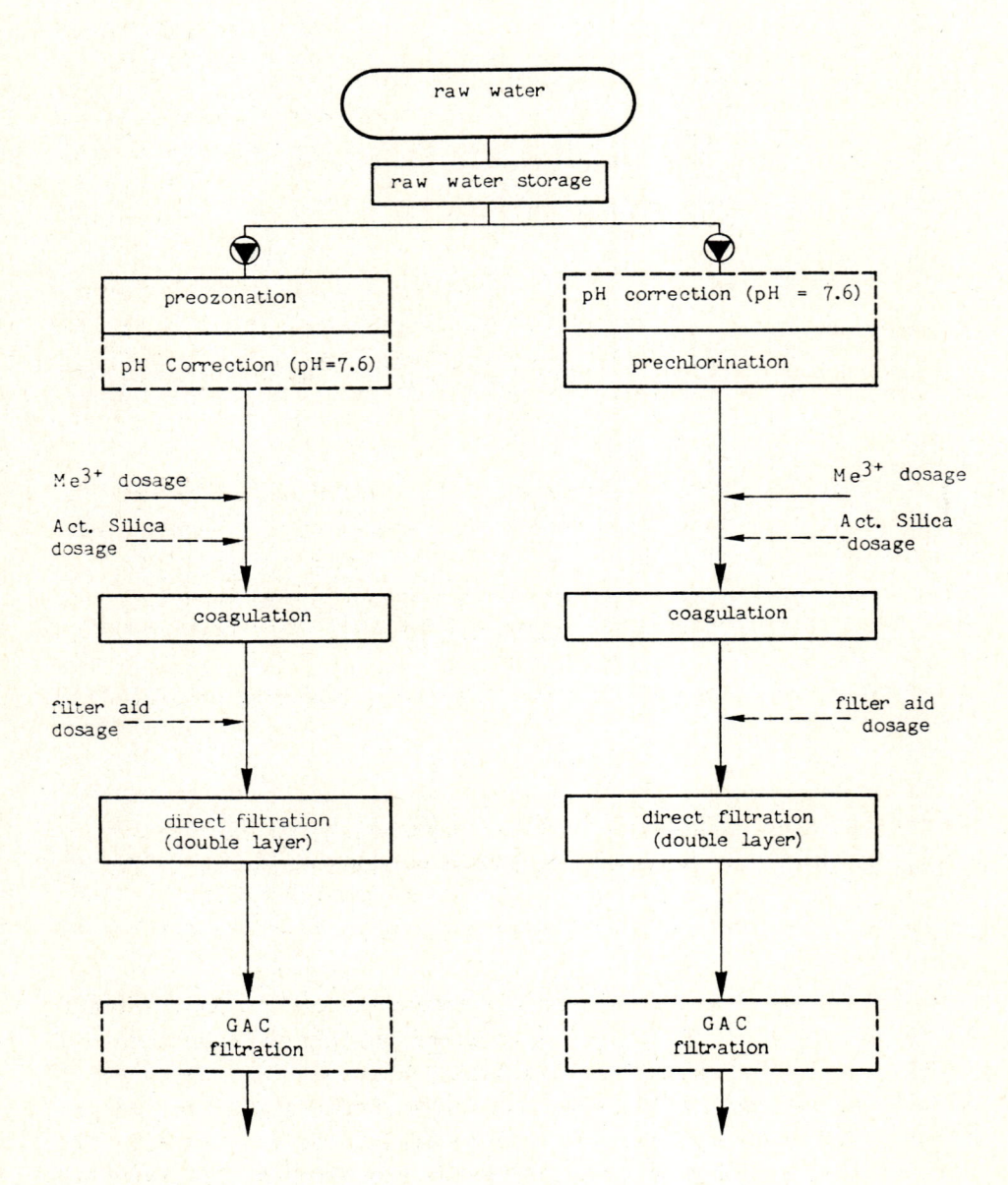

Fig. 1 : Experimental plant

treatment and their presence in relation to mutagenic activity in Salmonella typhimurium strain TA98.

EXPERIMENTAL
Water treatment

Surface water, derived from the river Meuse is treated with an ozone dose applied of 1 mg l^{-1} and is coagulated afterwards with trivalent metal (Me^{3+}) salts. This treatment is followed by direct filtration over a double layer filter bed using hydroanthracite and sand (Fig. 1).

Coagulants used were alum (eventually combined with activated silica and a polymeric filter aid), polyaluminiumchloride and ferric chloride. The double layer filter was operated at a filtration rate of 8,5 - 8,7 m/h.

Analysis of C$_1$-C$_3$ aldehydes

Low molecular weight aldehydes were analyzed as described in (7) with the exception that the 2,4-dinitrophenylhydrazine used was twice recrystallized from acetonitrile (HPLC grade). Separation of the 2,4-dinitrophenylhydrazones was done by reversed phase HPLC on a 25 cm 5 μ C$_8$ Ultrasphere column using a mixture of methanol-water (80-20).

Samples after filtration were all taken at minimum effluent turbidity.

Mutagenicity assays

Collection of samples and the Salmonella typhimurium microsomal assay on strains TA98 and TA100 were performed as described earlier. (8).

RESULTS AND DISCUSSION

Application of increasing ozone doses results in the formation of higher concentrations of low molecular weight aldehydes in all types of waters investigated (7). The increase in aldehyde levels correlates very well with the decrease in UV$_{254}$ absorbance for all types of waters tested as shown for the correlation coefficients : marshland water (-0.958), humic rich water (river Helle) (-0.973), surface water derived from storage reservoir (-0.984), prechlorinated, double layer filtered water (-0.984).
Similar correlation coefficients were found between aldehyde levels

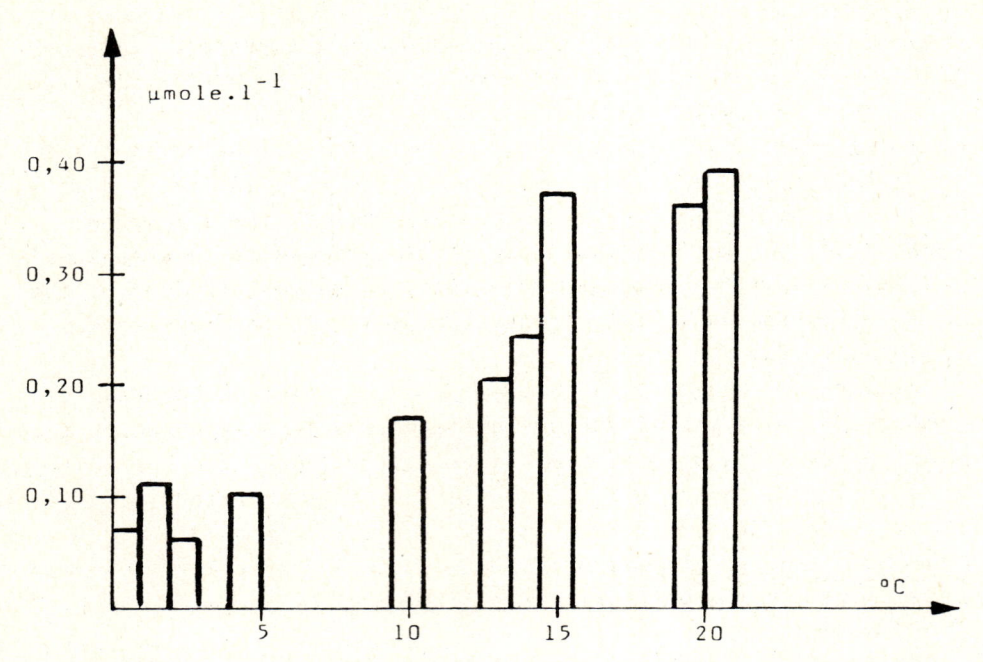

Fig. 2 : Seasonal variation of low molecular weight aldehydes, expressed as µmole aldehydes l^{-1}, after application of 1 mg l^{-1} O_3.

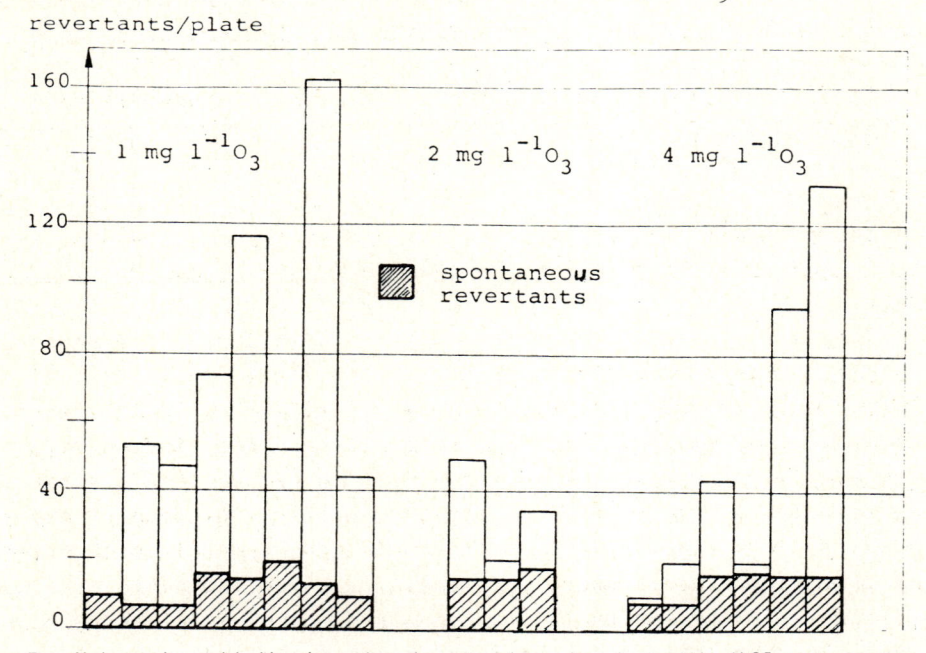

Fig. 3 : Mutagenic activity in extracts of waters treated with different ozone doses. All data represented were obtained on extracts corresponding with 1 liter samples in Salmonella typhimurium strain TA98.

and TOC in raw waters : river Helle (-0.936), marshland water (-0.947). No such correlations could be found on prechlorinated, filtered water when treated with ozone (-0.758).

The influence of seasonal variation on the formation of the aldehydes in ozonated surface water was studied by analyzing samples at varying water temperatures (1,5 - 20,5°C) at the same ozone dose applied (1 mg l^{-1}). A good correlation was found between aldehyde levels, expressed as μmole total aldehydes, the sum of the C_1-C_3 aldehydes, and water temperature (-0.952) for ten samples analyzed. (Fig. 2).

Mutagenic activity generated by preozonization in Salmonella typhimurium strain TA98 follows different patterns with respect to temperature and ozone dose applied : it is higher after application of 1 mg l^{-1} O_3 than at higher doses and increasing water temperature does not lead to increasing mutagenic activity (Fig. 3, 4).

Coagulation has little influence on the aldehyde levels irrespective of the coagulant used (Table 1).

TABLE 1
Influence of coagulation on low molecular weight aldehydes

Coagulant used	Before coagulation (μmole l^{-1})	After coagulation (μmole l^{-1})
Polyaluminiumchloride	0,20	0,20
Polyaluminiumchloride	0,29	0,20
Alum	0,37	0,36
Alum	0,39	0,53
Ferric chloride	0,36	0,25
Ferric chloride	0,04	0,07

The influence of rapid double layer filtration on the aldehydes present seems to vary strongly with temperature (Table 2). Aldehyde levels are not effected at temperatures below 10°C, removal is generally better above this temperature. This finding might lead to the conclusion that volatilization rather than filtration affects the aldehyde levels. However, within the Group of experiments performed above 10°C a better aldehyde removal is observed in those experiments where TOC removal is high indicating that under these circumstances filtration plays a role in aldehyde removal. No correlations can be found between aldehyde removal and changes in UV_{254} absorption before and after filtration (Table 2).

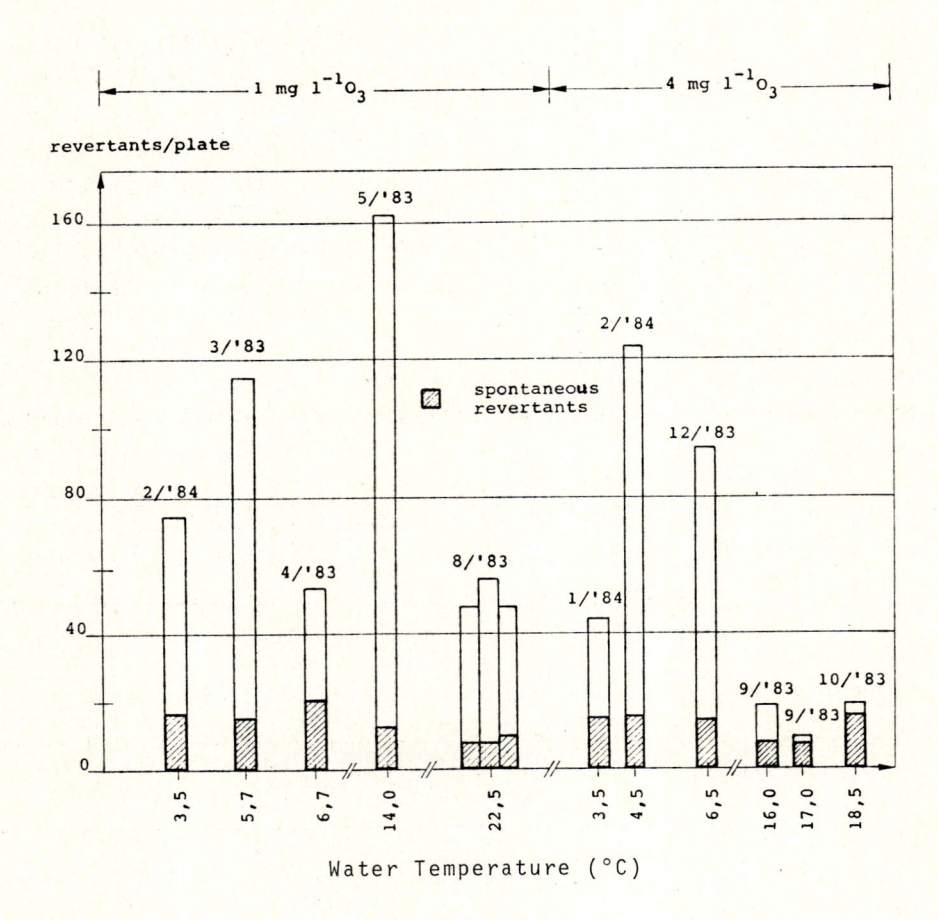

Fig. 4 : Seasonal variation of mutagenic activity on strain TA98 after application of 1 mg/l O_3 and 4 mg/l O_3 in 1 liter water samples.

Mutagenic activity in Salmonella typhimurium TA98 undergoes little influence from coagulation but is significantly reduced to marginal levels after filtration. (8). The experiments performed in this context were carried out at water temperatures below 10°C, therefore no direct analogy can be found between removal of aldehydes and removal of mutagenic activity.

Due to the different formation (as a function of the ozone dose applied) and removal patterns, low molecular weight aldehydes can not be used as surrogate parameters for mutagenic activity generated in Salmonella typhimurium TA98 during surface water preozonization. Wether higher molecular weight linear aldehydes which exhibit a different formation pattern under influence of ozone (6) can be used for this purpose deserves further investigation.

Neither low molecular weight linear aldehydes nor mutagenic activity in Salmonella typhimurium TA98 can be completely eliminated by coagulation in combination with rapid double layer filtration. Therefore, attention will have to be paid to other purification techniques e.g. aeration and activated carbon filtration.

TABLE 2

Influence of rapid double layer filtration on low molecular weight aldehydes, TOC and UV_{254}

Δ TOC (mg^{-1})[a]	ΔUV_{254} (m^{-1})[a]	Δ Aldehydes[a] (μmole l^{-1})	Temperature °C
-0,1	-0,4	-0,04	13,7
-0,1	-0,3	-0,08	13,7
-0,7	-0,3	-0,26	15,2
-1,5	-0,3	-0,32	20,5
-1,7	-0,0	-0,19	19,6
-0,1	-0,3	+0,01	2,0
-0,6	-1,4	+0,02	2,0
-0,6	-4,5	+0,06	4,5
-0,1	-3,3	+0,03	10

[a] Δ TOC, ΔUV_{254} and Δ aldehydes represent the changes in TOC, UV_{254} absorption and aldehydes levels caused by rapid double layer filtration.

REFERENCES

1. R.G. Rice, C.M. Robson, G.W. Miller and A.G. Hill, J. Amer. Water Works Ass., 73 (1981), 44.
2. D. Van der Kooy, J.P. Oranje and W.A.M. Hynen, Appl. Environ. Microbiol., 44 (1982), 1086.
3. F. Van Hoof, in R. Jolley, W. Brungs, J. Cotruvo, R. Cummings, J. Matice and V. Jacobs (Eds.), Water Chlorination - Environmental Impact and Health Effects, Vol. 4, Ann Arbor Science, Ann Arbor, 1983, p. 1211.
4. J.P. Duguet, A. Ellul, E. Brodard and J. Mallevialle, in Ozonation - Environmental Impact and Benefit, Proceedings of

the Symposium of the International Ozone Association, Brussels, September 12-13, 1983, p. 424.
5   R.E. Sievers, R.M. Barhley, G.A. Eiceman, R.H. Shapiro, H.F. Walton, K.J. Kolonko and L.R. Field, J. Chromatogr. Sci., 142 (1977), 745.
6   F. Zürcher, H. Bader and J. Hoigné, in H. Ott (Ed.), Analysis of organic micropollutants in water, D. Reidel, Dordrecht, 1980, p. 198.
7   F. Van Hoof, A. Wittocx, E. van Buggenhout and J. Janssens, Analytica Chimica Acta, 169 (1985), 419.
8   F. Van Hoof, J. Janssens, H. Van Dijck, accepted for publication in Chemosphere, 1985.

NEW DIRECTIONS IN OXIDANT BY-PRODUCT RESEARCH: IDENTIFICATION AND SIGNIFICANCE

R.F. CHRISTMAN, D.L. NORWOOD AND J.D. JOHNSON

Department of Environmental Sciences and Engineering, School of Public Health, University of North Carolina, Chapel Hill, NC 27514 (USA)

ABSTRACT

Exciting new research in the field of oxidant by-products in drinking water and wastewater is progressing along three main lines: 1) investigation into the mechanism of the aquatic humic/aqueous chlorine reaction, 2) correlation of by-product identity and yield with mutagenic activity, and 3) analysis of previously intractable organic by-products such as high molecular weight polar substances and N-chloroorganic compounds. Just as earlier research was stimulated by the development of the combined gas chromatograph/mass spectrometer, these new investigations have been fostered by the development of new analytical methods such as solid state C-13 nuclear magnetic resonance spectroscopy, extended mass range mass spectrometry, new ionization techniques such as fast atom bombardment and combined liquid chromatography/mass spectrometry. Application of these new techniques is allowing these and many other important public health questions to be addressed.

INTRODUCTION

Numerous research reports over the last decade have established that a wide variety of halogenated and non-halogenated by-products result from the exposure of aquatic organic material to chlorine. Much of this work has focused on chlorine/humic reactions (Johnson et al., 1982; Christman et al., 1983; de Leer et al., 1985; Coleman et al., 1984) although humic substances comprise only a fraction (Ca. 50%) (Malcolm, 1984) of the dissolved organic carbon in terrestrial streams. Other reports (Wachter and Andelman, 1984; Scully et al., 1984) have concentrated on chlorine reactions with algal exudates, and nitrogenous compounds. Exciting new research is progressing along three main lines: 1) investigation of the mechanism of the aquatic humic/aqueous chlorine reactions, 2) correlation of by-product identity and yield with mutagenic activity (Coleman et al., 1984; Horth et al., 1985), and 3) analysis of previously intractable organic by-products such as high molecular weight polar substances (Crathorne et al., 1984). Just as earlier research was stimulated by the development of the combined gas chromatograph/mass spectrometer (GC/MS), these new investigations have been catalyzed by the development of solid state C-13 nuclear magnetic resonance spectroscopy (NMR) extended mass range mass spectrometry, new ionization techniques (e.g., fast atom bombardment) and combined liquid chromatography/mass spectrometry (LC/MS).

Our purpose in this presentation is to comment on the impact these new techniques are having on our knowledge of the structures and reactivities of aquatic organic material.

MECHANISM OF THE HUMIC/CHLORINE REACTION

It is now clear that a variety of non-volatile aliphatic halogenated products result from the exposure of aquatic humic and fulvic acid fractions to chlorine (Johnson et al., 1982; Christman et al., 1983; de Leer et al., 1985; Coleman et al., 1984). Dominant among the hundreds of such products are the C_2 halogenated acids, principally di- and trichloroacetic acids (see Table 1 and Figure 1). Chloroform and trichloroacetic acid alone may account for as much as 50% of the total organic halogen in mixtures heavily chlorinated in the laboratory (Christman et al., 1983). These C_2 chlorinated acids have also been shown to be present in chlorinated municipal drinking water although they account for a smaller percentage of the total organic halogen (Norwood et al., 1984; Uden and Miller, 1983).

TABLE 1
Short chain chlorination products of aquatic humic and fulvic acids (Christman et al., 1983).

trichloromethane (chloroform)	dichloropropanedioic acid (dichloromalonic acid, DCM)
trichloroethanal (chloral)	butanedioic acid (succinic acid)
chloroethanoic acid (chloroacetic acid)	chlorobutanedioic acid (chlorosuccinic acid)
dichloroethanoic acid (dichloroacetic acid, DCA)	2,2-dichlorobutanedioic acid (alpha, alpha-dichlorosuccinic acid, DCS)
trichloroethanoic acid (trichloroacetic acid, TCA)	cis-chlorobutenedioic acid (chloromaleic acid)
trans-dichlorobutenedioic acid (dichlorofumaric acid)	cis-dichlorobutenedioic acid (dichloromaleic acid)
bromodichloromethane	2,2-dichloropropanoic acid
3,3-dichloropropenoic acid	2,3,3-trichloropropenoic acid

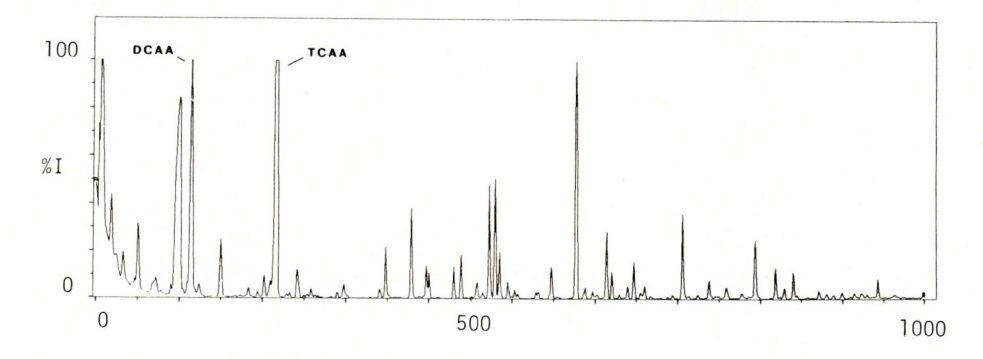

Fig. 1. A reconstructed total ion chromatogram of the ether extractable aqueous chlorination products of Singletary Lake fulvic acid.

It has been hypothesized (Rook, 1980) that chloroform is produced in these reactions by a phenolic ring rupture mechanism. This hypothesis has become increasingly attractive since many of the identified products of several research groups are consistent with this mechanism and have been shown (de Leer et al., 1985; Boyce and Hornig, 1983) to result from laboratory chlorination of model phenols. However, some products (e.g. dichlorosuccinic acid, and the unhalogenated aliphatic and aromatic acids) cannot be explained by this mechanism. Aromatic side chain attack appears to be the most promising of many alternative mechanisms to account for these products (Christman et al., 1984; de Leer et al., 1985).

Current work in our laboratory involves the application of solid state C-13 NMR and sequential chemical degradation to this problem. C-13 NMR as applied to humic materials has suffered from poor sensitivity due to the low relative abundance of C-13 (1.08%) and the limited solubility of humic materials in most NMR solvents. It is therefore desirable to perform C-13 NMR experiments on humic materials in the solid state. However, solid state C-13 NMR presents unique problems (Levy et al., 1980):

  a) reduced signal-to-noise ratio due to powerful (relative to liquid state) proton/C-13 dipolar coupling interactions.
  b) line broadening due to chemical shift anisotropy in the motionally restricted solid samples.
  c) poor sensitivity and lack of quantitative capacity due to inefficient C-13 spin-lattice relaxation.

The techniques used to minimize these problems include high power proton decoupling, magic-angle spinning (MAS) and cross-polarization (CP). For a comrehensive treatment of the CP/MAS technique the reader is referred to the monograph by Fukushima and Roeder (1981).

A CP/MAS C-13 NMR spectrum of Singletary Lake[*] fulvic acid is shown in Figure 2. The first step in spectral analysis is to assign the observed resonances to the types of carbon atoms present in the fulvic acid macromolecule. Our treatment of this data is patterned after the work of Hatcher et al., 1983. The spectrum in Figure 2 can be divided into the four regions summarized in Table 2. Region I, from 0 to 50 ppm, which contains a relatively intense resonance with maxima at 23.5 and 40.8 ppm is usually assigned to paraffinic carbons, i.e. carbons that are singly bonded only to other carbons. These may include methyl, methylene and methine carbons which appear to be very important in this particular aquatic fulvic acid.

TABLE 2

C-13 NMR chemical shift regions.

Region	Carbon Type	% Contribution
I (0 to 50 ppm)	methyl methylene methine (etc.)	29
II (50 to 110 ppm)	alcohol amine carbohydrate ether methoxyl acetal (etc.)	22
III (110 to 160 ppm)	olefinic aromatic phenolic	21
IV (160 to 220 ppm)	carboxyl ester amide aldehyde ketone	24

Region II, from 50 to 110 ppm, is assigned to aliphatic carbons substituted with heteroatoms such as oxygen. Signals from methoxyl carbons appear at approximately 55 ppm and a distinct shoulder may be noted in this region. Methoxyl carbons are often taken as indicators of lignin content due to their importance in lignin structure. The intense resonance centered at around 76 ppm with maxima at 74.3 and 78.8 ppm is assigned to carbohydrates and/or aliphatic alcohols. The presence of carbohydrates cannot be confirmed since a

---

[*]Singletary Lake is a freshwater lake in eastern North Carolina (USA).

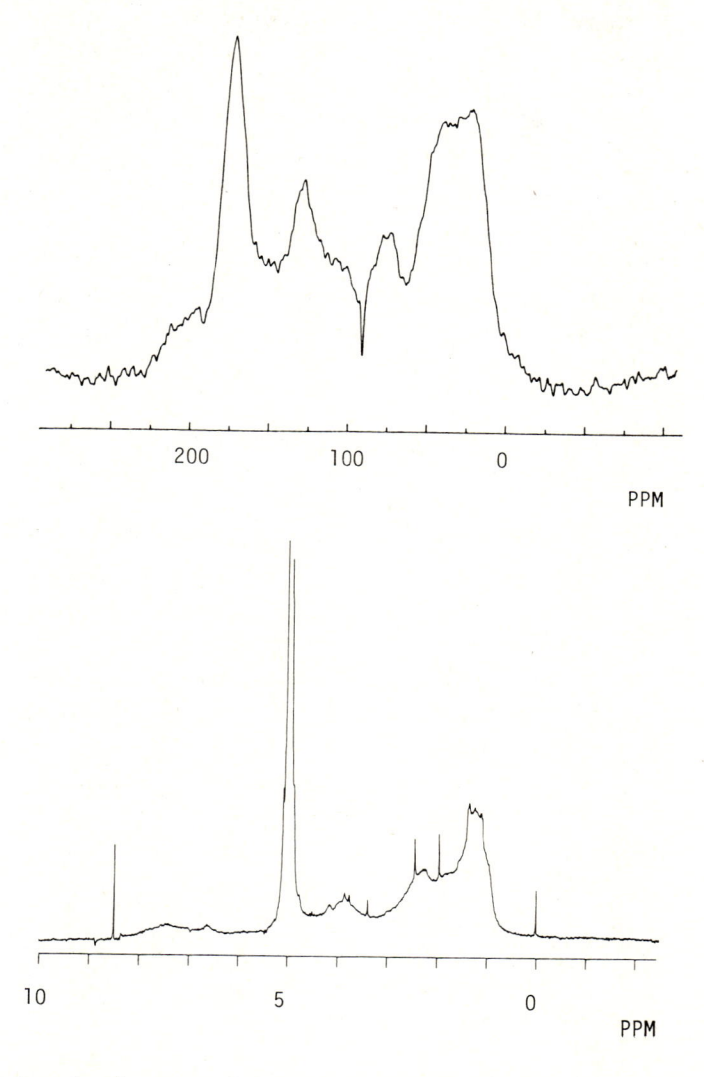

Fig. 2. CP/MAS (top) and solution proton (bottom) NMR spectra of Singletary Lake fulvic acid. The proton spectrum was acquired in D_2O/NaOD solution.

distinct peak for the anomeric carbon of polysaccharides (105 ppm) is not observed.

Region III, from 110 to 160 ppm, is assigned to aromatic and olefinic carbons. Olefinic carbons are generally not thought to be important in fulvic acid structures, therefore the resonance centered at 128.5 ppm is most likely due to aromatic carbons not substituted with heteroatoms. Oxygen substituted aromatic carbons of phenols should show signals centered at approximately 150 ppm. Note that a reasonably intense shoulder but no distinct peak is apparent at 150 ppm in the fulvic acid spectrum.

Region IV, from 160 to 220 ppm, is assigned to carboxyl, amide, ester, aldehyde and ketone carbons. The intense resonance centered at 173.4 ppm is probably due to carboxyl carbons although it is impossible to separate the ester contribution. Aldehyde and ketone carbons (190 to 220 ppm) are also present.

The second step in spectral analysis is to calculate the relative contributions by peak area integration of carbon atoms in each of the four spectral regions to the fulvic macromolecular structure, even though it must be stated that quantitative use of CP/MAS spectra is controversial. The relative contributions shown in Table 2 suggest a relatively low degree of aromaticity, i.e. 21%, or approximately three aromatic rings per 100 carbon atoms.

A typical proton spectrum of Singletary Lake fulvic acid also appears in Figure 2. Our interpretations have been influenced by Wilson's (1981) summary of chemical shift assignments for humic materials. Chemical shift assignments from Figure 2 are summarized in Table 3, and like the CP/MAS C-13 spectrum can be divided into four principal regions. Region I, from 0.4 to 1.7 ppm, includes protons on terminal methyl groups of methylene chains (0.8 - 1.0), methylene and methine protons (1.0 - 1.4), and methylene protons of alicyclic compounds (1.4 - 1.7). All of these protons should be at least two carbon atoms removed from aromatic rings or other polar functional groups. Peak area measurements indicate that these represent 38.3% of the total non-exchangeable proton which correlates well with Region I of the CP/MAS spectrum.

Region II, from 1.7 to 3.3 ppm, includes protons of methyl and methylene groups alpha to aromatic rings or carboxylic acid groups, and alpha and beta protons of indanes and tetralins. This region includes 36.3% of the non-exchangeable protons and is suggestive of a high degree of aromatic substitution and branching. The carbons bearing these protons would not be resolved from Region I in the CP/MAS spectrum.

Region III, from 3.3 to 4.6 ppm, represents protons on carbon attached to oxygen groups. It is usually interpreted as being indicative of carbohydrate content in humic material and corresponds to Region II in the CP/MAS spectrum. Table 3 indicates that 16.2% of the non-exchangeable proton is of this variety. The large peak centered at approximately 5 ppm in the proton spectrum is due to HOD and residual water.

Region IV, from 6.5 to 8.1 ppm, represents protons bound to aromatic rings. It is important to note that these include only 9.2% of the non-exchangeable proton whereas 21.0% of the carbon is probably present in aromatic rings. This result strongly suggests that aromatic rings present are heavily substituted. Olefinic protons should be observed in the 5.0 to 6.5 ppm region and although the water peak prevents the result from being conclusive, few if any appear to be present.

TABLE 3

Proton NMR chemical shift regions.

Regions	Proton Type	% Contribution
I (0.4 to 1.7 ppm)	methyl methylene methine (etc.)	38
II (1.7 to 3.3 ppm)	methyl methylene (α to aromatic rings or carboxyl groups) α and β protons of indanes and tetralins	36
III (3.3 to 4.6 ppm)	protons on carbons α to oxygen	16
IV (6.5 to 8.1 ppm)	aromatic	9

The CP/MAS spectrum of the humic acid fraction from Singletary Lake shows interesting differences from that of the fulvic acid fraction in the carboxyl and aromatic regions. The fulvic acid fraction appears relatively more acidic and less aromatic than the humic acid fraction. This observation correlates well with the observed physical properties (i.e., relative acid-base solubility). Similarly, it should be noted that the CP/MAS spectra for fulvic acids from two different sources are strikingly similar (Thurman and Malcolm, 1983).

It is attractive to ask an important experimental question, i.e., can phenolic moities present in natural polymers be observed with the CP/MAS technique. Figure 3 shows the CP/MAS spectra for a lignin sample from a species of woody tissue and it is apparent that the phenolic character of this natural polymer shows strongly in the 150-160 ppm and 50-60 ppm regions. However, these regions show only indistinct shoulders in the CP/MAS spectrum of Singletary Lake fulvic acid making it attractive to assume that lignin derived phenols constitute a relatively small part of the fulvic acid molecular structure.

Additional structural information can be obtained through the use of alternative pulse sequences with the CP/MAS technique. Wilson, et al. (1983) for example, have shown that it is possible to obtain spectra free of signals from CH and CH_2 protonated carbons. This is accomplished by exploiting

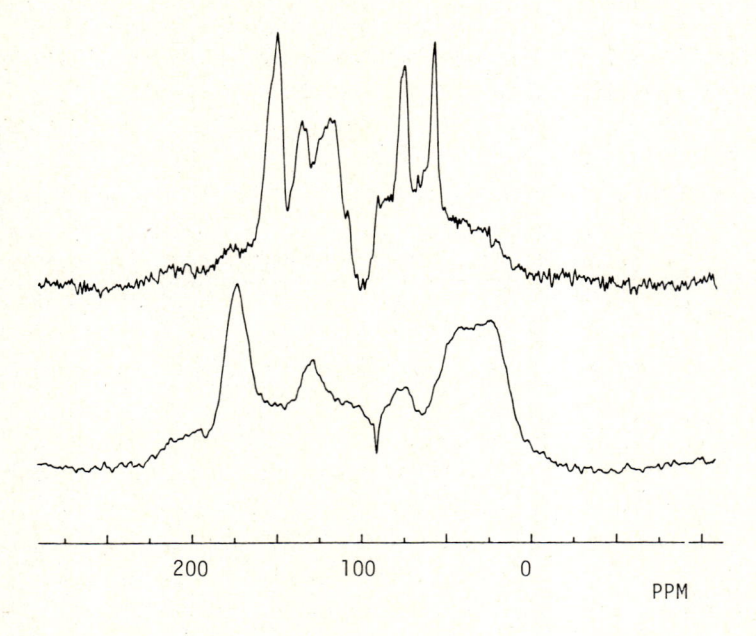

Fig. 3. CP/MAS spectra of a lignin (top) and Singletary Lake fulvic acid (bottom).

relaxation rate differences between carbon atoms under a variety of proton/C-13 dipolar interactions with an instrumental method known as "dipolar dephasing." With the exception of signals from groups which behave as if they were non-protonated (methyl and t-butyl methyl groups) the spectra show signals only from carbon which is actually non-protonated. These include carbon in carboxyl and amide groups, non-protonated aromatic carbon (e.g., phenol, arylether, bridgehead and substituted carbon), O-alkyl carbon and other non-protonated aliphatic carbon. Preliminary studies on the Singletary Lake fulvic acid utilizing the dipolar dephasing method indicate that small phenolic signals can be observed in the aromatic region once signals from protonated aromatic carbons are removed

It is worth noting that CP/MAS C-13 NMR is a technique currently in a rapid state of development, and the interpretations offered here may need to be revised as a result of future developments. For instance, the influence of free radicals and isolated paramagnetic centers on the measurements is presently difficult to evaluate.

Another way to demonstrate the presence of phenolic moities in aquatic fulvic acid is to employ milder degradative techniques which would permit identification of simpler phenolic fragments in the product mixture. Earlier degradation work with KMnO4 (Liao et al., 1982) did not show strong evidence of phenolic products. Some reports of chlorine degradation have shown chlorinated phenolic products (Kringstad et al., 1985) and some reports have failed (Christman et al., 1983) to show these products.

In our laboratory we have recently employed alkaline copper oxide degradation of aquatic fulvic acid using the method of Hedges and Ertel (1982). Figure 4 shows flame ionization detector (FID) traces of ether extractable CuO products from Singletary Lake fulvic acid. Also shown (bottom trace) is a chromatogram of a mixture of various phenols, structures of which are shown in Figure 5. Products identified and quantified are listed in Table 4 and additional identified products are listed in Table 5.

These data show that a wide range of phenols can indeed be produced from CuO degradation of aquatic fulvic acid, and that some of these phenols are lignin derived. Overall product yields are relatively small on a total carbon basis (Table 4) but several of the phenols dominate the product mixture. Of particular interest is the apparent abundance of 3,5-dihydroxybenzoic acid which has been shown repeatedly (Rook, 1980; Boyce and Hornig, 1983; Norwood et al., 1980) to produce chloroform in great yield upon aqueous chlorination. Other structures in Table 4 have also been shown to be reactive with chlorine (Norwood et al., 1980).

These findings coupled with the NMR data suggest that phenolic moities are present in aquatic fulvic acid in small amounts relative to total carbon. However, it must be noted that the m-dihydroxy moities are extremely reactive to chlorine and that the yields of chloroform from aquatic fulvic acids is relatively small on a total carbon basis.

CORRELATION OF BY-PRODUCTS WITH MUTAGENIC ACITIVITY

Chlorination by-products from natural waters (Cheh et al., 1980; de Greef et al., 1980; Horth et al., 1985) and from humic materials (Bull et al., 1982; Coleman et al., 1984) have been shown to be mutagenic and carcinogenic. It has, however, been difficult to correlate the products found from these laboratory and field experiments with the levels of mutagenic activity measured and the levels of activity in the chlorinated samples have been much higher than in the identified products. Recent evidence has shown that sample pretreatment and separation methods both in the initial isolation and final product identification steps can significantly decrease the mutagenic activity of the sample.

Cheh et al. (1980), and Wilcox and Denny (1984) have shown that a majority but not all mutagenic activity is destroyed in chlorinated samples treated with sulfite, thiosulfate and other reducing agents. This and other current literature suggests that an important group of mutagenic compounds is being destroyed by reducing agents used for dechlorination. Oxidants containing chlorine which

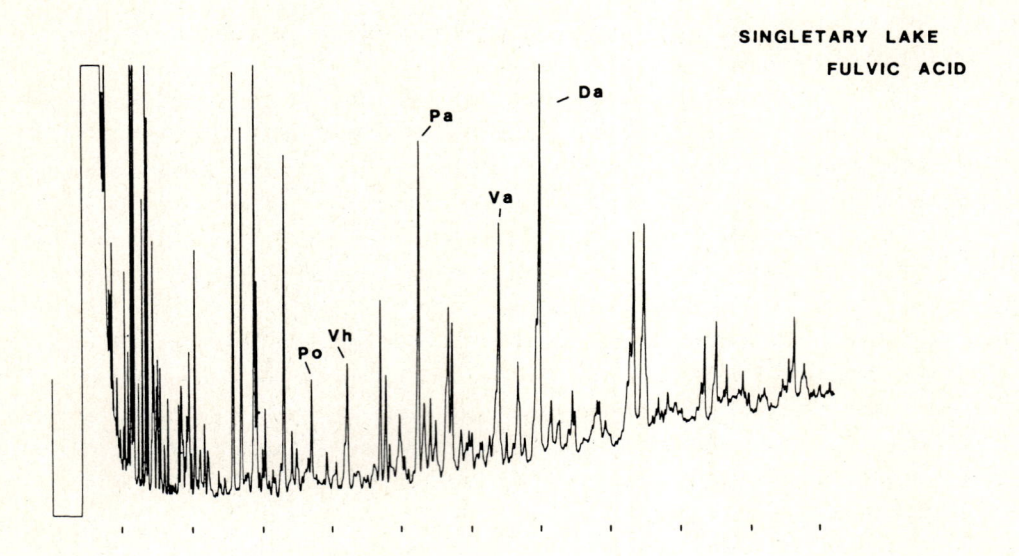

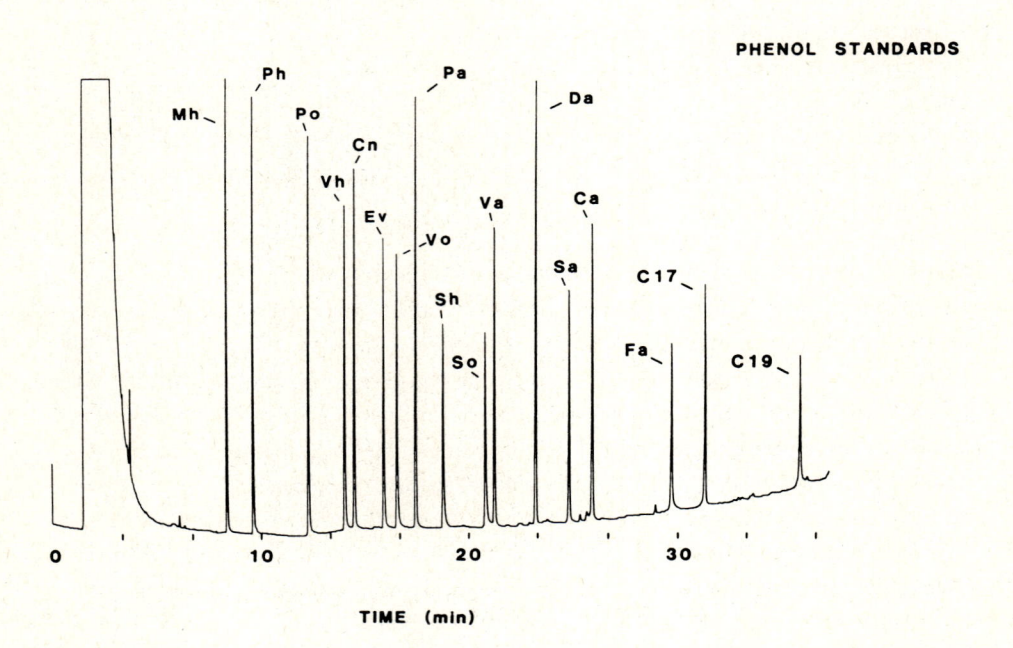

Fig. 4. Gas chromatograms of the ether extractable CuO products (as trimethyl-silyl derivatives) from Singletary Lake fulvic acid (top) superimposed on a mixture of standard compounds (bottom). Flame ionization detection was utilized.

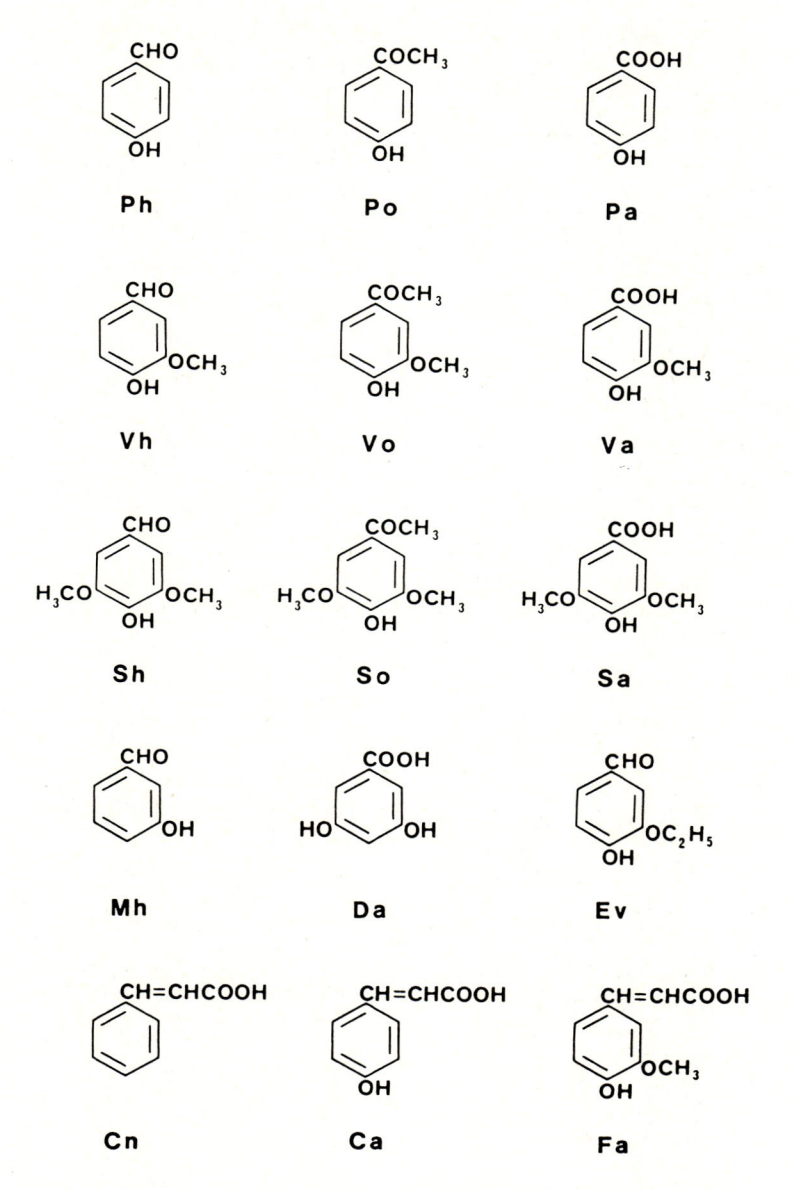

Fig. 5. Phenolic structures examined with the copper oxide degradation technique.

TABLE 4

[a]CuO oxidation products from Singletary Lake fulvic acid with yield measurements.

compound	yield (mg/gC)[b] A	B
p-hydroxybenzaldehyde (Ph)[c]	2.41[d]	1.45[d]
p-hydroxyacetophenone (Po)	0.433[d]	0.268[d]
p-hydroxybenzoic acid (Pa)	1.11[d]	0.694[d]
vanillin (Vh)	0.823[d]	0.484[d]
acetovanillone (Vo)[c]	trace[d]	0.292[d]
vanillic acid (Va)	0.650[d]	0.825[d]
syringaldehyde (Sh)[c]	0.212[d]	0.127[d]
acetosyringone (So)[c]	0.247[d]	0.174
syringic acid (Sa)[c]	0.298[d]	0.105[d]
trans-p-coumaric acid (Ca)[c]	0.151[d]	0.124
trans-ferulic acid (Fa)[c]	trace	0.124
3,5-dihydroxybenzoic acid (Da)	1.22[d]	0.936[d]
m-hydroxybenzaldehyde (Mh)	trace[d]	------
trans-cinnamic acid (Cn)	trace[d]	------
ethyl vanillin (Ev)	i.s.[d]	i.s.[d]
heptadecanoic acid (C17)	------	------
nonadecanoic acid (C19)	------	------

[a] based on work of Hedges, J.I. and J.R. Ertel, Anal. Chem., 54(2), pp. 174-178 (1982).
[b] $FA_o(A)$ = 59.3mg, $FA_o(B)$ = 61.5mg.
[c] significant error in yield measurement due to coeluting components.
[d] component identification confirmed by comparison of electron ionization mass spectrum and gas chromatographic retention index with those of an authentic standard.

TABLE 5

[a]Other identified CuO oxidation products of Singletary Lake fulvic acid.

---

lactic acid
alpha-hydroxyisobutyric acid
levulinic acid
3,5-dimethylphenol
benzoic acid
succinic acid
fumaric acid
resorcinol
salicylic acid
m-hydroxybenzoic acid
phthalic acid
isophthalic acid
terephthalic acid

---

[a]component identification confirmed by comparison of electron ionization mass spectrum and gas chromatographic retention index with those of an authentic standard.

are present in natural waters are generally thought to be chloramines. Nearly all the products of chlorination identified to date contain chlorine-carbon substitution, which is stable to reducing agents. Emphasis has also been placed on compounds which are relatively small and volatile even though recent evidence suggests that many mutagenic compounds isolated from water supplies may be polar and nonvolatile (Zoeteman et al., 1982; Kool et al., 1982). Ram and Malley (1984) along with other workers have shown that organic nitrogen compounds as well as ammonia form chlorine nitrogen substitution in which the oxidizing power of chlorine is retained by the structure, and it has also been shown that organic nitrogen compounds are abundant in municipal water supplies (Ram and Morris, 1980; Ram and Morris, 1981).

ANALYSIS OF POLAR BY-PRODUCTS AND N-CHLOROORGANIC COMPOUNDS

Owing to the foregoing findings, the development of methods of analysis for polar by-products and N-chloroorganic compounds is an important research frontier. The coupling of high performance liquid chromatography (HPLC) and mass spectroscopy (LC/MS) is leading to new opportunities for identification of previously intractable and unidentified by-products of chlorination such as high molecular weight, polar compounds. Crathorne et al. (1984) separated such compounds in water into HPLC fractions and analyzed them by fast atom bombardment (FAB) and field desorption (FD) ionization methods with a double-focusing mass spectrometer. New methods are now being used which allow directly coupled LC/MS, such as the moving belt and thermospray interfaces. The analysis of N-chloroorganic compounds presents an additional problem in that many of these compounds are not only polar and of high molecular weight but are also oxidants. Recently, Scully et al. (1984) have derivatized organic and inorganic N-chloramines for HPLC analysis of chlorinated municipal water and wastewater. The products formed are stable sulfonamides from the displacement of the chlorine from the N-chloro group by sulfinic acid derivatives. The use of HPLC separations especially with mass spectroscopy should provide much needed new identification and eventually quantification of the mutagenic products produced by chlorination.

REFERENCES

Boyce, S.D. and Hornig, J.F., 1983. Reaction pathways of trihalomethane formation from the halogenation of dihydroxy model compounds for humic acid. Environ. Sci. Technol., 17:202-211.

Bull, R.J., Robinson, M., Meier, J.R. and Stober, J., 1982. Use of biological assay systems to assess the relative carcinogenic hazards of disinfection by-products. Environ. Health Perspect., 46:215-224.

Cheh, A.M., Skochdopole, J., Koski, P. and Cole, L., 1980. Non-volatile mutagens in drinking water: production by chlorination and destruction by sulfite. Science, 207:90-92.

Christman, R.F., Norwood, D.L., Millington, D.S., Johnson, J.D. and Stevens, A.A., 1983. Identity and yields of major halogenated products of aquatic fulvic acid chlorination. Environ. Sci. Technol., 17:625-628.

Christman, R.F., Norwood, D.L., Seo, Y. and Frimmel, F.H., 1984. Oxidative degradation of aquatic humic material. presented at the 2nd International Humic Substances Society Meeting, Birmingham, England.

Coleman, W.E., Munch, J.W., Kaylor, W.H., Streicher, R.F., Ringhand, H.P. and Meier, J.R., 1984. Gas chromatography/mass spectroscopy analysis of mutagenic extracts of aqueous chlorinated humic acid. a comparison of byproducts to drinking water contaminants. Environ. Sci. Technol., 18:674-681.

Crathorne, B., Fielding, M., Steel, C.P. and Watts, C.D., 1984. Organic compounds in water: analysis using coupled-column high-performance liquid chromatography and soft-ionization mass spectrometry. Environ. Sci. Technol., 18:797-802.

de Greef, E., Morris, J.C., van Kreijl, C.F. and Morra, C.F.H., 1980. Health effects in the chemical oxidation of polluted waters. In: R.L. Jolley, W.A. Brungs and R.B. Cumming (Editors), Water Chlorination: Environmental Impact and Health Effects, Vol. 3. Ann Arbor Science Publishers, Ann Arbor, MI, pp. 913-924.

de Leer, E.W.B., Damste, J.S.S., Erkelens, C. and de Galan, L., 1985. Identification of intermediates leading to chloroform and C-4 diacids in the chlorination of humic acid. Environ. Sci. Technol., 19:512-522.

Fukushima, E. and Roeder, J.B.W., 1981. Experimental Pulse NMR, A Nuts and Bolts Approach. Addison-Wesley, Reading, MA, 539pp.

Hatcher, P.G., Berger, I.A., Dennis, L.W. and Maciel, G.E., 1983. Solid-state C-NMR of sedimentary humic substances: new revelations on their chemical composition. In: R.F. Christman and E.T. Gjessing (Editors), Aquatic and Terrestrial Humic Materials. Ann Arbor Science Publishers, Ann Arbor, MI, pp. 37-81.

Hedges, J.I. and Ertel, J.R., 1982. Characterization of lignin by gas capillary chromatography of cupric oxide oxidation products. Anal. Chem., 54:174-178.

Horth, H., Crathorne, B., Gwillian, R.D., Stanley, J.A., Steel, C.P. and Thomas, M.J., 1985. Techniques for the fractionation and identification of mutagens produced by water treatment chlorination. In: I.H. Suffet and M. Malayandi (Editors), Advances in Sampling and Analysis of Organic Pollutants from Water, Volume 2: Toxicity Testing/Analysis Interface. American Chemical Society, Washington, D.C., in press.

Johnson, J.D., Christman, R.F., Norwood, D.L. and Millington, D.S., 1982. Reaction products of aquatic humic substances with chlorine. Environ. Health Perspect., 46:63-71.

Kool, H.J., van Kreijl, C.F., de Greef, E. and van Kranen, H.J., 1982. Presence, introduction and removal of mutagenic activity during the preparation of drinking water in the Netherlands. Environ. Health Perspect., 46:207-214.

Kringstad, K.P., de Sousa, F. and Stromberg, L.M., 1985. Studies on the chlorination of cholorolignins and humic acid. Environ. Sci. Technol., 19:427-431.

Levy, G.C., Lichter, R.L. and Nelson, G.L., 1980. Carbon-13 Nuclear Magnetic Resonance Spectroscopy (Second Edition). John Wiley and Sons, New York, 338 pp.

Liao, W.T., Christman, R.F., Johnson, J.D., Millington, D.S. and Hass, J.R., 1982. Structural characterization of aquatic humic material. Environ. Sci. Technol., 16:403-410.

Malcolm, R.L., 1984. The geochemistry of stream fulvic and humic substances. In: Humic Substances: I. Geochemistry, Isolation and Characterization. Wiley and Sons, New York.

Norwood, D.L., Johnson, J.D., Christman, R.F., Hass, J.R. and Bobenrieth, M.J., 1980. Reactions of chlorine with selected aromatic models of aquatic humic material. Environ. Sci. Technol., 14:187-190.

Norwood, D.L., Christman, R.F., Johnson, J.D. and Hass, J.R., 1984. Determination of aqueous trichloroacetic acid utilizing isotope dilution mass spectrometry. J.-Am. Water Works Assoc., in press.

Ram, N.M. and Malley, J.P., 1984. Chlorine residual monitoring in the presence of N-organic compounds. J.-Am. Water Works Assoc., 76:74-81.

Ram, N.M. and Morris, J.C., 1980. Environmental significance of nitrogenous organic compounds in aquatic sources. Environment International, 4:397-405.

Ram, N.M. and Morris, J.C., 1981. Identification of nitrogenous organic compounds in aquatic sources by stopped-flow spectral scanning techniques. Journal of Liquid Chromatography, 4:791-811.

Rook, J.J., 1980. Possible pathways for the formation of chlorinated degradation products during chlorination of humic acids and resorcinol. In: R.L. Jolley, et al. (Editors), Water Chlorination: Environmental Impact and Health Effects, Vol. 3. Ann Arbor Science Publishers, Ann Arbor, MI, pp. 85-98.

Scully, F.E., Yang, J.P., Mazina, K. and Daniel, F.B., 1984. Derivatization of organic and inorganic N-chloramines for high-performance liquid chromatographic analysis of chlorinated water. Environ. Sci. Technol., 18:787-792.

Thurman, E.M. and Malcolm, R.L., 1983. Structural study of humic substances: new approaches and methods. In: R.F. Christman and E.T. Gjessing (Editors), Aquatic and Terrestrial Humic Materials. Ann Arbor Science Publishers, Ann Arbor, MI, pp. 2-23.

Uden, P.C. and Miller, J.W., 1983. Chlorinated acids and chloral in drinking water. J.-Am. Water Works Assoc., 75:524-527.

Wachter, J.K. and Andelman, J.B., 1984. Organohalide formation on chlorination of algal extracellular products. Environ. Sci. Technol., 18:811-817.

Wilcox, P. and Denny, S., 1984. The effect of dechlorinating agents on the mutagenic activity of chlorinated water samples. presented at the 5th Conference on Water Chlorination: Environmental Impact and Health Effects, Williamsburg, VA.

Wilson, M.A., 1981. Applications of nuclear magnetic resonance spectroscopy to the study of the structure of soil organic matter. J. Soil Sci., 32:167-186.

Wilson, M.A., Pugmire, R.J. and Grant, D.M., 1983. Nuclear magnetic resonance spectroscopy of soils and related materials. Relaxation of C-13 nuclei in cross polarization nuclear magnetic resonance experiments. Org. Geochem., 5:121-129.

Zoeteman, B.C.J., Hrabec, J., de Greef, E. and Kool, H.J., 1982. Mutagenic activity associated with by-products of drinking water disinfection by chlorine, chlorine dioxide, ozone and UV-irradiation. Environ. Health Perspect., 46:197-205.

CHLOROFORM PRODUCTION FROM MODEL COMPOUNDS OF AQUATIC HUMIC MATERIAL THE ROLE OF PENTACHLORORESORCINOL AS AN INTERMEDIATE

ED W.B. DE LEER and CORRIE ERKELENS
Laboratory for Analytical Chemistry, Delft University of Technology, Jaffalaan 9, 2628 BX Delft (The Netherlands)

ABSTRACT

The chlorination of pentachlororesorcinol was studied in view of its postulated role as an intermediate in the chlorination of resorcinol. Chlorination of resorcinol and pentachlororesorcinol produced several identical products, but the large differences in reaction rate, chloroform production, and products formed, point to a minor role of pentachlororesorcinol.

INTRODUCTION

Humic and fulvic acids are widely held responsible for the production of chloroform and other chlorinated compounds on chlorination of drinking water. Rook (ref.1) proposed that chloroform was produced from 1,3-dihydroxybenzene structures from the aromatic part of humic acid, but later other favourable precursors such as citric acid (ref.2), uracil (ref.3), and mannitol (ref.4) have been identified.

Recently, De Leer et al. (ref.5) identified a number of products on chlorination of terrestrial humic acid, which contained a trichloromethyl group. These compounds supported the hypothesis of Rook, because several of them were produced also on chlorination of 3,5-dihydroxybenzoic acid, a 1,3-dihydroxybenzene model compound of humic acid.

The mechanism for the production of chloroform from 1,3-dihydroxybenzene compounds is still under debate. Originally, Rook (ref.1,6) proposed a mechanism based on the production of pentachlororesorcinol (PCR) as an intermediate. (1)

$$\text{resorcinol} \xrightarrow{Cl_2} \text{intermediate} \longrightarrow {}^*CHCl_3 + \text{OTHER PRODUCTS} \tag{1}$$

Later (ref.7), he speculated on a different pathway, which included a radical-anion and 2,2-dichlorocyclohex-5-ene-1,3-dione as the key intermediates. However, the detection of the non polar reaction product 3,5,5-trichlorocyclopent-3-ene-

-1,3-dione (ref.6,8), was believed to be inconsistent with the mechanism proposed by Rook. The PCR postulation however, was supported by Boyce and Hornig (ref. 9) who explained the production of a number of reaction products on the basis of the PCR intermediate. To explain the production of other chlorination products they alternatively proposed tetrachlororesorcinol as an intermediate.

In this study, we tried to demonstrate the proposed role of PCR as an intermediate by studying its chlorination behaviour after an independent synthesis.

METHODS
General

PCR was synthesized according to Zincke and Rabinowitsch (ref.10). Other chlorinated resorcinols were synthesized according to known procedures or obtained commercially. The experimental details of the chlorination reactions of model compounds, the analytical procedures used for the GC and GC/MS analyses, and the determination of the yield of chloroform and the residual amount of chlorine have been described before (ref.5).

CO_2 determination

20 μL samples of the dechlorinated reaction mixture were injected in a tube oven at 150°C, filled with H_3PO_4 on glass. The CO_2 produced was measured with a non dispersive IR analyser. Corrections were made for CO_2 from the reagents.

Chloride determination

The reaction mixture was acidified to pH 2 with concentrated HNO_3 and N_2 was bubbled through the solution to eliminate the residual free chlorine. Chloride was then titrated with $AgNO_3$ with a Cl^- indicating electrode. Corrections were made for Cl^- from the reagent solutions.

Residual resorcinol determination

Residual resorcinol was determined by HPLC on a reversed phase C-18 column with H_2O-methanol (80:20) + 10^{-3} M H_3PO_4 as the eluent. UV detection was used.

RESULTS AND DISCUSSION

The synthesis of PCR, already described in 1890, presented no special problems. The product was shown to be pure by capillary GC and its EI mass spectrum showed a small molecular ion at m/z 280 (2%) and major fragments at m/z 170 (3Cl, 100%), 245 (4Cl, 75%), 217 (4Cl, 31%), 189 (4Cl, 22%), 142 (3Cl, 33%), and 107 (2Cl, 40%). Most fragments can be explained by successive eliminations of CCl_2=C=O, C=O, and/or Cl.

The chlorination of resorcinol was studied in more detail before starting on PCR. The results from the measurement of the chlorine demand (ref.11), the production of chloride (ref.7), chloroform (ref.8), and CO_2, and the consumption of resorcinol as a function of the chlorine dose are given in Fig. 1.

LEFT SCALE
- residual resorcinol (M)
- ▲ CO_2 production (M/M)
- ■ $CHCl_3$ production (M/M)

RIGHT SCALE
- ○ chlorine consumption (M/M)
- □ chloride production (M/M)

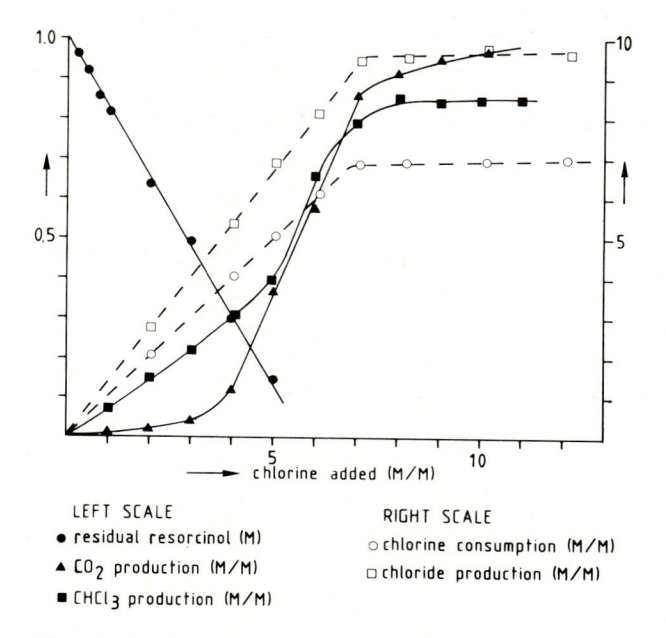

Fig. 1. Product formation and chlorine consumption on chlorination of resorcinol.

The results can be summarized as follows:
1. The ratio between the chlorine demand and the chloride production was independant of the chlorine dose. The large ratio (1:1.4) showed that chlorine is not used for substitution only (ratio 1:1), but also for oxidation.
2. 1 mole of CO_2 is produced when an excess (> 7 moles) of Cl_2 is used.
3. When resorcinol starts to react with chlorine, a series of rapid reactions occurs consuming 5.8 moles of Cl_2 per mole of resorcinol.
4. The production of $CHCl_3$ and CO_2 shows no linear relation with the Cl_2 dose, demonstrating that chlorinated intermediates must be produced which are slowly converted into $CHCl_3$ and CO_2.

Several investigators (ref.8,9,11) detected chlorinated resorcinols as intermediates. We studied the production of $CHCl_3$ and CO_2 of these intermediates and showed that up to 2,4,6-trichlororesorcinol the final $CHCl_3$ and CO_2 yield was the same as for resorcinol (see Table 1). However, when PCR was chlorinated the $CHCl_3$ yield was only 28% after 2 hrs and 45% after 22 hrs. Apparently, PCR was capable of producing $CHCl_3$ on chlorination as proposed by Rook, but the reaction rate and the final $CHCl_3$ yield was much lower than for resorcinol, excluding it as a major intermediate.

TABLE 1

Yield of chloroform and carbondioxide on chlorination of resorcinol derivatives

Compound	CHCl$_3$ yield (M/M)	CO$_2$ yield (M/M)
Resorcinol	0.78	1.02
4-Chlororesorcinol	0.76	1.02
2,4-Dichlororesorcinol	0.83	1.07
2,4,6-Trichlororesorcinol	0.78	-
Pentachlororesorcinol	0.28	-
Pentachlororesorcinol	0.45	0.95

Chlorine 9M/M; Reaction time 2 hrs; pH 7.2

Another striking difference was observed when we studied the other chlorination products with GC and GC/MS. The gaschromatogram of the chlorination products of resorcinol is given in Fig. 2. Several identical products were detected, but the differences outnumbered the similarities.

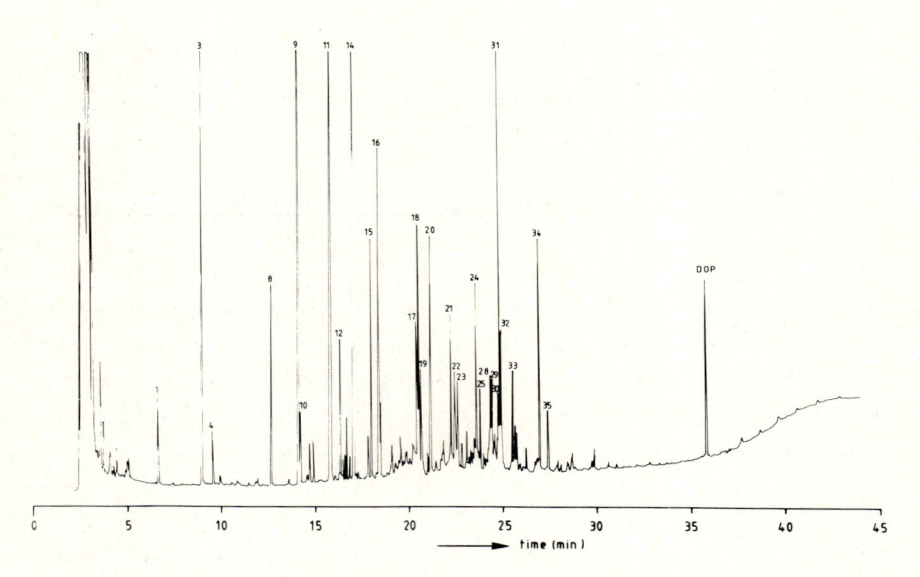

Fig. 2. Chlorination products of resorcinol. The peak numbers correspond with the compound numbers given in Table 2.

Most products were identified with GC/MS, recording both the electron impact and the chemical ionization mass spectra. The results are given in Table 2. The major chlorination product of resorcinol was shown to be 2-chlorobutenedioic acid (9) (ref.12), together with 3 isomers of chloropentenedioic acid (11,12,14). PCR produced a cyclopentenonecarboxylic acid (27) as the major reaction product,

TABLE 2

Major reaction products produced on chlorination of resorcinol (RES) and pentachlororesorcinol (PCR)

Compound Nr[1]	Assigned structure[2]	Confidence level[3]	Chlorination product of: RES	PCR
1	$CHCl_2-COOCH_3$	SC	+	++
2	$C_2H_3Cl_2-COOCH_3$	CT	-	+
3	$CCl_3-COOCH_3$	SC	++	+++
4	$CCl_2=CH-COOCH_3$	CT	+	+
5/6	$CHCl=CCl-COOCH_3$	SC	-	++
7	$CCl_3-CH_2-COOCH_3$	CT	-	+
8	$C_3H_3Cl_2-COOCH_3$	CT	+	-
9	$H_3COOC-CH=CCl-COOCH_3$	SC	+++	++
11/12/14	$H_3COOC-(C_3H_3Cl)-COOCH_3$	CT	+++	+
13	$H_3COOC-CCl=CCl-COOCH_3$	SC	-	++
15	$C_4H_4OCl_3-COOCH_3$	TE	+	-
16	$H_3COOC-(C_3H_3OCl)-COOCH_3$	CT	++	-
17/18	$H_3COOC-(C_3H_3Cl_3)-COOCH_3$	CT	++	-
19/20	$H_3COOC-(C_3HCl_3)-COOCH_3$	CT	+	+
22	$C_4H_2OCl_5-COOCH_3$	CT	+	-
23	$H_3COOC-(C_3H_2Cl_4)-COOCH_3$	CT	+	-
24/25	$CCl_3-(C_4H_3O_2Cl)-COOCH_3$	CT	+	-
26/27	$C_6H_4O_2Cl_3-COOCH_3$	CT	-	+++
29	$CCl_3-(C_4H_4O_3)-COOCH_3$	CT	+	-
31	$CCl_3-(C_4H_2O_2Cl_2)-COOCH_3$	CT	+	-
32	$CCl_3-(C_4H_2O_4)-COOCH_3$	CT	+	-
33	$CCl_3-(C_4O_3Cl_2)-COOCH_3$	CT	+	-
DOP	Dioctylphthalate		-	-

[1] The numbers correspond with the numbers given in the gaschromatogram (see Fig.2)
[2] The products are present as the free acids after the chlorination reaction, but analyzed as their methyl esters and in some cases methyl ethers.
[3] SC=Standard Confirmed (EI+CI+Reference), CT=Confident (EI+CI), TE=Tentative (EI)

which was not detected after chlorination of resorcinol. The production of 27 and its isomer 26 may be explained by hydrolysis of PCR, followed by a nucleophylic ring opening and an intramolecular substitution (2,3).

(2)

(3)

CONCLUSIONS

From the large differences in chlorination behaviour between resorcinol and PCR, it can be concluded that the original proposal for the chlorination mechanism of resorcinol is not correct.

The predominant reaction and most important side reaction in the chlorination of resorcinol is given in equation 4 and 5, but many other side products including chloroform precursors and higly oxidized products are formed.

PREDOMINANT REACTION

$$\text{resorcinol} + 7Cl_2 + 4H_2O \longrightarrow CHCl_3 + CO_2 + \underset{H}{\overset{Cl}{C}}=\underset{COOH}{\overset{COOH}{C}} + 10 HCl \qquad (4)$$

SIDE REACTION

$$\text{resorcinol} \longrightarrow \underset{H}{\overset{Cl}{C}}=\underset{CH_2-COOH}{\overset{COOH}{C}} + CO_2 \text{ or } CHCl_3 \qquad (5)$$

REFERENCES

1. J.J. Rook, Environ. Sci. Technol., 11 (1977) 478-482.
2. R.A. Larson and A.L. Rockwell, Environ. Sci. Technol.,13 (1979) 325-329.
3. J.C. Morris and B. Baum, In "Water Chlorination: Environmental Impact and Health Effects"; R.L. Jolley et al. (Eds.), Ann Arbor Science Publishers: Ann Arbor, 1978; Vol. 2, pp 29-48.
4. A.M. Crane, P. Kovacic and E.D. Kovacic, Environ. Sci. Technol., 14 (1980) 1371-1374.
5. E.W.B. de Leer, J. Sinninghe Damste, C. Erkelens and L. de Galan, Environ. Sci. Technol. 19 (1985) 512-522.
6. J.J. Rook, Thesis Wageningen, 1978.
7. J.J. Rook, In "Water Chlorination: Environmental Impact and Health Effects"; R.L. Jolley et al. (Eds.), Ann Arbor Science Publishers: Ann Arbor, 1980; Vol. 3, pp 85-98.
8. R.F. Christman, J.D. Johnson, J.R. Hass, F.K. Pfaender, W.T. Liao, D.L. Norwood and H.J. Alexander, In "Water Chlorination: Environmental Impact and Health Effects"; R.L. Jolley et al. (Eds.), Ann Arbor Science Publishers: Ann Arbor, 1978, pp 15-28.
9. S.D. Boyce and J.F. Hornig, Environ. Sci. Technol. 17 (1983) 202-211.
10. T. Zincke and S. Rabinowitsch, Ber. Dtsch. Chem. Ges. 23 (1890) 3767-3784.
11. D.L. Norwood, J.D. Johnson, R.F. Christman, J.R. Hass and M.J. Bobenrieth, Environ. Sci. Technol. 14 (1980) 187-190.
12. J. de Laat, N. Merlet and M. Dore, Water Res. 16 (1982) 1437-1450.

CHLORINATION OF HUMIC SUBSTANCES IN AQUEOUS SOLUTION : YIELDS OF VOLATILE AND MAJOR NON-VOLATILE ORGANIC HALIDES

B. LEGUBE, J.P. CROUE and M. DORE

Laboratoire de Chimie de l'Eau et des Nuisances, Université de Poitiers, 40, avenue du Recteur Pineau, 86022 Poitiers Cedex (France)

ABSTRACT

In this paper, we present the study of the effect of chlorination contact time, of the chlorine/carbon ratio and of the bromide concentration on the major halogenated products in the case of chlorinated humic substance solutions at neutral pH. The yields of chloroform, of trichloroacetic acid and of dichloroacetic acid increase with the increase in reaction time and in chlorine dosage. Various values for the yield are given for chlorinated commercial humic acid solutions and for chlorinated solutions of aquatic fulvic and humic acids. The simultaneous presence of trihalomethanes precursors and bromides in the solution leads to the formation of organo-brominated products. Total trihalomethanes increase with the increase of bromide concentration.

INTRODUCTION

Each time surface waters are chlorinated, significant concentration of trihalomethanes (THM) and total organic halides (TOX) are formed. Among the THM precursors present in drinking water, humic and fulvic acids play a very significant role. Recent works (Christman et al., 1983 ; Reckhow and Singer, 1984) have shown that a major part of non-volatile TOX in the case of chlorinated surface waters and of chlorinated solutions of humic substances could be due by trichloroacetic acid (TCAA) and by dichloroacetic acid (DCAA). Moreover, a recent survey (Masschelein, 1982) indicated that the bromide concentration can vary from 10 $\mu g\ l^{-1}$ to 2 500 $\mu g\ l^{-1}$ in Western European countries. The simultaneous presence of THM precursors and of bromide in water submitted to chlorination, results in the formation of organo-brominated products

Our research is based on a comparative study of the chlorination of different humic substances.

- Two commercial humic acids (FLUKA and ALDRICH)
- Humic and fulvic acids extrated from the water of a lake in the "Landes" region of FRANCE.

The formation of organo-halogenated products (THM, TCAA, DCAA) was studied as a function of chlorine contact time, of chlorine dosage and of bromide concentration.

EXPERIMENTAL

Aquatic fulvic and humic acids were extrated as indicated, following the method employed by Thurman and Malcolm (1981). The elementary analyses of humic and of fulvic samples are given in table 1.

TABLE 1
Elementary analyses of humic acid and of fulvic samples.

SAMPLES	% C	% H	% N	% O
Aldrich humic acid	36.94	4.24	0.46	38.76
Fluka humic acid	38.23	4.45	0.43	34.45
Landes fulvic acid	53.42	5.31	1.10	29.98
Landes humic acid	46.64	4.04	1.75	/

Chlorinations were carried out under the following conditions
- 5 mg l^{-1} humic substance ; pH = 7.5 ; phosphate buffer (I = 0.01) ; ultra-pure water (Milli RO then Milli Q water system Millipore)
- 10 mg l^{-1} chlorine dose (or 1 to 16 mg l^{-1})
- 72 hours reaction time (or 10 minutes to 400 hours)
- 20°C in the obscurity

Residual chlorine was measured by the iodometric method. After quenching the residual chlorine by sodium arsenite, THM were extracted with pentane at neutral pH, DCAA and TCAA were extracted with ether at acidic pH.

THM extracts were analyzed by gas chromatography (3 % SP 1 000 ; Electron capture detector) and chloroacetic acid extracts were analyzed after methylation by diazomethane using the same GC system.

RESULTS AND DISCUSSION

Organic halogenated formation as a function of time

Figure 1 represents the formation of $CHCl_3$, TCAA and DCAA for a chlorinated "Landes" fulvic acid as a function of time.

The results show that the major chlorination end-products began to be formed very rapidly at the start of the reaction (10 to 30 minutes), continued to be formed significantly for about three days and then slowly during an indefinite period of time (more than 400 hours). The same results could be observed with the chlorination of Fluka and Aldrich humic acids (Table 2).

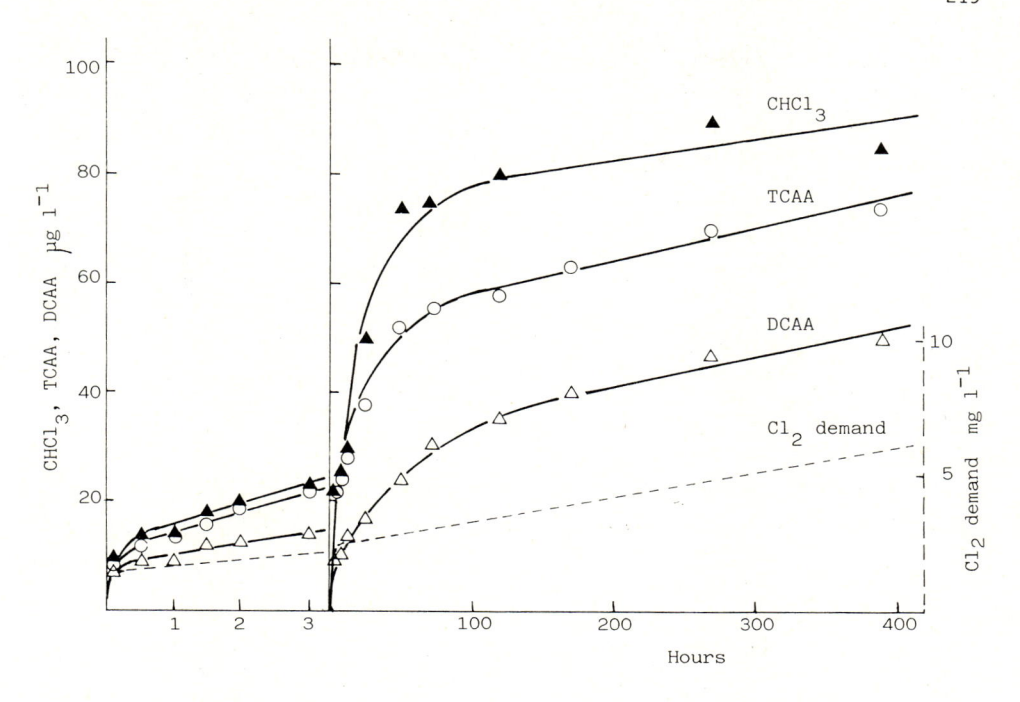

Fig. 1. Formation of $CHCl_3$, TCAA and DCAA as a function of time from the "Landes" fulvic acid.

TABLE 2

Formation of $CHCl_3$, TCAA and DCAA as a function of time (10 mg l^{-1} Cl_2). Yields in µg/mg TOC

TIME		Aldrich humic acid	Fluka humic acid	Landes fulvic acid
10 min	$CHCl_3$	23	35	3.5
	TCAA	30	14	3
	DCAA	5.5	5	3
72 h	$CHCl_3$	65	60	30
	TCAA	80	65	22
	DCAA	30	25	12
400 h	$CHCl_3$	76	73	33
	TCAA	114	90	28
	DCAA	54	53	19

These results seem to show that humic substances contain two or more THM precursor sites :
- THM precursors giving rapidly chloroform at neutral pH
- THM precursors reacting slowly with chlorine.

The latter could also be the chlorination intermediates of the former.

If we refer to the kinetic studies carried out on model compounds (De Laat et al., 1982) it would appear that, the part of the molecule of humic substance which reacts rapidly with chlorine is probably of an aromatic nature (i.e. resorcinol) while the slow formation of chloroform could be generated by the methyl ketone precursor sites (i.e. acetylacetone).

Effect of the chlorine/carbon ratio

The curves in figure 2 show the formation of $CHCl_3$, TCAA and DCAA for a chlorinated "Landes" fulvic acid solution against the chlorine dosage.

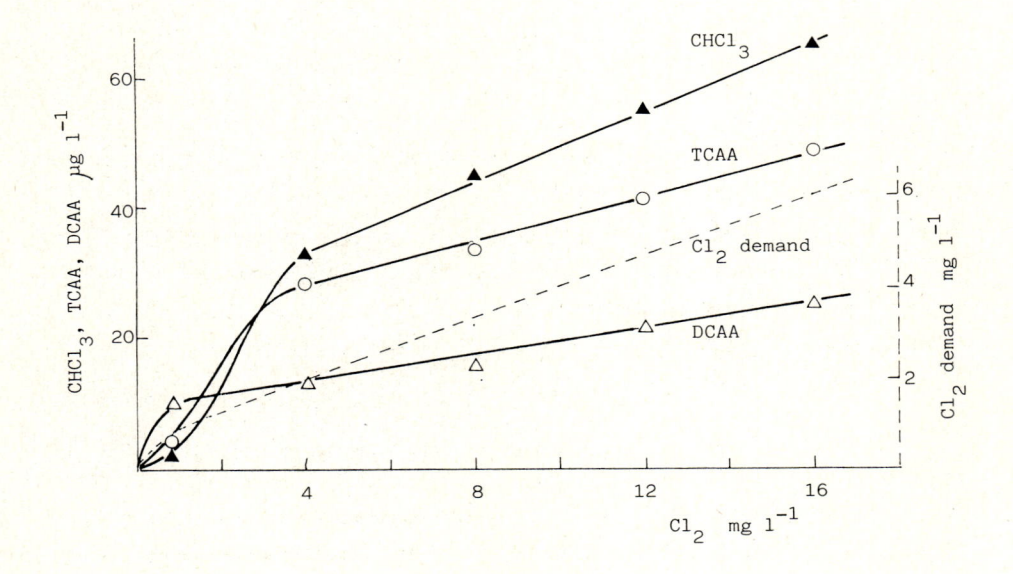

Fig. 2. Formation of $CHCl_3$, TCAA and DCAA as a function dosage from the "Landes" fulvic acid.

For a value of chlorine dosage of about 1.5 mg Cl_2/mg TOC, the slope of the curve decreases. The same observation could be made for the chlorination of the "Landes", Fluka and Aldrich humic acid solutions (table 3).

TABLE 3

Formation of $CHCl_3$, TCAA and DCAA as a function of the chlorine/carbon ratio (72 hours contact time). Yields in µg/mg TOC

mg Cl_2/ mg TOC		Aldrich humic acid	Fluka humic acid	Landes fulvic acid	Landes humic acid
0.5	$CHCl_3$	27	16	1	1
	TCAA	5.5	3	1.5	2
	DCAA	5.5	4	4	5
1	$CHCl_3$	47	30	10	9
	TCAA	20	15	9	10
	DCAA	12	15	5	7.5
3	$CHCl_3$	78	65	17	24
	TCAA	62	47	11	23
	DCAA	25	26	6.5	11

Effect of bromide concentration

Figure 3 shows the formation of THM, TCAA and DCAA for a chlorinated "Landes" fulvic acid solution as a function of the bromide concentration.

The similar change of chlorination end-products and the increase of total THM with the increase of bromide concentration could also be observed during the chlorination of Fluka and of Aldrich humic acid solutions.

The explanation of these results can be found in the literature :

- bromide is rapidly oxidized by chlorine ; $k_{HOBr} = 3.10 \ 10^3 \ l \ mol^{-1} \ s^{-1}$ at 20°C (Merlet et al., 1982)

- bromide reacts at the same rate as chlorine on model precursors like resorcinol and acetone (Merlet et al., 1982)

- the bromoform yield is greater than the chloroform yield at neutral pH for several aromatic and methyl ketone precursors with the exception of resorcinol (ref.6).

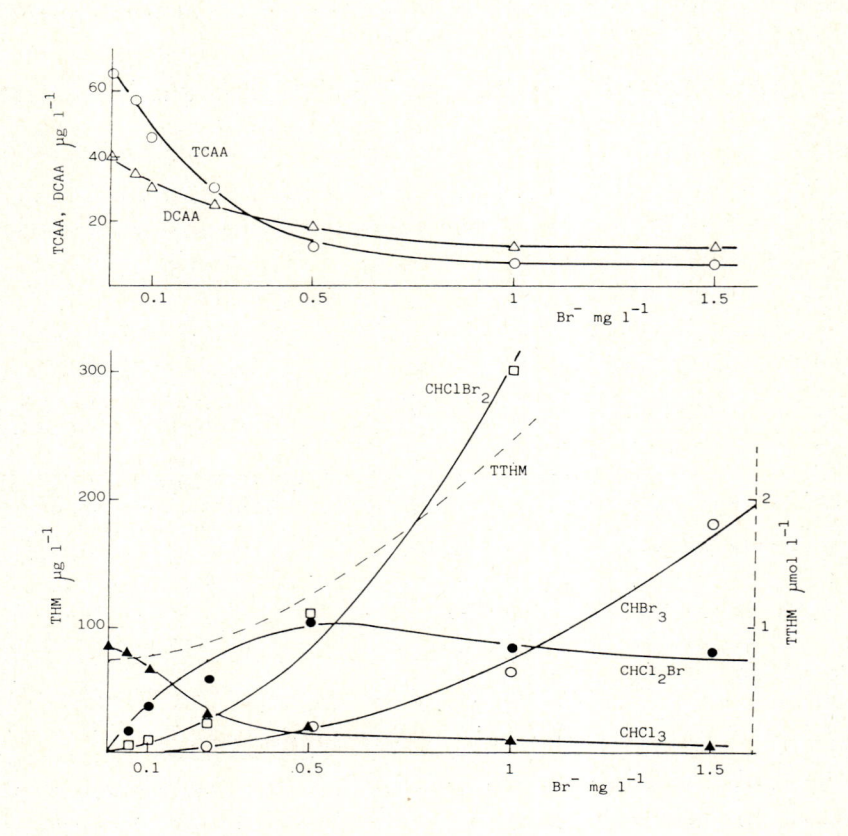

Fig. 3. Formation of THM, TCAA and DCAA as a function of the bromide concentration from "Landes" fulvic acid.

REFERENCES

Christman, R.F., Norwood, D.L., Millington, D.S., Johnson, J.D., and Stevens, A.A., 1983. Identity and Yields of Major Halogenated Products of Aquatic Fulvic Acid Chlorination. Environ. Sci. Techno., 17: 625-628.
De Laat, J., Merlet, N. and Dore, M., 1982. Chloration de composés organiques : Demande en chlore et réactivité vis-à-vis de la formation des trihalométhanes. Incidence de l'azote ammoniacal. Water Res., 16: 1437-1450.
Masschelein, W.J. and Denis, M., 1982. Sur la signification des bromures dans l'eau. Sciences de l'Eau, 1: 65-83.
Merlet, N., De Laat, J. and Dore, M., 1982. Oxydation des bromures au cours de la chloration des eaux de surface, incidence sur la production de composés organohalogénés. Sciences de l'Eau, 1: 215-231.
Reckhow, D.A. and Singer, P.C., 1981. The removal of organic halide precursors by preozonation and alum coagulation. J. Am. Water Works Ass., 76: 151-157.
Thurman, E.M. and Malcolm, R.L., 1981. Preparative isolation of aquatic humic substances. Environ. Sci. Techno., 15: 463-466.

CHLOROPICRIN FORMATION DURING OXIDATIVE TREATMENTS IN THE PREPARATION OF DRINKING WATER

N. MERLET, H. THIBAUD and M. DORE

Laboratoire de Chimie de l'Eau et des Nuisances, Université de Poitiers, 40, Avenue du Recteur Pineau, 86022 Poitiers cedex (France)

ABSTRACT

Chlorination of water can lead to the formation of chloropicrin. The numerous potential precursors (of various reactivities) observed during this study, confirm this hypothesis.

Combination of ozonation and chlorination can also lead to the formation of this compound, dangerous to health ; however, the conditions of the formation and particularly the impact of a nitration reaction in the gas phase are still not clearly defined.

INTRODUCTION

In the past few years, trichloronitromethane formation (chloropicrin) during the process of potable water production has often been underlined (Coleman et al., 1977 ; Allen et al., 1977 ; Elmghari et al., 1982 ; Becke, 1983). The production of this hazardous compound gives rise to various hypothesis incriminating either the chlorination step or the combined treatment method of preozonation and chlorination.

Because of the recognition of the different problems due to factors such as additional identified chloropicrin precursors, pH influence, etc..., concerning chloropicrin, our main objectives in this study are :

a) with regard to chlorination, to try and find some precursors for the production of chloropicrin and to determine the conditions of its formation

b) with regard to the combination of preozonation and chlorination, to define first the different ways liable to chloropicrin formation and second, to study the principal parameters of the reactions in question.

EXPERIMENTAL

The determination of chloropicrin was performed after pentane extraction by GC analysis with a ^{63}Ni electron capture detector. Residual chlorine was quenched by adding NH_4Cl to each sample (Na_2SO_3, usually employed, reduces chloropicrin as well).

All the experiments were carried out using high quality purified water ($0.05 <$ TOC < 0.2 mg l^{-1}, $\rho \simeq 18$ MΩ.cm).

RESULTS and DISCUSSION

Chlorination

Chlorination experiments carried out in batch reactors show that many organic compounds can lead to chloropicrin formation under chlorination conditions in the preparation of drinking water. This is demonstrated in the next table (table 1) for the following experimental conditions :

$|Cl_2| = 2 \cdot 10^{-3}$ mol l^{-1}
pH $= 7.1$ (phosphate buffer) ; t = 112 hours ; $\theta = 20°C$

TABLE 1
Chloropicrin yields from various precursors

	Initial concentration	Chloropicrin yields *
Natural Humic acid (0,22 µm filtrated solution)	TOC = 20 mg l^{-1}	1.45 µg/mg TOC
nitromethane	10^{-4} mol l^{-1}	45 %
o-nitrophenol	10^{-4} mol l^{-1}	5.7 %
m-nitrophenol	10^{-4} mol l^{-1}	53 %
p-nitrophenol	10^{-4} mol l^{-1}	7.2 %
picric acid	10^{-4} mol l^{-1}	1 %
nitrobenzene	10^{-4} mol l^{-1}	0.3 %
glycine	10^{-4} mol l^{-1}	0.01 %

* We suppose that a 100 % yield is obtained when one mole of initial compound leads to the production of one mole of chloropicrin.

Thus, the precursors converted by chlorination at pH 7.1 are not only nitro-compounds, but also an amino acid and a surface water heavily loaded with humic substances (TOC = 20 ppm).

pH effect. This study was carried out on nitromethane solutions (10^{-5} mol l^{-1} with a chlorine dose of 10^{-4} mol l^{-1} for different pH values (fig. 1).

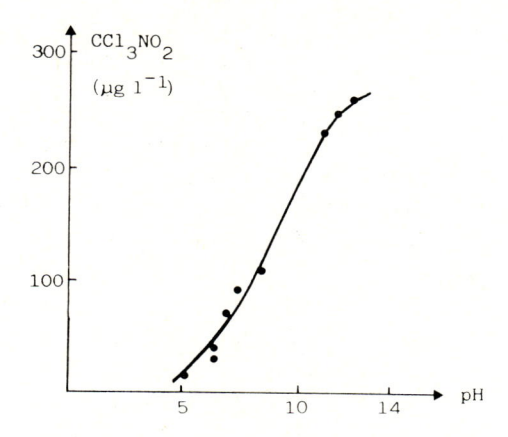

Fig. 1. Effect of pH on chloropicrin formation from nitromethane.

<u>Chloropicrin formation rate.</u> As shown in figure 2, the formation of chloropicrin from a very reactive precursor (nitromethane), appears to change slowly, even for highly concentrated solutions.

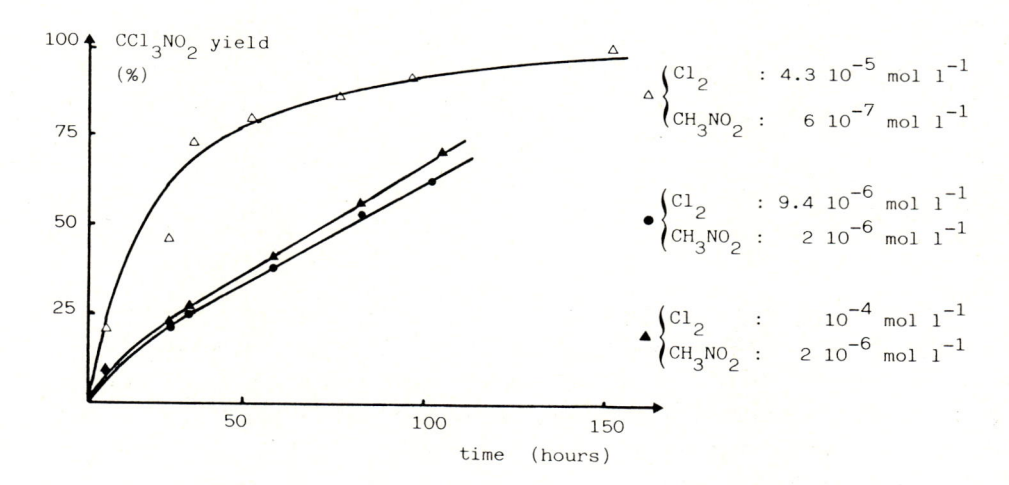

△ $\begin{cases} Cl_2 & : 4.3 \ 10^{-5} \ mol \ l^{-1} \\ CH_3NO_2 & : \ 6 \ 10^{-7} \ mol \ l^{-1} \end{cases}$

● $\begin{cases} Cl_2 & : 9.4 \ 10^{-6} \ mol \ l^{-1} \\ CH_3NO_2 & : \ 2 \ 10^{-6} \ mol \ l^{-1} \end{cases}$

▲ $\begin{cases} Cl_2 & : \ 10^{-4} \ mol \ l^{-1} \\ CH_3NO_2 & : \ 2 \ 10^{-6} \ mol \ l^{-1} \end{cases}$

Fig. 2. Rate of chloropicrin formation from nitromethane.

However, it could be concluded that this reaction can occur either during the treatment process, or more likely in the water supply systems because of the presence of residual chlorine.

These results show that under these conditions, chloropicrin formation is possible, not only on a nitrated organic compound :

$$R-NO_2 \xrightarrow{HOCl} CCl_3NO_2$$

but also by oxidation by chlorine of amino compounds (for instance aminoacids) in nitrated compounds (Elmghari et al., 1982) which would yield chloropicrin by further chlorination :

$$R-NH_2 \xrightarrow{HOCl} R-NO_2 \xrightarrow{HOCl} CCl_3NO_2$$

Therefore, the results thus obtained indicate it is most likely that formation of chloropicrin may occur from many potential precursors, whether aliphatic or aromatic, nitrated or aminated.

Combination of ozonation and chlorination

Though we find very few data on the ozonation/chlorination combination in the literature, it must not be neglected for two reasons :
- on the one hand, it could be that because of its oxidizing power, ozone is susceptible of causing the formation of nitrated compounds from certain aminated compounds :

$$R-NH_2 \xrightarrow{O_3} R-NO_2 \xrightarrow{Cl_2} CCl_3NO_2$$

- on the other, ozone generation by electric discharge in air is accompanied by the production of nitrogen oxides (Lapeyre et al., 1981), among which N_2O_5 (nitrogen pentoxide) may lead to NO_2^+, susceptible of reacting as a nitrating agent during the ozonation of organic compounds (Becke et al., 1981) :

$$\text{C}_6\text{H}_5\text{R} \xrightarrow{N_2O_5} \text{R-C}_6\text{H}_4\text{-NO}_2 \xrightarrow{HOCl} CCl_3NO_2$$

The aim of the present survey was :
- to verify the possibilities of N_2O_5 formation in ozonated air

– and to determine the conditions of N_2O_5 formation with the organic compounds present in the water to be treated.

Experimental. Two different reactors were used :
– In the first one (cf. fig. 3) ozonated air passed through the bubble column. For given ozone doses, samples were taken from the column and postchlorinated ($|Cl_2| = 5.10^{-4}$ mol l^{-1}, t = 2 h).
– In the second one, ozone was produced from oxygen (cf. fig. 3). Moreover, this reactor was fitted with an apparatus allowing the production of N_2O_5 (by dehydratation of HNO_3 on P_2O_5). Parallel circuits allowed air to be introduced into the system.

For increasing doses during treatment, samples taken from the reactor are chlorinated ($|Cl_2| = 5.10^{-4}$ mol l^{-1}, t = 16 h).

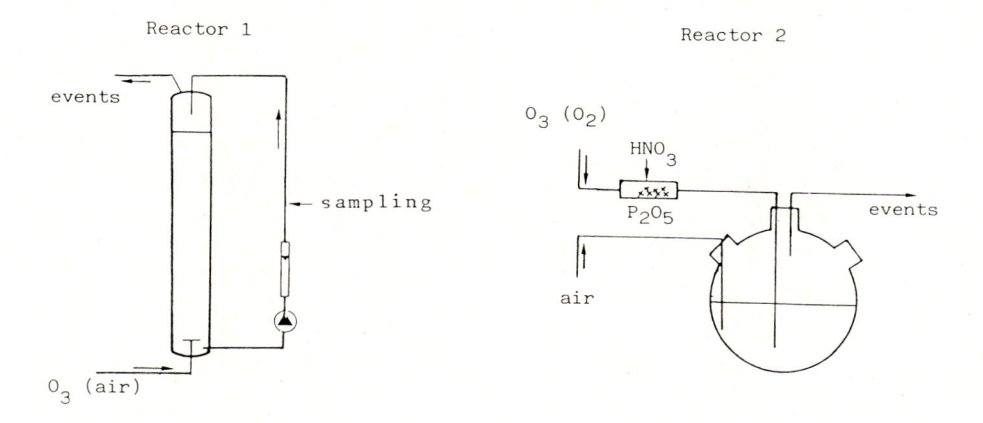

Fig. 3. Ozonation reactors.

N_2O_5 production during ozonation. The experiments confirmed the formation of nitrogen oxides by oxidation of atmospheric nitrogen during the generation of ozone from air under our experimental conditions, as was also found by Becke (1983). As was observed by Becke, nitration only occur in the gas phase, in the ozonation columns, at the ozonated air/water interface.

Thus, using reactor 2 with ozone produced from oxygen, we were able to show the formation of chloropicrin from impurities in the compressed air produced in the laboratory and used for ozone generating (fig. 4).

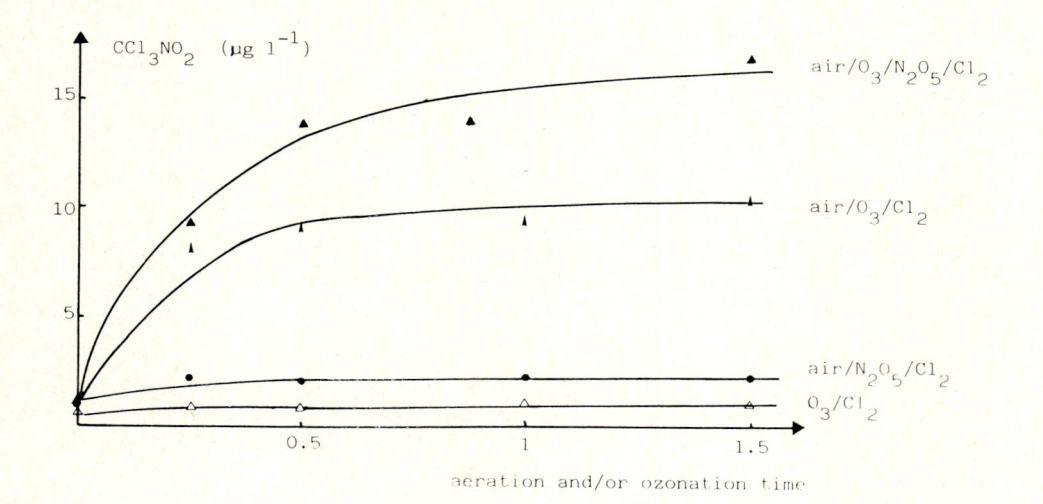

Fig. 4. Chloropicrin production for various processes.

Thus, in the combined ozonation/nitration system, the impurities present in the air lead to chloropicrin precursors ; ozonation seems to be necessary to obtain a higher concentration of chloropicrin.

We can conclude that as far as the ozonation of organic compounds in diluted synthetic solutions is concerned, the results obtained are at present too incomplete to allow a definite answer to the formulated hypothesis ; however, the observations in this survey indicate it is quite probable that formation of chloropicrin takes place in these systems.

REFERENCES

Becke, Ch., Maier, D., 1981. Nebenproduckte bei der Ozonerzeugung. Wasser Berlin, 5: 859-874.
Becke, Ch., 1983, Untersuchungen zur Bildung von Trichloromethan bei der Anwendung oxidativer Wasseraufbereitungsverfahren. Ph. D. thesis Univ. Karlruhe, 132 pp.
Coleman, W.E., Lingg, R.G., Melton, F.C., Kopfler, F.C., 1977. The occurence of volatile organics in five drinking water supplies using gas chromatography/mass spectrometry. In: L.H. Keith (Editor), Identification and analysis of organic pollutants in water, pp 305-327.
Elmghari, T.M., Laplanche, F., Venien, F., Martin, G., 1982. Ozonation des amines dans l'eau. Water Research, 16: 223-229.
Keith, L.H. et al., 1977. Identification of organic compounds in drinking water from thirteen U.S. cities. In: L.H. Keith (Editor), Identification and Analysis of Organic pollutants in water, pp 329-373.
Lapeyre, R.M., Peyrous, R. 1981. Produits gazeux (NO, NO_x, O_3) et noyaux de condensation créés par des décharges électriques entre un point et un plan. Environ. Technol. Letters, 2: 29-38.

EVALUATION OF DIFFERENT TREATMENT PROCESSES WITH RESPECT TO MUTAGENIC ACTIVITY IN DRINKING WATER

H.J. Kool, J. Hrubec, C.F. van Kreijl and G.J. Piet

National Institute of Public Health and Environmental Hygiene
P.O. Box 150, 2260 AD LEIDSCHENDAM, THE NETHERLANDS

ABSTRACT

Treatment processes which are applied in The Netherlands during the preparation of drinking water have been evaluated with regard to introduction and removal of organic mutagens as well as halogenated organics. It appeared that the most efficient processes in reducing mutagenic activity were activated carbon filtration and artificial dune recharge. In general these processes were also the most efficient in removing halogenated organics. Using low doses of chlorine dioxide (< 1 mg ClO_2/l) for safety disinfection of drinking water, no change or substantial less mutagenic activity than by chlorination (1 mg Cl/l) was found. This counts too for the formation of halogenated organics. Transport chlorination of stored river Meuse water was able to introduce or activate mutagenic nitro organics which have not been found previously. Ozone treatment under field conditions showed mostly a tendency to decrease the activity of organic mutagens. It was also shown that dependent on the water quality and treatment conditions a slight increase of mutagenic activity occurred, but this activity would be reduced by increasing the ozone dose. It seems possible to optimalize the ozone treatment conditions regarding the level of ozone dose and the contact time to avoid an increase of mutagenic activity. Futhermore it was shown that when a mutagenic raw water source was used a proper combination of treatment processes is able to produce drinking water in which no mutagenic activity could be detected under the test conditions. Finally it is stated that before far-reaching decissions with respect to use mutagenicity data for a selection of water sources or treatment processes will be made, more information on the relation mutagenic activity from drinking water and effects on human health should become available.

INTRODUCTION

In The Netherlands the presence of 280 organic compounds in drinking water was recorded during a drinking water survey in 20 cities and almost 190 of these have been identified among which mutagens and (suspect) carcinogens (Zoeteman, 1978).

Limitations of time, manpower and scientific information have not permitted an indepth evaluation of most of the compounds recently found in drinking water and therefore relatively little is known about their toxic effects including their carcinogenic potential. In addition to this it is recognized that in fact the non purgeable fraction which comprises 90-95% of the total organics in the water have not been identified (NAS, 1977, 1980, 1981).

The development of the so-called short-term tests for genotoxic compounds, in particular the Ames Salmonella/microsome assay (Ames et al., 1973) has greatly influenced th toxicological examination of drinking water over the past 7 years.

At present the bacterial mutagenicity tests in particular the Salmonella Microsome assay (Ames test) is considered to be a useful screeningstest for predicting mutagenicity and possible carcinogenicity of organic chemicals (McCann et al. 1975; Purchase, 1982). As a result the screening of drinking water (as well as the raw water sources) for the presence of organic mutagens, has become an important part of the toxicological studies. To date a considerable and still growing number of papers has been published on this subject and most of these have been reviewed by Loper (1980). It was shown that in addition to the chemically identified rather volatile mutagens and carcinogens in drinking water, there is now ample evidence for mutagenic activity among the far more abundant nonvolatile organics. Although some positive data were obtained in other microbiological assays and with a cell transformation assay, most of the experimental evidence came from the Ames Salmonella/microsome assay.

Testing of complex organic concentrates prepared from (drinking) water in other short-term tests beside the Ames test is rather time consuming, since in general fractionation procedures are required to obtain non toxic fractions (Kool et al. 1984 a). Therefore application of the Ames assay in testing "raw" organic concentrates prepared from (drinking) water is rather a convenient method for evaluating different drinking water treatment processes with respect to mutagenicity.

This paper presents recent information on the changes of mutagenic activity in drinking water treatment processes obtained in field studies. These processes will be evaluated with respect to the removal or the introduction of mutagenicity during the treatment. Some chemical parameters have been measured also to see how they correlate with the observed mutgenic activity. Finally results with regard to the characterization and identification of a particular group of organic mutagens will be presented and discussed.

MATERIALS AND METHODS

XAD resin.

Amberlite XAD-4 resin was obtained from KIWA Research Institute Nieuwegein, The Netherlands. Control by GC analysis, and storage of the resins in methanol have been described previously (Kool et al., 1981 a).

XAD concentration procedure.

Water samples were taken before and after the studied treatment process.

To obtain about 7000 fold concentration, 150 L of water was passed over columns containing 20 cm^3 XAD-4/8 at a flow rate of maximal 4 bed vol/min and at a constant temperature of 15 °C. Elution of the adsorbed neutral fraction of the organic constituents was carried out with the appropriate volume (> 1 bed volume) of either dimethylsulfoxide (DMSO) or acetone. The XAD filtrate was adjusted to pH 2 with HCl, and readsorbed on XAD-4/8. Subsequent elution of this acid fraction was carried out with DMSO or acetone. For lower or higher concentration factors, correspondingly smaller or larger volumes of water were passed through the XAD column until the desired water/eluate ratio (v/v) was obtained.

Bacterial strains

The Salmonella typhimurium strains TA98 and TA100 were used (Ames et al., 1975). They were stored frozen at -80 °C in nutrient broth containing 10% DMSO.

Ames Salmonella/microsome assay

Mutagenicity testing of the organic concentrates of drinking water was carried out according to the plate incorporation assay (Ames et al., 1975)). The induction of microsomal enzymes with Aroclor 1254 and the preparation of the rat liver homogenates (S9) have also been described previously (Ames et al., 1975). In the S9 mix, 0.075 ml of liver homogenate was added per ml of mix. All water concentrates were tested in 3-5 replicates, and the results were considered significant when a 2-fold increase above the background and dose-response effects were observed. The deviation of the mean was usually below 20%. Routine controls to check for the presence of factors affecting bacterial growth were incorporated as described previously (Kool et al., 1981 a).

Chemical analyses

The analysis of water quality parameters in the unconcentrated water samples was carried out according to standard procedures, described previously. Adsorbable organic chlorine (AOCl) was determined by microcoulometry after adsorption on activated carbon and pyrolysis (Sanders, 1980). Total organic carbon (TOC) was analyzed with a Beckman TOC analyser (Tocomaster model 915 B). For a direct headspace analysis of very volatile halogenated compounds (VOCl) the procedure described by Piet et al. (1978) was used.
Trihalomethanes (THM) were analyzed with a Carlo Erba 2900 gaschromatograph containing a capillary column coated with OV/225, dia 0.5 mm, length 50 m, and equipped with an automatic headspace sampler, model 250.

INFLUENCE OF WATER TREATMENT ON MUTAGENIC ACTIVITY

In The Netherlands, groundwater contributes for about two thirds to the total

volume of the drinking water production and surface water for about one third.
The river bank infiltrate - a mixture of river and groundwater - also
contributes for a small part to the drinking water supply of the country.
During recent servey's in all three types of raw water sources, mutagenic
activity could be detected (Kool et al., 1982, 1984 b, 1985 a, b). However the
polluted rivers Rhine and Meuse showed the highest mutagenic activity (Van
Kreijl et al., 1980). The results of the testing the river water concentrates,
in particular river Rhine concentrates, demonstrated that the mutagenic activity
predommently required metabolic activation, while results with concentrates of
drinking water prepared from this river showed that the mutagenic activity in
the Ames test in many cases was most pronounced without metabolic activation
(Kool et al., 1982 a, 1984 b). These results indicated that during drinking
water treatment a change in the bacterial mutagenicity had occurred. The
indications of introduction of mutagenic activity during drinking water
treatment was supported by the fact that several pilot plant studies had shown
an increase in mutagenicity in the Ames test after chlorination (Cheh et al.,
1980; Zoeteman et al., 1982). These results were confirmed by the later
experiments under field conditions (Dolara et al., 1981; Kool et al., 1981 b,
1984 c; van der Gaag et al., 1982).
In this part some recent information on the changes of mutagenic activity by
different processes used in the drinking water treatment practice in The
Netherlands will be presented.

EFFECTS OF STORAGE ON RIVERWATER

The most important sources of surface water in The Netherlands are the rivers
Rhine and Meuse. To tide over periods of accidental pollution of the river water
and to egalize fluctuations in river water quality a storage of rawwater before
the treatment is necessary. This storage is realized in open reservoirs and by
means of underground storage in the dunes along the Northsea coast. The
influence of the storage in an open reservoir is demonstrated in fig. 1.
The figure shows a drastic decrease of mutagenic activity after storage the
Rhine water in an open storage reservoir. Although variation in the mutagenic
activity in the Rhine was observed throughout the year (Slooff et al, 1984), the
mutagenic activity in the storage reservoirs remained on a constant level and
was
always found to be much less then the activity in the river Rhine (not shown).
On the other hand it appeared that storage of river Meuse water did not show up
a significant reduction of mutagenic activity. Nevertheless a small reduction
after storage was also observed (not shown). That a long term storage of river
water in open reservoirs results in an effective removal of some groups of
organic micropollutants such as PAH's, phenols and aromatic amines has been

reported previously (Oskam, 1980; Knoppert et al., 1980).
The changes of mutagenic activity during artificial groundwater recharge in dunes are illustrated in fig. 2, showing the results of a study in two water works.

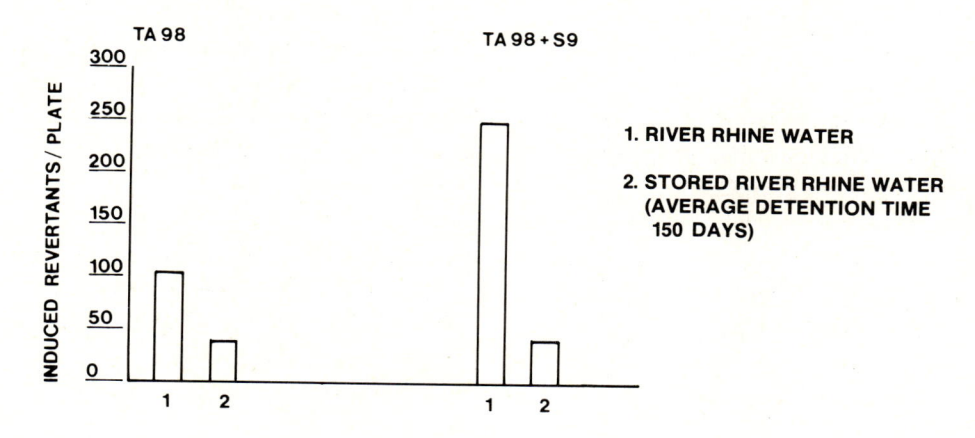

Figure 1. Effect of an open storage on mutagenic activity in water.
Sampling - 900 and 1800 fold concentration of Rhine respectively stored Rhine water in a water work (average detention time 150 days) on XAD-4/8 elution with DMSO and subsequent testing the DMSO concentrate in the Ames assay as described in Method. Each value represents the average of 3 plates. The results correspond to 450 ml of water per plate.

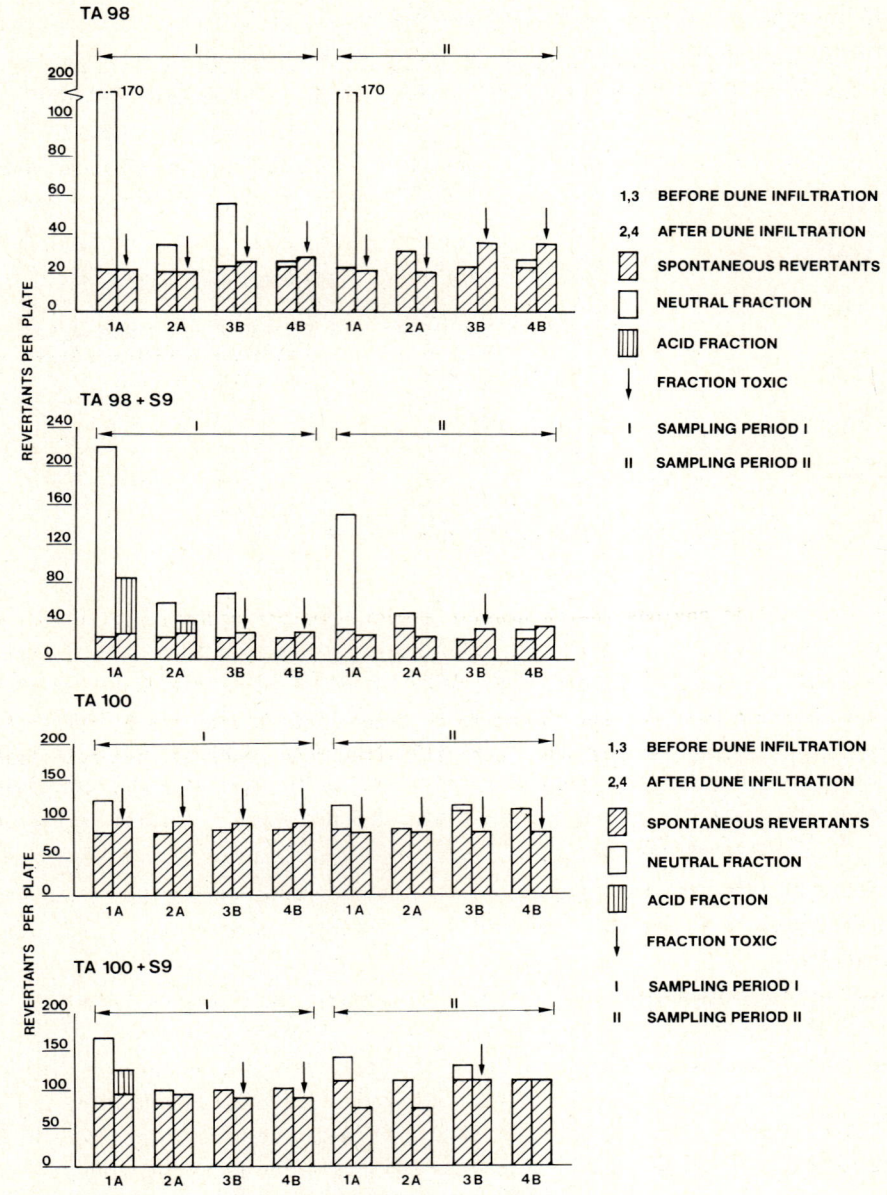

Figure 2. Effect of dune recharge on mutagenic activity.
Sampling, 7500 fold concentration of (treated) water in two water works before and after dune filtration (water work A average detention time 2 months, water work B average detention time 3 months) on XAD-4 elution with DMSO and subsequent testing the DMSO concentrate in the Ames assay as described in Methods. Each value represents the average of 3 plates. The results correspond to 3.5 litre of water per plate.

Table 1. Effect of dune recharge on chemical parameters.

Sampling site		Before treatment			After treatment		
		TOC mg C/l	VOCl nmol HAL/l	AOCl μmol HAL/l	TOC mg C/l	VOCl nmol HAL/l	AOCl μmol HAL/l
Water work	A*						
PERIOD	I	2.6	30	0.8	2.0	<6	0.2
	II	3.2	7	0.8	1.5	<6	0.6
Water work	B**						
Period	I	4.2	14	1.0	3.9	<6	0.3
	II	2.1	<6	1.0	1.8	<6	0.2

* Watersource the river Rhine (average detention time 2 months)
** Watersource the river Meuse (average detention time 3 months)

From fig. 2 it is obvious that a drastic almost complete reduction of direct and promutagenic activity with both strains occurred after dune filtration in both water works. It appeared here too that all the acid fractions without addition of S-9 and a few with S-9 mix showed toxic properties towards the bacteria. A substantial reduction of mutagenic activity after dune recharge, has been found in our previous study (Kool et al., 1984 c). The results of a survey of the mutagenicity changes during the artificial recharge of a majority of the Dutch dune water works carried out by the KIWA (1984) demonstrated the reduction of the activity after the recharge also. The observed reduction in this study was generally greater at the pH 7 (neutral) XAD concentrates than at the pH2 (acid) concentrates.

Removal of mutagens seems to be a paralel of removal of different organic micropollutants during the dune recharge (table 1), which is ascribed to biological and sorption processes during the water percolation through the ground. A number of analytical studies indicates a quite large concentration reduction of some relevant groups of micropollutants. For instance a very high and completed transformation of trihalomethanes, organic chlorine - and phosphorous pesticides, haloacetonitriles and nitroaromatics has been found (Piet and Zoeteman, 190; KIWA, 1984; Hrubec et al., 1985a).

Effect of oxidation and disinfection processes

Chlorination

In The Netherlands in the last years chlorine application has been reduced as

much as much as possible and most of the drinking water does not receive a chlorine treatment. The use of chlorine in general is related to surface water treatment. Even there a part of drinking water derived from surface water is distributed without chlorination which often is the case when bank infiltration or artificial groundwater recharge in dunes is applied. According to results of a great number of field-, pilot plant- and laboratory studies, chlorination of (drinking) water in general will increase or even will generate mutagenic activity in water (Loper, 1980; Kool et al., 1982 b, 1985 a). Some representative results with respect to the changes of mutagenic activity and halogenated organics in water derived from surface water obtained from Dutch water works after a chlorine treatment have been published previously (Kool and Hrubec,
1985 b) and are shown in fig. 3 and table 3.

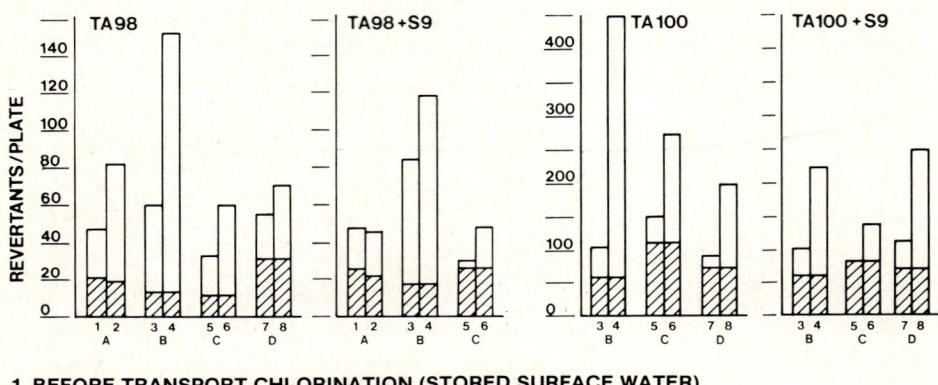

1. BEFORE TRANSPORT CHLORINATION (STORED SURFACE WATER)
2. AFTER TRANSPORT CHLORINATION (1-2 MG CHLORINE/L)
3. BEFORE PRÉ CHLORINATION } 1.8 MG CHLORINE/L
4. AFTER PRÉ CHLORINATION
5. BEFORE POST CHLORINATION
6. AFTER POST CHLORINATION } 0.15 MG/L CHLORINE PRESENT
7. BEFORE POST CHLORINATION } AFTER A CONTACT TIME OF 20 MIN.
8. AFTER POST CHLORINATION

▨ SPONTANEOUS REVERTANTS

Figure 3. Effect of transport-, pre and post chlorination on mutagenic activity in drinking water treatment.
The sampling: Treated water in three water works was concentrated before and after a chlorine treatment (A, B and C raw water source the river Meuse, D raw water source the river Rhine) on XAD-4/8, elution with DMSO and subsequent testing the DMSO concentrate in the Ames assay as described in Methods. Each value represents the average of 3 plates. The result correspond to 1.5 litre (A) and 3.5 litre (B, C, D) of water per plate.

Table 2. Effect of chlorine treatments on chemical parameters.

Sampling site	Type of chlorine treatment		Before treatment			After chlorine treatment			
		TOC mgC/l	THM µg/l	EOCL nmol HAL/l	AOCL µmolHAL/l	TOC mg/l	THM µg/l	EOCL nmol HAL/l	AOCL µmol HAL/l
Water work A	Transport 1.8 mg/l Cl$_2$	3.8	ND	-	0.7	3.7	35	50	1.2
B	" 2.5 mg/l Cl$_2$	3.1	ND	30	1.2	3.2	100	120	4.7
A	Pre chlorination	3.5	0.1	30	0.4	3.6	31	80	3.5
B	Post "	1.8	4.7	25	0.3	1.8	10	75	1.8
C	" "	2.5	-	25	1.0	2.3	36	45	1.4

ND Not detectable
- Not measured

The data in table 2 and fig. 3 confirmed the previously mentioned fact that a chlorine treatment increases the mutagenic activity as well as the level of halogenated organics in the treated water. However from the studies no clear influence of the quality of the treated water, the dosage of chlorine and the water temperature on the level of the mutagenic activity could be deduced.

Ozonation

Ozon which is a powerful disinfectant and oxidant has become accepted as an alternative to chlorination, since it posses some important advantages, a.o.: it improves taste and odour of water, has a relatively high redoxpotential, oxidizes organics to more biodegradable compounds and has a microflocculation effect. On the other hand ozonation can stimulate bacterial regrowth in the distribution systems and may generate hazardous compounds. A number of drinking water studies on mutagenicity changes during ozonation have been carried out in the last years and have been reviewed recently (Kool et al., 1985 a). These studies give no clear picture, because a reduction as well as an increase of mutagenic activity in water after ozonation was found. In a pilot plant study an ozone treatment of stored Rhine water it was shown that mutagenic activity was completely removed (Zoeteman et al., 1982) while a substantial decrease of mutagenic activity was observed in other drinking water studies (Dolara, et al., 1981, Kool et al., 1981 b, 1985 c, Hrubec and Kool, 1985 b, Bourbigot et al., 1983). However several reports document an increase of the mutagenicity after ozonation (Dolara et al., 1981; Van der Gaag et al., 1982; Bourbigot et al., 1983; Duget et al., 1983; Van Hoof et al., 1983). This has been mostly ascribed to the formation of hydrophile direct acting mutagens. In the recent study of van Hoof et al. (1985) a formation of direct lipophilic frame shift mutagens was detected also. To investigate the mutagenicity change under the conditiones of the ozon application in the drinking water treatment practice an orientation field study in two Dutch waterworks has been performed. The results are shown in figure 4 and table 3. Fig. 4 shows that in all cases but one no effect or a decrease in mutagenic activity was observed.

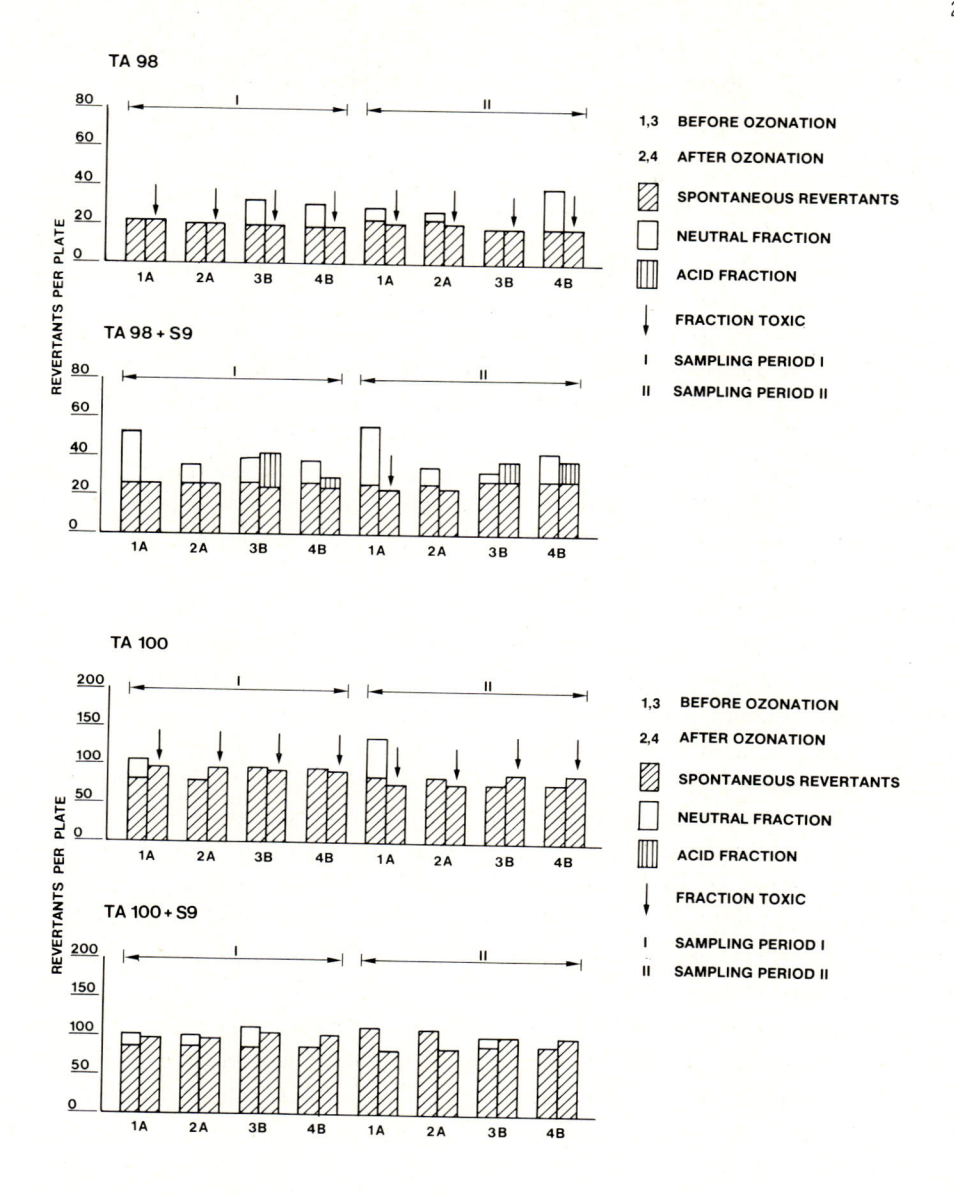

Figure 4. Effect of an ozone treatment on mutagenic activity.
Sampling, 7500 fold concentration of (treated) water in two water works before and after ozonation (water work A 2 mg ozone/l contacttime 10 min.; waterwork B 0.8-1.4 mg ozone/l contacttime 30-60 min.) on XAD-4, elution with DMSO and subsequent testing the DMSO concentrate in the Ames assay as described in Methods. Each volume represents the average of 3 plates. The results correspond to 3.5 litre of water per date.

Table 3.

Effect of an ozone treatment on chemical parameters

Sampling site	Before treatment			After treatment		
	TOC	VOCl	AOCl	TOC	VOCl	AOCl
Water work A*	mg C/l	nmolHAL/l	µmolHAL/l	mg C/l	nmolHAL/l	µmol HAL/l
Period I	2.0	<6	0.2	1.9	<6	<0.1
II	2.1	<6	0.2	1.9	<6	<0.1
Water work B**						
Period I	0.9	<6	<0.1	1.0	<6	<0.1
II	1.0	<6	0.3	0.9	<6	<0.1

* Water source the River Meuse; dose 2 mg O_3/l, contact time 10 minutes
** Water source groundwater; dose 0.8-1.4 mg O_3/l, contact time 30-60 minutes

Only the second period in water work B an increase in direct mutagenic activity with strain TA98 occurred while in the other cases no effect or a reduction in activity was observed when mutagenic activity was present. It also appeared that the acid fraction showed toxic properties towards the test bacteria (without S-9 mix) and therefore it is not possible to evaluate these results.

The changes of mutagenic activity during the treatment in waterwork A have also been studied by the KIWA. The study as far as the influence of the ozonation concerns, gave no consistent results. Great differences were found especially in the activity with strain TA 100 with pH2 concentrates while in the sample taken during the first period the activity with this strain after ozonation highly increased, in the samples from the second period remained unchanged.

The mentioned results confirmed the supposition that the changes of mutagenic activity is highly dependent on the proces conditions. In a few recent studies (Bourbigot et al., 1983; Duget et al., 1983; Kool et al., 1985 b) was recently found that the activity decreased when the level of ozone dose was high and the contact time long.

Nevertheless our present knowlegde on the variety of factors influencing changes of the mutagenic activity during the ozonation is still insufficient to make a reliable evaluation of this process.

Concerning the chemical parameters it appeared that the ozone treatment did not influence the level of volatile halogenated organics but the level of AOCl decreased after ozonation (table 3). An increace of AOClY, by ozonation was reported previously (Bruchet et al., 1983; Kool and Hrubec, 1985 b). In this context it should be mentioned that there a possibility exits that the decrease of AOCl level is not caused by a removal, but by a conversion of adsorbable to nonadsorbable halogenated compounds.

Chlorine dioxide.

In contrast with chlorine, chlorine dioxide is currently hardly applied as disinfectant in water treatment in The Netherlands. One of the main reasons therefore is, that application of chlorine dioxide introduces toxic reaction products like chlorite and chlorate in the treated water. Chlorine dioxide is however considered a promising alternative to chlorine because it causes no taste and odour problems, is fairly stable in water and its disinfection capacity is rather independant from the pH. Furthermore, it is recognized that chlorine dioxide has a low potential in producing halogenated organics. Several pilot plant studies which compare chlorine dioxide with chlorine with respect to mutagenicity and formation of halogenated organics have been carried recently in The Netherlands. The investigations were focused on the use of chlorine dioxide for transport desinfection of Rhine and Meuse water after storage in open reservoirs and for safety disinfection of drinking water. It was demonstrated that treatment of stored river water with high doses of chlorine dioxide (up to 15 mg ClO_2/l) (De Greef et al, 1980; Zoeteman et al., 1982; Hrubec et al., 1985 c) resulted in an increase of direct acting mutagenicity with strain TA-98.

The safety disinfection of drinking water from different sources of surface and groundwater with lower doses of chlorine dioxide (< 1 mg ClO_2/l) in most cases did not increase the mutagenic activity. Safety disinfection with 1 mg/l chlorine under the same conditions resulted mostly in an increase of direct mutagenic activity (Kool et al., 1985 a, b; Hrubec et al. 1985 c).

In general no increase of trihalomethanes, volatile organic halogene (VOCl) during the disinfection with chlorine dioxide, was detected. However after disinfection with chlorine dioxide in some cases an increase of adsorbable organic halogens (AOCl) was observed (Kool et al., 1985 a, b). An representative example is shown in table 4.

Table 4. Effect of a chlorine dioxide treatment om the level of halogenated organics in water

Type of water	Before treatment			After treatment 1 mg/l CLO$_2$	
	THM µg/l	AOCl µmol HAL/l	TOC mgC/l	THM µg/l	AOCl µmol HAL/l
1. Stored Rhine water	ND*	0.7	3.1	NO	0.7
2. Store Meuse water	ND	0.8	2.2	ND	1.0
3. Groundwater	ND	0.1	0.8	ND	0.1
4. Mixture of ground** and surface water	0.3	0.2	7.5	0.3	0.4
5. Groundwater	ND	0.3	3.1	3.2	0.3

* ND-Not detectable
** Bank infiltrated Rhine water

It seems that an analogy exists between the changes of the mutagenic activity by chlorine dioxide treatment and with these by ozonation. By both processes the activity, dependent on the condition of the treatment, might be increased as well as decreased. Due to a lower oxidation potential of chlorine dioxide than of ozone, the changes seems to be less pronounced by the chlorine dioxide treatment.

Slow sand filtration.

Recently slow sand filtration - one of the oldest drinking water treatment processes - is receiving an increased attention as an alternative for chemical treatment. Due to effective removal of pathogens and precursors of trihalomethanes by application of slow sand filtration in water treatment systems a limitation of the use of chlorine and of trihalomethanes forming can be achieved. In The Netherlands slow sand filtration is widely applied especially in the posttreatment of water after the dune recharge. Therefore we have followed the changes of mutagenic activity and removal of halogenated organics by slow sand filtration in two dune waterworks.

The results obtained (fig. 5 and table 5) showed, that after this treatment, when mutagenic activity in the influent water was present, the activity decreased. The mutagenic activity of the influents, however was so marginal, that no firm conclusion may be drawn from these results. No increase was observed after the treatment however. The acid fractions without S-9 showed toxic properties though and therefore these data are difficult to be evaluated.

The small changes of chemical parameters (table 5) are not surprising since in both studied water works the slow sand filtration follows the dune recharge which is more or less thought to have qualitatively the same purification effect as slow sand filtration.

Activated carbon filtration.

Activated carbon filtration is generally recognized as the best available method for removal of many potential dangerous organic micropollutants.
A number of pilot plant and field studies under various process conditious have shown a reduction of mutagenic activity by activated carbon filtration (Kool et al., 1982; Van der Gaag et al., 1982; Hrubec, 1983; Monarca et al., 1983; Loper et al., 1984; Kool et al., 1984 c;). The observations of inefficient reduction of mutagenic activity (Denkhaus et al., 1980; Pendygraft et al., 1979) could be explained by a break through of the organic mutagens due to a too long operation time.
In fig. 6 and table 6 the changes of the mutagenic activity and some chemical parameters after activated carbon filtration in three waterworks are shown. These results show that even after the breakthrough of TOC the filters were able to reduce the mutagenic activity drastically or even completely. A significant reduction of non volatile halogenated organics was observed also (table 6).

Discussion.

The results of the presented survey's as well as the literature data indicate that only some of studied processes give a clear picture with respect of the changes of mutagenic activity.

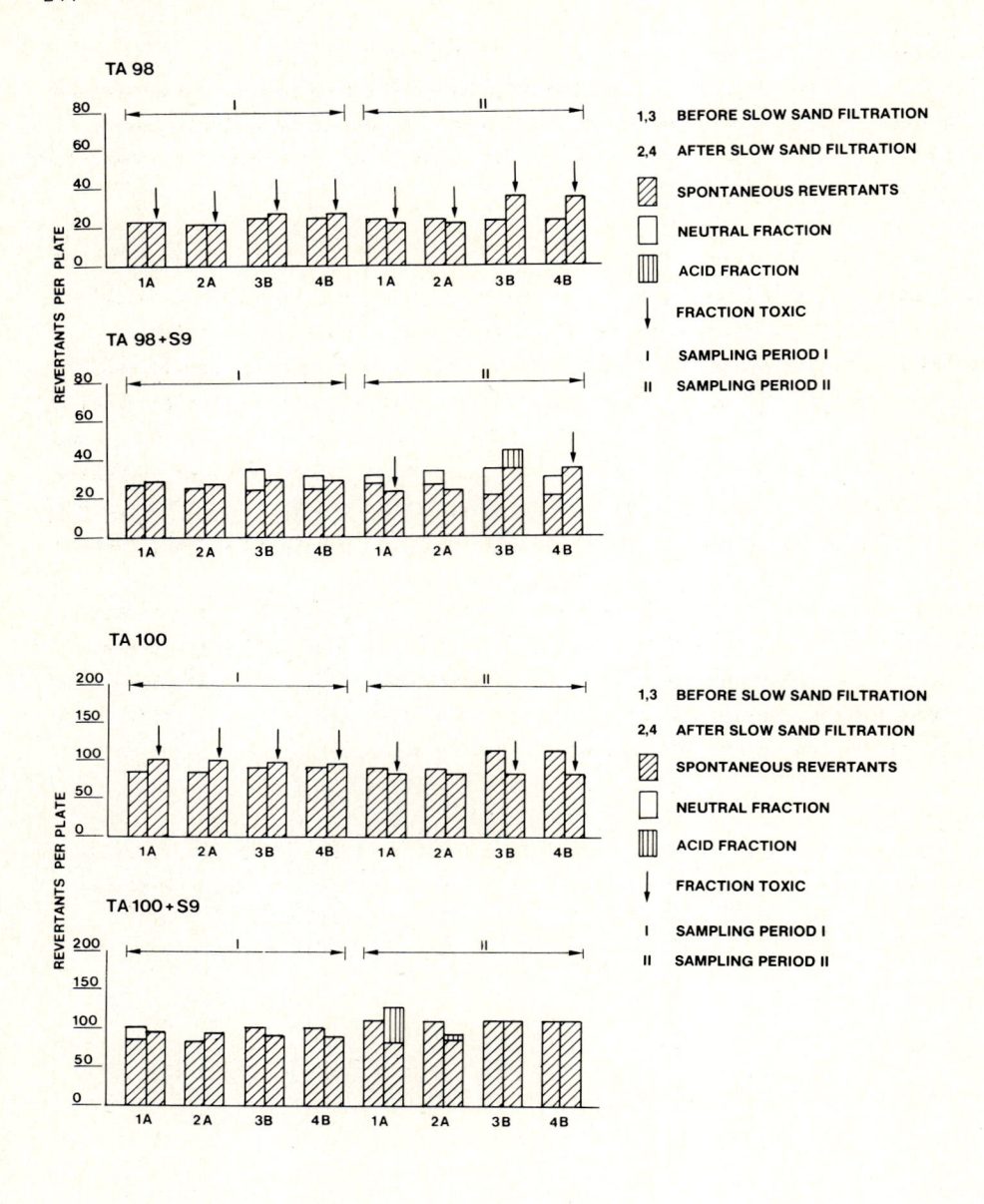

Figure 5. Effect of slow sand filtration on mutagenic activity.
Sampling, 7500 fold concentration of treated water in two water supplies before and after slow sand filtration (20-30 cm/hr) on XAD-4, eluation with DMSO and subsequent testing the DMSO concentrate in the Ames assay as described in Methods. Each value represents the average of 3 plates. The results correspond to 3.5 litre of water per plate.

Table 5. Effect of slow sand filtration on chemical parameters

Sampling site	Before treatment			After treatment		
	TOC mg C/l	VOCl nmol HAL/l	AOCl μmol HAL/l	TOC mg C/l	VOCl nmol HAL/l	AOCl μmol HAL/l
Water work A*						
Period I	1.6	27	0.2	1.6	<6	0.2
II	2.0	23	0.4	2.0	<6	0.4
Water work B**						
Period I	4.6	26	0.1	4.2	<6	0.1
II	2.1	<6	<0.1	1.4	<6	<0.1

* Water source the river Rhine
** Water source the river Meuse

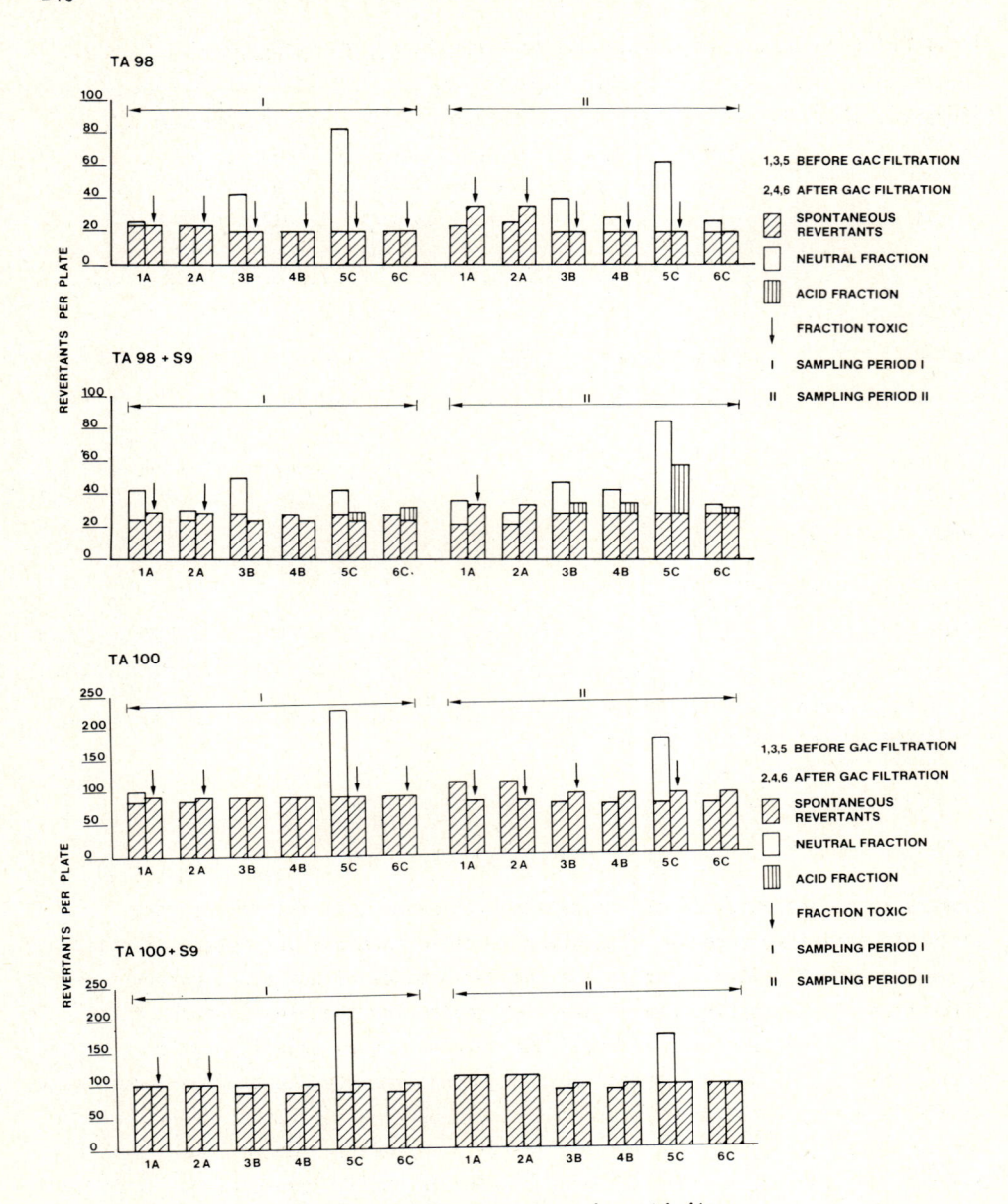

Figure 6. Effect of GAC filtration on mutagenic activity.
Sampling, 7500 fold concentration of treated water in three water works before and after GAC filtration (A contacttime 8-9 minutes, B contacttime 15 minutes, C contacttime 15-20 minutes) on XAD-4, elution with DMSO and subsequent testing the DMSO concentrate in the Ames assay as described in Methods. Each value represents the average of 3 plates. The results correspond to 3.5 litre of water per plate.

Table 6. Effect of granular activated carbon filtration on chemical parameters

SAMPLING SITE	CONTACT TIME (minutes)	BEFORE TREATMENT TOC mg C/l	VOCl µmolHAL/l	AOCl mmolHAL/l	AFTER TREATMENT TOC mg C/l	VOCl nmolHAL/l	AOCl µmolHAL/l
water work A*							
period I	8-9	5.4	< 6	0.7	5.4	< 6	0.3
II		3.0	12	0.3	3.1	< 6	0.1
water work B**							
period I	15	1.4	< 6	<0.1	0.8	< 6	< 0.1
II		2.0	< 6	0.3	1.6	< 6	< 0.1
water work C***							
period I	15-20	2.0	1300	4.4	2.0	890	1.4
II		2.2	1710	4.2	1.8	1060	2.3

* water source - the river IJssel
** water source - the river Meuse
*** water source - Lake IJsselmeer

From the processes which reduce the activity, the most efficient and reliable appears to be activated carbon filtration. It seems that the data on the mutagenic activity, together with other specific chemical parameters, might become an import criterion to assess the performance of activated carbon filtration. These testing can be used for monitoring the breakthrough of the adsorbent bed (i.e. an exhaustion of the carbon removal capacity) with regard to organic micropollutants of health concern. Long time storage of river water in open reservoirs, like groundwater recharge, under the conditions characteristic for the dune recharge in The Netherlands also appear for remove mutagens to a high extent. However due to lack of knowledge on the influence of the factors and the processes influencing the transformation of the mutagens during the open storage and during the percolation through the ground a generalization of the Dutch experiences seems to be premature.

Slow sand filtration in the studied waterworks showed a tendency to decrease mutagenic activity. However due to marginal activity present in the influent water of the filters a firm conclusion on the effect of the filtration is not possible.

Concerning oxidation and desinfection processes only a reliable conclusion on mutagenicity changes by chlorination is possible.

Chlorine use for transport, pre- and post chlorination, in general, increases or introduces direct and promutagenic activity. In our studies no clear correlation between the level of mutagenic activity and the quality of treated water, the dosis of chlorine and the reaction time was found.

Ozone treatment under the field conditions showed mostly a tendency to decrease the activity of mutagens, although here too in one case an increase in activity was found. So it is likely that, dependent on water quality and treatment conditions different results with respect to mutagenicity changes by ozonation can can be obtained. It seems that a realistic possibility exists to optimalize the ozon treatment condition, regarding the level of ozone dosis and the contact time, to avoid an increase of mutagenic activity.

As indicated in an extended pilot plant study, transport disinfection of river water with high doses of chlorine dioxide might increase or introduce mutagenic activity. By the use of low doses of chlorine dioxide for safety disinfection of drinking water no change of activity or substantial less activity than by chlorination was found.

From the data available from our studies no reliable correlation between the change of the mutagenic activity in all water types and of the measured chemical parameters could be derived although AOX for chlorinated (drinking) water may be a reasonable surogate.

The reported and previous studies (Kool et al., 1982 a, 1984 b) indicate also, that with a proper combination of treatment processes drinking water can be produced, having no detectable mutagenic activity (Ames test) under the test conditions.

Furthermore it is suggested that data on mutagenic activity might be used as a quality criterion for selction of raw water sources for drinking water supply. For this purpose, when the data from different laboratories have to be compare, a standardizaton of the test and concentration procedure is inevitable.

Finally it should be stated, that before far-reaching policy decissions could be made on the use of the mutagenicity data, for a selection of the water sources on the treatment processes, more information on the relation of the mutagenicity data and effects on human health should become available. In connection with this, a comparison of the risk from drinking water with the risks via other exposure route (food, air) should be made.

Based on the present level of knowledge, it is however obvious, that in the mean time, where ever it is possible, those treatment processes should be applied which do not increase or introduce mutagens/carcinogens in drinking water.

FRACTIONATION AND IDENTIFICATION OF ORGANIC MUTAGENS

In addition to mutagenicity testing a number of studies on characterization of the organic mutagens present in complex organic mixtures prepared from water have been carried out in the past several years. Coleman et al. (1980) investigated what kind of organic compounds could be identified in a mutagenic concentrate of Cincinnati tapwater. More than 700 organic compounds could be detected in an Ames test positive concentrate from which 460 could be identified. This result clearly shows that a more sophisticated coupled bioassay/chemical fractionation procedure, in a sence that the major part of the organics which is not responsable for mutagenic activity will be separated from the organic mutagens, is necessary to identify the biological active organics in a complex mixture.

At present the most generally used fractionation method suitable for fractionation of non volatile organics is high performance liquid chromatography (HPLC), because alteration of organics during the separation is unlikely to occur and derivatization can be avoided. Nevertheless the efficiency of HPLC columns is not enough for a desirable separation of a high number of organic compounds in water. Recently Wilson Tabor and Loper (1980) carried out initial partitioning by liquid/liquid extractions, followed by repeated high performance liquid chromatography (HPLC) for separation into smaller subfractions. Active subfractions (positive in the Ames test) were analysed by GC/MS and consequently for peak identification. The structure of this mutagenic compound has been identified and this compound is probably derived from the herbicide diallate widely used in that period (Wilson Tabor, 1982).

Extended studies on identification of nonvolatile organic compounds using HPLC have been carried out by the researchers of the WRC (Crathome et al., 1979, 1984). The most promising results have been achieved by fractionating methanol extracts of freeze-dried concentrates with reverse phase HPLC, followed by field desorption mass spectrometry (FD-MS). Another promising identification research area is the investigation to seperate electrophic compounds from the rest, because direct acting mutagens and carcinogens generally possess electrophilic properties. This approach was applied by Cheh et al. (1983) in drinking water. Recently, studies on the identification of organic mutagens has been carried out in our institute (Kool et al., 1984 a).

Characterization of the XAD-4/8 concentrates prepared from drinking water of several Dutch cities showed that besides the mutagens adsorbed on the XAD at neutral pH, a class of mutagens existed which adsorbed at pH 2-3 (Kool et al., 1981 b).
Furthermore it was demonstrated that by elution of XAD 4/8 with acetone substantially more mutagenic activity was recovered than using diethyl ether,

which is widely used for analytical purposes (GC/MS). From the mentioned characteristic behaviour of the organic mutagens can be derived that a substantial part of mutagens consists of somewhat with more polar substances. The compounds identified by GC/MS in ether eluates will in general not be identical with the organic mutagens.

The organic mutagens concentrated with the XAD procedure proved to be not gaschromatographable, since in a routine preparative GC-procedure less than 10% of the mutagenic activity (Ames test) could be recovered. To find out whether the heating step in the GC analysis may have caused this poor recovery, an experiment was set up in which a mutagenic drinking water concentrate was heated to 250° C, just like in a routine GC-analysis. The results showed that after heating a mutagenic drinking water concentrate mutagenic activity with strain TA98 and TA100 is completely lost. This indicates that the organic mutagens in the concentrate show probably thermolabile properties under GC conditions (Kool et al. 1984 a).

In previous studies, when the concentrates were fractionated by thin layer chromatography, the mutagenic activity was found predominantly in one distinct zone. Furthermore, using gelfiltration it was observed that the zone with the highest mutagenic activity in chlorinated and in chlorinated drinking water consisted of organic compounds with a molecular weight of 100 to 300 (Kool et al, 1981 b; 1982 a, 1984 a).

Additional fractionation with Sephadex LH20 using stepwise elution with isopropanol and dioxane/water was carried out to see whether the majority of the mutagenic activity could be separated from the bulk of the organics.
These results reported recently (Kool et al., 1984 a) showed that this stepwise isopropanol-dioxane/water (7:3) elution was able to separate most of the mutagenic activity with strain TA98 from the TA100 activity and the bulk of organic matter as detected by 263 nm absorption. Further fractionation of these LH20 fractions by means of repeated linear gradient HPLC-analyses (analytical and semi-preparative) in combination with the Salmonella mutagenicity test showed that the TA98 $\pm$ S9 mutagenic activity was found predominantly in two fractions and this phenomenon appeared to be good reproducable. Further attempts to separate the mutagens in these fractions from the remaining non-mutagenic organics so far were unsuccessful.

Because the identification of the mutagens by GC/MS has not been succesful up till now, attempts has been made to use specific enzyme deficient strains of Salmonella typhimurium for characterization purposes. A first indication of the nature of some of these compounds was recently obtained by testing organic drinking water concentrates with the nitroreductase deficient (NR^-) strains of Salmonella, isolated by Rosenkranz et al. (1983).
As shown in fig. 7 the direct mutagenic activity with strain TA100 of three LH20

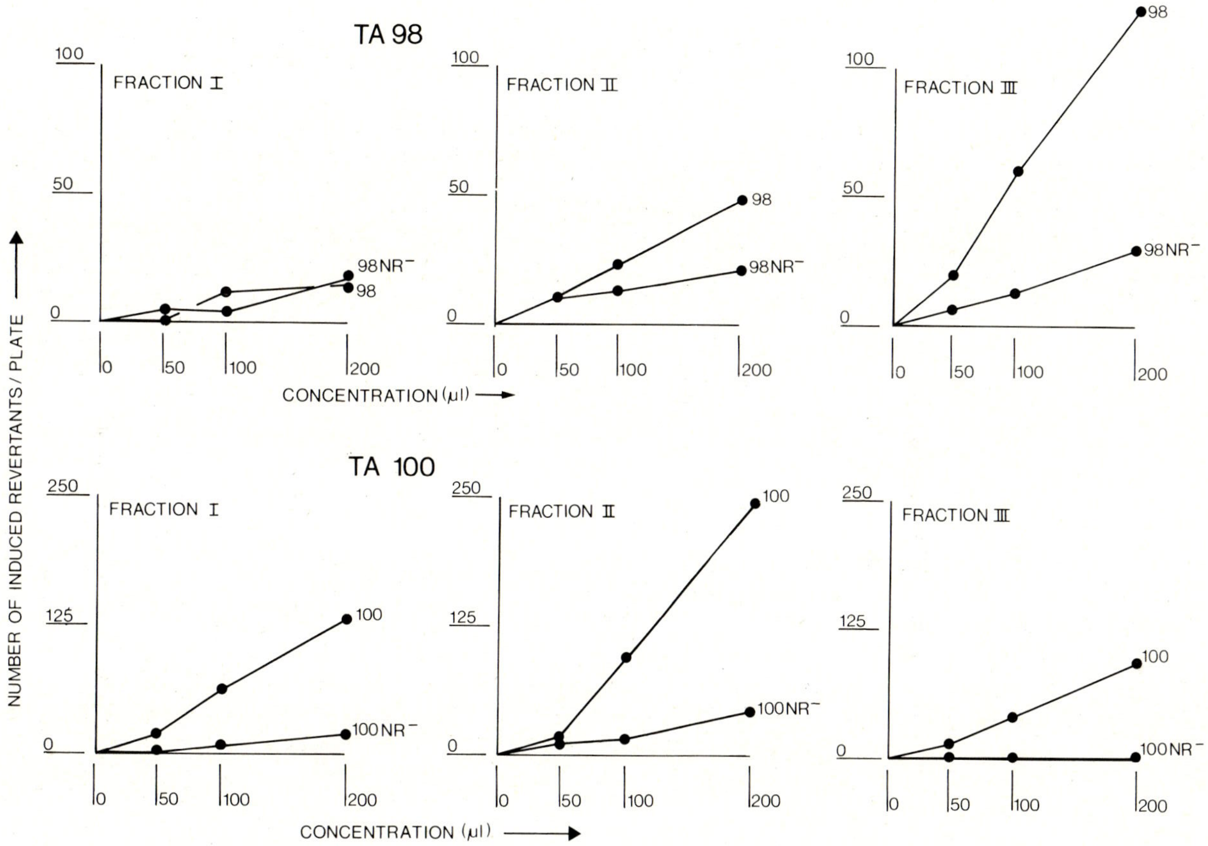

Figure 7. Mutagenic activity of a fractionated LH20 drinking water concentrate detected with nitro-reductase-deficient bacterial strains.

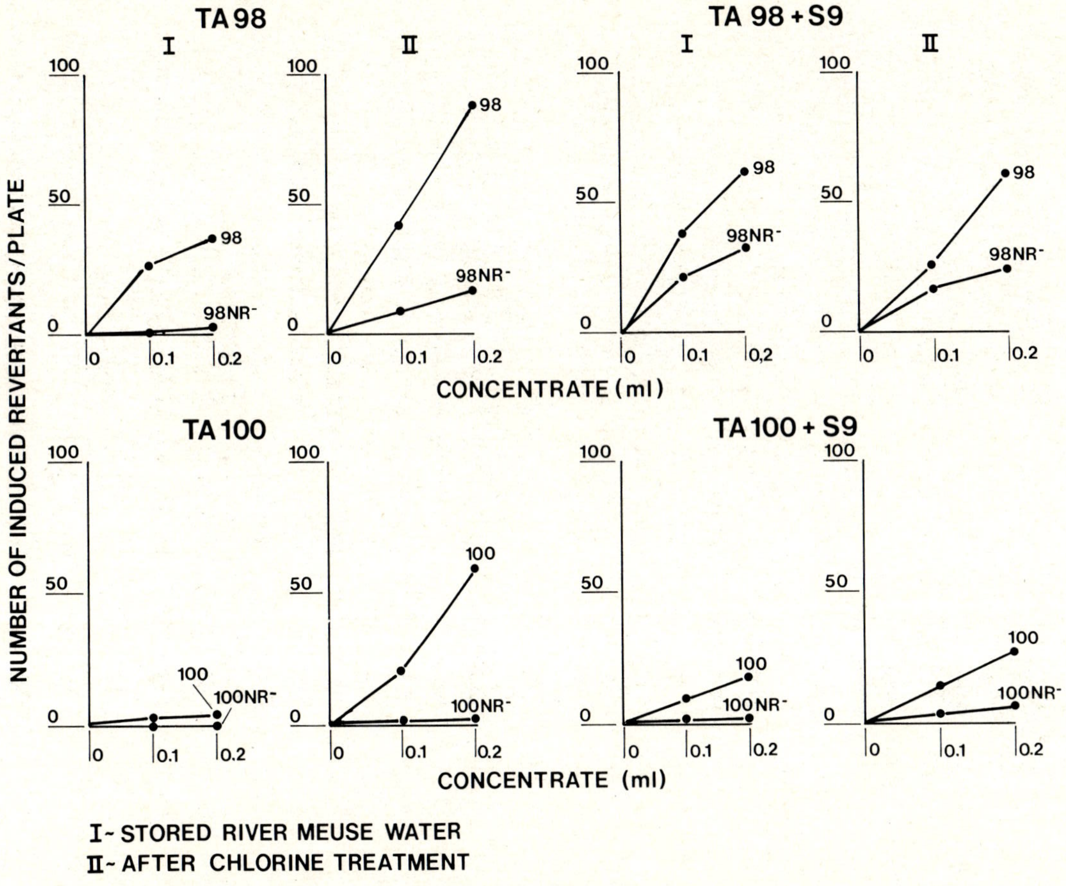

Figure 8. Influence of chlorine treatment on mutagenic activity detected with nitro-reductase deficient strains TA98 nr.- and TA100 nr-.

fractions (see reference 9) is almost completely abolished when the reductasedeficient strain TA100 NR⁻ is used. Since hardly any reduction of promutagenic activity with strain TA100 was observed (not shown) and the fact that control experiments demonstrated the capability (∼ 70%) of the 2 enzyme-deficient strains (TA98NR⁻, TA100NR⁻) to detect non-nitro reductase dependent mutagens, the results with the 3 fractions are suggestive for the presence of nitro-organics. As for the direct mutagens (strains TA98) in fraction I and II no or a small reduction was observed and only in fraction III a significant reduction occurred, this result is also suggestive for the presence of nitro-organics.

The presence of mutagenic nitroaromatic compounds in the environment is known, since these compounds are by-products of incomplete combustion processes and their presence is thus mainly a result from man made activities (59). In drinking water, however, these kinds of compounds have not been found in sufficient amounts to produce such a mutagenic effect in the Ames test. For this reason, it was investigated whether these activities with the nitro-reductase deficient strains are introduced during drinking water treatment for instance during a chlorine treatment (fig. 8).
activity in drinking water treatment.
Fig. 8 shows that stored river Meuse water also contains nitroreductase dependent mutagens (see TA98). This indicates that sufficient amounts of mutagenic nitro-organic are present in the river Meuse to show an effect in the Ames test. After a chlorine treatment (1.5 mg/l Cl_2) it is clear that new direct TA98 mutagens, not nitroreductase dependent are introduced, because a significant increase of activity is observed with both strains TA98 and TA100. The generated direct mutagenic activity with strain TA100 after the chlorine treatment however, is completely abolished with TA100NR⁻ indicating that clearly mutagenic, nitro-organics are introduced during a chlorine treatment.
Regarding the promutagenic activities with all 4 strains, no differences before and after chlorination were observed. Further investigations in this respect are underway.

ACKNOWLEDGEMENTS

The authors wish to thank the water works for their co-operation. Also, the assistance of S. Persad, R. Kerkhoff, W. Pool and the technical assistance of W. Willemse and M. Middelburg is gratefully acknowledged.

REFERENCES

Ames, B.N., Durston, W.E., Yamasaki, E. and Lee, F.D., (1973)
 Carcinogens are mutagens: A simple test system combining liver homogenates for activation and bacteria for detection. Proc. Natl. Acad. Sci., USA, 70: 2281-2285.

Ames, B.N., McCann, J. and Yamasaki; E. (1975). Mutagenicity Test, Methods for Detecting Carcinogens and Mutagens with Salmonella/Mammalian-Microsome. Mutat. Res. 31: 347-364.

Bourbigot, M.M., Paquin, J.L., Pottenger, L.H., Blech, M.F. and Hartemann, Ph. (1983) Etude du caratère mutagene de l'eau dans une filière de production à ozonation etage. Aqua 3: 99-102.

Bruchet, A., Mallevialle, J. (1983). Etude de cas d'elimination de composes organics dans de usimes de production d'eau potable. J. Francais d'Hydrologie, 14: 31-43.

Cheh, A.M., Stockdopole, J., Koski, P. and Cole, L., (1980). Non-volatile mutagens in drinking water: production by chlorination and destruction by sulphite.
 Science, 207: 90-92.

Cheh, A.M., Carlson, R.E., Hildebrand, J.R., Woodward, C. and Pereira, M.A. (1983). Contamination of purified water by mutagenic electrophiles. In Water Chlorination, Environmental Impact and Health Effects" Vol. 4. Eds. Jolley R.L., Brings, W.A., Cotruvo, J.A., Cumming, R.B., Hatice, J.S., Jacobs, V.A. Ann Arbor Sci, Ann Arbor, Mich., 1221-1235.

Coleman, W.E., Melton, R.G., Kopfler, F.C., Barone, K.A., Aurand, T.A.A. and Jellison, M.G. (1980). Identification of organic compounds in a mutagenic extract of a surface drinking water by a computerized gaschromatography/mass spectrometry system (GC/MS/com.). Environm. Sci. Technol., 14: 576-588.

Crathorne, B., Watts, C.D. and Fielding, M. (1979). The analysis of nonvolatile organic compounds in water by high performance liquid chromatography. J. Chromatogr., 185: 671-690.

Crathorne, B., Fielding, M., Steel, C.P. and Watts, C.D. (1984). Organic compounds in water. Analysis using coupled-column high-performance liquid chromatography and soft-ionization mas spectrometry. Environ. Sci. Technol., 18: 797-802.

De Greef, E., Morris, J.C., Van Kreijl, C.F. and Morra, C.F.H. (1980). Health effects in the chemical oxidation of polluted water. In: Water Chlorination Environmental Impact and Health Effects, vol. 3. Eds. Jolley R.L., Brings W.A. and Cumming R.B. Ann Arbor. Mich. 913-924.

Denkhaus, R., Grabow, W.O.K. and Prozesky, O.A. (1980). Removal of mutagenic compounds in waste water reclamation system evaluated by means of the Ames Salmonella/Microsome assay. Progr. Water Technol., 12: 571-589.

Dolara, P., Ricci, V., Burrini, D. and Griffini, O. (1981). Effect of ozonation and chlorination on the mutagenic potential of drinking water. Bull. Environm. Contam. Toxicol., 27: 1-6.

Duguet J.P., Ellal A, Brodard E. and Mallevialle J. (1983). Evolution de la mutagenese des eaux au cours d'un traitement d'ozonation. Paper presented on I.O.A. Brussel 1983.

Hrubec, J. (1983) Van Kreijl, C.F., Morra, C.F.H. and Sloof, W. Treatment of Municipal Waste Water by reverse osmoss and activated carbon-removal of organic micropollutants and reduction of toxicity. Sci. Total Environ., 27: 71-88.

Hrubec, J. Luijten, J.A., Luijten, W.C.M.M. and Piet, G.J. (1985 a). Quality changes during groundwater recharge, in press.

Hrubec, J. and Kool, H.J. (1985 b). Onderzoek betreffende enkele gezondheidsaspecten van de toepassing van chloordioxide, ozon en UV bij de bereiding van drinkwater. Paper presented at the 38th International Conference of CEBEDEAU Brussel, June 10-12, 1985.

Hrubec J., 't Hart M.J., Kool H.J. and Marsman, P. (1985 c). Onderzoek betreffende de toepassing van alternatieve desinfectiemethoden bij de bereiding en distributie van drinkwater.

RIVM-rapport nr. 840132001. RIVM P.O. Box 10, Leidschendam, The Netherlands.
KIWA (1984). Kwaliteit van oppervlaktewater bij kunstmatige infiltratie in de Nederlandse kustduinen; Hygiënische aspecten.
Mededeling 81, Ed. Stuijfzand, P.J. KIWA N.V. Rijswijk, The Netherlands.
KIWA (1983). Neveneffecten van de chloring. Mededeling nr. 74, Ed. Kruithof, J.C., Nieuwegein, The Netherlands.
Knoppert, P.L., Oskam, G. and Vreedenburgh, R.G.H. (1980). An overview of European water treatment practice. J.A.M. Wat. Works Ass., 72: 592-599.
Kool, H.J., Van Kreijl, C.F., Van Kranen, H.J. and De Greef, E. (1981 a). The use of XAD resins for the detection of mutagenic activity in water I. Studies with surface water: Chemosphere 10: 85-98.
Kool, H.J., Van Kreijl, C.F., Van Kranen, H.J. and De Greef, E. (1981 b). Toxicity assessment of organic compounds in drinking water in The Netherlands. Sci. Total Environm. 18: 135-153.
Kool, H.J., Van Kreijl, C.F., De Greef, E. and Van Kranen, H.J. (1982 a). Presence, introduction and removal of mutagenic activity during the preparation of drinking water in The Netherlands.
Environm. Health Persp., 46: 207-214.
Kool, H.J., Van Kreijl, C.F. and Zoeteman, B.C.J. (1982 b). Toxicity assessment of organic compounds in drinking water. CRC Crit. Rev. Environm. Control, 12: 307-357.
Kool, H.J., Van Kreijl, C.F. and Verlaan de Vries M. (1984 a). Concentration, fractionation and characterization of organic mutagens in drinking water. Paper presented at Environmental Chemistry Division Symposium: Concentration Techniques for Collecting and Analyses of Organic Chemicals for Biological Testing of Environmental Samples. Philadelphia, 26-31 August 1984.
Kool H.J., Van Kreijl, C.F. and Van Oers H. (1984 c). Mutagenic activity in drinking water in The Netherlands. A survey and a correlation study. Toxicol. Environm. Chem., 7: 111-129.
Kool, H.J. and Van Kreijl C.F. (1984 c). Formation and removal of mutagenic activity during drinking water preparation. Water Res., 18: 1011-1016.
Kool H.J., Van Kreijl, C.F. and Hrubec, J. (1985 a). Mutagenic and carcinogenic properties of drinking water. In: Water Chlorination Environmental Impact and Health Effects vol. 5 Eds. Jolley R.L. et al. Lewis Publishers (in press), 187-205.
Kool, H.J. and Hrubec, J. (1985 b). The influence of an ozone chlorine and chlorine dioxide treatment on mutagenic activity in (drinking) water. Paper presented on the International Ozone Symposium, Wasser Berlin, April 23-24, Berlin 1985.
Loper, J.C., (1980) Mutagenic effects of organic compounds in drinking water. Mutation Res., 76: 241-268.
Loper, J.C., Rosenblum, L. and Tabor M.W. (1984). Granular activated carbon removes both mutagens and mutagens forming potential chlorinated drinking water. Paper presented at 5th Conference on Water Chlorination: Environmental Impact and Health Effects, Williamsburg, Virginia, June 3-8, 1984.
McCann, J., Choi, E., Yamasaki, E. and Ames, B.N. (1975b). Detection of carcinogens as mutagens in the Salmonella/microsome test: Assay of 300 chemicals. Proc. Nat. Acad, Sci., 72: 5135-5139.
Monarca, S., Meier, J.R. and Bull, R.J. (1983). Removal of mutagens from drinking water by granular activated carbon. Water Res., 17: 1015-1026.
NAS, (1977). National Academy of Sciences Drinking Water and Health volume 1 and 2. Washington DC.
NAS, (1980). National Academy of Sciences Drinking Water and Health volume 3. National Academy press. Washington DC.
NAS, (1981). National Academy of Sciences Drinking Water and Health volume 4. National Acadamy Press. Washington DC.
Oskam, G. (1980). Berging van oppervlaktewater in open bekkens. H_2O, 13: 189-197.

Pendygraft, G.N., Schlegel, F.E. and Huston, M.J. (1979). Organics in drinking water: maximum contaminant level as an alternative to the GAC treatment requirement. J. Am. Water Works Assoc., 71: 174-183.

Piet, G.J., Slingerland, P., De Grunt F.E., Van der Heuvel, M.P.M. and Zoeteman, B.C.J. (1978). Determination of very volatile halogenated organic compounds in water by means of direct headspace analysis. Anal. Letters A11 5: 437-448.

Piet, G.J. and Zoeteman, B.C.J. (1980). Organic waterquality changes during sand-, bank- and dune infiltration of surface water in The Netherlands. J. Am. Water Wks. Ass., 72: 400-404.

Purchase, I.H.P., (1982). An appraisal of predictive tests for carcinogenicity. ICPEMC Working Paper 2/6. Reviews in Genetic Toxicology.

Rosenkranz, H.S., and Mermelstein, R. (1983). Mutagenicity and genotoxicity of nitroarenes. Mutation Res., 114: 217-267.

Sanders, R. (1980). Verbesserung des Pyrohydrolyse Verfahrens. Veroffentlichungen des Bereichs und des Lehrstuhls für Wasserchemie, Karlsruhe Heft 15, 128-162.

Slooff, W. and Van Kreijl, C.F. and De Zwart, D. (1984). Biologische parameters en oppervlaktewater (meetnetten). H_2O 17: 2-5.

Van der Gaag, M.A., Noordsij, A. and Oranje J.P. (1982). "Presence of Mutagens in Dutch Surface water and Effects of water Treatment processes for Drinking Water Preparation", in progress in Clinical and Biological Research, Vol. 109, Mutagens in Our Environment 277-286. M. Sorse and H. Vaino, Eds. New York: Alan R. Liss, Inc.

Van Hoof, F. (1983). Influence of ozonation on direct acting mutagens formed during water chlorination. In "Water Chlorination, Environmental Impact and Health Effects" Vol. 4. Eds. Jolley, R.L., Brings, W.A., Cotruvo, J.N., Cumming, R.B. Ann Arbor Sci, Ann Arbor Mich. 1211-1220.

Van Hoof, F. (1985). Formation of mutagenic activity during surface water preozonation and its removal in drinking water treatment. Chemosphere (in press).

Van Kreijl, C.F., Kool, H.J., De Vries, M., Van Kranen, H.J., and De Greef, E. (1980). Mutagenic activity in the rivers Rhine and Meuse in The Netherlands. Sci. Total Environm., 15: 137-147.

Wilson Tabor, M. and Loper, J.C. (1980). Separation of mutagens from drinking water using coupled bioassay/analytical fractionation. Intern. J. Environm. Anal. Chem. 8: 197-215.

Wilson Tabor, M. (1982). Structure elucidation of 3-(2-chloroethoxy)-1,2-dichloropropene a new promutagen from an old drinking water residue. Paper presented at the 12th Ann. Symposium on the Analytical Chemistry of Pollutants, April 14-16, 1982, Amsterdam.

Zoeteman, B.C.J. (1978). Sensory assessment and chemical composition of drinking water. Thesis State University of Utrecht. The Netherlands.

Zoeteman, B.C.J., Hrubec, H., De Greef, E. and Kool, H.J. (1982). Mutagenic activity associated with by-products of drinking water disinfection by chlorine, chlorine dioxide dioxide, ozone and U.V. irradiation. Environm. Health, Persp., 46: 197-205.

MUTAGENIC ACTIVITY IN HUMIC WATER AND ALUM FLOCCULATED HUMIC WATER TREATED WITH ALTERNATIVE DISINFECTANTS

P. Backlund[1], L. Kronberg[1], G. Pensar[1] and L. Tikkanen[2]

[1] Department of Organic Chemistry, Åbo Akademi, SF-20500 Turku, Finland

[2] Technical Research Centre of Finland, Food Research Laboratory, SF-02150 Espoo, Finland

ABSTRACT

Mutagenic activity in Salmonella typhimurium strains TA 100, TA 98 and TA 97 has been determined for humic water and alum flocculated humic water, treated with the alternative disinfectants chlorine, ozone, chlorine dioxide, ozone/chlorine and chlorine/chlorine dioxide.

The most pronounced activity was found for chlorine treated water tested on strain TA 100 without metabolic activation (S9 mix). Ozone treatment prior to chlorination did not alter the activity, while treatment with chlorine in combination with chlorine dioxide reduced the activity to a level somewhat over the background. No mutagenic response was detected in waters treated with ozone or chlorine dioxide alone. In presence of S9 mix all water extracts studied were non-mutagenic.

INTRODUCTION

Municipal drinking water is generally supplied from surface water in nearby rivers or lakes. In Finland humic substances comprise by far the major portion of the organic material present in surface waters (TOC 10-30 mg/l), while the contribution of anthropogenic compounds usually is small.

During water purification, using conventional treatment practices (flocculation, sedimentation and filtration), 70-80 % of the humic material is removed from the water. The remaining organics are subjected to the oxidizing effect of water disinfectants.

During the last decade there have been several reports on mutagenic activity in concentrates of chlorine treated waters (Cheh et al. 1980, Kool et al. 1982, Marouka et al. 1983). The influence of other disinfectants on the mutagenic activity of water has, however, been studied to a more limited extent. Zoeteman et al. (1982) have reported an increase of direct acting mutagens in surface water after treatment with chlorine dioxide (strains TA 100 and TA 98). In the same study ozone treatment of raw water was found to reduce mutagenic activity in strain TA 98 both with and without metabolic activation. Increased mutagenic activity has been reported in water containing soil fulvic acid after treatment with ozone in combination with chlorine (Kowbel et al. 1984).

In the present study mutagenic activity in Salmonella typhimurium strains TA 100, TA 98 and TA 97 was determined for humic water (HW) and alum flocculated humic water (FHW), treated with the alternative disinfectants chlorine, ozone, chlorine dioxide, chlorine/chlorine dioxide and ozone/chlorine. The resulting response patterns of humic water and alum flocculated water were compared in order to test the hypothesis that mutagenic compounds are produced as a result of reactions between disinfectants and humic substances (Meier et al. 1983, Kronberg et al. 1985).

MATERIALS AND METHODS

Non-mutagenic natural humic water (see Table 1) was collected from a lake (Savojärvi) situated in a marsh region in the southwestern part of Finland, and treated according to the scheme shown in Fig. 1.

Alum flocculation was performed in the laboratory by adding an aqueous solution of $Al_2(SO_4)_3 \cdot 18\ H_2O$ to humic water (pH 5.9) while stirring. After settling the clear water was decanted for further treatment.

TABLE 1
Total organic carbon contents of water samples studied and disinfectant dosages applied.

	TOC - value (mg/l)	Cl_2- dosage (mg/l)	O_3- dosage (mg/l)	ClO_2- dosage (mg/l)	Cl_2/ClO_2- dosage (mg/l)	O_3/Cl_2- dosage (mg/l)
Humic water	21.0	21.0	10.0	21.0	10.5/10.5	10.0/21.0
Alum flocc. humic water	6.4	6.5	2.9	6.5	3.25/3.25	2.9/6.5

```
                HUMIC WATER
          ┌──────────┬──────────┐
          │          │       ── pH 5.9
          │     ┌────┴──────────┐
          │     │ Alum flocculation │
          │     └────┬──────────┘
          │── pH 7.0 │
          │   ┌──────┴──────────┐
          │   │ Treatment with   │
          │   │ Cl₂, O₃, O₃/Cl₂, │
          │   │ ClO₂, or Cl₂/ClO₂│
          │   └──────┬──────────┘
          │────── 60 h ─────────│
          │────── pH 2.1 ───────│
    ┌───┐ │   ┌─────────────┐   │ ┌───┐
    │ X │ │   │Ethyl acetate│   │ │ X │
    │ A ├─┤   └──────┬──────┘   ├─┤ A │
    │ D │ │   ┌──────┴──────┐   │ │ D │
    │4/8│ │   │Concentrates of│  │ │4/8│
    └───┘ └───┤  HW and FHW  ├───┘ └───┘
              └──────┬──────┘
                ┌────┴─────┐
                │ AMES TEST│
                └──────────┘
```

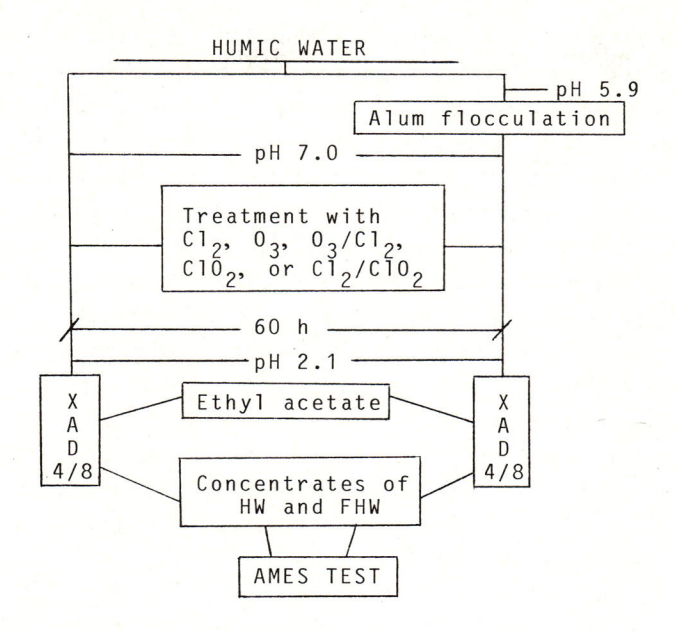

Fig. 1. Scheme of water treatment.

Prior to disinfection pH of the water samples was adjusted to 7.0 by adding 4 M NaOH and a potassium phosphate buffer.

Chlorine- and chlorine dioxide treatments were carried out at room temperature using freshly made stock solutions in distilled water.

The chlorine solution was prepared by bubbling chlorine gas (generated by adding fuming HCl to $KMnO_4$, with subsequent cleaning and drying) into a solution of 2.3 % NaOH until a pH of 7-8 was obtained. The concentration of chlorine in the solution was determined by the starch-potassium iodide technique.

The chlorine dioxide stock solution was prepared by adding a solution of HCl (9 %) to aqueous 7.5 % $NaClO_2$. The resulting chlorine dioxide gas was swept off from the mixing bottle by nitrogen and collected in buffered (pH 7.0) distilled water. The concentration of chlorine dioxide in the solution was determined by the starch-potassium iodide method.

Ozonation was performed using a laboratory ozonator (Herrman-Labor Lo-50-1) with a maximum capacity of 7 g O_3/h. Oxygen was used as feed gas. The ozone concentration in the gas stream was determined both before and after ozonation by the starch-potassium iodide technique.

The ratio of TOC to the amount of chlorine or chlorine dioxide added to the water was 1:1, resulting in a residual concentration of 0-0.2 mg/l after a reaction time of 60 h at room temperature in the dark. Noack et al. (1978) have shown that treatment of water containing humic acid with a chlorine/chlorine dioxide mixture of 1:1 (w/w) minimizes the combined formation of chloroform and chlorite. For this reason the chlorine/chlorine dioxide treatment was carried out by adding a mixture of 1:1 of the two disinfectants at a ratio of TOC to total disinfectant of 1:1. The ozone dosage applied to humic water and alum flocculated humic water was 10 mg/l and 2.9 mg/l respectively. In the combined ozone/chlorine treatment, ozonation was performed 2 h prior to chlorination. We have noticed in earlier experiments that pre-ozonation of water containing humic substances does not significantly influence the chlorine consumption of the water. For this reason the chlorine dosage applied after ozonation was equal to that applied when using chlorine alone, and, consequently, resulting in a residual chlorine concentration of 0-0.2 mg/l after 60 h reaction time.

Prior to concentration on XAD 4/8, pH of the water was adjusted to 2.1 (4 M HCl). The adsorbed organics were eluted by ethyl acetate and, prior to assay, the solvent was replaced by DMSO. The concentration procedure used has previously been described in detail (Kronberg et al. 1985). The largest equivalent volumes of water extracts tested were 50 ml and 200 ml for humic water and flocculated humic water respectively.

Mutation tests were performed according to the method of Ames (Ames et al. 1975) with minor modifications (von Wright et al. 1978). The Salmonella typhimurium tester strains TA 100, TA 98 and TA 97 were provided by Dr. B.N. Ames, University of California. When metabolic activation was used, liver homogenate (S9mix) was prepared from the livers of male Sprague-Dawley rats induced with Na-phenobarbital and β-naphtoflavone (Matsushima et al. 1979).

The test results are means of duplicate plates and each test was repeated at least once.

RESULTS

Mutagenic activity was found in chlorinated water concentrates when tested on strains TA 100, TA 98 and TA 97 without enzyme activation (see Fig. 2). The highest responces were noted for strain TA 100.

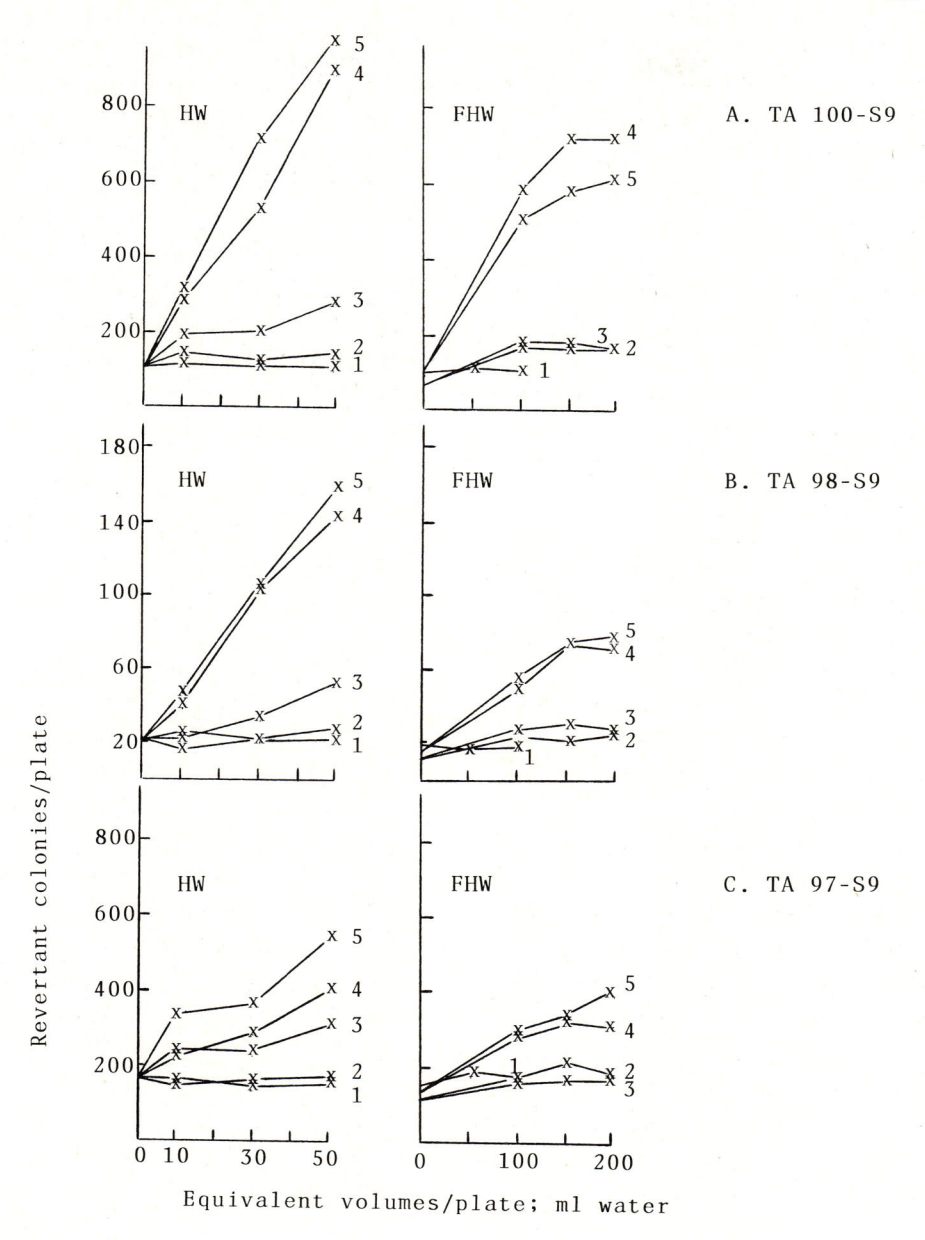

Fig. 2. Mutagenic responce of strains TA 100 (A), TA 98 (B) and TA 97 (C) (-S9mix) for humic water and alum flocculated humic water treated with 1. ozone, 2. chlorine dioxide, 3. chlorine/chlorine dioxide, 4. chlorine, 5. ozone/chlorine.

Ozone treatment prior to chlorination did not alter the activity significantly. However, when using extremely high ozone dosages (33.2 mg/l) a reduced activity was noted (see Fig. 3).

The combined chlorine/chlorine dioxide treatment reduced the activity to a level only somewhat over the background. This reduction (as compared to the activity obtained after chlorine treatment) was greater than the reduction due to the lower chlorine dosage applied in the combined treatment (see Fig. 4).

No mutagenic activity was detected in waters treated with ozone or chlorine dioxide alone (see Fig. 2).

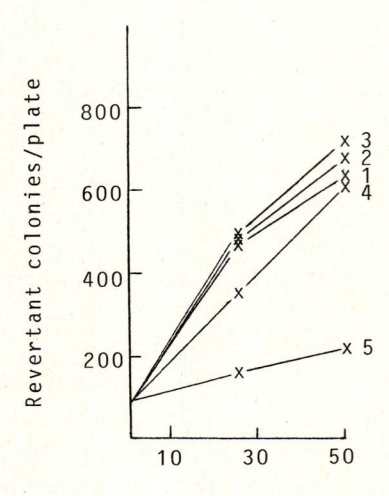

Fig. 3. Influence of pre-ozonation on mutagenic activity (TA 100-S9mix) of humic water treated with chlorine. Chlorine dosage 21.0 mg/l.
1. 0 mg O_3/l, 2. 5.7 mg O_3/l, 3. 11.5 mg O_3/l,
4. 23.5 mg O_3/l, 5. 33.2 mg O_3/l.

In presence of metabolic activation all water extracts studied were non-mutagenic (not shown).

The mutagenic response patterns for concentrates of humic water and alum flocculated humic water were similar, but the activities in humic water were more pronounced (see Fig. 2).

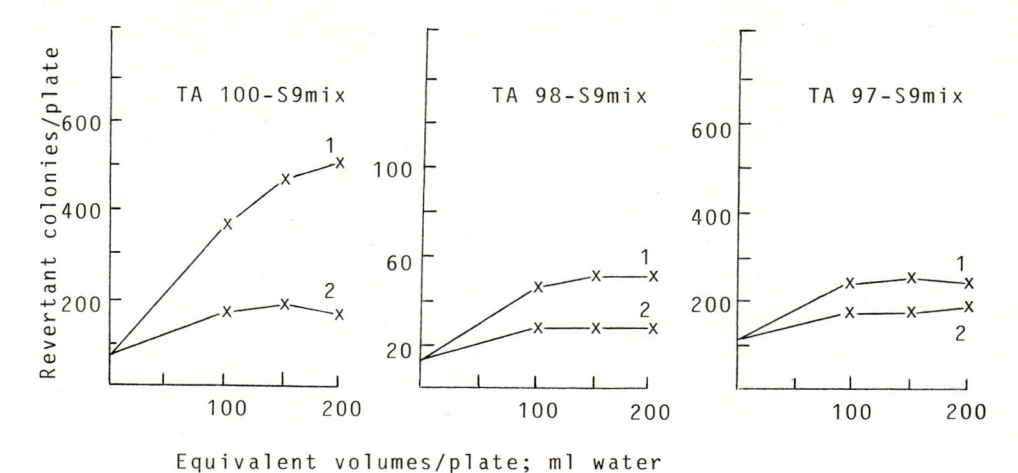

Fig. 4. Mutagenic response of strains TA 100, TA 98 and TA 97 (-S9mix) for alum flocculated humic water treated with
1. chlorine (3.25 mg/l), 2. chlorine/chlorine dioxide (3.25 mg/l/ 3.25 mg/l).

CONCLUSIONS

The results show that chlorine treated water containing humic substances exhibits mutagenic activity. This is in agreement with several previous studies (Meier et al. 1983, Kringstad et al. 1983, Kronberg et al. 1985). Ozone treatment prior to chlorination is not able to alter the activity produced during chlorination significantly at ozone dosages normally used in water treatment practices. Treatment with ozone or chlorine dioxide alone does not produce detectable amounts of mutagenic compounds.

The results indicate that a combined chlorine/chlorine dioxide treatment might be a better alternative than a combined ozone/ chlorine treatment in order to minimize mutagenic activity produced during disinfection of water containing humic substances.

As stated before alum flocculation is capable of removing only 70-80 % of the humic material present in water. The similar responses patterns obtained for the mutagenic activity in humic water and alum flocculated humic water support the assumption that mutagenic compound(s) are products of reactions between the disinfectant(s) and humic substances.

ACKNOWLEDGEMENT

We wish to thank Bjarne Holmbom, Rainer Ekman (Laboratory of Forest Products Chemistry, Åbo Akademi) and Gösta Wahlroos (Turku Water Works) for valuable comments and practical help during the study and the Academy of Finland for financial support.

REFERENCES

Ames, B.N., McCann, J. and Yamasaki, E., 1975. Methods for detecting carcinogens and mutagens with the Salmonella/mammalian-microsome mutagenicity test. Mutat. Res. 31:347.

Cheh, A.M., Skochdopole, J., Koski, P. and Cole, L., 1980. Non-volatile mutagens in drinking water: Production by chlorination and destruction by sulphite. Science 207:90-92.

Kool, H.J., van Kreijl, C.F., de Greef, E. and van Kranen, H.J., 1982. Presence, introduction and removal of mutagenic activity during the preparation of drinking water in the Netherlands. Environ. Health Perspect. 46:215-227.

Kowbel, D.J., Malaiyandi, M., Paramasigamani, V. and Nestmann, E.R., 1984. Chlorination of ozonated soil fulvic acid: Mutagenicity studies in Salmonella. Sci. Total Environ. 37:171-176.

Kringstad, K.P., Ljungquist, P.O., deSousa, F. and Strömberg, L. M., 1983. On the formation of mutagens in the chlorination of humic acid. Environ. Sci. Technol. 17:553-555.

Kronberg, L., Holmbom, B. and Tikkanen, L., 1985. Mutagenic activity in drinking water and humic water after chlorine treatment. Vatten 41:106-109.

Matsushima, T., Sawamura, M., Hara, K. and Sugimura, T., 1979. A safe substitute for poly-chlorinated biphenyl as an inducer of metabolic activation system. -In vitro metabolic activation in mutagenesis testing (de Serres, F.J., Fouts, J.R., Bend, J.R. and Philpot, R.M. eds.). Elsevier/North-Holland, Amsterdam.

Marouka, S. and Yamanaka, S., 1983. Mutagenic potential of laboratory chlorinated river water. Sci. Total Environ. 29:143-154.

Meier, J.R., Lingg, R.D. and Bull, R.J., 1983. Formation of mutagens following chlorination of humic acid. A model for mutagen formation during drinking water treatment. Mutat. Res. 118:25-41.

Noack, M.G. and Doerr, R.L., 1978. Reactions of chlorine, chlorine dioxide and mixtures thereof with humic acid: An interim report. Water Chlorination: Environ. Impact Health Eff., Proc. Conf. 1977. 2:49-58.

von Wright, A., Niskanen, A. and Pyysalo, H.J., 1978. Mutagenic properties of ethylidene gyromitrin and its metabolites in microsomal activation tests and in host mediated assay. Mutat. Res. 38:117.

Zoeteman, B.C.J., Hrubec, J., de Greef, E. and Kool, H.J., 1982. Mutagenic activity associated with by-products of drinking water disinfection by chlorine, chlorine dioxide, ozone and UV-irradiation. Environ. Health Perspect. 46:197-205.

A COMPARISON OF METHODS FOR CONCENTRATING MUTAGENS IN DRINKING WATER - RECOVERY ASPECTS AND THEIR IMPLICATIONS FOR THE CHEMICAL CHARACTER OF MAJOR UNIDENTIFIED MUTAGENS

B. Wigilius[1], H. Borén[1], G.E. Carlberg[2], A. Grimvall[1] and M. Möller[2]
[1]Department of Water in Environment and Society, Linköping University, S-581 83 Linköping, Sweden
[2]Center for Industrial Research, P.B. 350, Blindern, 0314 Oslo 3, Norway

ABSTRACT

A comparison of techniques for concentrating mutagenic compounds in drinking water has shown that XAD-2 adsorption and dichloromethane extraction have acceptable and almost identical enrichment properties, while purging at an elevated temperature is inappropriate in this context. Quantitatively, the most important drinking water mutagens could only be adsorbed (extracted) after acidification of the water, and even then recovery was far from complete. Recovery experiments with known mutagens from pulp mill effluents have shown that none of the major chlorination-stage mutagens identified thus far can explain the mutagenic activity of extracts from neutral or acidified chlorinated drinking water.

INTRODUCTION

During the last decade there have been numerous reports on mutagenic substances in chlorinated drinking water. Extracts of organic compounds obtained by resin adsorption, liquid-liquid extraction and other concentration techniques have been shown to be mutagenic in various short-term mutagenicity tests (Loper, 1980; Cheh et al., 1980; Kool et al., 1982a,b; Nestman et al., 1979; Gruener and Lockwood, 1980 and Grabow et al., 1981a,b). From experiments with laboratory chlorination of humic material, it is known that naturally occurring organic substances are important precursors of mutagenic compounds (Meier et al., 1983 and Kringstad et al., 1983a). The chemical character of the mutagens themselves has been described in terms of polarity and volatility, but their exact identity is still unknown (Kool et al., 1982a,b). Furthermore, the efficiency of concentration techniques presently used is only partially known.

The first part of the present study deals with experiments performed to recover Ames mutagens (strain TA 100) enriched from chlorinated drinking water by resin adsorption, liquid-liquid extraction and a purging technique. The results of these experiments form a basis for discussion of the chemical character of mutagenic compounds formed when drinking water is chlorinated. In the second part of the study, the recovery experiments were extended to water samples spiked with known mutagens from chlorobleaching of pulp. The resulting recovery patterns were compared in order to test the hypothesis that chlorinated drinking water contains almost the same mutagenic compounds as spent chlorination liquor from pulp bleaching (Kringstad et al., 1983a).

MATERIALS AND METHODS

All samples of chlorinated drinking water were taken from a water works with non-mutagenic raw water. The $KMnO_4$-value after alum flocculation, rapid sand filtration and slow sand filtration was approximately 12 mg/l expressed as $KMnO_4$. The chlorine dose used was 0.7 mg/l, and chlorination was performed at pH = 8.7.

Our procedure for XAD-adsorption has previously been described in detail (Wigilius et al.). In order to minimize the leakage of impurities from the resin the original solvent sequence methanol-water has been replaced by the sequence methanol-ether-water (James et al., 1981). The liquid-liquid extraction was performed by simply mixing the water sample with dichloromethane (ratio 12:1) and stirring for half an hour. The purging technique we used has also been described previously (Borén et al., 1982). By replacing Grob's original closed-loop system with an open system, purging can be performed at a higher temperature (60°C), with a corresponding increase in the recovery of more polar compounds (Borén et al., in press).

Throughout the study, mutagenicity refers to a positive response in Ames bacterial assay, strain TA 100, without metabolic activation. Before testing for mutagenicity the drinking water extracts were evaporated to a small volume and then dissolved in DMSO. Recovery of model compounds has been calculated using capillary gas chromatography.

RECOVERY EXPERIMENTS WITH MUTAGENS IN AUTHENTIC DRINKING WATER SAMPLES

Resin adsorption and liquid-liquid extraction are by far the most widely used techniques for the concentration of mutagens from drinking water. In this study we have compared adsorption on the unpolar resin XAD-2, liquid-liquid extraction with dichloromethane and a modified version of Grob's closed-loop stripping technique. In order to be able to observe any incomplete adsorption with the XAD-technique, we used as many as four moderately sized columns run in series (25 ml resin for a 20 l water sample). For the same reason repeated liquid-liquid extractions of each water sample were performed.

Figures 1a and 1b show typical yields of mutagenic substances from the first XAD column and the first dichloromethane extraction. For neutral water samples, recovery was slightly higher when using liquid-liquid extraction than when using

XAD-adsorption; opposite results were obtained after acidifying the water samples. However, as in previous comparisons of these two techniques (Grabow et al., 1981a; Maruoka & Yamanaha, 1983), the differences between XAD-adsorption and liquid-liquid extraction were not very pronounced.

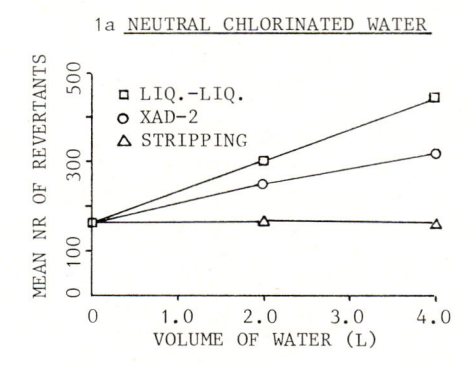

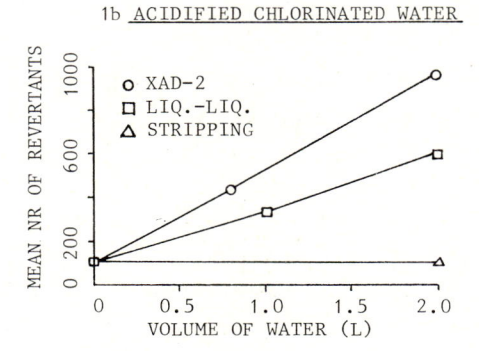

Fig. 1a. Comparison of the effectiveness of liquid-liquid extraction, XAD-2 and stripping in recovering mutagens from neutral water.
Fig. 1b. Comparison of the effectiveness of liquid-liquid extraction, XAD-2 and stripping in recovering mutagens from acidified water.

The extracts obtained by stripping were all non-mutagenic. Our experiments showed that there were three reasons why this technique fails to enrich mutagens: some of the mutagens were destroyed when the water samples were heated to and held at 60°C for two hours; the most important mutagens could not be removed from the water by stripping; certain suspected drinking water mutagens, e.g. tetra-, penta- and hexachloroacetone, were irreversibly adsorbed to the charcoal filter.

Figures 2a and 2b illustrate the dramatically increased yield of mutagens after acidification of chlorinated drinking water samples. If the mutagens from the second dichloromethane extraction or the second XAD column are also included (see the latter part of this section), the increase due to acidification approaches a factor ten. Similar results have been obtained by Kronberg et al. (1985). The fact that certain mutagens could only be adsorbed on XAD after acidification was also noticed by Kool et al. (1981). Grabow (1981a) on the other hand, reported a lower yield of mutagens after acidification.

Extracts from four XAD columns run in series were tested separately in order to obtain more detailed information about the adsorption of mutagenic compounds on XAD-2. When neutral chlorinated drinking water was tested only the first XAD

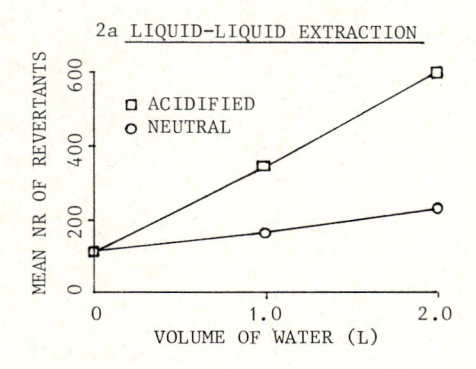

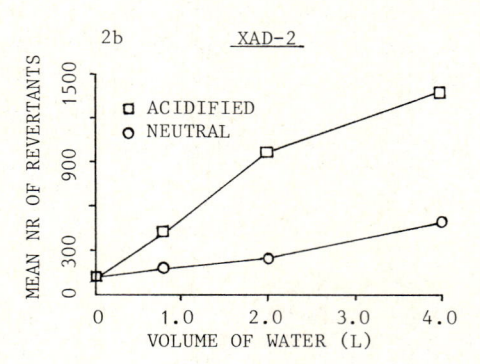

Fig. 2a. Comparison of recovery of mutagens from neutral and acidified water by the use of liquid-liquid extraction.
Fig. 2b. Comparison of recovery of mutagens from neutral and acidified water by the use of XAD-2 adsorption.

column gave rise to a positive response in the Ames test. Acidification (pH 2.1) increased the mutagenic activity from the first column approximately five fold, but at the same time there was a considerable breakthrough of mutagenic compounds (see Fig. 3). In fact, the extract from the second column showed about half the mutagenicity of the extract from the first column. Similar experiments with repeated dichloromethane extractions gave almost identical results.

What conclusions can be drawn from the above results about the chemical character of the mutagenic compounds? First of all, it is obvious that the compounds accounting for the large increase in mutagenicity when the water is acidified must have acidic properties. When the pH-value is decreased, the affinity of such compounds to the XAD resin increases, which explains the relatively high mutagenicity from the first column after acidification. Since acids are rather polar even in protonized form, the adsorption on the first XAD column will be incomplete, thus giving a positive response from the second column as well. When concentrating organic compounds from neutral or basic water samples, the yield of acids is very low. Therefore if the acids were the only mutagenic compounds in chlorinated drinking water, we could, in the neutral case, expect either a negative response from both the first and second column or a low but equal mutagenicity from both columns. We have, however, observed a positive response from the first column and no significant response from the second. Consequently, the chlorinated drinking water samples must contain at least two groups of mutagens: one neutral, relatively unpolar group and one acidic group. The same arguments can be used and the same conclusions can be drawn when considering the results from the dichloromethane extractions.

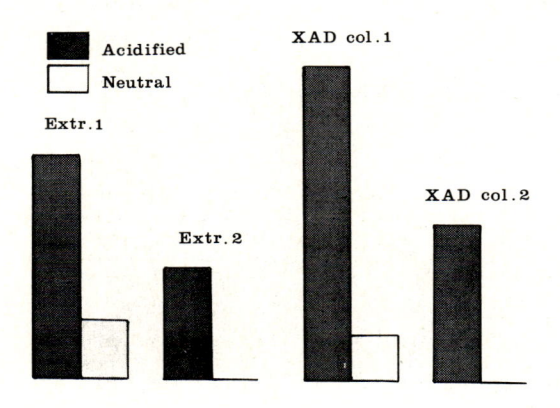

Fig. 3. Recovery pattern of mutagens from experiments with repeated extractions and XAD-columns run in series.

In principle, a lower yield of mutagens from non-acidified water samples can also be explained by alkaline instability of important mutagens (Kringstad et al., 1983b). In our tests, however, this explanation was rejected since storing the water sample for one day before acidification did not alter the mutagenicity pattern.

RELATIONSHIP BETWEEN DRINKING WATER MUTAGENICITY AND BLEACHERY EFFLUENT MUTAGENS

Several authors have suggested that the mutagenicity of chlorinated drinking water has, at least partly and for certain types of raw water, the same chemical origin as the mutagenicity of spent liquor from chlorobleaching of pulp (Kringstad et al., 1983a and Rapson et al., 1980). We have tested this hypothesis by comparing the recovery patterns of some important bleachery effluent mutagens with the mutagenicity patterns reported in section three. Inspired by the work of Kringstad and co-workers (1983a) we first considered 2-chloropropenal and a number of chloroacetones (see Table 1).

Recoveries and detection limits for the above mentioned compounds were determined by gas chromatography in experiments where the model compounds were first added to water and then enriched by XAD adsorption, dichloromethane extraction and stripping, respectively. As can be seen in the qualitative results shown in

Table 1, the dicholoroacetones and 2-chloropropenal can be enriched by XAD adsorption and dichloromethane extraction, while the stripping technique is less successful for these compounds. The more substituted chloroacetones are not satisfactorily enriched by any of the concentration methods.

TABLE 1
Chemical recovery of bleaching effluent mutagens compared to mutagenic response from chlorinated drinking water. The model compounds are 2-chloropropenal and five chloroacetones.

	Mutagenicity		Recovery					
	Neutral water	Acidif. water	2-Cl-pr.	1,1-di	1,3-di	1,1,3,3-tetra	penta	hexa
XAD 1st	+	++	+	+	+	(+)	−	−
XAD 2nd	−	+	+	+	+	−	−	−
Extr. 1st	+	++	+	+	+	(+)	−	−
Extr. 2nd	−	+	NT	+	+	(+)	−	−
Stripping	−	−	−	−	−	−	−	−

NT = not tested

By comparing the recovery from the first extraction with that of the second (the first and second XAD columns), we could further show that the extraction (adsorption) efficiency for all model compounds was less than 50%. Due to instability of the model compounds and other losses, the recovery from the first extraction (the first XAD column) was even lower. In fact, the highest values were obtained for 2-chloropropenal and 1,1-dichloroacetone, the first dichloromethane extraction resulted in a recovery of approximately 30%, while XAD adsorption gave slightly lower values. Replacing dichloromethane with ether in the liquid-liquid extraction did not cause any dramatic changes in recovery.

In section three we concluded that acidic compounds are the most active mutagens in chlorinated drinking water. The selected bleachery effluent mutagens are all neutral, and thus cannot be responsible for the main mutagenicity in acidified drinking water.

A closer analysis of the results has shown that the model compounds are not important mutagens in neutral water samples either. In fact, if 2-chloropropenal or any of the dichloroacetones were responsible for the mutagenic activity of the first dichloromethane extraction (the first XAD column), we should have obtained a positive response from the second extraction (the second XAD column) as well. The fact that none of the model compounds could be detected when the drinking water extracts were analysed by gas chromatography gives further evidence in the same direction. Our concentration methods were not successful in enriching the more substituted chloroacetones (alkaline lability is one explanation for this (Douglas et al., 1982)), but as long as our conclusions are restricted to the mutagenicity we have observed, such compounds play a minor role.

Another important bleaching effluent mutagen is 3-chloro-4-(dichloromethyl)-5-hydroxy-2(5H)-furanone (Holmbom et al., 1984). From our results we cannot exclude the possibility of this compound being an important drinking water mutagen but recent research by Kronberg et al. (1985) has indicated that it can be excluded. Thus, all the candidates suggested as being responsible for the mutagenicity of chlorinated drinking water have been rejected. There are certainly qualitative similarities between the formation of mutagenic compounds during chlorobleaching of pulp and during drinking water chlorination (Kringstad et al., 1983a), but quantitatively the mutagenicity seems to be of entirely different origin.

CONCLUSIONS

Our study has given further evidence that two major groups of mutagens are formed during the chlorination of drinking water. The neutral mutagens are easily extracted from the water with dichloromethane or adsorbed on XAD-2. The acidic mutagens can only be enriched after acidification of the water, and even then satisfactory recovery requires relatively large XAD-2 columns or repeated or continuous liquid-liquid extractions. The most natural parallels between the mutagenicity of chlorinated drinking water and pulp mill effluents have all been rejected. Further research on this point should primarily deal with compounds similar to the hydroxy-furanone mentioned above.

ACKNOWLEDGEMENT

We wish to thank Lars Strömberg, Swedish Forest Products Research Laboratory, for valuable comments and for supplying the mutagenic model compounds and the National Swedish Environmental Protection Board for financial support.

REFERENCES

Borén, H., Grimvall, A., Sävenhed, R. 1982. Modified stripping technique for the analysis of trace organics in water. - J. Chromatogr. 252:139-146.
Borén, H., Grimvall, A., Palmborg, J., Sävenhed, R., Wigilius, B. (in press). Optimization of the open stripping system for the analysis of trace organics in water.
Cheh, A.M., Skochdopole, J., Koski, P., Cole, L. 1980. Nonvolatile mutagens in drinking water: Production by chlorination and destruction by sulfite. - Science 207:90-92.

Douglas, G.R., Nestmann, E.R., McKague, A.B., Kamra, O.P., Lee, E.G.-H, Ellenton J.A., Bell, R., Kowbel, D., Liu, V., Pooley, J. 1982. Mutagenicity of pulp and paper mill effluent: A comprehensive study of complex mixtures. - In: Waters, M. et al. (eds) Application of short-term bioassays in the analysis of complex environmental mixtures III, Plenum Press.

Grabow, W.O.K., Burger, J.S., Hilner, C.A. 1981a. Comparison of liquid-liquid extraction and resin adsorption for concentrating mutagens in Ames Salmonellamicrosome assays on water. - Bull. Environm. Contam. Toxicol. 27:442-449.

Grabow, W.O.K., vanRossum, P.G., Grabow, N.A., Denkhaus, R. 1981b. Relationship of the raw water quality to mutagens detectable by the Ames Salmonella/microsome assway in a drinking water supply. - Water Res. 15:1037-1043.

Gruener, N., Lockwood, M.P. 1980. Mutagenic activity in drinking water. - Am. J. Publ. Health, 70:276-278.

Holmbom, B., Voss, R.H., Mortimer, R.D., Vong, A. 1984. Fractionation, isolation and characterization of Ames mutagenic compounds in kraft chlorination effluents. - Environ. Sci. Technol. 18:333-337.

James, H.A., Steel, C.P., Wilson, I. 1981. Impurities arising from the use of XAD-2 resin for the extraction of organic pollutants in drinking water. - J. Chromatogr. 208:89-95.

Kool, H.J., vanKreijl, C.F., vanKranen, J.H., deGreef, E. 1981. Toxicity assessment of organic compounds in drinking water in the Netherlands. - In: van Lelyveld, H., Zoeteman, B.C.J. (eds) Water Supply and Health, Elsevier, Amsterdam, pp. 135-153.

Kool, H.J., vanKreijl, C.F., Zoeteman, B.C.J. 1982a. Toxicology assessment of organic compounds in drinking water. - CRC Critical Reviews in Environmental Control 12(4):307-357.

Kool, H.J., van Kreijl, C.F., deGreef, E., van Kranen, H.J. 1982b. Presence, introduction and removal of mutagenic activity during the preparation of drinking water in the Netherlands. - Environ. Health Perspect. 46:207-214.

Kringstad, K.P., Ljungquist, P.O., deSousa, F., Strömberg, L.M. 1983a. On the formation of mutagens in the chlorination of humic acid. - Environ. Sci. Technol. 17:553-555.

Kringstad, K.P., Ljungquist, P.O., de Sousa, F, Strömberg, L.M. 1983b. Stability of 2-chloropropenal and some other mutagens formed in the chlorination of softwood kraft pulp. - Environ. Sci. Technol. 17:468-471.

Kronberg, L., Holmbom, B. 1985. (Personal communication), Åbo Academy, Finland.

Loper, J.C. 1980. Mutagenic effects of organic compounds in drinking water. - Mutat. Res. 76:241-268.

Maruoka, S., Yamanaha, S. 1983. Mutagenic potential of laboratory chlorinated river water. - Sci. Tot. Environ. 29:143-154.

Meier, J.R., Lingg, R.D., Bull, R.J. 1983. Formation of mutagens following chlorination of humic acid. A model for mutagen formation during drinking water treatment. - Mutat. Res. 118:25-41.

Nestman, E.R., LeBel, G.L., Williams, D.T., Kowbel, D.J. 1979. Mutagenicity of organic extracts from Canadian drinking water in the Salmonella/mammalian microsome assay. - Environ. Mutagen. 1:337-345.

Rapson, W.H., Nazar, M.A., Butsky, V.V. 1980. Mutagenicity produced by aqueous chlorination of organic compounds. - Bull. Environ. Contam. Toxicol. 24:590-596.

Wigilius, B., Borén H., Grimvall, A., Sävenhed, R. (In press). A systematic approach to XAD-asorption for the concentration and analysis of trace organics in water at the ng/l level.

THE QUALITY OF DRINKING WATER PREPARED FROM BANK-FILTERED RIVER WATER IN THE NETHERLANDS

A.NOORDSIJ , L.M.PUYKER and M.A.VAN DER GAAG
The Netherlands Waterworks' testing and research Institute, KIWA Ltd.
Research Department,P.O.B.1072,3430 BB Nieuwegein, The Netherlands.

ABSTRACT

The quality of raw water and drinking water was analysed at 18 sites in the Rhine delta in an integrated program with hydrological,chemical and toxicological parameters. About 600 different organic compounds have been detected,of which about 200 could be identified,and 400 more have been classified on the basis of their mass spectra. Lipophilic compounds with a log P(octanol/water) greater than 3.2 are adsorbed during the first few meters of the bank infiltration. More hydrophilic compounds are often not adsorbed and not removed by the currently used water purification systems,such as aeration and sand filtration.

The presence of organohalogens was demonstrated with surrogate parameters,such as extractable and adsorbable organohalogen. All the samples tested were mutagenic in the Ames test with strain TA98 + S9 mix. Only part of the mutagenic activity was related to compounds originating from Rhine water.

KEYWORDS
River bank filtration;drinking water quality;organic micro pollutants; organo chlorine compounds;mutagenicity

INTRODUCTION

About 70 % of the drinking water in the Netherlands is prepared from ground water. A combination of aeration and sand filtration is usualy applied for the purification of the ground water. These treatment technics remove iron,manganese, methane and ammonia. Some of these plants are located near the rivers and may abstract bank filtered river water together with ground water.

Circa 50 of such plants are located in the Rhine delta, along the rivers Lek, Waal, Oude Rijn and IJssel. Some 25 of these process varying amounts of river bank filtered ground water, representing about 7% of the total production of drinking water in the Netherlands.

At some of these plants the conventional treatment systems were not able to produce drinking water free of taste and odour.

Zoeteman et al.(1978) studied the quality of drinking water prepared from river bank filtered ground water. They showed the presence of a number of organic substances in river bank ground water after residence times of one to twelve months, such as halogenated alkanes, dichloro- to hexachlorobenzenes, lindanes, polychloro biphenyls, bis(chloro-iso-propyl)ethers, various nitro aromatics and anilines. Trichloromethane, tetrachloroethylene, dichlorobenzenes and chloroethers in particular were found in relatively high concentrations.

Kool (1983) also demonstrated that mutagenic substances were present in drinking water prepared from river bank filtered ground water.

Infiltration of Rhine water into the river banks has also been subject of study in Switzerland and Germany (Kussmaul,1977; Schwarzenbach,1983). The results of these investigations agree with those of Zoeteman, although the situations studied in Switzerland and Germany involved shorter residence times of the infiltrated water than in the Netherlands.

In the period 1982 - 1984 KIWA and RIVM (the National Institute for Public Health and Environmental Hygiene) cooperated in an extensive investigation of the quality of drinking water prepared from river bank filtered ground water. This study involved an integrated approach for the measurement of chemical and toxicological parameters (Noordsij et al.1983;Van der Gaag et al.1982), complemented by a large number of standard analyses. Many disciplines were involved: analytical chemistry,toxicology, microbiology, hydrology, geohydro- chemistry, process technology and mathematics. Details of this project are described in KIWA report no. 89(1985)

This paper discusses in particular the results of the organic-chemical and toxicological investigations and their significance for the assessment of the quality of the drinking water.

METHODS

The research program included:

-isolation of volatile organic compounds by means of the gas stripping technique, (Purge), followed by microcoulometric POCl measurement and identification of individual compounds with the help of the gaschromatograph/mass spectrometer.

-solvent extraction of very lipophilic compounds followed by gas chromato- graphy (PCBs and chlorinated pesticides) and HPLC (polycyclic aromatics) and the surrogates EOCl, EON, EOS and EOP.

-adsorption of lipophilic and weakly hydrophilic compounds on the resin XAD-4 at successive pH values of 7 and 2(Noordsij et al.1983),followed by:

.identification of individual compounds with GC/MS

.group parameter determination X_7OCl, X_7ON, X_7OS, X_7OP, X_2OCl, X_2OS and X_2OP (Veenendaal,1984; Puyker et al. 1981)

.molecular weight distribution with Gel Permeation Chromatography

.salmonella microsome mutagenicity test (Ames test) with strains TA 98 and TA 100 (Maron and Ames,1983) without and with addition of S9-mix at six dose levels (Van der Gaag et al.1982).

—determination of on activated carbon adsorbable organo chlorine (AOCl)
—determination of the content of dissolved organic carbon DOC.

PROGRAM OF INVESTIGATIONS

Both the raw river bank filtered ground water and the drinking water were investigated at some twenty sites in the Rhine delta. Samples of river water and of ground water not subject to the influence of rivers were also taken. Two of these ground water sites were located near the river, the two other sites were selected on their DOC-content, one with a low DOC(0.7 mg/l) and one with a relatively high DOC(4.3 mg/l). In cooperation with the RIVM measurements were also carried out in a flow path perpendicular to the river Lek at Opperduit, at distances of 6, 15, 45, 220 and 675 m from point of infiltration.

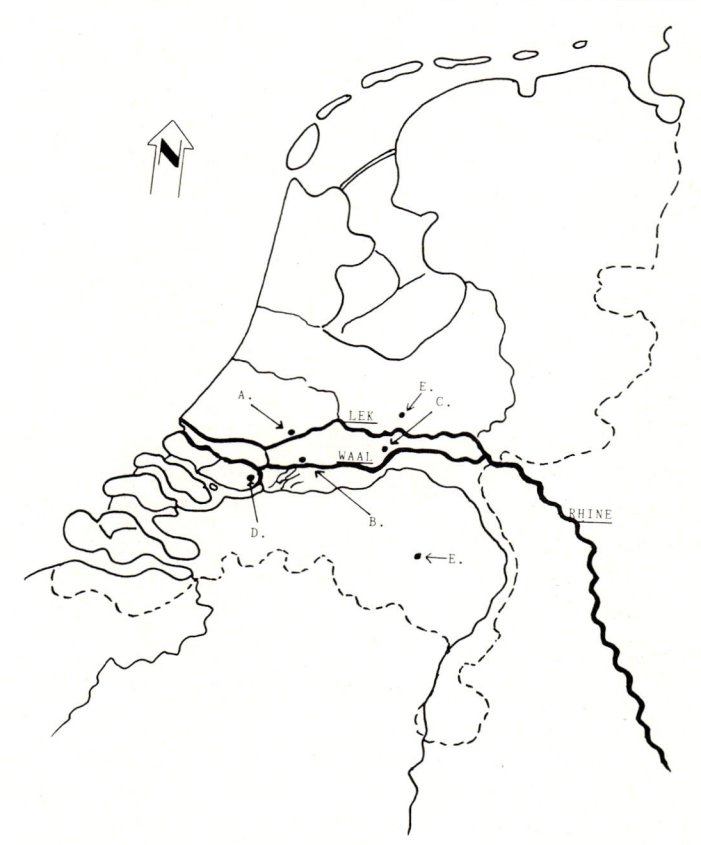

Fig.1. Some of the investigated plants in the river Rhine delta where river bank filtered water is used for the preparation of drinking water. For declaration of letters see text.

The results from a selected number of sampling sites are presented here. They include: river water from Lek and Waal
- 5 sampling wells at the experimental site Opperduit (Site A)
- a production plant with shallow wells and conventional ground water treatment (aeration and Rapid sand Filtration) complemented with Granular Activated Carbon filtration (Site B)
- a production plant with deep wells and conventional ground water treatment (Site C)
- a ground water production plant located near the river, with conventional treatment (Site D)
- two ground water plants, one with low DOC, and the other one with high DOC in the abstracted ground water (Site E)

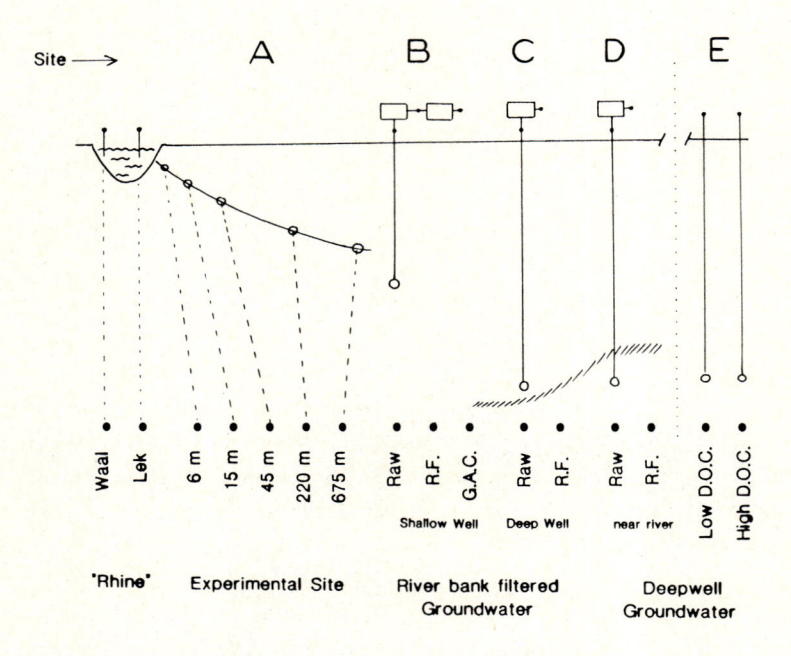

Fig.2. The different sampling points and their location in relation to the river.

RESULTS OF THE INVESTIGATION

Volatile compounds

The volatile compounds found in the various water samples are summarised in Table 1. Concentration of for example the halogenated compounds is very low in the drinking water samples.

TABLE 1

Concentrations of volatile organic compounds, analysed with gasstrip-GC/MS

Volatile compounds (µg/liter)	"Rhine" River Waal	"Rhine" River Lek	Flowpath at Opperduit 6m Exp.Site	15m Exp.Site	45m Exp.Site	220m Exp.Site	675m Exp.Site	Riverbank filt.water Shallow Raw	Shallow Rapid Filtrate	Gr.Act.Carbon	Deep Raw	Deep Rapid Filtrate	Ground water Near river Raw	Rapid Filtrate	Low DOC	High DOC
Aromatic compounds	0.1	1.0	0.6	0.2	0.2	0.1	0.1	0.1	0.1		0.1		0.1	0.1	0.1	
"Oxygen" compounds	6.0	4.9					1.2	2.4	2.0		2.7	0.1				
Chloroalkanes(enes)	2.9	2.1		0.3	0.7	6.0	1.9	4.7	0.6		0.1					
Bis(Cl-i-propyl)ethers	0.1			0.2	0.4	3.1	2.4	1.3	0.5				0.1	0.1		
Chlorobenzenes	0.1	0.4	0.5	1.1	1.1	1.4	0.6	0.2								

Polycyclic Aromatic Hydrocarbons (PAHs), chloropesticides and PCBs.

PAHs were found only in river water and in the raw water from Opperduit in the well nearest to the river. (Rivers: 0.1 and Opperduit 6m: 0.7 µg/l)

Chlorinated pesticides and polychloro biphenyls were not detected (< 0.01 µg/l)

Lipophilic and weakly hydrophilic compounds

Individual compounds. With GC/MS about 600 different substances were detected in the XAD-isolated fractions, of which 180 could be identified. An additional 380 substances could be classified on the base of their mass spectrum. Table 2 summarises the results for compounds or groups of compounds and the pH at which the isolation took place. The peak heights for each substance in the various chromatograms are indicated in millimeters. Experience learns that under these conditions of analysis a peak of 100 mm corresponds with a quantity of several tenths of a microgramme of substance per liter of water at an isolation yield of nearly 100%. With the more hydrophilic (polar) compounds much lower isolation recoveries must be reckoned with, down to below 20%. The measured peak heights of the polar compounds then correspond with several microgrammes per liter of water. It is clear that a true quantitative analysis is difficult to realize.

From table 2 it can be seen that there is great variation in behaviour from one substance or group to another. The origin of the various compounds is also different. The components which contain chlorine, nitro compounds, anilines and sulphones in particular are xenobiotic. The various terpenoids and low molecular dicarbonic acids may however be of natural origin. The alifatic oxygen compounds whose structures could not be determined may also be of natural origin, or may be degradation products of river pollutants.

It is difficult to compare concentrations of organic pollutants in the river with those in the bank filtered ground water, because in the river an instantaneous measurement is made of a concentration which may be varying with time, whereas in the river bank ground water considerable smoothing of variations in concentration will have taken place.

TABLE 2

Compounds and groups of compounds, isolated with XAD in different water types. Peak heights in the MS-chromatograms in millimeters.(Summary)

(Groups of) compounds peakheights in mm.	pH XAD-isolation	"Rhine" River Waal	"Rhine" River Lek	Flowpath at Opperduit Exp.Site 6m	Exp.Site 15m	Exp.Site 45m	Exp.Site 220m	Exp.Site 675m	Riverbank filt.water Shallow Raw	Rapid Filtrate	Gr.Act.Carbon	Deep Raw	Rapid Filtrate	Ground water Near river Raw	Rapid Filtrate	Low DOC	High DOC
Alkylbenzenes and indanes	7	21	25	650	93	48	64	13	28	8			8				8
Anilines	7	444	240	233	145	85	138	38	40	14			42				5
Nitro benzenes	7	100	147	42	56	19											
Nitro anilines	7	30	42														
Chloro benzenes	7			105	130	97			14	16							
Chloro anilines	7			151	168	121	170	142	138	53							
Alkyl acetyl-anilines	7		27	58	77	78	85	3	35	55							
bis(Chloro-iso-propyl)Ethers	7			22	39	55	371	290	258	136		16	6	6	7		
Trialkylphosphates	7	74	50	22	17	30	20	179	64	70							
Alkyl phenols	7	124	97	148	330	221	143	56	77	49			8				
Arom.aldehydes	7	64	25	273	66	81	68	64	64	8		18				8	13
N-butylbenzene-sulphonamide	7	20		32	40	15	5	10	6	13							
Trimethyloxindole	7	20	10	70	70	52	63	24	42	49							
p.p'bis(dimethyl-amino)benzophenone	7	13	13	16	30	60	62	5	8	10							
Alkoxy-alkanes	7	22	5	150	240	97	115	95	25	54		12					
Diacetoneglucose	7	240	190	270	240	200	300	200	180	180		38	35	12	15		
Alif.CHO compounds	7	161	155	202	155	74	180	130	62	86	8	13			19	5	13
Alif.CHO compounds	2	129	195	325	249	119	265	222	385	264	47		27	7	27	13	29
Terpenoids M.W.152-156	7	15		99	83	46	8	24	30	23					8		
Terpenoids M.W.152-156	2	115	95	82	81	43	8	12	11	6							
Terpenoids M.W.170	7	262	290	83	93	38	150	60	92	84							
Terpenoids M.W.170	2	98	160	12	13	9	12		6	5							
Arom.carbon acids	2	200	380	87	41	26	38	25	28	32	11	11	8	15	14		
Monochloro phenoxy-alkanoic acids(MCPA)	2	55	25	63	37	9	12	59	154	70							
Dichloro phenoxy-alkanoic acids(2,4-D)	2	90	200	17	25		25	18	80	45							
Sulphones	2	28	30	63	70	51	55	33	50	32							10
Propoxy ethers (n x 58)	2			109	75	59	173	72	160	128			23				14
"Oxalic"acids	2	200	135	85	78	64	78	80	51	8		9	16	51	32		7
Fatty acids	2	469	789	19	14	14	33	23	78	74	102	30	42		54	23	24

Group parameters

From the Gel Permeation Chromatography investigation it appears that the greater part of the with XAD isolated organic material has a molecular weight higher than 400. For this reason the analysis of components by GC/MS provides limited information: Only the low molecular fraction of the isolated organic material can be analysed with GC/MS, of which about 30% could be identified.

These limitations do not apply to the group parameter experiments in which a chemical or physical characteristic is measured completely, either in the water sample (UV; DOC) or in combination with a specific isolation technique such as extraction (E), gasstripping or purging (P) and adsorption on XAD at pH 7 (X_7), at pH 2 (X_2) or on activated carbon (A).

TABLE 3

Results of the various group parameter determinations

Group-parameter		River Waal "Rhine"	River Lek	Exp.Site 6m	Exp.Site 15m	Flowpath at Opperduit Exp.Site 45m	Exp.Site 220m	Exp.Site 675m	Raw Shallow Riverbank filt.water	Rapid Filtrate	Gr.Act.Carbon	Raw Deep	Rapid Filtrate	Raw Near river Ground water	Rapid Filtrate	Low DOC	High DOC
DOC	mg/l	3.8	4.0	2.3	2.5	2.2	2.7	3.3	3.0	2.6	1.1	1.6	1.3	6.3	5.7	0.7	4.3
AOCl	µg/l	32	35	25	12	6	9	10	15	14	<5	<5	<5	6	20	<5	<5
X_7OCl	µg/l	7.0	8.2	5.0	4.3	3.0	3.0	2.1	3.0	2.6	<0.3	1.4	2.3	1.3	4.4	<0.3	<0.3
X_2OCl	µg/l	-	-	6.3	4.7	2.4	3.0	2.7	2.9	3.1	0.3	1.2	1.5	3.2	20	0.5	0.5
EOCl	µg/l	1.8	1.1	1.9	2.2	2.1	1.0	1.0	1.9	0.5	0.5	0.3	0.1	0.7	0.9	<0.2	<0.2
POCl	µg/l	1.7	3.1	-	-	-	-	-	1.5	1.2	<0.5	0.8	-	0.5	0.7	<0.5	<0.5
X_7ON	µg/l	27.5	34.5	14.5	13.0	9.0	9.5	7.5	-	-	-	4.1	5.5	7.2	6.2	1.1	5.5
EON	µg/l	2.0	3.2	1.1	1.1	1.1	0.7	0.8	-	-	-	0.1	<0.1	4.0	3.2	<0.5	<0.5
X_7OS	µg/l	20	25	19	19	9	17	17	19	13	1	5	4	17	9	2	8
X_2OS	µg/l	30	20	23	21	16	26	25	20	19	<0.2	12	13	42	29	4.5	-
EOS	µg/l	<1	<1	<1	<1	<1	<1	<1	-	1	-	<1	<1	<1	<1	6	-
X_7OP	µg/l	0.5	0.4	0.1	0.1	<0.1	0.2	0.3	-	-	-	<0.1	<0.1	0.2	0.2	<0.1	<0.1
X_2OP	µg/l	0.5	0.2	0.2	0.2	0.1	0.1	0.1	-	-	-	0.1	<0.1	0.1	0.1	0.3	-
EOP	µg/l	0.2	<0.1	0.3	0.1	<0.1	<0.1	<0.1	-	0.1	-	<0.1	<0.1	0.2	0.2	<0.1	0.2

- = not measured

The physical character of the compounds indicated with a group parameter is defined by the method of isolation used. With extraction and gasstripping techniques substances of a lipophilic nature are isolated, and with XAD also weakly hydrophilic substances. With activated carbon the compounds mentioned and also more polar compounds are isolated from the water sample.

In figures 3 to 6 some group parameters are reproduced in the form of histograms (see figure 2 also). It is evident that passage through soil and conventional purification are more effective in lowering surrogates representative of non polar compounds (EO-parameters) then those of more polar compounds (X and A).

The DOC is a measure of the total of organic substances dissolved in water, both lipophilic and hydrophilic. As it is primarily the more hydrophilic compounds which are present in the river bank filtered ground water there is no significant decrease in the DOC content after passage through soil or conventional treatment.

Only after filtration of the water through activated carbon a clear reduction of the DOC-value is observed.

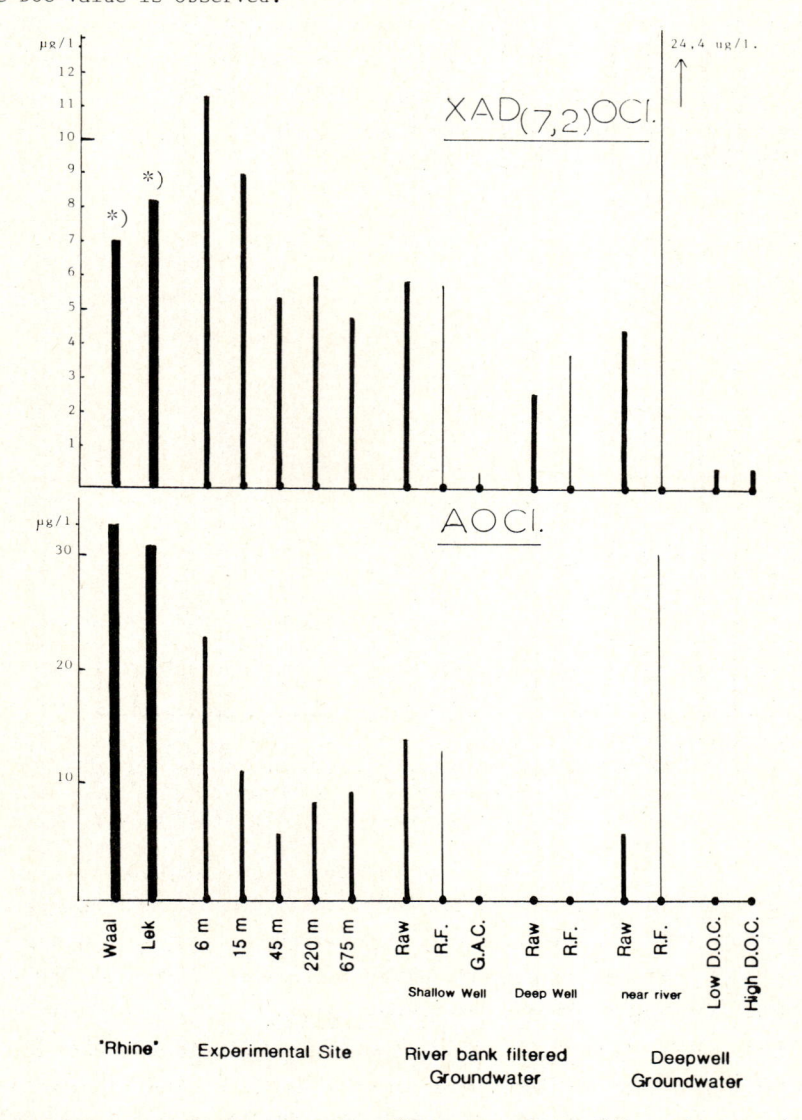

Fig.3 and 4. Organo halogen parameters: $X_7OCl + X_2OCl$ in XAD-isolates and Activated Carbon adsorbable Organo halogen compounds AOCl. The high values found at sample point RF(Ground water near river) are caused by desinfection with chlorine. (* = X_2OCl not measured)

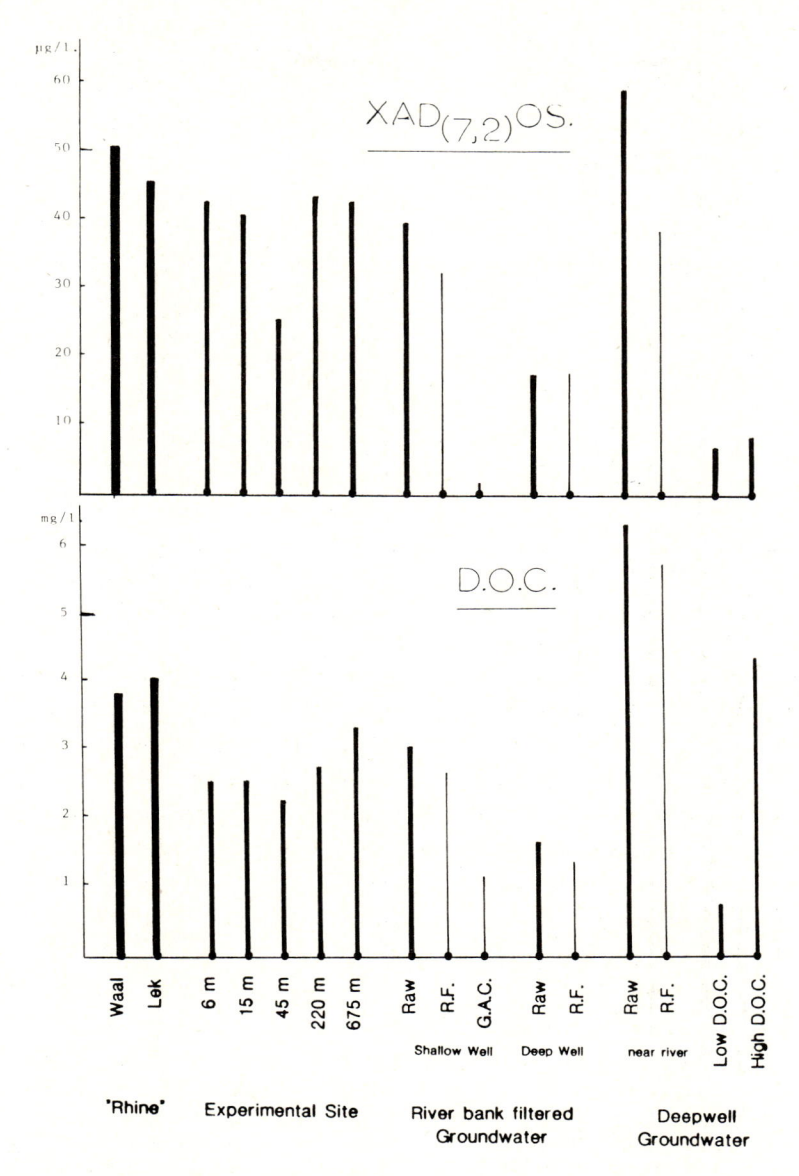

Fig.5 and 6. Organo-sulphur content in XAD-isolates and concentrations of dissolved organic carbon DOC. Both parameters probably of natural origine.

Relationship between mobility and lipophilic nature

Study of the different group parameters and the individual compounds shows that the hydrophilic organic substances are more easily carried along with the infiltered water than the lipophilic substances are. The degree by which an organic compound is retained in an aquifer depends on both the structure and composition of the ground, and the physical nature of the organic compound.

Schwarzenbach (1983) describes in his study of the behaviour of organic compounds during infiltration of river water through the river bank retardation factors of 2 to 4000 for compounds with a $logP_{(octanol-water)}$ of 1 to 6.

His calculations were based on three types of soil with different organic carbon contents and different fractions of sorbing material. Table 4 gives a summary of the retardation factors determined and calculated by Schwarzenbach in the different ground layers and in dependence on the $logP_{(O-W)}$

TABLE 4
Retardation factors in three river bank layers for organic compounds with different $logP_{(Octanol-Water)}$ as calculated by Schwarzenbach(1983).

layer	logP → 2	3	3.5	4	6
I a)	5-10	20-30	40-60	100-150	2000-4000
II	< 3	ca 10	20-25	40- 60	1000-2000
III	< 2	< 3	< 3	< 5	< 10

  a) I : river sediment, org.C content in sorbing fraction 1-2%
     II : Distance to river 0.1-5m; org.C in sorbing fraction 0.1-1%
     III : Distance to river > 5m; org.C in sorbing fraction < 0.1%

The situation studied by Schwarzenbach is not directly comparable to the Rhine delta investigated by KIWA. Moreover, during passage through soil more processes take place than the sorption on organic soil material, processes such as sorption on inorganic constituents, biological degradation and possibly transport of lipophilic compounds by combination with more mobile hydrophilic material. Nevertheless the approach of Schwarzenbach can be helpfull to interpret the results of this investigation.

The $logP_{(O-W)}$ values were determined as far as possible for the organic compounds identified in the river bank filtered water (Leo et al.1971). It showed that compounds which reach the abstraction wells have a $logP_{(O-W)}$ lower than 3.5. Substances with a $logP_{(O-W)}$ of 4 and higher, such as polycyclic aromatics, chlorinated pesticides and PCBs are not found in river bank filtered ground water.

Figure 7 shows the results for the aromatics isolated with XAD (alkyl benzenes and two-ring structures), and figure 8 the results for alkyl phenols. The decrease in concentration of aromatics during passage through the first part of the soil and the simultaneous increase in the phenol content may be caused by oxidation of the aromatics in the river sediment. The mobility of these aromatic compounds is lower than that of the volatile aromatics and higher than that of the poly-aromatic hydrocarbons. This agrees with the $logP_{(O-W)}$ values: toluene/xylenes with logP 2.7-3, naphthalene/anthracene 3.5-4.5 and PAHs approximately 6.

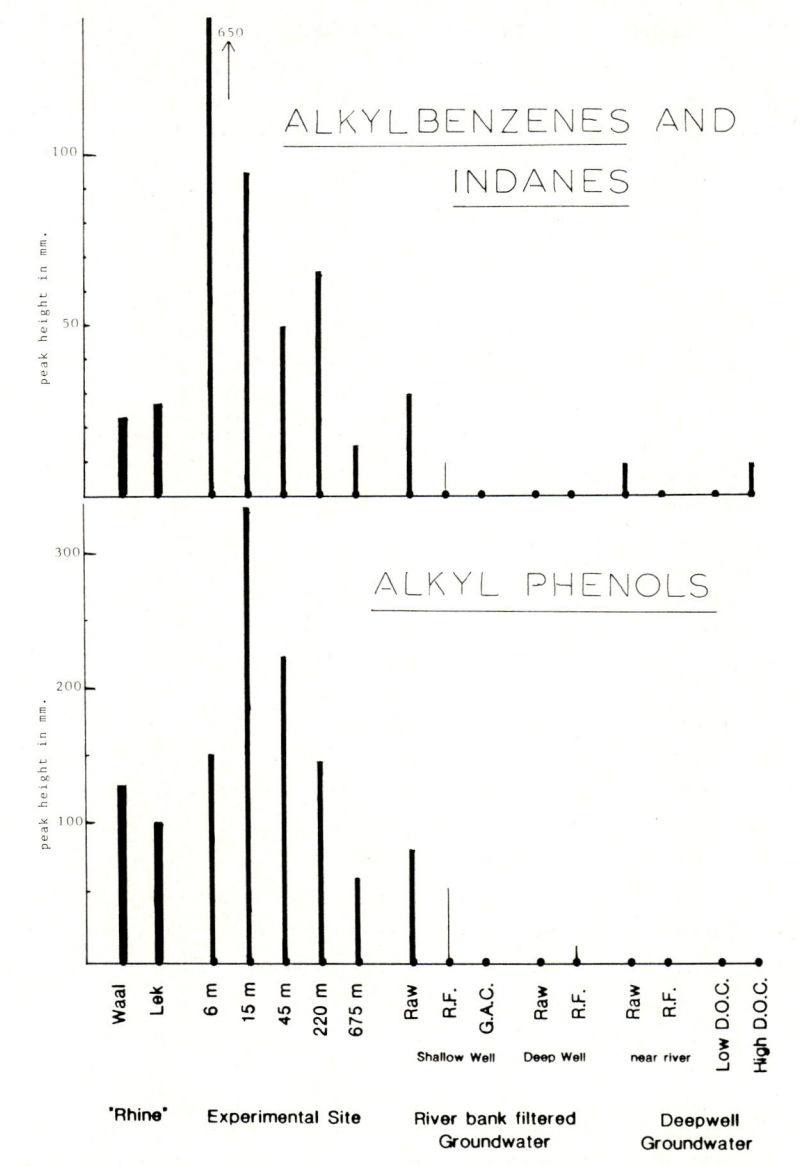

Fig.7 and 8. Concentration patterns of aromates and phenols in XAD-isolates.

Figure 9 gives an **exam**ple of a compound with a low $\log P_{(O-W)}$, i.e. -0.6: diacetone-glucose. This substance is present in river water as a pollutant and appears to be very persistent during passage through soil and conventional ground water treatment. Figure 10 shows a high correlation between the occurrence of this compound and the tritium content of the river bank ground water. This last parameter (^{3}H) is a measure of the amount of river water younger than 22 years in the abstracted ground water.(Stuyfzand,1985)

There is no particular reason to consider this substance as harmful,but it illustrates that persistent hydrophilic compounds from the river can reach the drinking water. The investigation showed that compounds such as trimethyloxindole and p.p'-bis(dimethylamino)benzophenone, both products from the manufacture of dyes,chloro-anilines and tri-alkyl(thio)phosphates all have a behaviour comparable to that of diacetone glucose.

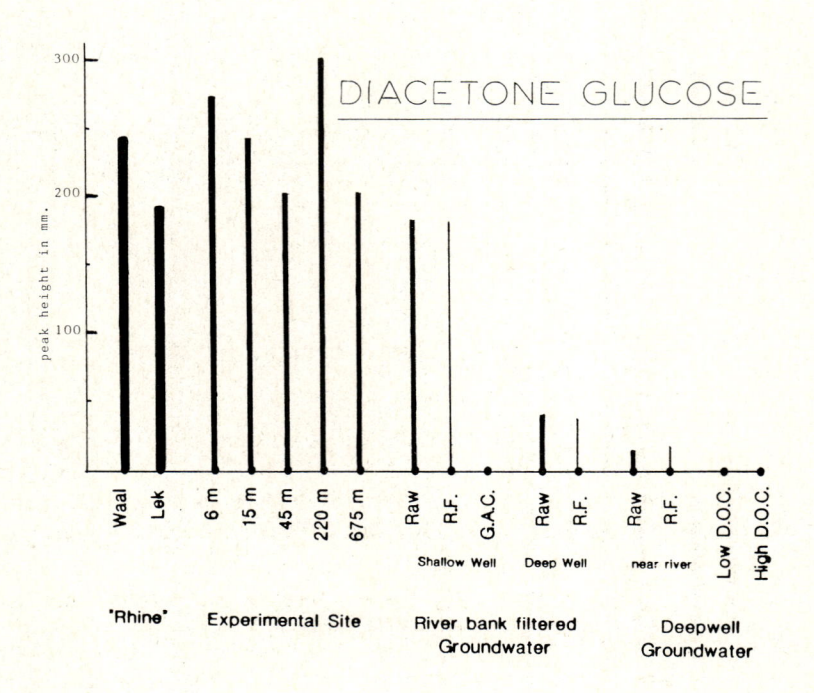

Fig.9. Concentration pattern of Diacetone glucose in XAD-isolates.No significant decrease is observed neither during soil passage, nor during conventional water purification. Carbon filtration looks very effective.

Figure 11 shows the correlation between trimethyloxindole and tritium in the investigated drinking water samples after aeration and rapid sand filtration.

Measuring point (a) concerns the effluent from the activated carbon filter, in which no trimethyloxindole was found but in which the tritium content was the same as in the influent.

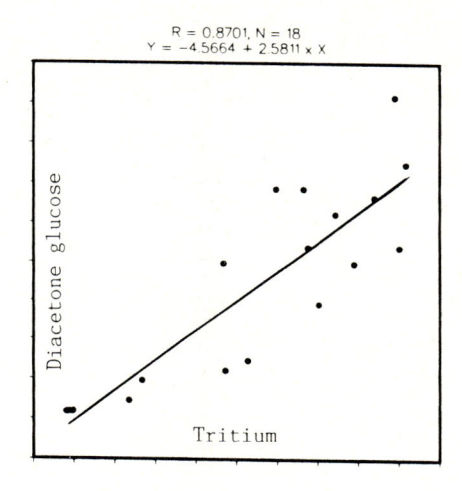

Fig.10. Relation between diacetone glucose content and tritium content in river bank ground water.

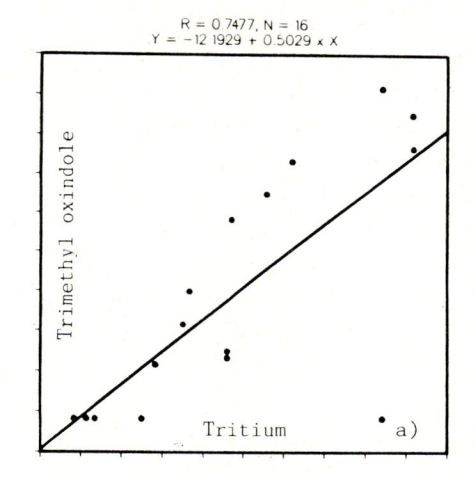

Fig.11. Relation between trimethyl-oxindole content and tritium content in drinking water produced from river bank ground water.

Variation of concentration during infiltration

Bis(Chloro-iso-propyl)ethers have been found in river bank filtered ground water several times, but at present they hardly occur in river water. A few years ago however these substances were present in river water in much higher concentrations. The results of the analysis at Opperduit illustrate the relation between the residence time and the concentration of the chloro-ethers found at different distances from the river. The pattern of concentrations measured with the gasstripping technique is identical to that of the measurements carried out as part of the investigations by RIWA(Co-operating Rhine and Meuse Water Supply Companies in the Netherlands and Belgium).

TABLE 5

Comparison of investigation results at five sample wells at Opperduit 1983 and RIWA measurements in the river Lek since 1977 (averiged values). Concentration of bis(Chloro-iso-propyl)ethers in microgram per liter.

Opperduit 1983		RIWA measurements in the Lek	
distance		year	
675 m	2.35 µg/l	1977	3 µg/l
220 m	3.05	1978	3
45 m	0.43	1981	1.3
15 m	0.2	1982	0.5
6 m	< 0.1	1983	< 0.2

This example shows that the concentration of a pollutant in river bank ground water measured at this moment is not merely dependent on the residence time of the river water and the retardation factor of the compound, but also on the sometimes highly variable concentration of it in the river water at the moment of the bank infiltration.

Figure 12 gives the results of the chloro-iso-propyl ethers as measured after XAD-isolation. These rather polar and very persistent compounds have already penetrated far into the aquifer and are even encountered in ground water where no influence of river water is assumed (site D). They cannot be eliminated by the conventional treatment, carbon filtration is very effective (site B).

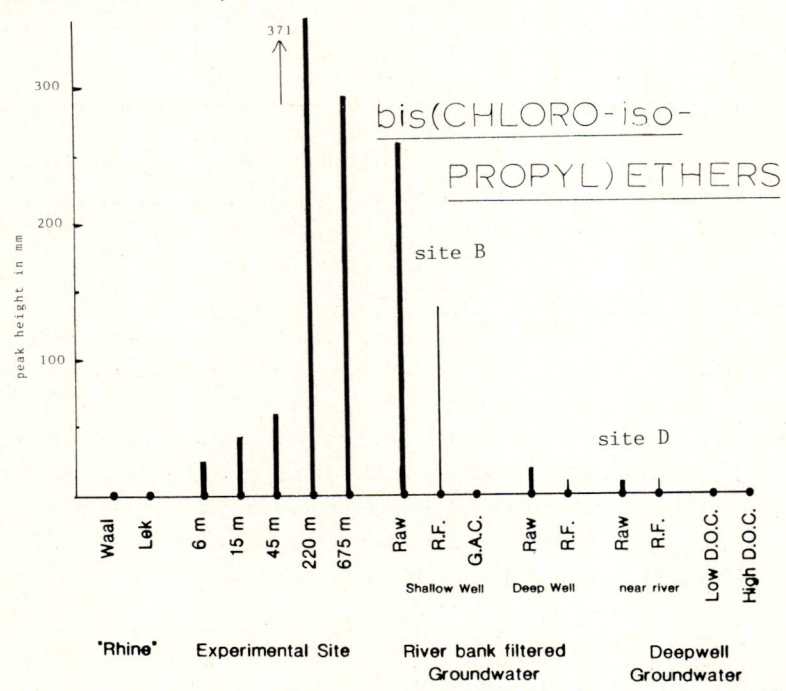

Fig.12. Concentration pattern of Chloro-iso-propyl ethers in XAD-isolates.

Compounds whiche are present in water in dissociated form.

Compounds which are dissociated at pH 7 such as mono- and dichlorophenoxy-alkanoic acids (MCPA and 2,4-D) can be isolated with XAD at pH 2. These compounds have been used as herbicides for as long as 40 years. Figures 13 and 14 show the concentration patterns in the various samples and illustrate that these compounds are mobile in the aquifer. (logP of MCPA =2.0, and of 2,4-D = 2.8). The monochloro compounds occur primarily in groundwater, the dichloro compounds in river water.

It is possible that dehalogenation occurs during passage (Zepp,1974).

Purification by aeration and sand filtration has little effect on these compounds. Filtration over activated carbon removes these substances completely.

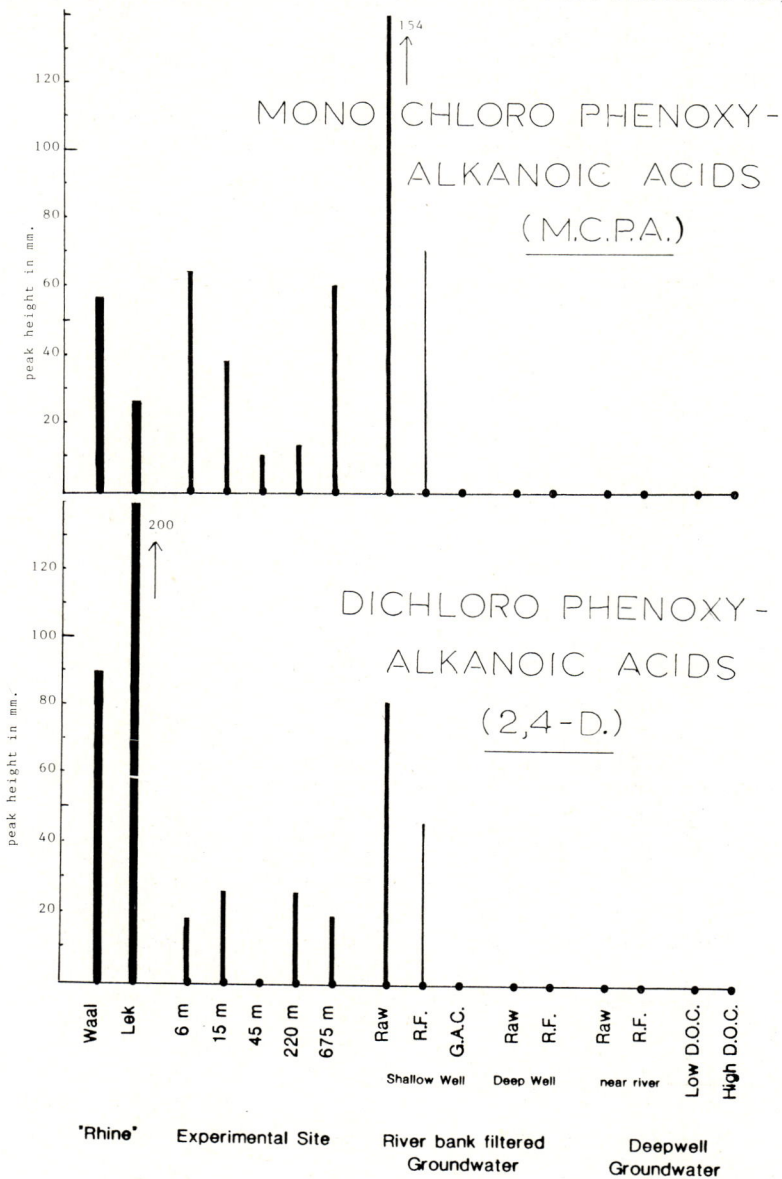

Fig. 13 and 14. Concentration patterns of chloro-phenoxy-alkanoic acids in XAD-isolates.

Mutagenicity

The mutagenicity of infiltrated Rhine water is lowered during the first few meters of bank infiltration. At the experimental site (A) in Opperduit the mutagenic activity of the water at a depth of 6 meter was considerably lower than the mean mutagenicity of river Lek water in the preceeding year. At a greater distance from the river the mutagenicity of the XAD pH7 fraction is still lower, but no further change was observed in the pH2 fraction.

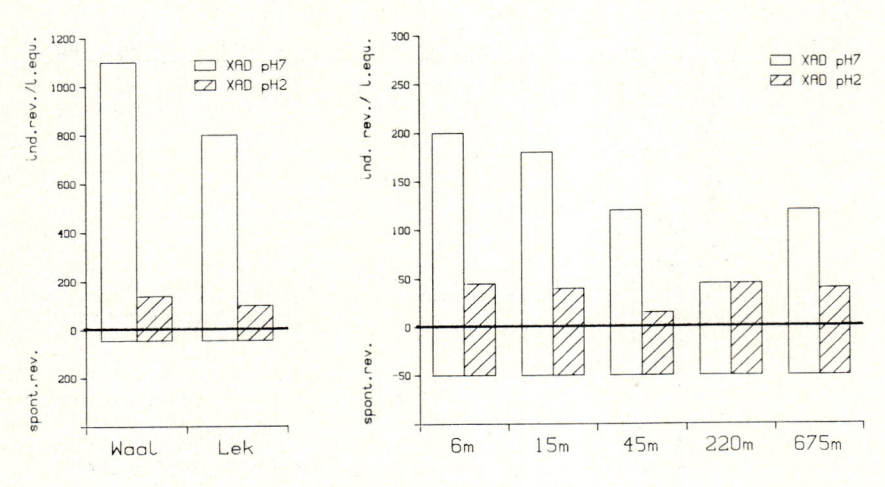

Fig.15. Mutagenic activity of river water in the Rhine delta and of bank filtered water at Opperduit. At left: mutagenic activity of water of the Waal and mean mutagenicity of water of the Lek in 1983. At right: mutagenicity of ground water in Opperduit at different distances from the Lek.

All raw ground waters assayed in this study were mutagenic in the Ames test.

This mutagenic activity was quite specific, as it was only detected with tester strain TA98 in the presence of S9 mix. With a few exceptions no mutagenic activity was observed for strain TA100 and for strain TA98 without S9 mix. There was however a difference between the mutagenicity of ground water with and without bank filtrate(figure 16.) When recently infiltrated river water was present in the raw water, mutagenicity for TA98+S9 mix was detected both in the XAD pH7 and pH2 fractions. The samples containing bankfiltrate older than 22 years, deep ground water or anaerobic ground water with no influence from the river were only mutagenic in the pH7 fraction.

The current treatment processes for ground water (aeration and rapid sand filtration) removed only a part of the mutagenic activity from the raw water.

GAC filtration which was introduced in site B to remove taste and odour , also reduced the mutagenic activity to the detection level for more than two years (figure 16).

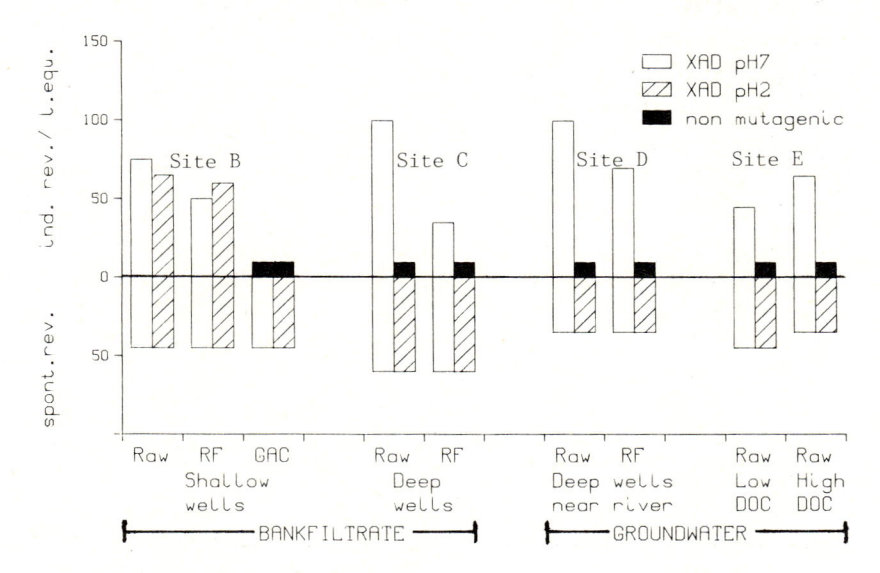

Fig.16. Ground water from site B, with a high content of bank filtrate was mutagenic for TA98 +S9 mix in both XAD fractions. Deep bank filtrate more than 22 years old (site C) was only mutagenic in the pH7 fraction. This mutagenic activity was similar to that of deep ground water in the Rhine delta area (site D) and anaerobic ground water with low and high DOC levels (site E).

<u>Does Rhine water contribute to the mutagenicity of bank filtered ground water?</u>

Rhine water has an influence on the mutagenicity of bank infiltrated water in the delta area in the Netherlands. The mutagenic activity of the XAD pH2 fraction was related with the tritium content of the sampled waters (figure 17). It is not clear if the mutagenicity of the pH2 fraction originates directly from the Rhine water or if it was introduced by transforming processes in the ground.

There is only few evidence on the chemical identity of these mutagens. The organohalogen-surrogates correlate poorly with the mutagenic activity, indicating that the major part of these compounds in this type of water is not mutagenic in the Ames test (figure 17)

The contribution of Rhine water to the mutagenicity of the pH7 fraction is not so evident. Similar mutagenicity was observed in ground water types from "uncontaminated" sites. The results from the experimental site at Opperduit showed that a part of the mutagenicity of the pH7 fraction in bank filtrate may be attributed to the influence of Rhine water. The origin of the major part of the mutagenicity in the pH7 fractions in bank filtrate as well as in uncontaminated ground waters will need further investigation. This research will focus on the possible influence of anaerobic conditions in the ground on this parameter. Studies on artificial recharge with Rhine water in coastal dune areas have shown that an increase of mutagenicity can happen in anaerobic sites in the dunes (Kool et al.,1984; van der Gaag,1985).

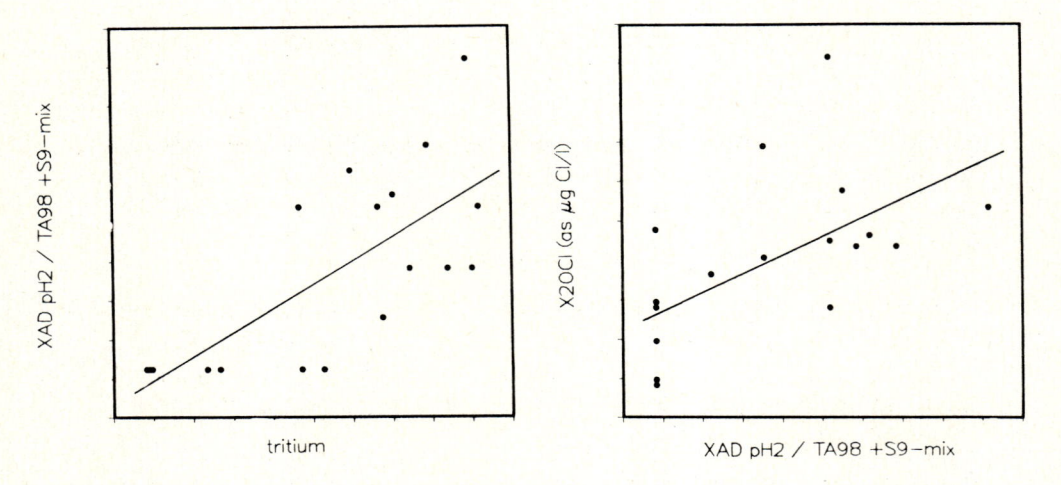

Fig.17. Relations between mutagenicity of the pH2 for TA98 +S9 mix and the tritium content (left,n=18, r=0.72) respectively organohalogen content (right ,n=18, r=0.59).

What is the meaning of these results for the quality of the drinking water, prepared from bank filtered ground water?

Only a part of all the organic substances present in water can be analysed by the techniques applied. The compounds which are poorly isolated because they are hydrophilic, are the most persistent in the soil and during the "classical" treatment processes aeration and sand filtration.

The AOCl measurements also indicate that xenobiotic compounds are present in river bank filtered ground water which cannot be detected by the other analytical techniques. The high values for XOS and XON correlate with the DOC-values and may be from natural origin. The individual component investigation gave no indication of high concentrations of sulphur and nitrogen compounds.

A large number of compounds from industrial origine has been detected in the drinking water. However the concentrations of these compounds were very low.

Twenty of the 180 identified compounds have been evaluated for carcinogenic effects, and 96 have been assayed in short term mutagenicity assays, mainly in the Ames test. The mutagenicity observed in XAD fractions in the Ames test could not be explained by the mutagenic properties of the identified compounds. The mutagenic activity from the samples therefore has to be caused either by compounds which are not analysed with GC/MS (possibly with a molecular weight higher than 400), or by the identified compounds which have not yet been assayed for their mutagenic properties. From the relation with the XOCl values it appears that organo halogens are not predominantly responsible for the mutagenic activity. Further study will be needed to characterize the mutagens, and to assess possible implications for health.

Taste and odour. A number of compounds found in the water such as chlorobenzenes, chloro-anilines, aromatic sulphur compounds and terpene-like materials can affect odour and taste of the drinking water. Determination of taste values provided no direct correlation between these values and the concentrations of these substances. Taste only correlated with the percentage of young river water in the abstracted ground water.

CONCLUSIONS AND RECOMMENDATIONS

Ground water in the delta of the Rhine contains various amounts of bank filtered river water. Organic substances from Rhine water were detected in this type of water. Primarily (moderately) hydrophilic compounds with a $logP_{(Oct-water)}$ lower than 3.5 were able to infiltrate in the aquifers. The presence of infiltrated Rhine water has a negative influence on taste and odour of the drinking water prepared from it.

The mutagenic activity in drinking water prepared from bank filtered ground water was only partly related to the presence of river water. It is possible that mutagens can be formed under anaerobic conditions in the ground, but this theory needs further investigation. Little is known about the chemical identity of these mutagens. Only a minor number of mutagenic substances could be identified.

The current treatment processes were not able to remove the major part of the hydrophilic pollutants which had allready percolated through the aquifer.

Additional filtration over granular activated carbon gave excellent results, both for removal of taste and odour and for lowering the concentrations of the organic pollutants.

The presence of these hydrophilic organic contaminants still is a problem which needs further study. The analytical methods are not adequate for an optimal investigation of this group of compounds, and will have to be improved or developed.

The investigations described in this paper are part of the research program performed by KIWA Ltd. commissioned and financed by The Netherlands Waterworks Association VEWIN, in close cooperation with the waterworks involved in river bank filtration and with the RIVM, the National Institute for Public Health and Environmental Hygiene.

REFERENCES

Kool, H.J. and C.F. van Kreijl (1984). Formation and removal of mutagenic activity during drinking water preparation.Water Res. 18: 1011-1016

Kussmaul, H.(1977). Behaviour of persistent organic compounds in bank-filtrated Rhine water. Second international symposium on aquatic pollutants: Transformation and biological effects. Ed. O.Hutzinger, L.H.van Lelyveld and B.C.J.Zoeteman. Pergamon Press, Oxford.

Leo, A.,C. Hansch and D. Elkins.(1971). Partition coefficients and their uses. Chem.Revs.71:525-616.

Maron,D.M. and B.N.Ames (1983). Revised methods for the Salmonella mutagenicity test. Mutation Res. 113 :173-215.

Noordsij,A.,J. van Beveren and A. Brandt (1983). Isolation of organic compounds from water for chemical analysis and toxicological testing. Intern.J.Environ. Anal. Chem. 13: 205-217.

Puyker,L.M., G.Veenendaal and H.M.J.Janssen (1981). Determination of organo-phosphorus and organo-sulphur at the sub-ng-level for use in water analysis. Fres.Z.Anal.Chem.306:1-6.

Schwarzenbach, R.P.,et al. (1983). Behaviour of organic compounds during the infiltration of river water to ground water. Field studies. Environ. Sci. Technol.,17:472-479.

Stuyfzand,P.J. (1985). Hydrology, recognition and dating. In: Drinking water from river bank filtered ground water. KIWA report nr. 89: 2,47-2,54.(In Dutch)

Van der Gaag,M.A.,A.Noordsij and J.P. Oranje (1982). Presence of mutagenens in Dutch surface water and effects of water treatment processes for drinking water preparation. In: "Mutagens in our environment"(M.Sorsa and H.Vainio eds.) p.277-286. Alan R. Liss Inc., New York.

Van der Gaag,M.A. (1985). Some toxicological aspects. In:"Micropollutants and artificial dune recharge." KIWA report nr. 81, P.J.Stuyfzand ed.:260-283. (In Dutch).

Veenendaal,G.(1984) Group parameters, technique and application. H2O,17 :314-317. (In Dutch).

Zepp, R.G.,et al. (1974). Chemical and photochemical alteration of 2,4-D-esters in the aquatic environment. Proc. of the IUPAC-third international congress of pesticide chemistry, Helsinki.

Zoeteman,B.C.J. (1978). Sensory assessment and chemical composition of drinking water. Dissertation, Van der Gang B.V., 's-Gravenhage.

MUTAGENICITY TESTING OF WATER WITH FISH: A STEP FORWARD TO A RELIABLE ASSAY

M.A. VAN DER GAAG and J.F.J. VAN DE KERKHOFF[*]
The Netherlands Waterworks' testing and research institute, KIWA Ltd.
Research Department, P.O. Box 1072, 3430 BB Nieuwegein (The Netherlands)

ABSTRACT

A step forward has been set towards the realization of a reliable "sister-chromatid" exchange (SCE) assay with fish for the detection of mutagens in water. This test can detect mutagenic contaminants in water without prior concentration steps. A healthy breeding stock of the tropical fish Nothobranchius rachowi has been set up in our laboratory. This fish is particularily suited for the SCE assay because of its low number of chromosomes. The major improvement consisted of the development of a reliable staining technique for the differentiation of sister-chromatids. The assay can now be used for research purposes. With a few improvements it can become a rapid low-cost test for the screening of effluents on contamination with DNA-damaging agents.

INTRODUCTION

Exchanges between sister-chromatids of a chromosome are an indication of DNA-damage. Nine years ago Kligerman and Bloom (1976) introduced the SCE test in fish (Kligerman, 1979 and 1982). The potential of this "in vivo" assay was demonstrated by the detection of the induction of SCE's in fish exposed to unconcentrated Rhine water (Alink et al., 1980; Hooftman and Vink, 1981). The further development of the assay was not so successfull since suitable fish were difficult to obtain. It was not possible to breed the Mudminnow (Alink et al., 1980) in the laboratory. Nothobranchius rachowi (Hooftman, 1981) from commercial breeders were infected with parasites (van der Gaag et al, 1983). The unreliable sister-chromatid differentiation (SCD) staining technique however was the major problem (Kligerman and Bloom, 1976; van der Gaag et al, 1983). Further research on the assay was therefore abandoned by many laboratories.

BREEDING OF HEALTHY FISH

Healthy test animals are a "conditio sine qua non" for toxicological testing. A suitable fish for the SCE "in vivo" assay is Nothobranchius rachowi. It has 16 fairly large chromosomes, an exceptionally low number for a fish. The rate of

---

[*] This project was financed by the Netherlands Waterworks Association (VEWIN).

cell divisions is also high. This is important, because two cell replication cycles are needed to obtain sister-chromatid differentiation. In its natural environment in Eastern Africa, Nothobranchius lives in small pools during the wet season. The species is fully adapted to the seasonal changes in this part of the world. It grows very rapidly, and can reach its adult stage within two months after hatching. It survives the dry season as eggs. The embryonal development can be blocked for as long as 9 months in this period. In the laboratory, the adult fish can live for more than a year. The eggs are kept in humid peat in small polyethylene bags. We were able to control a number of factors which determine the length of the embryonic blockade. If necessary, it can be reduced to about four weeks (unpublished results). This controlable embryonic stage has many advantages. The breeding stock can be reduced to a small number of fish, lowering the maintenance costs. The embryos can be hatched when needed. It is also possible to mail the eggs over long distances without problems. We expect to publish the details of the breeding procedures within a few months.

A RELIABLE STAINING TECHNIQUE

The SCE assay is based on a technique which makes it possible to stain both chromatids of one chromosome with a different intensity. This differential staining of sister-chromatids is obtained by incorporating 5-bromo-2-deoxyuridine (BrdU) in the DNA as an analogue for thymidine (figure 1). A number of techniques are available for the SCD staining. Most are based on the fluorescence-plus-Giemsa protocol (FPG) introduced by Perry and Wolff (1974).

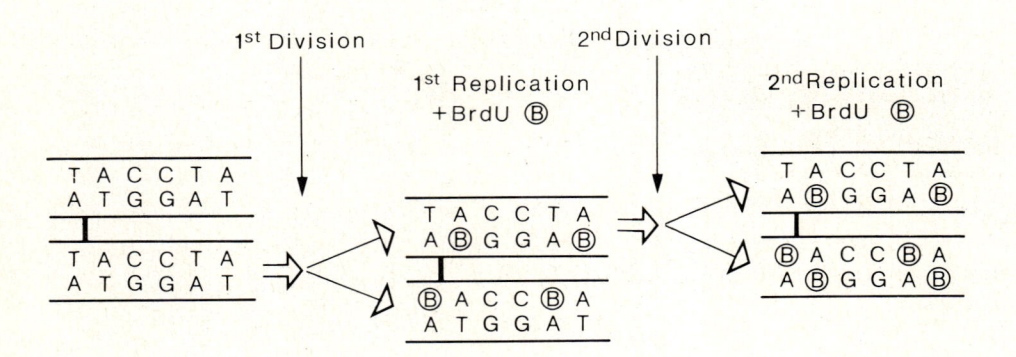

Fig. 1. Originally (at left), the DNA in the chromatids consists of sequences of A-T (adenine-thymine) and C-G (cytosine and guanine) base-pairs. After one replication round (middle), bromodeoxyuridine (B) has been incorporated in one strand of each chromatid instead of thymidine. After two replication rounds (at right), thymidine has been fully replaced by bromodeoxyuridine in one chromatid. The other chromatid still has one of the original DNA strands with thymidine. This will make the differential staining possible. Exchanges between sister-chromatids are seen as a harlequin pattern.

In fish however, the differential staining of sister-chromatids succeeded only in about 20% of the test animals (Kligerman and Bloom 1976; van der Gaag et al, 1983). With a few modifications we have obtained a rapid and effective SCD-procedure (van de Kerkhoff and van der Gaag, 1985). The major improvement consisted of a post treatment of the slides with 5N HCl (Gonzales-Gil and Navarette, 1982). The result is a high quality contrast between sister-chromatids (figure 2).

In our protocol, the fish are first exposed to a 100 mg/l BrdU solution at 26-27°C for 48 hours. During the last 10 hours, colchicine is added to arrest the cell division in the metaphase. After killing the fish, the gills are removed and allowed to swell for 30 minutes in a hypotonic solution of 0.4% KCl. The tissue is then fixed in methanol-acetic acid (3:1). Slides are prepared according to the solid tissue technique (Kligerman and Bloom, 1977). The slides have to be stained on the day of preparation, according to the following protocol: treat for 15 minutes with a 5 µg/ml solution of Hoechst 33258 in PBS buffer. Then mount under coverslips in the same solution, and expose to two 8 watt black light bulbs for 10 minutes at a distance of 2 cm. The slides are subsequently treated for 15 minutes with 5N HCl, and finally stained with a 2% Giemsa solution (pH 6.8) for 7-8 minutes.

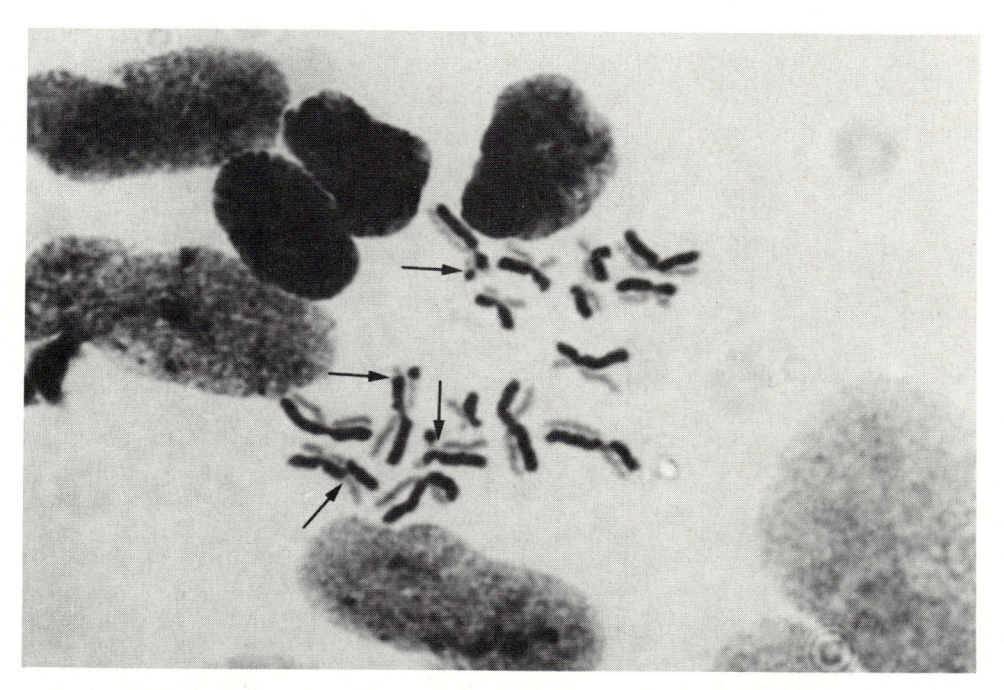

Fig. 2. Metaphase of a gill cell, consisting of 16 chromosomes. The BrdU substituted chromatids stain palely in comparison to the thymidine-chromatids. The arrows indicate the sites of sister-chromatids exchanges.

FIRST RESULTS

Base line SCE-frequencies

The base line SCE frequencies obtained with this protocol are quite low compared to other "in vivo" tests with aquatic organisms. They average at 0.08 SCE/chromosome (0.9 SCE/metaphase), with a 95% confidence interval ranging from 0.04 to 0.11. Only the mussel larvae SCE-assay (Harrison and Jones, 1982) showed a lower level (0.02-0.06 SCE/chromosome). The BrdU concentration in water was 30 times lower in the mussel larvae assay than in our test. The higher BrdU levels do not appear however to cause much higher SCE frequencies. It is probable that only a part of the BrdU present in the water has been incorporated in the DNA. This could be caused either by a limited uptake of BrdU through the gills, or a rapid metabolization by the fish.

Induction of SCE's by cyclophosphamide

Cyclophosphamide (CP) is a well known mutagen which is only positive in short term mutagenicity assays after metabolization by liver enzyme systems. Van der Hoeven and coworkers (1982) showed that Nothobranchius is capable of converting CP into its active metabolite. In a preliminary experiment we confirmed the mutagenic effect of CP. It also became evident that the timing of the exposure to this agent is of great influence on its response in the assay (figure 3). In this respect the results in Nothobranchius are very similar to those of a time course experiment with CP in mice (Charles et al., 1984).

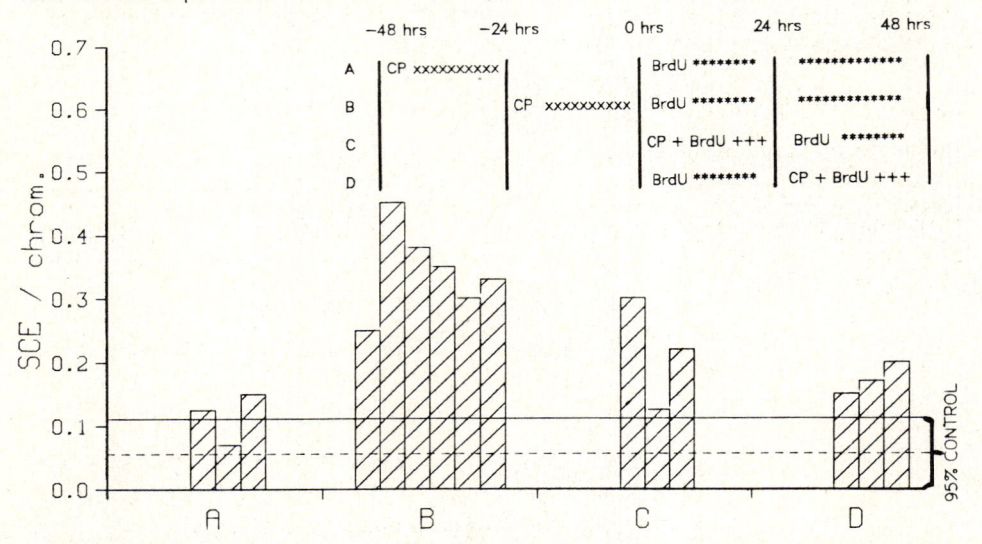

Fig. 3. If the exposure to CP takes place too long before the BrdU incorporation (group A), the SCE level is back to base line frequencies. The highest response is measured if the exposure to CP happens immediately before the incorporation of BrdU (group B). Decreasing SCE levels are noted if the exposure to CP is done simultaneously with the BrdU incorporation.

PROBLEMS AND PERSPECTIVES
Duplication of results

Further experiments with CP exposed a number of new problems. The variation in the response to CP appeared to be larger than was expected from the initial experiments. The mean number of SCE's from fish exposed to 25 mg CP before BrdU incorporation varied between about 0.10 and 0.50 SCE/chromosome. Three factors which could cause this variation are now under investigation. Temperature could be important, because the metabolic activity of fish is temperature dependent and CP has to be metabolized into its active form. The number of metaphases counted in each fish was limited to ten, because of initial difficulties in obtaining sufficient second division metaphases. This number may be too small, as some counts in fish with sufficient cells demonstrated a variation in mean SCE frequencies in groups of ten metaphases from a control fish. In one of these fish, the counts in four groups gave a series of mean values (0.030; 0.076; 0.076 and 0.100) with a threefold difference. Finally, the individual variation in the response of these animals to a chemical can not be estimated at this time.

How to obtain a sufficient number of metaphases?

A problem mentioned above, is the difficulty in obtaining consistently a high number of dividing cells in all fish. The time needed to score the SCE frequency in ten metaphases varied from about 15 minutes to more than an hour. As with the variation of response to chemicals, determination of temperature optimum might solve a part of this problem (table 1).

TABLE 1
Influence of temperature on cell division

Temper. (°C)	Number of fish	Number of metaphases*	Percentage of cells in 1st cycle	2nd cycle	≥3rd cycle
22	6	-	88-94	6-12	0
25	22	+/+	38-84	16-52	0-14
28	9	+/++	32-84	10-56	0-28

* -: very few dividing cells; ++: many dividing cells

Perspectives

The final step towards a routine assay will include the definition of a standardized protocol. This will have to take into account the statistical interpretation of the results and the economical feasibility of the assay.

At this moment, the SCE assay with Nothobranchius is only suited for research purposes. It is not yet standardized as an assay for the monitoring of water quality. We hope to make further progress in the coming year, leading to a final protocol. The costs of the assay are still difficult to estimate at this moment, as the definition of the statistical conditions will be of influence on the protocol. If the test turns out to be sufficiently sensitive, it will become a rapid and quite cheap assay for the monitoring of mutagenic contaminations in effluents and in river water, and possibly also for the analysis of water treatment systems. Its costs could be comparable to those of an Amestest, with the advantage that no preconcentration techniques are needed, and that the test is carried out "in vivo" in vertebrates.

REFERENCES

Alink, G.M., E.M.H. Frederix-Wolters, M.A. van der Gaag, J.F.J. van de Kerkhoff and C.L.M. Poels (1980). Induction of sister-chromatid exchanges in fish exposed to Rhine water. Mutation Res. 78, 369-374.

Charles, J.L., D. Jacobson-Kram, J.F. Borzelleca and R.A. Carchman (1984). The kynetics of in vivo sister chromatid exchange induction in mouse bone marrow cells by alkylating agents: Cyclophosphamide. Environ. Mutagen. 5, 825-834.

Gonzales-Gil G. and M.H. Navarrete (1982). On the mechanism of differential Giemsa staining of BrdU-substituted chromatids. Chromosoma 86, 375-382.

Harrison F.L. and I.M. Jones (1982). An in vivo sister-chromatid exchange assay in the larvae of the mussel Mytilus edulis: response to 3 mutagens. Mutation Res. 105, 235-242.

Hooftman, R.N. (1981). The induction of chromosome aberrations in Nothobranchius rachowi (Pisces: Cyprinodontiae) after treatment with ethyl methane sulphonate or benzo(a)pyrene. Mutation Res. 91, 347-352.

Hooftman, R.N. and G.J. Vink (1981). Cytogenetic effects on the eastern mudminnow, Umbra pygmaea, exposed to ethyl methane sulphonate, benzo(a)pyrene or river water. Ecotoxicol. Environm. Safety, 5, 261-269.

Kligerman, A.D. (1979). Induction of sister-chromatid exchanges in the central mudminnow following in vivo exposure to mutagenic agents. Mutation Res. 64, 205-217.

Kligerman, A.D. (1982). Fishes as biological detectors of the effects of genotoxic agents. In: "Mutagenicity. New horizons in genetic toxicology". J. Heddle (ed.), p. 435-356.

Kligerman, A.D. and S.E. Bloom (1976). Sister chromatid exchanges in adult mudminnows (Umbra limi) after in vivo exposure to 5-bromodeoxyuridine. Chromosoma 56, 101-109.

Kligerman, A.D. and S.E. Bloom (1977). Rapid chromosome preparation form solid tissues of fishes. J. Fish. Res. Bd. Canada 34, 266-269.

Perry, P. and S. Wolff (1974). New Giemsa method for the differential staining of sister chromatids. Nature 251, 156-158.

Van de Kerkhoff, J.F.J., and M.A. van der Gaag (1985). Some factors affecting optimal deffential staining of sister-chromatids "in vivo" in the fish Nothobranchius rachowi. Mutation Res. 143, 39-43.

Van der Gaag, M.A., G.M. Alink, P. Hack, J.C.M. van der Hoeven and J.F.J. van de Kerkhoff (1983). Genotoxicological study of Rhine water with the killifish Nothobranchius rachowi. Mutation Res. 113, 311.

Van der Hoeven, J.C.M., I.M. Bruggeman, G.M. Alink and J.H. Koeman (1982). The killifish Nothobranchius rachowi, a new animal in genetic toxicology. Mutation Res. 97, 35-42.

ALTERNATIVE METHODS FOR CHLORINATION

F. Fiessinger, J.J. Rook and J.P. Duguet
Laboratoire Central Lyonnaise des Eaux, 78230 Le Pecq (France)

ABSTRACT

Existing disinfectants are oxidative agents which all present negative effects on subsequent treatment processes. None of them has decisive advantages over chlorine, although chlorine-dioxide and chloramines might at times be preferable. Optimum treatment practices will improve the removal of organic precursors before final disinfection which could then consist in a light chlorine addition. A philosophy of radical change in water treatment technology encompassing physical treatment without chemicals such as membrane filtration, solid disinfectants is presented.

INTRODUCTION

Since its introduction into water treatment in the beginning of this century, chlorine has held a predominant position as reliable disinfectant because of its broad range biocidal effectiveness, its reasonable persistence in treated waters, its ease of application and control and its cost effectiveness. In addition chlorine is the only chemical agent that is able to oxidize ammonia readily. Chlorine is also used for controlling the proliferation of algae during the warm periods in uncovered coagulation and sedimentation basins. Unfortunately, when organic matter is present chlorination results in the formation of undesirable halogenated compounds ; i.e. total halogenated compounds (TOX) and more particularly the Trihalomethanes (THM). The use of modern analytical techniques has led to the identification of a large part of them (Christman et al., 1983, Bruchet et al.1984, Coleman et al. 1984, Duguet et al. 1984b) as illustrated in Figure 1. Some of these compounds like dichloroacetonitrile, several chlorinated ketones, chloroform are known to be mutagenic or toxic (Simmon et al. 1977, Bull 1980).

The effects of water chlorination on mutagenic activity have been studied. Many authors, (Kool, 1984), conclude that chlorination increases mutagenicity (see Fig.2) but the nature of organics and chlorination conditions have a great influence. Thus Cognet (1984) found that during chlorination of treated Seine water, variation of mutagenicity during one year was not statistically significative.

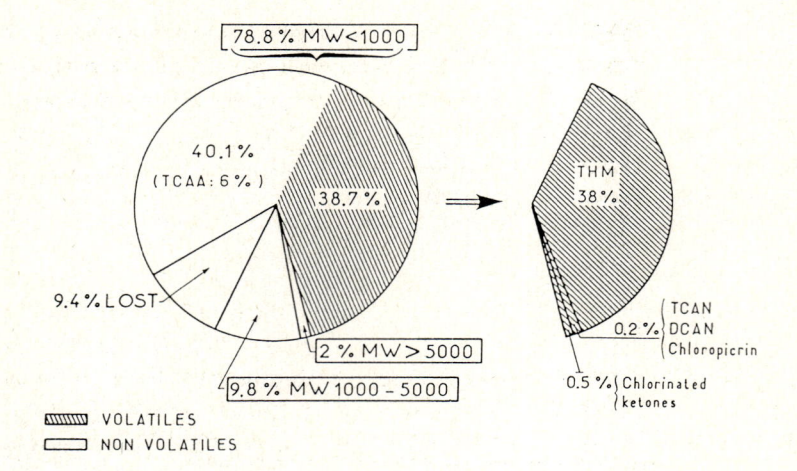

Fig. 1. Diagram of molecular weight distribution of TOX after prechlorination of a reservoir water (Cholet, France) and speciation of the volatile fraction.

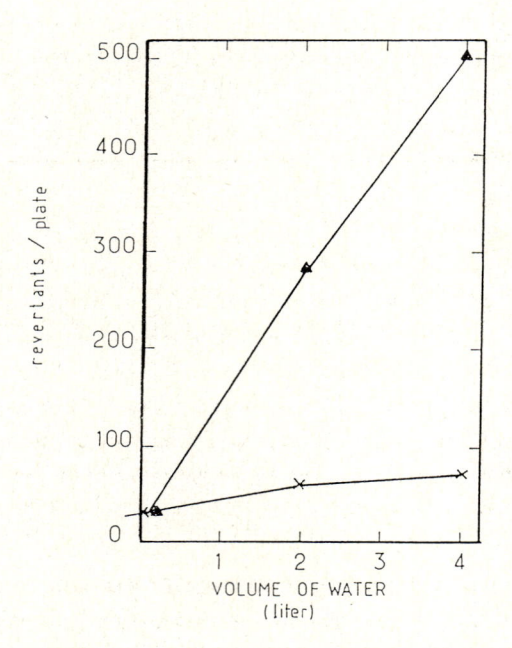

Fig. 2. Comparison of mutagenic activity on Salmonella thyphimurium strain TA 98(-S9), before (x) and after prechlorination/Dissoved air flotation ▲ of a surface water (Moulle, France)

Another risk is related to the direct toxicity of chlorine gas during its road transport and its storage as liquid chlorine, in plants located near densely populated areas. This risk can be avoided by using sodium hypochlorite instead of liquid chlorine, though at a 20 to 25 % increase in costs (Gomella,1980). In situ generation of chlorine, through electrolysis might also be applied.

ALTERNATIVE DISINFECTANTS

Although chlorine is a good disinfectant, the potential health risks of halogenated by-products formation have caused the entire subject of drinking water disinfection to be reconsidered. The main alternative disinfectants in use are ozone, chlorine dioxide, chloramines and to a less extend ultraviolet light, hydrogen peroxide, permanganate, other halogens and silver ions. Gomella (1980) and Fiessinger (1981) summarized the properties of several disinfectants with respect to their possible reactivities with water constituents. An extended summary of oxidant properties based on these surveys is shown in table 1.

TABLE 1
Comparison of Various Disinfectants

	Cl_2	ClO_2	O_3	$KMnO_4$	NH_2Cl	H_2O_2
Iron and manganese	+	++	+++	+	-	+
Ammonia	+++	-	-	-	-	-
THM formation	+++	-	-	-	-	-
THM precursors removal		+	++	+	-	-
Formation of mutagens or toxic substances	+	+-	+-	-	?	?
	+	+-	+-	-	?	
Enhanced biodegradability	+	++	++	+	-	?
Taste removal	-	+	++	+	-	+
Disinfection	+	++	+++	+-	+-	+

- no effect   ++ effect
+ little effect   +++ very effective

For the evaluation of the different properties and effects of an alternative disinfectant it must be distinguished where it is applied in the treatment process,(see Fig.3) either in preoxidation, as an intermediate step or in post-disinfection, i.e. before or after the bulk of the organic matter, especially the trihalomethane formation potential (THMFP) has been removed.

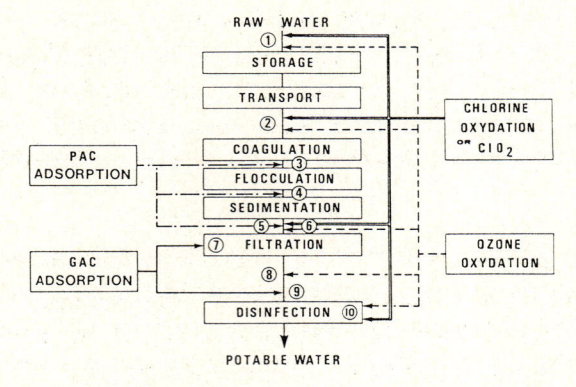

Position of oxidation treatments in a general surface water treatment process.

Fig. 3. Influence of light on residual ClO_2 and chlorite concentration

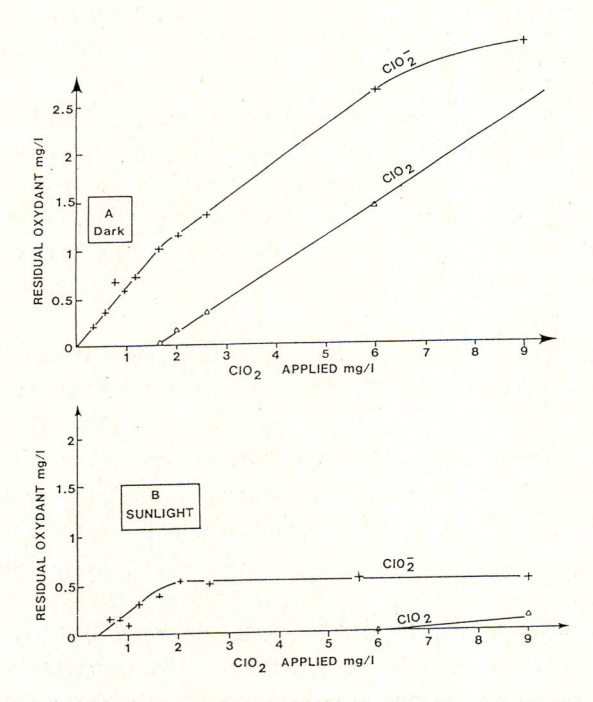

Fig. 4. Possible positions of oxidation treatments in surface water treatment process

Preoxidation

Traditionally preoxidation is performed, at the beginning of the treatment process, to ensure good hygienic conditions throughout the treatment and to control algal growth in flocculation basins. Thus growth can be limited by covering the basins provided that ammonia would be removed in subsequent treatment steps.

__Permanganate__. The application of permanganate in pretreatment oxidizes iron, manganese and destroys taste and odor causing substances. It can reduce THMFP but at the pretreatment dosage usually applied the reduction of chloroform formation is relatively modest (Singer et al.,1980). Permanganate is not a very effective disinfectant and a possible residual of manganese precipitation in the distribution system are two reasons why permanganate is rarely used in water treatment.

__Chlorine dioxide__. The great advantages of the use of chlorine dioxide in pretreatment are algicidal effect and negligible formation of halogenated by-products (Stevens 1982). However, chlorine dioxide produces polar compounds such as aldehydes, ketones and acids (Rav Acha, 1984). The main inorganic by-products is chlorite which is reported to be toxic.

Using mouse skin initiation promotion essays, among mice when concentrates of water disinfected with chlorine dioxide were applied, no effect was observed. However, short term toxicity of chlorine dioxide and more specifically its inorganic reaction products may present a higher risk than chlorine or ozone (Bull 1980).

Chlorite (ClO_2^-) and chlorate (ClO_3^-) in high concentrations (100 mg/l) have been found to produce methemoglobinemia in animals (Komorita, 1985). These uncertainties have made health authorities reluctant to allow the application of chlorine dioxide in many countries. In some countries standards as low as 0.1 mg/l chlorites have been edicted. In order to maintain these inorganic by-products at such a low level, the reagents and the by-products concentrations have to be constantly monitored which cause. Many analytical problems are not yet solved (Masschelein, 1984). The disinfective efficiency of chlorine dioxide is reported to be superior to that of chlorine (Hoff, 1981, Longley ,1981). On the other hand when ClO_2 is applied in open basins the photodecomposition of ClO_2 necessitates the application of an excessive dose (Fiessinger, 1981) (see Fig.4).

The costs for applying chlorine dioxide are 3 to 4 times than those for chlorine (Gomella, 1980).

Ozone. The preozonation of natural organic matter does not lead to its direct removal as is reflected by nearly unchanged TOC values. The primary effect is to modify the chemical nature of the molecules towards increase of polarity. The increased polarity will in turn favor adsorbability in subsequent coagulation filtration and adsorption processes (Kuehn, 1981). A clear example of improved turbidity removal induced by oxidative pretreatment was found for preozonated Seine water (Fiessinger, 1981) (see Fig. 5). The same improvement however, could be achieved by a slighty increased alum dose, such that in this case preozonation was an expensive means for saving on flocculant costs.

Reckhow and Singer (1984) have shown that preozonation of certain lake waters reduced the formation potentials of THM and TOX significantly (see fig.6). Doses exceeding 1.2 mg O3 per mg TOC per liter hampered the removal of THM and TOX precursors in alum coagulation,. Jekel however reported improved removal of turbidity caused by humic acid coated mineral particles along with improved particle agglomeration, when preozonation was applied in ratio's below 0.8 mg O3 per mg TOC (Jekel, 1983). Increase of that ratio had no further effect.

The polymerizing effect of ozone on small sized micropollutants may present more interesting possibilities. Duguet and al.(1985a)) using a dichlorophenol synthetic organic compounds found that ozonation induced polymerization to hexamers and insoluble polymers, which will improve removal in coagulation filtration.

As with other oxidants, ozonation also leads to formation of organic by-products mostly aldehydes, ketones and carboxylic acids (Mallevialle, 1980) Especially aldehydes have been quantified (Van Hoof et al., 1985).

Ozonation by-products may also induce mutagenicity in the treated water which can be effectively removed by activated carbon (Van Hoof,1983). It is also possible to diminish the production of mutagens during ozonation by prolonged contact time or dosage (Duguet et al.,1984a) (see Fig.7). Ozonation of organic matter will naturally increase biodegradability,(Peel and Benedek, 1983) which can be considered (see Fig. 8) as an advantage as well as an disadvantage. This material must be removed biologically before the water leaves the treatment plant since it may interfere with post disinfection. A principle disadvantage of the biodegradation taking place in biologically operated carbon filters is that a complete ecology of higher plankton will develop and foul the filters. Practical experience has shown that the filtered waters may contain heavy loads of zooplankton during summer periods (Rook, 1983).

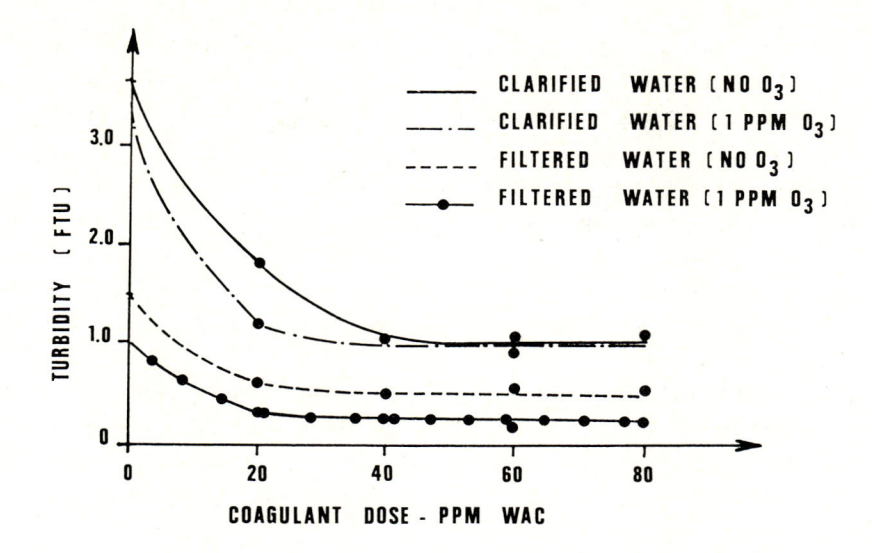

Fig. 5. Effect of preozonation on clarification and filtration of a surface water (Croissy, France)

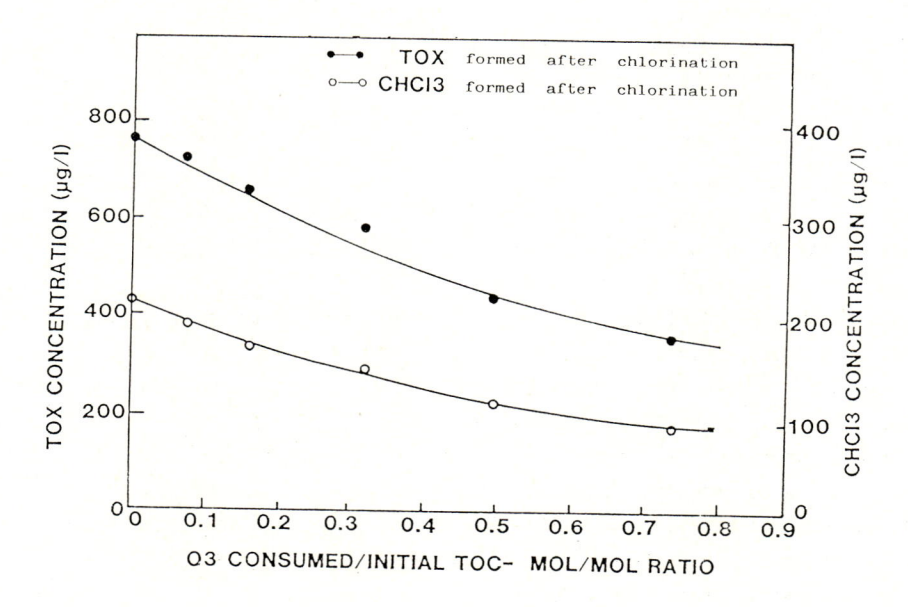

Fig.6 Effect of preozonation on TOX and CHCl₃ formations during subsequent chlorination of black lake fulvic acid (from Reckow, Singer, 1984)

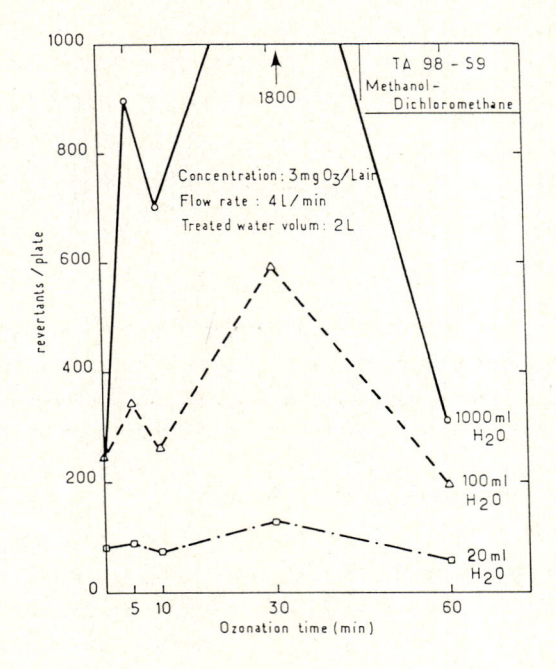

Fig. 7. Effect of ozonation conditions on mutagenic activity of a ground water at three sample volumes (Le Pecq, France)

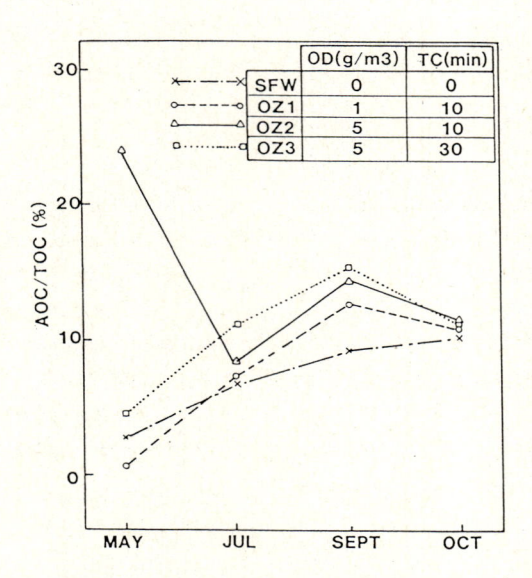

Fig. 8. Effects of ozone dose (OD) and contact time (TC) on the biodegradability of a filtered water (SFW)

In our view the full advantage of preoxidation can only be obtained when used in combination with an effective means of removing biodegradable subtances, such as slow sand filtration. Walker observed in a comparative study (1984) of preozonated slow sand filters a doubled removal of TOC, 35 % of the initial 3.7 mg/l TOC versus 16 % in the non-ozonated control filter.

Post Disinfection

The reason for the use of disinfectants after the complete water treatment is to kill or inactivate microorganisms still present to protect the distribution system from regrowth and safeguard hygienic quality. This necessity is even more pronounced if prechlorination is abandoned. In this case post-disinfection may be the only hygienic barrier. The choice of disinfectant must be made on the basis of germicidal activity and persistence for maintaining a residual in the network to prevent any further contamination. Experimental data show the following order of decreasing germicidal efficiency $O_3 > ClO_2 > HOCl > OCl^- > NHCl_2 > NH_2Cl$. Ozone is the best disinfectant but its half life is not sufficient to maintain a residual (see Fig. 9). Moreover the application of ozone may produce some biodegradable matter resulting in regrowth.

For chlorine dioxide this problem exists to a less extent. Besides it has the advantage of its long half life ensuring a residual throughout the network. At the final point of the treatment organic matter is at minimum. The formation of chlorite by the organic matter reduction of ClO_2 is than limited such that the acute toxicity may be considered to be less significant.

Some combinations of oxidants like chloramine + hydrogen peroxide have been shown to improve control of bacterial regrowth in a treated surface water for which chlorine disinfection was insufficient (Germonpré, 1985). More studies should be done to verify the efficiencies of such combinations (H_2O, $O_3 + H_2O_2 + UV$, $O_3 + UV$). Disinfection using UV involves different mechanisms such as direct UV action and formation of high energetic radicals which have short half lifes, but the problem of persistence remains. This short survey shows that replacement of chlorine by another disinfectant is not quite an easy task. The ideal disinfectant may have four main properties :
- germicidal effect
- persistent residual
- no precursor of toxic by-products
- low cost.

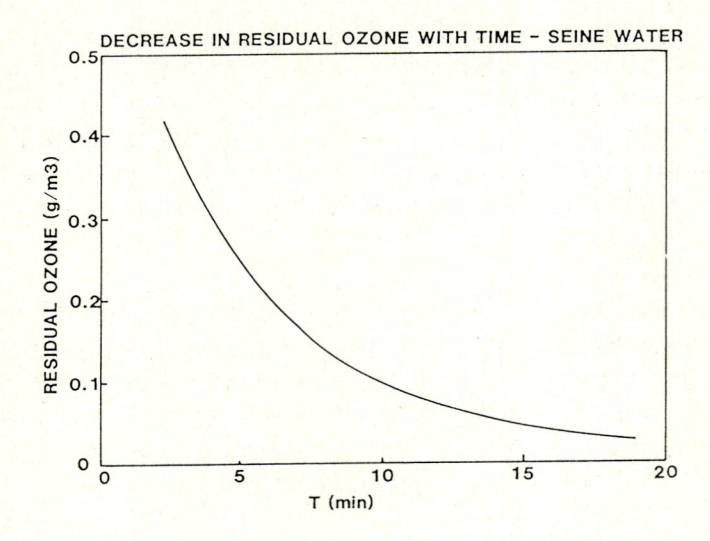

Fig. 9. Decrease in residual ozone with time (Seine water, France)

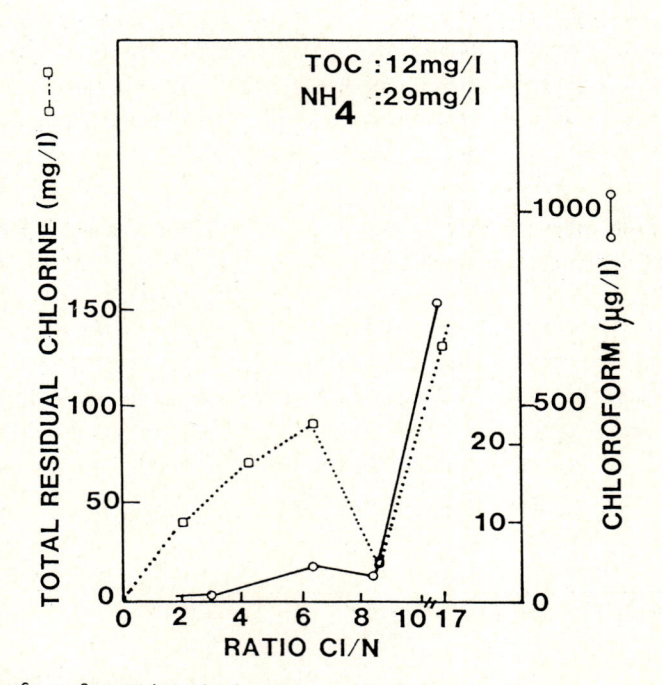

Fig. 10. Chloroform formation during conventinal breakpoint chlorination curve

Chlorine still remains the best available means of disinfection. However optimal conditions for its application must be determined with respect to a minimizing by-product formation to ensure hygienic safeguard. It is still preferable to reduce the ammonia content by biological treatment in order to avoid high chlorine demands.

ALTERNATIVE STRATEGIES FOR A BETTER USE OF CHLORINE

With the objective to reduce halogenated compounds such as THM's, two treatment strategies can be followed : the chloramination and the use of chlorine after a maximal reduction of organics and specifically mainly the precursors of chlorinated compounds. The development of new adsorbents such as activated alumina and resins may further improve the removal of precursors

Chloramination

As illustrated in figure 10 in natural water containing ammonia, the trihalomethanes are formed at a chlorine dose corresponding to the destruction of chloramines of free appearance of free chlorine.

Although the chloramines are weak disinfectants they have algal inhibiting properties. With the objective to reduce halogenated compounds, the pretreatment may be realized with chloramines. In this case, the chlorine dose must be adjusted to the maximum of chloramine with or without addition of ammonia. On line measurement of chloramines is not easy and the process control is difficult to realize.

The main problems related to chloramines are tastes and odors and some health effects. Although monochloramine is the predominant form at a pH of 8, the correlated traces of di an trichloramines have offensive odors as found by Krasner (1984) (Table 2).

TABLE 2
Sensory Threshold Values

Compounds	Threshold Aroma	(mg/l as Cl2) Flavor
Hypochlorous acid	0.28	0.24
Hypochlorite ion	0.36	0.30
Monochloramine	0.65	0.48
Dichloramine	0.15	0.13

(from : Krasner 1984)

The health effects of chloramines and the knowledge of by-products formed (organic chloramines ...) need more research, thus chloramines have recently been found to cause hemolytic anemia in patients undergoing kidney dialysis (Komorita, 1985). To solve the problems of taste, odors and health effects chloramination may be followed by a chloramine removal from water by reduction on activated carbon which should be carefully monitored to produce a water of desired quality. This alternative chlorination has a promising development; thus one of the largest water utilities in the USA has changed from chlorine to chloramines for the control of THM formation.

This control may be also realized by the use of chlorine after the reduction of organics to a great extent.

OPTIMIZATION OF ORGANICS REMOVAL BEFORE CHLORINATION

As illustrated by figure 11 the use of chlorine only at the end of the water treatment permits a great reduction of THM formation. A low THM level may be obtained by the optimization of each treatment step. Coagulants such as Al^{3+} and Fe^{3+} remove significant concentrations of TOC and THM precursors which, in actual practice tends to range from about 40 % to 70 %. Singer (1983) found that the removals of THMFP concentrations tend to be higher than the corresponding reduction in TOC. Some improvements in the utilization of metal coagulants can result in a better reduction of precursors but the combination of ozone with an adsorbant like activated carbon can retain a large part of organics from clarified water. This unit operation is already used routinely in drinking water production in Europe. However, activated carbon is a rather non specific adsorbant which while a good principle, frequently leaves a fraction of organics in the water which may give rise to halogenated compounds during post-chlorination. Improvements of the yield of the adsorption will be feasible by a research of new adsorbants of different nature combined with an optimized oxidation like e.g. ozone coupled with hydrogen peroxide (Duguet et al, 1985b) (see Fig. 12). Ozonation in all cases produces polar products which are less adsorbable by activated carbon. This is illustrated by shifts in the adsorption isotherms shown in Figure 13. In a study reported by Duguet et al. (1985b) the k-value of the Freundlich isotherm (TOC) was diminished by a factor 3 after preozonation. With activated alumina, however, an increased adsorption was observed, enhanced by the combined action of O_3 and H_2O_2. This case is an example of an optimization of organics removal which can lead to more effective reduction of the formation of toxic compounds during final chlorination.

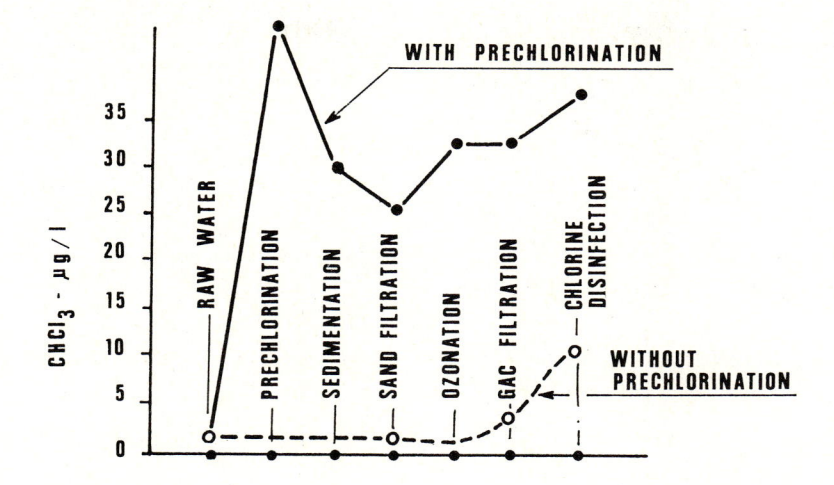

Fig.11. Changes in chloroform concentrations during treatment with and without prechlorination

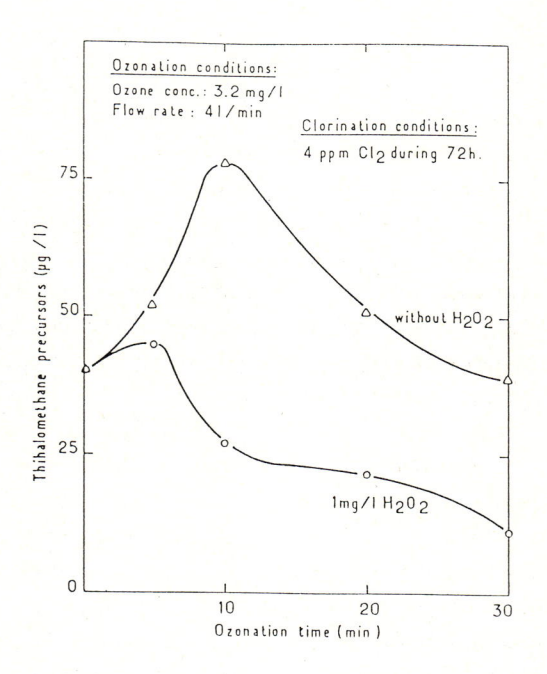

Fig.12. Change in the THM precursors in a lake water (Cholet) during ozonation - effect of H_2O_2

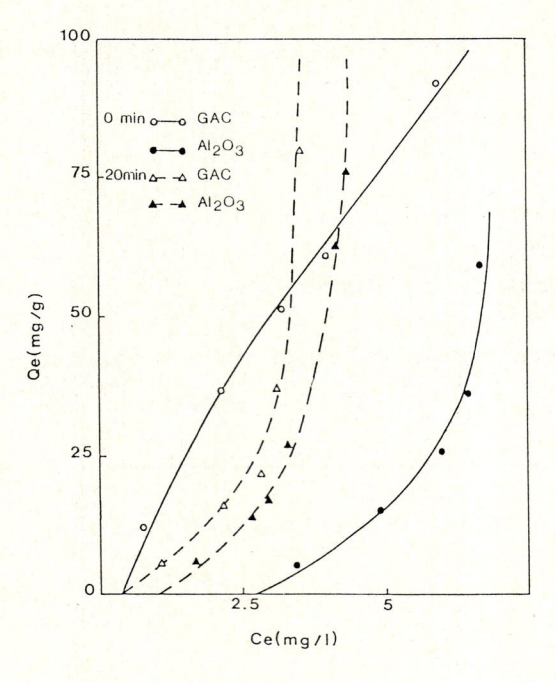

Fig. 13. Effect of ozonation on the isotherms for adsorption of a lake water (Cholet) on alumina and activated carbon.

An additional advantage of having removed organic matter as far as possible is the reduction of chlorine consumption in the distribution system resulting in lower doses needed for maintaining a desired residual. Depending on the state of network multiple injections of low doses of chlorine at several crucial points may be necessary.

CONCLUSIONS

- There is no satisfactory alternative for chlorination:
    . Prechlorination should be abandoned
    . Algal growth may be controlled with chloramines, or coverage of open
    . sedimentation basins
    . Ammonia can be removed biologically
    . Chlorine dioxide and chloramines may constitute satisfactory interim alternatives

. Ozone should be used as a complement not for a replacement.

- Combination of oxidants might present interesting synergistic effects in particular for oxidation O3 + UV.

- Chlorine can be maintained as a final disinfectant if organic precursors are sufficiently removed. Process sequences of oxidation, adsorption and biodegradation should be carefully designed.

- Disinfection practices should be revised in the light of new germs, Giardia, Legionella, ...

- Disinfection mechanisms should be further investigated.

- Megatrends
. New disinfectants
   Immunological applications
   Immobilized disinfectants, Ag ...
   nascent chlorine, electrical treatment

. Improved removal techniques
   micro and ultrafiltration
   improved coagulants

REFERENCES

Bruchet, A., Tsutsumi, Y.; Duguet, J.P. and Mallevialle, J., 1984. Characterization of total halogenated compounds along various water treatment processes. Presented to the Fifth Conference on Water Chlorination. Environmental Impact and Health Effects. Williamsburg, Virginia.
Bull, R.J., 1980. Health effects of alternate disinfectants and their reaction products. J. Awwa, 5: 299-303.
Christman, R.F., Norvood, D.L., Millington, D.S., Johnson J.D. and Stevens, AA 1983. Identification and yields of major halogenated products of aquatic fulvic acid chlorination. Env. Sci. tech., 17 : 625-628.
Cognet, L., Duguet, J.P., Courtois, Y., Bordet, J.P. and Mallevialle, J., 1984. Use of MRR for Ames test in a practical operating system. Congress of the Division of Environmental Chemistry. ACS. Philadelphia, 26-31.
Coleman, W.E., Munch, J.W., Kaylor, W.H., Streicher, R.P., Ringhand, H.P. and Meier, J.R. 1984. Gas chromatography/mass spectroscopy analysis of mutagenic extracts of aqueous chlorinated humic acid. A comparison of the by-products to drinking water contaminants. Env. Sci. Techn., 18: 674-681.
Duguet, J.P., Ellul, A., Brodard, E. and Mallevialle J., 1984a. Evolution de la mutagénèse au cours d'un traitement d'ozonation, IOA Congress, Brussels.

Duguet, J.P., Tsutsumi, Y., Mallevialle, J. and Fiessinger, F., 1984b. Chloropicrin in potable water : conditions of formation and evolution along various treatment processes. Fifth Conference on Water Chlorination : Environmental Impact and Health Effects, Williamsburg, Virginia.

Duguet. J.P., Dussert, B., Bruchet, A., Mallevialle, J. and Richard Y., 1985a. La polymérisation des matières organiques des eaux, comparaison des effets de l'ozone sur le 2-4 dichlorophenol. IOA Meeting, Wasser Berlin.

Duguet, J.P., Brodard, E., Dussert, B. and Mallevialle, J., 1985b. Improvement in the effectiveness of ozonation of drinking water through the use of hydrogen peroxide. Pre-publication Ozone Science and Engineering.

Fiessinger, F., Richard, Y., Montiel, A. and Musquere, P.,1981 Advantages and Disadvantages of Chemical oxidation and disinfection by ozone and chlorine dioxide. Sc. of the Total Environment, 18:245-260.

Germonpré R., personal communication.

Gomella, C., Musquere, P., 1980. La désinfection des eaux par le chlore, l'ozone et le dioxide de chlore. 13ème Congrès AIDE, Paris.

Hoff, J.C. and Geldreich, E.E., 1981. Comparison of the biocidal efficiency of alternative disinfectants. J. Awwa, 1: 40-44.

Jekel, M.R., 1983. The benefits of ozone treatment prior to flocculation processes. Ozone : Science and Engineering, 5/21-35.

Kool, H.J. and Van Kreikl, C.F., 1984. Formation and removal of mutagenic activity during drinking water preparation. Water Res. 8: 1011-1016.

Komorita, J.D., Snoeyink, V.L. 1985. Technical note : Monochloramine removal from water by activated carbon. J. Awwa, 1 : 62-64.

Krasner, S.W. and Barret, S.E., 1984. Aroma and flavor characteristics of free chlorine and chloramines. Water Quality Technology Conference, Am. Wat. Works Ass., Denver, Colorado.

Kuehn, W. and Sontheimer, H., 1981. Treatment Improvement on deterioration of water quality. Sc. of the total Environment, 18: 219 - 223.

Longley, K.E., Moore, B.E. and Sorber, C.A. 1981. Disinfection efficiencies of chlorine and chlorine dioxide in a gravity flow reactor. J. W.P.C.F., 54 : 140 - 145.

Mallevialle, J., 1980. Produits de réaction identifiés au cours de l'ozonation. l'ozonation des eaux, Manuel pratique (Editor W.J. Masschelein, IOA, Paris, 78-99.

Masschelein, W.J., 1984. Methods for controlling chlorine dioxide operation. Part. 2, J. Awwa, 3 : 80-82.

Peel, R.G. and Benedek, A., 1983. Biodegradation and adsorption within activated carbon adsorbers. J. W.P.C.F., 55: 1168-1173.

Rav., Ch. Acha., 1984. Review Paper : The reactions of chlorine dioxine with aquatic organic materials and their health effects. Water Res. 18: 1329-1341.

Reckhow, J.J., Singer, Ph.C. 1984. The removal of organic halide precursors by preozonation and alum coagulation. J. Awwa, 4: 151-157.

Rook, J.J. 1983. Comparison of removal of halogenatedcompounds by carbon in pilot filters. Proc. 181st Meeting Am. Chem. Soc., Atlanta. Advances in Chemistry series, Washington DC, 202/ 455-479.

Simmon, V.F., Kauhanen, K. and Tardiff, R.G. 1977. In : Progress in genetic toxicology, D. Scott, B.A. Bridges, F.H. Sobles, Elsevier. North Holland Biomedical Press : Amsterdam, The Netherlands, 249-258.

Singer, P.C., Borchardt, J.H. and Colthurst, J.M., 1980. The effets of permanganate pretreatment on trihalomethane formation in drinking water. J. AWWA, 10: 573-578.

Singer, P.C. 1983, Applicability of coagulants and alternative oxidants for controlling THM formation. Presented at the Seminar on Strategies for the control of trihalomethanes at the Awwa Annual Conference, Las Vegas, Nevada.

Stevens, A.A., 1982. reaction products of chlorine dioxide. Env. Health Persp. 46: 101-110.

Walker, R.A. 1984. Use of ozone for enhancing organics removal, EEC Workshop COST Project 641, Barcelone, 86-89.

Van Hoof, F., 1983. Removal and formation of mutagenic activity by ozone. Proc. IOA Symposium, "Environmental Impact and benefit", Brussels, 410-423.
Van Hoof, F., Wittocx, A., Van Bruggenhout, E. and Janssens, J., 1985. Determination of aliphatic aldehydes in water by high pressure liquid chromatography. (in press).

IDENTIFICATION AND ASSESSMENT OF HAZARDOUS COMPOUNDS IN DRINKING WATER

J K Fawell and M Fielding

Water Research Centre, Henley Road, Medmenham, PO Box 16 Marlow, Buckinghamshire, SL7 2HD, United Kingdom

ABSTRACT

The identification of organic chemicals in drinking water and their assessment in terms of potential hazardous effects are two very different but closely associated tasks. In relation to both continuous low-level background contamination and specific, often high-level, contamination due to pollution incidents, the identification of contaminants is a pre-requisite to evaluation of significant hazards. Even in the case of the rapidly developing short-term bio-assays which are applied to water to indicate a potential genotoxic hazard (for example Ames tests), identification of the active chemicals is becoming a major factor in the further assessment of the response.

Techniques for the identification of low concentrations of organic chemicals in drinking water have developed remarkably since the early 1970s and methods based upon gas chromatography-mass spectrometry (GC-MS) have revolutionised qualitative analysis of water. Such techniques are limited to "volatile" chemicals and these usually constitute a small fraction of the total organic material in water. However, in recent years there have been promising developments in techniques for "non-volatile" chemicals in water. Such techniques include combined high-performance liquid chromatography-mass spectrometry (HPLC-MS) and a variety of MS methods, involving, for example, field desorption, fast atom bombardment and thermospray ionisation techniques. In the paper identification techniques in general are reviewed and likely future developments outlined.

The assessment of hazards associated with chemicals identified in drinking and related waters usually centres upon toxicology - an applied science which involves numerous disciplines. The paper examines the toxicological information needed, the quality and deployment of such information and discusses future research needs.

Application of short-term bio-assays to drinking water is a developing area and one which is closely involved with, and to some extent dependent on, powerful methods of identification. Recent developments are discussed.

INTRODUCTION

Drinking water usually contains a complex mixture of organic chemicals that covers a remarkably diverse range of chemical type. The substances are of natural and anthropogenic origin although often a distinction between the two is difficult. Anthropogenic substances can be introduced into raw waters by industrial and domestic effluents, surface drainage, spillage etc and by less obvious routes such as synthesis during disinfection and leaching from distribution mains.

We are able to make such comments because of the dramatic development and application of sensitive qualitative analytical techniques during the last decade or so. The driving force behind this development was mainly concern over potential health risks, such as carcinogenicity, that may be associated with exposure to small amounts of organic chemicals. Initially, many studies aimed at an assessment of these risks to health had a very basic approach, namely to identify the organic chemicals in drinking water and assess any associated hazards by consultation of relevant toxicity data. Very quickly researchers in this field became aware that the newly acquired ability to show the presence of low concentrations of a variety of anthropogenic organic chemicals in drinking water was countered, to a large extent, by an inability to evaluate the practical significance. In essence, the required toxicity data and basic knowledge did not exist. Additionally, although the analytical techniques developed were exceedingly powerful compared to previous methods, they were limited and overlooked a major fraction of organic chemicals in drinking water - the so-called non-volatiles (Garrison, 1976, Fielding and Packham, 1977).

More recently, and to some extent because of these limitations, an additional approach has been adopted in many studies (Packham et al., 1981, Kool et al., 1981, Loper et al., 1978). This involves the deployment of genotoxicity tests, such as the bacterial mutagenicity assays, in conjunction with chemical analysis to focus attention on the identification of organic chemicals in drinking water of possible maximum concern. Such an approach enables resources to be concentrated and used more effectively and, consequently, it has considerable merit.

The identification of organic chemicals in drinking water and their assessment in terms of hazard involves two very different disciplines but, clearly, they are very closely related tasks. The paper is neither intended to catalogue all the research reported in the literature nor provide an inventory of all chemicals identified in drinking water. However, it is intended to consider recent progress in these fields generally and to discuss future requirements and developments.

IDENTIFICATION OF ORGANIC SUBSTANCES IN DRINKING WATER

With the exception of grossly contaminated drinking water, concentrations of organic chemicals (usually less than $1 ug.l^{-1}$) are such that direct application of identification techniques is usually impossible and, consequently, some form of isolation/concentration process is required. Usually the mixture of organic chemicals isolated is so complex that considerable separation, invariably by some form of chromatography, is also needed prior to application of instrumental techniques capable of providing structural information. Thus, the

overall identification technique deployed usually consists of:
  i.   isolation/concentration (not necessarily as one step)
  ii. separation (of the components in the complex mixtures isolated)
  iii. detection and structural information

Stages ii. and iii. are dealt with later. Various methods for isolating and concentrating organic chemicals from drinking water (and related waters) exist but it is not the intention of this paper to cover these in detail. A number of validated methods have emerged based upon solvent extraction, adsorption (usually by XAD resin) followed by solvent elution of the adsorbent, headspace and related methods. Good examples of such methods can be found in the literature (Keith, 1981) and reference to a so-called Master Analytical Scheme (Garrison, 1981, is worthwhile since this provides validated techniques for the following categories of organic substances in drinking water:

<u>Organic substances</u>

  i.   volatile (purgeable) organics
  ii. neutral water soluble organics
  iii. weak acids, bases and neutrals
  iv. extractable semi-volatile strong acids
  v.   volatile strong acids
  vi. strong amines

Most well-validated methods tend to be inherently more suitable for non-polar or moderately polar substances. However, potential methods for non-volatile and more polar substances have been reviewed (Crathorne et al., 1982) but, in general, validated methods for substances not amenable to gas chromatography-mass spectrometry are scarce.

<u>Gas chromatography-mass spectrometry</u>

Modern GC-MS provides a powerful and indispensible technique for the identification of organic chemicals in drinking and related waters. Used in conjunction with suitable isolation/concentration methods, such as those mentioned above, GC-MS combines a powerful separation method with an excellent technique for providing structural information. Without the early availability and refinement of GC-MS the remarkable growth in studies on potential health effects of organic substances in drinking water would have been impossible.

Capillary GC columns and suitable instrumentation are widely available and capable of separating organic matter extracted from almost any drinking water into a multitude of constituents. In the case of drinking water derived from moderately polluted sources, hundreds of separated components can be detected with relative ease. Modern mass spectrometers are entirely compatible with capillary column GC and scan fast enough to cope with the considerable numbers of separated components. Application of this techniques is now virtually routine and examples abound in the literature (James, et al., 1981, Coleman, 1980B, Keith, 1976, 1981). However, there are some important limitations to the techniques which need to be recognized since they ultimately affect our knowledge on organic chemicals in drinking water and consequently our understanding of any associated health hazards:

i. high capital and running cost
ii. GC-MS "unknowns"
iii "too much" data
iv. quantitative data
v. limitation to volatile compounds

The very high cost of GC-MS (and most MS systems) means that relatively few instruments are available for drinking water analysis and most drinking waters have never been examined. To some extent the situation could improve with the emergence of cheaper and simpler systems marketed as GC detectors and intended for automated, routine operation (Stafford et al., 1983).

Although GC-MS often detects large numbers of organic chemicals in drinking water, over 10% of these, depending on the degree of contamination, cannot be identified and to a large extent little impact upon these unknowns has been made in the last few years. They remain unknown largely because of the absence of reference mass spectra. It must be borne in mind that a major factor in the success of GC-MS in drinking water analysis has been the emergence of a large, reasonable quality, reference mass spectra data bank. Although such data banks contain about 40,000 mass spectra they are still inadequate. In principle it is possible that many of these unknowns could be identified by more advanced GC-MS approaches (see below) but the time and resources needed are often difficult to justify.

A curious criticism of GC-MS sometimes encountered is that it produces too much data; hundreds of chemicals being identified in a single drinking water. These are present at low concentrations and the problem concerns the burden of evaluating the potential health risks for each of the substances. The problem is touched upon later but it seems unreasonable to criticize an analytical technique for displaying the nature and complexity of organic chemicals in drinking water.

The main ability of mass spectrometric techniques is to identify organic substances in water and as such they are unrivalled. Often, after having performed this task there is an immediate demand for quantitative information from the same analytical data. As mentioned already, GC-MS often detects and identifies multitudes of organic chemicals and consequently it is hardly surprising that the efficiency of recovery during the analysis of many of these is not known precisely. To provide better quantitative data deuterated internal standards covering a range of chemical class can be added to the water sample (James et al., 1980, Garrison, 1981). This approach can provide reasonable estimates of a wide range of organic chemicals, quite sufficient for any initial assessment of significance. Good quantitative data for an identified substance considered to be potentially significant requires a technique optimized for the particular substance. GC-MS methods employing stable isotopically labelled substances are particularly effective for such analyses.

The most important drawback of GC and consequently GC-MS is a limitation to substances with sufficient volatility. While this enables substances with boiling points up to 500°C and molecular weights up to 700u to be identified, it is now well recognized that over 80% of the organic matter in drinking and related waters is not amenable to GC-MS even after derivatisation of substances to improve volatility. Some of this non-volatile matter consists of substances, such as humic material, which are intractable to most instrumental methods of structure elucidation, although some insight into their structure has been derived from indirect approaches (Christman et al., 1982, 1983). However, accumulating evidence (Crathorne et al., 1984a, 1984b) points to the fact that, in addition to these substances, the non-volatile matter consists of a mixture of organic chemicals at least as complex as those revealed by GC-MS-based methods. Most probably this mixture consists of discrete chemicals, non-volatile due to polarity, thermal instability and high molecular weight. Identification of these substances is discussed later.

It was explained above that at least 10% of the organic chemicals in drinking water detected by GC-MS remain unidentified. Recent advances in MS instrumentation, particularly improved magnets, mean that capillary GC-MS at high resolution ($\geq 3000 \frac{m}{\Delta m}$) can be carried out successfully and with reasonable sensitivity. The use of capillary column GC requires fast scanning by the MS and, until recently, fast scanning was incompatible with high resolution MS operation. At relatively high resolution, the masses of molecular and fragment ions can be determined with sufficient accuracy (to within a few ppm) to limit the calculated possible empirical formulae of molecular ions and fragments to manageable proportions. This can greatly assist the elucidation of structure and the identification of the unknowns. As yet little information has been published on the systematic use of this approach to identification of organic substances in drinking water. The technique is relatively demanding and consequently will often be reserved for GC-MS 'unknowns' of some priority, for example substances highlighted by application of genotoxicity tests. However, routine GC-high resolution MS is now a distinct possibility.

Modern GC-MS systems allow the rapid alternation between electron impact ionisation (EI) and chemical ionisation (CI) which enables EI and CI spectra on the same GC peak. However, no systematic application to this useful facility to water analysis has been published.

High-performance liquid chromatography and MS

As mentioned already, drinking and related waters contain a substantial proportion of organic chemicals that are not amenable to GC-MS due to insufficient volatility and/or thermal instability. HPLC is a powerful and extremely versatile separation technique that operates at ambient temperatures with liquid mobile phases and consequently does not suffer from this limitation.

Its combination with MS to provide structural information on separated substances has been an obvious goal for over a decade. However, technical difficulties inherent in combing two basically incompatible processes has meant that a practical, routine system has not yet appeared. In essence, the major obstacle has been the disparity between the commonly used flow rates in HPLC (more than 1 $ml.min^{-1}$) and the maximum liquid flow tolerated by a MS (less than 50 $ul.min^{-1}$).

However, the emergence of practical microbore columns with flow rates of less than 100 $ul.min^{-1}$ promises to overcome this obstacle to a large extent The vast majority of GC-MS analyses have employed EI. In the case of non-volatile and thermally unstable chemical substances EI has two major drawbacks; i) substances need to be volatilized prior to ionisation and ii) the molecular ion is often absent or of very low intensity in the mass spectrum which usually precludes identification. While the first limitation can be overcome to a limited extent, in the case of environmental samples the latter leads to complex mass spectra which are difficult to interpret. Ideally, for mass spectrometry of a mixture of non-volatile and thermally unstable compounds, a MS technique is required that can ionise compounds in the unvaporised state with transfer of little excess energy.

This can provide mass spectra consisting of intense molecular (or related) ions with few fragment ions. Several MS ionisation methods are potentially available including desorption chemical ionisation, plasma desorption, laser desorption, field desorption and fast atom bombardment. The latter two techniques, which have been applied to drinking water, are mentioned later. A drawback of such "soft" ionisation techniques is the lack of fragment ions in the mass spectrum and consequently structural information to relate to the molecular ions. With suitable MS instrumentation accurate mass measurement of the molecular ions can provide information on empirical formulae and, although a number of theoretical chemical structures are usually possible, in some instances this alone is sufficient to identify the substance. However, structural information can be obtained (again with suitable MS instrumentation) from selected molecular ions by collisional activation dissociation (CAD) methods. Basically, molecular ions can be selected to undergo controlled fragmentation which provides information resembling that obtained from conventional EI-MS.

Techniques employed for isolation of non-volatile and thermally unstable substances are referred to later in the section but, as a generalisation, it is becoming clear that due to the extreme variety of chemical compounds present or potentially contaminating drinking water, a variety of well-validated techniques is needed.

The application of HPLC and MS methods to the identification of these substances in drinking and related waters has been very limited to date. It has been approached in two ways: i) off-line HPLC-MS and ii) on-line (coupled) HPLC-MS. The former approach was adopted in the authors' own laboratory based on a desire to use a variety of HPLC techniques including gradient elution and 'soft' ionisation MS methods - and an awareness of the limitations of on-line HPLC-MS systems available at the time.

Off-line HPLC-MS

Off-line HPLC-MS is based upon separation of extracts of water by HPLC and collection of fractions of interest (usually decided by reference to UV detection) from preparative-scale columns in order to provide sufficient material for study. The resulting fractions (and often unseparated extracts) are examined directly by MS. There is little published information on the application of this approach to drinking water (or indeed any water type). In the authors' laboratory extensive use of this approach has been made (Crathorne, 1982, 1984, 1984b) Overall the method involved isolation of organic chemicals by techniques based upon freeze drying/solvent extraction and adsorption on/elution from XAD resin, and separation by normal and reversed-phase coupled HPLC columns (to achieve increased resolution of components). Selected fractions (and unseparated extracts) were examined by FD- and FAB-MS. Peak matching was employed to obtain empirical formulae of detected molecular ions and CAD was employed, where appropriate, to provide structural information. Due to insufficient sensitivity, the latter technique was rarely feasible in practice. Both FD- and FAB-MS demonstrated the presence of a complex mixture of organic chemicals up to at least 1750u molecular weight (the limit of the instrumentation deployed). FAB-MS tended to give less information overall that FD-MS (in terms of substances detected) but was much more suited to routine use and showed noticeably more specifity towards surface active substances. The substances positively and tentatively identified by this approach (Crathorne et al., 1984b) include polychlorinated terphenyls, various anionic, cationic and non-ionic surfactants, uron-type pesticides and some antibiotics. In general, FD- and FAB-MS tended to give complimentary data.

Off-line HPLC and FD-MS (and also IR and Fourier transform-nuclear magnetic resonance spectroscopy (FT-NMR)) have been applied to XAD derived extracts of tapwater (Shinohara et al., 1980). Samples were separated into acids, bases and neutrals, separated further by column chromatography and then HPLC. The conclusion reached after application of FD-MS and FT-NMR was that the main substance encountered was polyvinylacetate.

HPLC and FD-MS was used to identify a variety of poly(oxyethylene)alkylphenyl ethers in river water (Otsuki and Shiraishi, 1979). Samples were concentrated by enrichment on a C_{18} accumulator column. A similar approach, but using freeze drying, identified polyethylene glycols, long-chain fatty acids and other substances in polluted river water (Shiraishi et al., 1981).

An analytical method based upon off-line HPLC and FD-MS was optimised for identification of a range of biocides in river water (Schulten and Stober, 1980). Samples were extracted with chloroform, submitted to column chromatography followed by HPLC and application of FD-MS to appropriate fractions. The range of substances (phenyl ureas, carbamates and thiocarbamates) positively identified by detection of appropriate molecular ions and observation of characteristic chlorine isotope ratios were buturon, chloroxuron, chlorotoluron, diuron, linuron, methabenzthiazuron, metabromuron, metoxuron and monolinuron.

Although there have been relatively few systematic applications of off-line HPLC-MS to the identification of organic substances in drinking water and related water the results to date clearly show the presence of a wide variety of substances up to at least 1750 molecular weight. The majority of the detected substances are non-volatile and while many have been identified most have not. Although in principle a great many more could be characterized by existing methodology, the effort and resources would be considerable with this type of approach. In the examples referred to the idea of a "target compound" approach emerges. The basis of this approach has been outlined (Crathorne et al., 1984). The approach requires the selection of specific compounds or classes of compound which could pose a health risk if present in drinking water. Criteria for selection of target substances depends on a variety of interests but in the reference cited above the following are given;

- a) non-volatility
- b) widespread use
- c) produced in large amounts
- d) of potential toxicological concern
- e) likely to enter the water cycle.

Thus, the target compound approach has two attractive features;
i) it optimizes and simplifies powerful but tedious qualitative techniques and
ii) it focusses on substances of potential importance.

On-line HPLC-MS

The success of GC-MS and an awareness of its limits convinced many research workers investigating the nature of organic matter in drinking water that a combined HPLC-MS system would be an extremely useful technique. However, it has been a promising technique for about a decade and many interesting interfaces for combining HPLC with MS have come and gone. Little application of commercially available systems to the identification of organic chemicals in drinking and related water has taken place. In a very readable review of LC-MS developments in the last ten years (Arpino, 1985) it is suggested that most of the HPLC-MS interfaces in laboratories were purchased as an accessory with new MS instrumentation but subsequently never used. The benefits of on-line HPLC-MS over off-line HPLC-MS have been clearly stated (Karger and Vouros, 1985);

- i. convenience, especially when analysing complex mixtures
- ii. much faster analysis
- iii. reduced sample losses, especially at trace levels
- iv. ease of deconvolution of partially resolved components due to the selectivity of the MS selector
- v. reliable evaluation of peak purity

In the last two years considerable optimism in the HPLC-MS field has been aroused which is evident in an excellent recent review of HPLC-MS developments

(Lederer, 1985). In particular, the emergence of the thermospray interface (see later) and the increasing impact of microbore HPLC has led to a resurgence of activity. The more recent systems can, with selected examples, show remarkable results, with good sensitivity, for a variety of non-volatile and thermally unstable compounds of biological, pharmaceutical and industrial relevance. However, environmental applications, let alone water analyses, are very few and it is consequently difficult to estimate practical detection limits and likely interferences etc and ultimately potential for application to drinking water. The impact on environmental analysis of a versatile routine HPLC-MS system with good sensitivity would be considerable and possibly exceed that of GC-MS. Consequently, it is useful to consider the more promising HPLC systems and the few relevant applications. However, for a more detailed review of HPLC-MS and applications the reader is directed to the review of Games and McDonnell (1984) and Levsen (1982).

The moving belt interface involves a continuous moving belt onto which is deposited the effluent from the HPLC column, prior to evaporation of solvent and vaporisation of solutes into the ion source of the MS. Early experience highlighted very restricted HPLC conditions and restrictions on the type of substances examined due to the necessity to vaporise solutes into the ion source. The system has been successfully applied to the identification of aldicarb, aldicarb sulphoxide and aldicarb sulphone in groundwater with detection limits of below 1ppb (Wright et al., 1982) and identification of binaphthyl sulphone (which was not detected by GC-MS) in an effluent (Thruston and Maguire, 1981). Some preliminary findings of the application of this system to groundwater potentially contaminated by landfill leachate show that compounds were detected (but not positively identified) which were overlooked by GC-MS methods (Games et al., 1983). Levsen (1985) reported the application of a recent commercial moving belt interface to the characterisation of various alkylphenolethoxylates and biodegradation products in river water using EI and CI. Currently improved methods of depositing HPLC column effluent onto the belt, incorporation of microbore HPLC and utilization of new ionisation methods applied directly to the belt, such as FAB, could lead to a reliable and versatile interface.

Direct liquid introduction (DLI) involves allowing a portion of the column effluent into the ion source of the MS where CI is effected by the solvent. With conventional HPLC only a tiny portion of the flow is used and hence sensitivity is correspondingly reduced. HPLC-DLI-MS has been used by Levsen et al. (1983) for the identification of phenylurea herbicides in river water. Microbore HPLC was used which enabled all the column effluent to enter the MS. In a spiked water extract a range of phenyl urea herbicides were detectable at low concentrations. Application of this technqiue to the solvent extract of freeze-dried river water enabled the tentative identification of several phenylureas, although high-resolution MS measurement was considered necessary for confirmation of identity. The value of a target compound approach was considered important in the study.

However, some workers refer to severe operational problems with DLI and it has been suggested that many users have now switched to thermospray techniques (Arpino, 1985).

Recently a new interface/ionisation technique, Thermospray (TSP), has evolved and in the last couple of years interest in the technique has increased remarkably (Blakley and Vestel, 1983, McDowell et al., 1985, Catlow, 1985, Chapman, 1985). The particular advantage of TSP is that the conventional HPLC flow rate (1 ml.min^{-1} or more) can be tolerated by the MS without too much difficulty. In addition aqueous HPLC mobile phases are prefered. The exact nature of the ionisation which occurs in TSP is not fully understood although its basis is that a supersonic jet of mobile phase vapour from the HPLC column is produced (although complete evaporation is avoided). The presence of ammonium acetate in the mobile phase appears essential in order to achieve ionisation of the sample probably by a form of chemical ionisation. The use of microbore columns with aqueous ammonium acetate introduced at the end of the HPLC column (to produce the required flow rate of 1ml/min or so) could allow a variety of HPLC techniques to be deployed (Bruins, 1985). Whether or not the level of detection achievable with environmental samples is sufficient should become apparent in the near future. Although the interface is mechanically relatively simple and impressive results have been reported for selected organic chemicals, one cannot yet judge its practical applicability for water and evironmental analysis in general.

ASSESSMENT OF ORGANIC SUBSTANCES IN DRINKING WATER

When assessing the significance to health of organic micropollutants in drinking water, we are faced with a very complex question. Not only are there large numbers of compounds present but they are present individually in small quantities. There are three possible approaches to the problem;

1. Toxicological evaluation of individual compounds which have been identified.
2. <u>In vitro</u> bio-assays, in particular short-term tests for genotoxicity to detect the presence of unidentified, biologically-active compounds.
3. Epidemiology.

All three approaches have their own advantages and limitations and so may all contribute to the overall assessment. For example when using epidemiology and bio-assays the nature of the compounds need not be known whereas the toxicological evaluation of individual compounds first requires that those compounds have been identified.

Toxicological evaluation of identified organic compounds

The toxicity of a compound is its intrinsic capacity to cause injury. The hazard associated with that same compound is the capacity of that compound to cause injury under the circumstances of exposure. The first step in any assessment is to gather and critically evaluate the toxicological data available on the compounds identified. It is highly unlikely that the concentrations of organics observed in drinking water will give rise to acute toxicity, either individually or in mixtures with other compounds. However, since the period of exposure is very long it is necessary to search for data on chronic exposure. The most obvious possible effects from long-term exposure to low doses are mutagenicity and carcinogenicity but data is required on reproductive toxicity, which may be associated with mutagenicity, and metabolism and pharmacokinetics. At present our knowledge of other aspects of toxicology such as behavioural toxicity and immunotoxicity including allergy are very limited and thus remain possible problem areas. In the case of allergy, for example, though it is unlikely that the observed concentrations would sensitise, the consequences of low dose challenge in sensitised individuals are not clear.

When using the available toxicological data to assess hazards it is important to remember that there are many factors which can influence toxicity or the expression of toxicity in animals and man. These will differ from species to species. Arguably the most important of these is pharmacokinetics through which absorption, metabolism and excretion of compounds can be taken into account. The compound must enter the body and be transported to the target organ or tissue in the corect form and quantity to cause damage. However, metabolism by cellular enzymes may take place, often in a series of steps, leading to detoxification or activation. Many of the factors which influence toxicity actually do so by affecting absorption, metabolism and excretion of toxicants.

With so many uncertainties and the uncertainties of extrapolation discussed below it is clear that the requirement from the analyst is identification of the compounds present and only an approximate concentration.

When extrapolating from experimental animals to man the confounding effects of these factors must be carefully considered. However, judgements often must be made before all the information and understanding we need are available. The basic difference in the method of extrapolation depends very much on the acceptance of the theory that there is no concentration at which genotoxic agents are not capable of inducing cancer and that the dose response curve goes through the origin. For non-genotoxic agents the use of arbitrary safety or uncertainty factors, as used by WHO/FAO to calculate acceptable daily intakes or as defined by the U.S. National Academy of Sciences (1980) (Table 1) are often used. More recently mathematical models have been introduced to establish exposure levels for genotoxic carcinogens which result in "acceptable risk" to the population. Such models are useful in providing estimates within orders of magnitude, unfortunately it is still early days and many uncertainties remain so that they must be applied with caution.

TABLE 1 'SAFETY" or "UNCERTAINTY" FACTORS. AFTER NAS (1980)

	FACTOR
GOOD CHRONIC OR ACUTE HUMAN DATA + CHRONIC OR ACUTE DATA IN OTHER SPECIES	10
GOOD CHRONIC OR ACUTE TOXICITY DATA IN ONE OR MORE SPECIES	100
LIMITED OR INCOMPLETE ACUTE OR CHRONIC DATA	1000

The use of these models has been summarized by David Hoel (1979) who said:

"...recently interest has been directed to statistical models of dose response for estimating low dose risk based on high dose data. Curve fitting has been commonplace using functions such as probit, logit, multi-hit, multi-stage etc. Some of these functions have a tradition in bio-assay work, while others attempt crudely to describe biological mechanisms. For example, the multi-stage model assumes the carcinogenesis process is represented by a direct acting carcinogen interacting with DNA as in a single cell somatic mutation theory. This simple model does not include consideration of DNA repair, the immune surveillance system, genetic or environmental susceptibilities in the population or pharmacokinetics."

The additional problems with the use of these models at this time are the quality of the animal data to which they are applied and their applicability to some of the major compounds of interest, such as chloroform and carbon tetrachloride which appear to have an epigenetic mechanism.
However, as has been made clear individual micropollutants do not occur on

their own and the difficulties of studying mixtures of chemicals have not yet been mentioned. With organic micropollutants we are dealing with a very complex mixture which is largely uncharacterised and exposure to these water-borne organic chemicals is superimposed on exposure to other complex mixtures, often in much larger concentration, from the rest of our environment; air, food, industrial exposure. It is highly probable that additive and possibly synergistic and antagonisitic effects do occur in this mixture of many chemicals. However, although studies are now underway, our knowledge of the toxicology of mixtures is at best limited and at worst virtually non-existent.

With regard to those volatile micropollutants which have been identified in drinking water and referred to earlier, it is not practicable to generalise about the availability of information on specific compounds. By and large more data are available on acute than chronic toxicity - including carcinogenicity. Compounds such as chloroform, trichloroethylene, tetrachloroethylene and the aromatic hydrocarbons have been well studied. Other compounds including some of the aliphatic hydrocarbons and halogenated hydrocarbons have not been well studied whilst for others no information exists at all.

Although there are a number of known mutagens (Table 2), carcinogens (Table 3) and promoters together with a substantial proportion of the identified compounds being known to be toxic at much higher concentrations than those observed in water, there is no evidence at present to suggest that any individual compound is likely to pose a significant risk to the health of consumers. There are some instances where specific pollution has taken place, such as groundwater contamination by some halogenated solvents such as trichloroethylene. In these instances, (and with trihalomethanes in some surface derived drinking waters) the concentrations are much higher than usually found in drinking water and though it is difficult to identify any substantial risk it is conceivable that a low level effect could accrue from long term exposure to these higher concentrations.

The groups of compounds of most concern are perhaps the halogenated aliphatic hydrocarbons, the halogenated ethers and some miscellaneous halogenated compounds. It is possible that at least in some of these cases there could be an additive action which would, in effect, be equivalent to a higher concentration of an individual compound. However, the evidence we have to date on the volatile organic micropollutants which includes the risk extrapolations for carcinogenesis by the United States Environmental Protection Agency (1979) and the work of WHO on the Guidelines for Drinking Water Quality (1984) indicates that any risk from the concentrations normally encountered will be of a low order. It must be emphasised that this applies only to the volatile fraction which accounts for a small proportion of the organic

Table 2 Some compounds mutagenic in bacteria identified in drinking water

styrene	1,5-dibromopentane
1-methylnaphthalene	trichloroethylene
acenaphthalene	bis(2-chloroethyl)ether
fluoranthene	dichloroacetonitrile
9-methylfluorene	bromochloroacetonitrile
1-methylphenanthrene	chloral
9-methylphenanthrene	trichloronitromethane
2-methylphenanthrene	bromodichloromethane
bromochloromethane	bromoform
dibromomethane	dibromochloromethane
bromoethane	dimethylphthalate
iodoethane	diethylphthalate
1,2-dichloroethane	dibutylphthalate
tetrachloroethane	di-2-ethylhexylphthalate
1-bromopropane	3-butandione
1-bromobutane	butan,2,3-dione
	dichloropropene

Table 3 Some known and/or suspected animal/human carcinogens identified in drinking water

benzene	bis(2-chloroethyl)ether
styrene	trichloroaniline
tetrachloromethane	clofibrate
1,2-dichloroethane	polychlorinated terphenyls
1,1,1-trichloroethane	2,4,6-trichlorophenol
1,1,2-trichloroethane	chloroform
tetrachloroethane	di-2-ethylhexylphthalate
hexachloroethane	1,4-dioxane
trichloroethylene	carbazole
chlorobenzene	

matter present in drinking water.

Epidemiology

The advantage of epidemiology is that it looks at human populations and takes account of exposure from both volatile and non-volatile organic micropollutants. In practice, the complications of studying the epidemiology of disease in relation to environmental agents are also very great. This is especially true of retrospective studies since the number of unknown factors is considerable, particularly in relation to exposure to the agent under study. There has been no evidence to suggest that organic micropollutants will produce novel effects so one is looking for an inceased incidence of disease which is already found in the population, that is, effects superimposed on the "normal" pattern of disease. Therefore, all other contributory factors such as socio-economic status, diet, occupation, smoking and alcohol consumption must be considered.

In addition exposure to the environmental agent or agents under study may be difficult to measure and is likely to be very varied in the study population leading to a wide variation in response.

Results from epidemiological studies involving the consumption of organic micropollutants from water have tended to be inconsistent. However a number of studies of cancer mortality in the United States and Europe (Crump and Guess, 1982) and studies of cancer incidence (Beresford, 1983) have shown some association between drinking water from surface water sources, which are generally higher in micropollutants than groundwater, and cancer. The target organs appear to be the bladder, urinary tract and gastro-intestinal tract, particularly stomach, rectum and colon. However, it must be emphasised that the risk appears to be small and more definitive studies are still required, for example case control studies using living subjects.

Application of in-vitro bio-assays

The advantage of bio-assays is that, in principle, they can be used to test whole water samples so taking into account all of the organic material present. Many *in vitro* and *in-vivo* assays have been used. These have been reviewed by Loper (1980) and more recently by Kool et al., (1983). The most frequently used assays are those based on the *Salmonella typhimurium* strains, developed by Ames and his co-workers to detect point mutations. Mutagenicity in bacterial assays is also considered to be predictive of carcinogenic potential since with individual compounds there is a good correlation between positive results in these assays and carcinogenicity in laboratory animal studies.

Initial studies were carried out with unconcentrated water samples but the difficulties encountered with artifacts and with interpretion of the results have made this a less attractive proposition (Forster et al., 1983, Harrington et al., 1983). The great majority of studies have consequently used some means of concentrating the organics in the samples. These techniques have been reviewed (Jolley, 1981) but the methods most favoured are either those based on adsorption on XAD resins with subsequent elution using solvents or freeze drying followed by solvent extraction of the residue. There are groups using these techniques in many countries and the literature is now substantial. Although some differences in detail can be observed a number of broad conclusions can be drawn from this work;

a) Surface waters, after chlorination, are almost invariably mutagenic to *Salmonella typhimurium* TA100 and TA98.
b) This activity is usually reduced by the addition of rat liver microsome S9 preparation which simulates metabolism.
c) Some surface waters are mutagenic to strain TA98 before chlorination and

this is usually increased by the addition of S9.
d) Groundwaters usually show less activity than surface waters but there are exceptions.
e) In laboratory studies chlorine dose often correlates with mutagenicity.
f) In the field there are no consistent correlations with analytical data such as TOC, THMs, AOX.
g) The chlorine derived mutagens appear to be mostly non-volatile.

The presence of bacterial mutagens does not, however, confirm a hazard to consumers. Rather it indicates the presence of substances which possibly constitute a hazard to consumers. To confirm that any hazard exists requires that the material can also be shown to be positive in tests based on mammalian cells in tissue culture or *in vivo* systems. The health authorities of different countries make different recommendations as to the amount of testing required but the U.K. Committee on Mutagenicity have published guidelines (1981) based on a four tier system. These are i) point mutations in bacteria, ii) chromosome damage in mammalian cells *in vitro*, iii) point mutations in mammalian cells *in vitro* or recessive lethals in *Drosophila melanogaster* and iv) chromosomal or germ cell damage in the intact animal. A number of groups have carried out studies with water extracts using these higher test systems. However, the tests are much more complex than bacterial tests and the results to date have been rather sporadic. Three groups have reported positive results with concentrated drinking waters in assays to show chromosome abberrations and sister chromatid exchange, although the significance of this latter test is in doubt (Athanaisiov and Kryptopoulos, 1983, Guerrero and Rounds, 1979, van Kreijl et al., 1983). Three different groups have also reported the finding of positive results in mammalian point mutation assays with drinking water (Gruener and Lockwood, 1979, Pottenger et al., 1983, Tye, 1983).

In addition there have been positive results with drinking water extracts in mammalian cell transformation assays (Loper et al., 1978, Lang et al., 1980, Pelon et al., 1980) which are considered to be indicative of genotoxic activity and possible carcinogenicity.

The evidence, therefore, indicates that there is a likelihood that mutagens in chlorinated drinking water pose some risk to the consumer. The size of this risk is, however, much more difficult to determine. The methods available which have been used to estimate the size of the risk are epidemiology (with all its attendant difficulties regarding measurement of exposure), long term animal carcinogenicity studies with water concentrates or identification of the mutagens and subsequent toxicological evaluation of the known compounds.

Long-term animal studies have been carried out by Kool et al (1983) and these have proved negative. However, the difficulties in carrying out such a study with an unknown mixture, especially in ensuring a consistent test substance over a long period and in defining realistic dose levels are very great and the most reassurance which can be obtained from this is that the risks are not of a large order of magnitude.

Identification of organic chemicals detected by in vitro bio-assays

The work described above shows that genotoxicity can be readily detected in drinking and related waters. While one approach to assessing the significance of such findings is the application of additional assays to better predict the likely effects in man, for example, those involving higher cell systems, an additional approach is to identify the chemical or chemicals responsible. Such an approach has been successfully used in other fields, a classic example being the identification of the carcinogen benzo[a]pyrene in coal-tar pitch. If the identity of a genotoxic agent is known then several pathways to assessing its significance to health are available. For example, more appropriate bio-assays with the pure substance, examination of available toxicity data and accurate determination of exposure levels. In addition, the source of the substance can be located if necesary by rapid, specific analyses. Identification of such active substances should always be considered since they might be easily, and quickly, characterised.

The basis of most approaches used to date involves fractionation of the organic constituents in a "positive" drinking water concentrate by chemical/physical means and/or chromatography (thin-layer chromatography (TLC), HPLC, column chromatography) combined with re-testing of fractions to locate the genotoxic chemicals. The basic aim has been to separate active from non-active chemicals in the complex mixture to enable identification techniques to focus upon the active chemicals. Since the bulk of the genotoxicity studies reported to date relate to mutagenicity detected by bacterial assays, such as the Ames test, discussion will concentrate on identification of mutagens.

Many studies include initial detailed analysis, usually by GC-MS methods, of (unseparated) mutagenic drinking water concentrates. This usually leads to the following conclusions. The detected and identified components are numerous (for example, Coleman et al. (1980) reports 700 detected and 460 identified components in tapwater). Among these, known mutagenic organic chemicals can be recognized, but several studies, including our own, have found that these, even if tested for mutagenicity as a mixture, do not account for more than a few percent of the mutagenicity detected in mutagenicity assays. However, the general testing of individual identified substances (mostly unobtainable as pure substances) and the resolution of the unknown substances detected is a daunting prospect - which emphasises the need to focus attention on specific mutagenic fractions. Several studies on drinking and related waters have used the overall approach referred to above - involving XAD resin concentration and TLC/HPLC fractionation (Kool, et al., 1981), and XAD concentration, freeze drying/solvent extraction and HPLC (Horth et al., 1984) and application of HPLC to a carbon/chloroform extract (Tabor and Loper, 1983). No study has yet identified mutagens that appear to account for the bulk of the mutagenicity detected by mutagenicity tests and consequently it is difficult to compare studies. Most are successful in that several mutagenic sub-fractions which in total account for most of the original mutagenicity can be detected. These sub-fractions still appear to be quite complex and scrutiny of these, usually by GC-MS, gives no clear indication of the major mutagens. Many researchers conclude that the chemical mutagens in drinking water are non-volatile (or at least not readily amenable to GC-MS) and are numerous (Coleman et al., 1984, Kool et al., 1982, Horth et al., 1984. One study (Tabor, 1983) led to the identification of a new mutagen, 3-(2-chloroethoxy)- 1,2-dichloropropene, although this was encountered in an old carbon/chloroform extract and does not seem applicable to contemporary drinking water mutagenicity. The evidence to date suggests that the major drinking water mutagens are non-volatile, numerous, and at very low concentrations (i.e. quite potent).

In the case of mutagenicity produced by chlorination, which appears to involve reaction of chlorine with natural precursors in the raw water, an approach to characterizing the active mutagens is to study the reaction of potential precursors with chlorine in the laboratory. The potentially simpler products, and the ability to produce larger quantities of material for study - combined with knowledge of the precursors - could facilitate identification of mutagens produced by water treatment chlorination. To date, results of studies on humic substances (Coleman et al., 1984, Becher et al., 1985, Horth et al., 1984) and amino acids, purines, pyrimidines, nucleosides and nucleotides (Horth et al., 1984) have been reported. Although many interesting products can be identified and humic substances and certain amino acids produce mutagenicity similar to that found after water treatment chlorination, no major mutagens have been identified. Here too there is speculation that there are many mutagens involved, probably difficult to analyse and quite potent (very low quantities present).

Further progress in this area could depend on the following:

i. application of more powerful HPLC combined with new MS ionisation methods (so-called soft ionisation techniques).

ii. extraction techniques specific for mutagens and/or specific derivatives to "label" mutagens (here one notes the work of Cheh and Carlson, (1981) and Jacks et al., (1983) with derivatives specific for electrophilic, direct acting mutagens)

iii. clues from related areas of study. For example, the identification of an extremely potent mutagen (tentatively identified as 3-chloro-4-(dichloromethyl)-5-hydroxy-2(5H)-furanone) produced during chlorination of effluents from pulp mills (Holmbom, 1984).

CONCLUSIONS

Drinking water contains a very complex mixture of organic chemicals. Until quite recently the available methods have allowed the identification of only a minor fraction of the chemical impurities present. Even so hundreds of compounds have been identified. Our knowledge of the nature of organic contamination of drinking water is, therefore, severely limited and, consequently, so is our ability to assess its significance to health. However, methods for the identification of the unknown substances are becoming available and some of these have shown that the unknown non-volatile fraction is at least as complex as the known fraction. Continued development and application of such techniques is essential but since such methods are slow (at least for the foreseeable future) the target compound approach (referred to in the text) in which organic substances of potential concern are identified, has considerable attractions.

The ability to identify a multitude of organic chemicals in drinking water causes some problems since the information is embarrassingly difficult to evaluate in terms of possible risks to health. Toxicological information on many of the identified compounds does not exist and is limited for many others. For those compounds for which there is toxicological data from animal studies, there remain difficulties and uncertainties in applying such data to determine what would be the effect on humans of the very low concentrations found in drinking water. Estimates by other organisations (USEPA and WHO) have indicated that the risks are likely to be of a low order.

Research is required on the fundamental aspects of toxicology relating to low doses, long-term exposure and complex mixtures. The understanding of the mechanisms of toxicity is particularly important. There is a need for improving the basis of mathematical models for extrapolating from high to low doses and from species to species to assist in interpreting animal experiments in relation to the exposure of man to environmental levels of chemicals. To this end, more fundamental mechanistic studies in toxicology are required to provide information to support the mathematical models.

Because of the problems mentioned above, the detection of relevant toxic responses in drinking water by means of reasonably simple bio-assays is of vital importance. Most of the work to date in this area has involved _in vitro_ bacterial mutagenicity assays. The work has shown that most chlorinated surface waters are active in those assays. Some surface waters are active but most groundwaters are less mutagenic than surface waters or are not mutagenic. Mutagenicity in _in vitro_ assays in mammalian cells has also been observed but only a limited number of studies have been carried out.

Identification of mutagens (and other toxic agents detected by bio-assays) in drinking water is of crucial importance in the assessment of any associated hazards to health. At the present time, although many mutagens have been identified, they do not seem to amount for much of the mutagenicity detected. Fractionation of mutagenic extracts still produces complex fractions and consequently more powerful fractionation methods are required. Coupled conventional HPLC columns and microbore HPLC columns offer some promise although in the latter case the lower sample capacity is a major obstacle. The mutagens appear to be non-volatile and therefore application of the newer qualitative techniques mentioned in the text is necessary. Research into extraction methods selective for mutagens and specific derivatives to 'label' mutagens in order to eliminate non-mutagenic compounds could be rewarding.

REFERENCES

Arpino, P.J., 1985. Ten years of liquid chromatography-mass spectrometry. J. Chromat., 323, pp.3-11.

Athanasiou, K. and Kryptopoulos, S.A., 1983. Mutagenic and clastogenic effects of organic extracts from the Athenian drinking water. Science of the Total Environment. Vol.27, pp. 113-120.

Becher, G., Caslberg, G.E., Cjersing, E.T., Hongslo, J.K., Monarca, S., 1985. High performace size exclusion chromatography of chlorinated natural humic water and mutagenicity studies using the microscale fluctuation assay. Environ. Sci. Tech., 19, pp 422-426.

Beresford, S.A., Fielding M. and Packham, R.F., 1981. Health related studies of organic compounds in relation to re-use in the U.K. In: H.van Lelyveld and B.C.J. Zoeteman (Editors), Water Supply and Health, Elsevier, Amsterdam, pp 167-186.

Beresford, S.A.A., 1983. Cancer incidence and reuse of drinking water. Am. J. Epidemiol, Vol 117, No 3. pp 258-268.

Blakley, C.R. and Vestel, M.L., 1983. Thermospray interface for liquid chromatography/mass spectrometry. Anal. Chem., 55, pp. 750-754.

Bruins, A.P., 1985. Developments in interfacing microbore high-performance liquid chromatography with mass spectrometry (a review). J. Chromat., 323, pp 99-111.

Catlow, D.A., 1985. Applications of thermospray ionisation liquid chromatography - mass spectrometry to compounds of pharmaceutical interest. J. Chromat., 323, pp 163-170.

Cheh, A.M. and Carlson, R.E., 1981. Determination of potentially mutagenic and carcinogenic electrophiles in environmental samples. Anal. Chem., 53, pp 1001-1006.

Chapman, J.R., 1985. Liquid chromatography - mass spectrometry interfacing on a magnetic section mass spectrometer. J. Chromat., 323, pp 153-161.

Christman, R.F., Liao, L., Johnson, J.D., Millington, D.S. and Hass, J.R., 1982. Structural characterizations of humic material. Environ. Sci. Tech., 16, pp 403-410.

Christman, R.F., Norwood, D.L., Millington, D.S., Johnson, J.D. and Stevens, A.A., 1983. Identity and yields of major halogenated products of aquatic fulvic acid chlorination. Environ. Sci. Tech., 17, pp 625-628.

Coleman, W.E., Melton, R.G., Kopfler, F.C., Barone, K.A. Durand, T.A. and Jellison, M.G., 1980. Identification of organic compounds in a mutagenic extract of a surface drinking water by a computerised gas chromatography/mass spectrometry system (GC/MS/COM). Environ. Sci. Tech., 14, pp 576-588.

Coleman, W.E., Munch, J.W., Kaylor, W.H., Strelcher, R.P., Ringland, H.P. and Meier, J.R., 1984. Gas chromatography/mass spectroscopy analysis of mutagenic extracts of aqueous chlorinated humic acid. A comparison of the byproducts to drinking water contaminants. Environ. Sci. Tech., 18, pp 674-681.

Crathorne, B. and Watts, C.D., 1982. Identification of non-volatile organic compounds in water. In A. Bjorsetts and G.Angeletti, Analysis of Organic Micropollutants in Water, Reidel, London, pp 159-173.

Crathorne, B., Fielding, M., Steel, C.P. and Watts, C.D., 1984a. Identification of non-volatile organics in water using field desorption mass spectrometry and high performace liquid chromatography. In: A. Bjorseth and G. Angeletti (Editors), Analysis of Organic Mircropollutants in Water, Reidel, Dordrecht, 339 pp.

Crathorne, B., Fielding, M., Steel, C.P. and Watts, C.D., 1984b. Organic compounds in water: analysis using compled-column high-performance liquid chromatography and soft-ionization mass spectrometry. Environ. Sci. Tech., 18, pp 797-802.

Crump, K.S. and Guess, H.A., 1982. Drinking water and cancer: a review of recent epidemiological findings and assessment of risks. Ann. Rev. Public Health, 3, p 339-357.

Fielding, M. and Packham, R.F., 1977. Organic compounds in drinking water and public health. J. Inst. Wat. Eng. and Scient., 31, pp 353-375.

Forster, R., Green, M.H.L., Gwilliam, R.D., Priestley, A and Bridges, B.A. 1983. Use of the fluctuation test to detect mutagenic activity in unconcentrated samples of drinking waters in the United Kingdom. In: R.L.Jolley et al., (Editors), Water Chlorination Environmental Impact and Health Effects Vol 4, Book 2, Environment Health and Risk. Ann. Arbor. Science, Michigan.

Games, D.E., Lant, M.S., Westwood, S.A., Foster, M.G. and Meresz, O., 1983. Capillary gas chromatographic mass spectrometric and microbore liquid chromatographic mass spectrometric studies of a test well sample from a landfill site. Biomed. mass spectrom., 10, pp 338-342.

Games, D.E. and Mc Dowall, M.A., 1984. Recent progress in LC/MS. In: G. Angeletti and A. Bjorseth (Editors), Analysis of Organic Micropollutants in Water, Reidel, Dordrecht, pp 68-76.

Garrison, A.W., 1976. Technical Paper No 9, International Reference Centre for Community Water Supply, The Hague.

Garrison, A.W., Alford, A.L., Craig, J.S., Ellington, J.J., Haebener, A.F., Mc. Guire, J.M., Pope, J.D. and Shackleford, W.M., 1981. The master analytical scheme : an overview of interim procedures. In: L.H. Keith (Editor), Advances in the Identification and Analysis of Organic Pollutants in Water, Vol. 1, Ann. Arbor. Science, Michigan, pp 17-30.

Gruener, N. and Lockwood, M.P. 1979. Mutagenicity and Transformation by Recycled water. J. Tox Environ. Health, Vol. 5. pp 663-670.

Guerrero, R.R. and Rounds, D.E. 1979. The use of sister chromatid exchange analysis to detect ambient mutagens in drinking water. In vitro, Vol. 15, pp 171-172.

Harrington, T.R., Nestmann, E.R. and Kowbel, D.J. 1983. Suitability of the modified fluctuation assay for evaluating the mutagenicity of unconcentrated drinking water. Mutation Research 120, pp 97-103.
Hoel, D.G. 1979. Statistical approaches to toxicological data. Environmental Health perspectives Vol. 27, pp 267-271.
Holmbom, B., Voss, R.H., Mortimer, R.D. and Wong, A., 1984. Fractionation, isolation, and characterization of Ames mutagenic compounds in Kraft chlorination effluents. Environ. Sci. Tech., 18, pp 333-337.
Horth, H., Crathorne, B., Gwilliam, R.D., Stanley, J.A., Steel, C.P., Thomas, M.J., 1984. Techniques for the fractionation and identification of mutagens produced by water treatment chlorination. In I.H. Suffet and M. Malayandi (Editors), Advances in Sampling and Analysis of Organic Pollutants from Water, Vol. 2, Toxicity Testing/Analysis Interface. Advances in Chemistry series, American Chemical Society. (In press).
Jacks, C.A., Gute, J.P., Neisers, L.B., Van Sluis, R.J. and Baird, R.B., 1983. Health effects of water reuse : characterization of mutagenic residues isolated from reclaimed, surface, and groundwater supplies. In: R.L. Jolley et al., Water Chlorination: Environmental Impact and Health Effects, Vol. 4(2), Ann. Arbor. Science, Michigan, pp 1237-1248.
James, H.A., Fielding, M., Gibson, T.M., and Steel, C.P., 1980. Quantitative aspects of the determination of organic micropollutants in raw and portable water. In: Advances in Mass Spectrometry, Vol. 8, Heydon, London, pp 429-1435.
James, H.A., Fielding, M., Gibson, T.M., McLoughlin, K. and Steel, C.P., 1981. Organic micropollutants in drinking water, TR159, Water Research Centre, Marlow, Bucks., England.
Jolley, R.L. 1981. Concentrating organics in water for biological testing. Environ. Science and Tech. 15, 8, pp 874-880.
Karger, B.L. and Vouros, P., 1985. A chromatographic perspective of high-performance liquid chromatography-mass spectometry. J. Chromat., 323, pp 13-32.
Keith, L.H. (Editor), 1976. Identification and Analysis of Organic Pollutants in Water, Ann. Arbor. Science, Michigan, pp 718.
Keith, L.H. (Editor), 1981. Advances in the Identification and Analysis of Organic Pollutants in Water, Vols. 1 and 2, An. Arbor. Science, Michigan, 1170 pp.
Kool, H.J., Van Kreyl, C.F., van Kranen, H.J. and De Greef, E., 1981. Toxicity assessment of organic compounds in drinking water in the Netherlands. In: H. van Lelyveld and B.C.J. Zoeteman (Editors), Water Supply and Health, Elsevier, Amsterdam, pp 135-153.
Kool, H.J., van Kreijl, C.F., de Greef, E. and van Kranen, H.J., 1982. Presence, introduction and removal of mutagenic activity during the preparation of drinking water in the Netherlands. Environ. Health Perspectives, 46, pp 207-214.
Kool, H. 1983. Organic mutagens and drinking water in the Netherlands. A study on mutagenicity of organic constituents in drinking water in the Netherlands and their possible carcinogenic effects. Dissertation, Agricultural University Wagenigen, RID-mededeling 83-4 Leidschendam, 116pp.
Kool, H.J., van Kreijl, C.F., and Zoeteman, B.C.J. 1983. Toxicology assessment of organic compounds in drinking water. CRC Critical Reviews in Environ Control, 12, 4, pp 307-357.
van Kreijl, C.F., De Vries, M., Van Kranen, H., Kool, H.J. and De Greef, E. 1983. Properties of mutagenic activitiy in river Rhine water in the Netherlands. Mutation Research, Vol. 113, p 314.
Lang, D.R., Kurzepa, H., Cole, M.S. and Loper, J.C. 1980. Malignant transformations of BALB/3T3 cells by residue organic mixtures from drinking water, J. Environ. Path. Tox., Vol. 4, pp 41-54.
Lederer, M. (Editor), 1985. Third Symposium on Liquid Chromatography - Mass Spectrometry and Mass Spectrometry - Mass Spectrometry. J. Chromat., 323. 171 pp.

Levsen, K., 1982. Kepplung eines hochleistungschromatographen mit einem massenspecktrometer. In: A. Bjorseth and G. Angeletti (Editors), Analysis of Organic Micropollutants in Water, Reidel, Dordrecht, pp 149-158.

Levsen, K., Schafen, K.H. and Freudenthal, J., 1983. Determination of phenylureas by on-line liquid chromatography-mass spectrometry. J. Chromat., 271, pp 51-60.

Levsen, K., Wagner-Redeker, W., Schafer, K.H. and Dobberstein, P., 1985. On-line liquid chromatography-mass spectrometry analysis of non-ionic surfactants. J. Chromat., 323, pp 135-141.

Loper, J.C., Lang, D.R., Smith, C.C., Schoeny, R.S., Kopfler, F.C. and Tardiff, R.G., 1978. In vitro mutagenesis and carcinogenesis testing of residual organics in drinking water. In: R.O. Hutzinger et al (Editors)., Aquatic pollutants. Transformation and Biological effects, Pergamon, pp 405-417.

Loper, J.C., Lnag, D.R., Schoeny, R.S., Richmond, B.B., Gallagher, P.M. and Smith, C.C., 1978. Residue organic mixtures from drinking water show in vitro mutagenic and transforming activity. J. Tox. Environ. Health Vol. 4, pp 919-938.

Loper, J.C. and Tabor, W.M., 1980. Separation of mutagens from drinking water using coupled bioassay/analytical fractionation. Intern. J. Environ. Anal. Chem., 8, pp 197-215.

Loper, J.C., 1980. Mutagenic effects of organic compounds in drinking water. Mutation Research 76. pp 241-268.

McDowell, M.A., Schmelzeisen-Redeker, G., Guiessmann, V., Levsen, K and Rollgen, F.W., 1985. Studies with a laboratory-constructed thermospray liquid chromatograph-mass spectrometer interface. J. Chromat., 323, pp 127-133.

Otsuki, A. and Shiraishi, H., 1979. Determination of poly (oxyethylene) alkylphenyl ether nonionic surfactancts in water at trace levels by reversed phase adsorption liquid chromatography and field desorption mass spectrometry. Anal. Chem., 51, pp 2329-2332.

Pelon, W., Beasley, T.W. and Lesley, D.E., 1980. Transformation of the mouse clonal cell line R846-DP8 by Mississippi river raw and finished samples from South Eastern Louisiana. Environ. Science. Tech. Vol. 14, pp 723-726.

Pottenger, L.K., Helias, Z., Hartemann, P. and Bourbigot, M.M. 1983. Mutagenicity and promoting activity of water concentrates from a drinking water treatment process. Poster Presentation at the 13th EEMS conference, Montpellier.

Shinohara, A.K., Eto, S., Hori, T., Koga, K., and Akiyama, J., 1980. Identification of non-volatile neutral organic compounds in tap water. Environ. Intern., 4, pp 31-37.

Shiraishi, H., Yasuhura, A., Tsuji, M. and Okuno, T., 1981. Analysis of organic substances in highly polluted raw water by mass spectrometry. Environ. Sci. Tech., 15, pp 570-573.

Shulten, H.R. and Stober, I., 1980. Combined application of high-pressure liquid chromatography and field desorption mass spectrometry for the determination of biocides of the phenylurea- and carbamate-type in surface water. Sci. Total. Environ., 16, pp 249-262.

Stafford, G.C., Kelley, P.E. and Bradford, D.C., 1983. Advanced ion trap technology in an economical detector for G.C. American laboratory, June (1983).

Tabor, M.W. 1983. Structure elucidation of 3-(2-chloroethoxy)-1,2-dichloropropene, a new promutagen from an old drinking water residue. Environ. Sci. Tech., 17, pp 324-329.

Thruston, A.D. and Maguire, J.M., 1981. Biomed. Mass Spectrom., 8, 47 pp.

Tye, R.J., 1983. Comparison of Ames and Mouse Lymphoma L51784. Test Results on Extracts Derived from Water Samples. Poster presented at UKEMS Conference, Edinburgh.

U.K. Department of Health and Social Security, 1981. Guidelines for the testing of chemicals for mutagenicity. Report on Health and Social Subjects 24. HMSO London. 95 pp.

United States Environmental Protection Agency, 1979. Water quality criteria Federal Register 44. pp 15926-15981, pp 43660-43697, pp 56627-56657.

United States National Academy of Sciences, 1980. Drinking Water and Health, Vol. 3. National Academy Press, Washington DC. 415 pp.

WHO, 1984. Guidelines for drinking-water quality Vol.1. Recommendations. World Health Organisation, Geneva. 130 pp.

Wright, L.H., Jackson, M.D. and Lewis, R.G., 1982. Determination of aldicarb residues in water by combined high-performance liquid chromatography/mass spectrometry. Bull. Environ. Contam. Toxicol., 28, pp 740-747.

FRACTIONATION OF MUTAGENIC COMPOUNDS FORMED DURING CHLORINATION OF HUMIC WATER.

L. KRONBERG[1], B. HOLMBOM[2] and L. TIKKANEN[3]
[1]Department of Organic Chemistry, Åbo Akademi, SF-20500 Turku, Finland.
[2]Laboratory of Forest Products Chemistry, Åbo Akademi, SF-20500 Turku, Finland.
[3]Technical Research Centre of Finland, Food Research Laboratory, SF-02150 Espoo, Finland.

ABSTRACT

Mutagenic compounds formed during chlorination of humic water were fractionated by reversed phase high performance liquid chromatography, size exclusion chromatography and thin layer chromatography, in sequence. Following each chromatographic separation the mutagenic compounds were found in one fraction predominantly. A remarkable purification of active compounds was achieved without serious losses of activity, thus enabling further purification and mass spectrometric studies of the main mutagens.

INTRODUCTION

The presence of Ames-mutagenic compounds in chlorine treated drinking water has been noticed in several studies (Cheh et al., 1980; van der Gaag et al., 1982 and Kool et al., 1982). In a recent study (Kronberg et al., 1985) we compared the mutagenic activity in extracts of chlorinated drinking water, purified from surface water containing high amounts of humic subtances (TOC = 14 mg/l), with the activity in extracts of a laboratory chlorinated natural humic water. Similar mutagenic activity found in both types of water supports the assumption that mutagenic compounds in drinking water are products of reactions of chlorine with humic substances (Meier et al., 1983). Identification of the main mutagenic compounds would be an important step towards correct assessment of their effects on human health. The pronounced mutagenic activity found for extracts of acidified chlorine treated humic water makes these extracts suitable for fractionation and identification of mutagens.

A multistep fractionation scheme developed for isolation of

mutagens is presented. Mutagenic extracts were fractionated by reversed phase high performance liquid chromatography (RP-HPLC), high performance size exclusion chromatography (HPSEC) and thin layer chromatography (TLC), in sequence.

MATERIAL AND METHODS

Water sample.

Natural humic water (TOC = 25 mg/l) was collected from a lake (Savojärvi) situated in a marsh region in the southwestern part of Finland.

Chlorination and concentration procedure, and mutagenicity determination.

Chlorination and concentration procedures, and methods for determination of mutagenicity (Ames-test) have been described in detail previously (Kronberg et al., 1985). A volume of 60 l of humic water (pH 7) was treated with 21 mg Cl_2/l (contact time 90 h). Extracts were obtained by passing acidified (pH 2.1) water over a column (2 x 40 cm) of a 1:1 mixture of XAD-4 and XAD-8. Elution in reverse direction was carried out with 360 ml ethyl acetate (see Fig. 1). Throughout the work mutagenicity was determined by Ames tester strain TA100 without metabolic activation. Each test was performed on triplicate plates and repeated at least once.

Chromatographic methods.

High performance chromatographic separations (RP-HPLC and HPSEC) were carried out at ambient temperature using two Altex 110A pumps, an Altex Model 153 UV-254 nm absorbance detector, a Rheodyne Model 7120 injector and a solvent programmer. Separations were performed on a Nucleosil 7 C_{18} reverse phase column (10 x 250 mm) and on a TSK G3000SW size exclusion column (7.5 x 300 mm). The RP-HPLC column was eluted by the use of a linear gradient from 30% methanol in water (pH 3 with ortophosphoric acid) to 100% methanol over a 30 min period at a flow rate of 4 ml/min. The HPSEC column was eluted with 0.01M sodium acetate (pH 7) at 1 ml/min. Fractions of separated material were collected during chromatography.

TLC separations were carried out on 20 x 20 cm plates coated with 0.25 mm silica gel (Macherey-Nagel, DC-Fertigplatten SIL G-25 UV_{254}). The solvent system consisted of acetonitrile, chloroform and acetic acid (5:5:0.1 by vol.). Separated material was collected by scraping off the adsorbent in marked zones from the plate.

Reconcentration of fractions.

RP-HPLC fractions containing more than 50% methanol were diluted with 0.01M HCl (pH 2.1) in order to get solutions with a methanol content of 50% or less. Organics in each fraction were extracted with methylene chloride. HPSEC fractions were acidified to pH 2.1 by the addition of 0.1M HCl and reconcentrated on small columns of XAD-4/8 (1 x 10 cm). The collected TLC adsorbent was eluted with water and after acidification the eluate was reconcentrated on XAD-4/8. Reconcentrated organics were eluted from XAD-4/8 with 25 ml of ethylacetate.

RESULTS

Extracts of chlorine treated humic water gave mutagenic response for strain TA100 (see Fig. 1). The mutagenic compounds were found, after each chromatographic separation in one fraction predominantly (see Fig. 2). Mutagenic activity recovered in the last fractionation step (TLC fraction 3) accounted for 25 - 30% of the original activity.

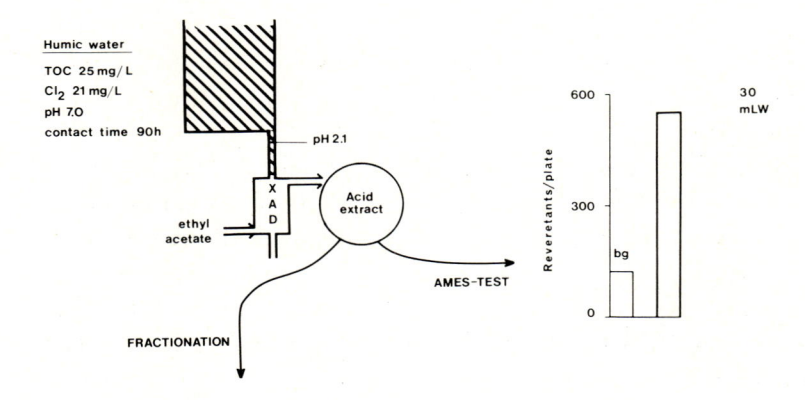

Fig. 1. Chlorination and concentration procedure for humic water. Mutagenic activity and equivalent volume of water tested are shown at the right side of the figure. bg = spontaneous revertants

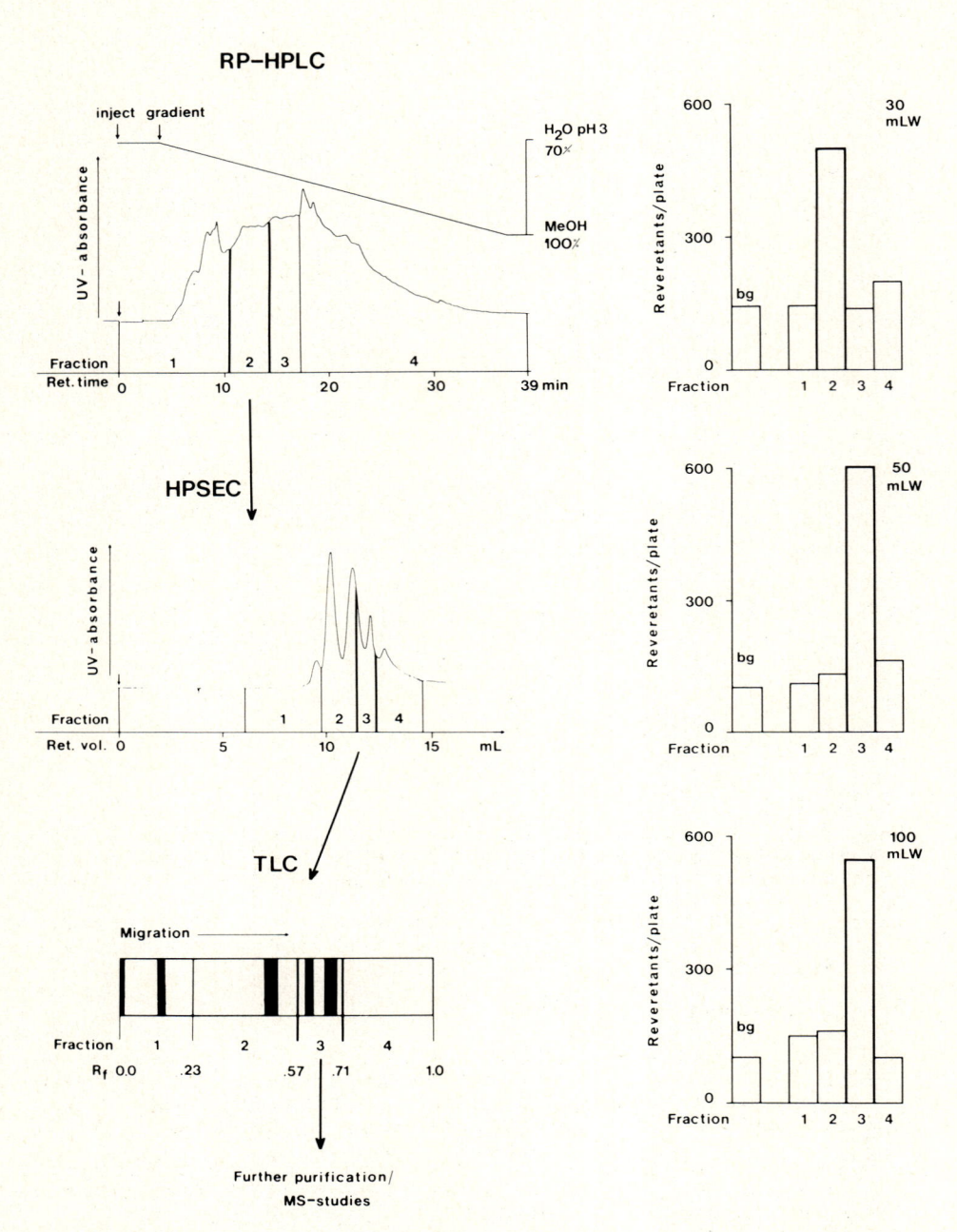

Fig. 2. Chromatographic separation and fractionation of mutagenic compounds in chlorine treated humic water. Mutagenic activity in fractions and equivalent volume of water tested are shown at the right side of the figure. bg = spontaneous revertants.

DISCUSSION

This work shows that it is possible, by using different chromatographic methods, to concentrate and purify mutagenic compounds in extracts of chlorine treated humic water. Mutagens are more readily adsorbed to XAD-4/8 resins under acid conditions than under neutral, indicating that the main active compounds are organic acids (Kronberg et al., 1985). Their chromatographic characteristics both in RP-HPLC and TLC separations are close to those of bensoic acid, used as a reference substance (retention time = 11.0 min; Rf = 0.63). HPSEC fractionation shows that the active compounds are eluted among substances of intermediate to low molecular weight.

Although a remarkable purification of mutagenic compounds is achieved through the three-step fractionation scheme presented, further fractionation has to be carried out before the active compounds are pure enough for a structural elucidation carried out by spectrometric techniques.

ACKNOWLEDGEMENT

We wish to thank Göran Pensar, Åbo Akademi, and Gösta Wahlroos, Turku Waterworks, for valuable advice and comments. The work was financially supported by the Academy of Finland.

REFERENCES

Cheh, A.M., Skochdopole, J., Koski, P. and Cole, L., 1980. Nonvolatile mutagens in drinking water: Production by chlorination and destruction by sulfite. Science., 207: 90-92.

van der Gaag, M.A., Noordsij, A. and Oranje, J.P., 1982. Presence of mutagens in Dutch surface water and effects of water treatment processes for drinking water preparation. Mutagens in Our Environment. Alan R. Liss, Inc., 150 Fifth Ave., New York, NY 10011, pp 277-286.

Kool, H.J., van Kreijl, C.F., de Greef, E. and van Kranen, J., 1982. Presence, introduction and removal of mutagenic activity during the preparation of drinking water in the Netherlands. Environ. Health Perspect., 46: 207-214.

Kronberg, L., Holmbom, B. and Tikkanen, L., 1985. Mutagenic activity in drinking water and humic water after chlorine treatment. Vatten, 41: 106-109.

Meier, J.R., Lingg, R.D. and Bull, R.J., 1983. Formation of mutagens following chlorination of humic acid. A model for mutagen formation during drinking water treatment. Mutat. Res., 118: 25-41.

LIQUID CHROMATOGRAPHIC DETERMINATION OF THE FUNGICIDE IPRODIONE IN SURFACE WATER, USING ON-LINE PRECONCENTRATION

C.E.GOEWIE and E.A.HOGENDOORN
National Institute of Public Health and Environmental Hygiene
Laboratory of Organic Chemistry
P.O.Box 1, 3720 BA Bilthoven (The Netherlands)

ABSTRACT

A new method for on-line sample preparation of surface water extracts for the reversed phase HPLC analysis of the fungicide iprodione at the ppb level is presented. Water samples are extracted with dichloromethane and after concentration and evaporation to dryness, taken up in a mixture of acetonitrile/water. 2-Ml aliquots are injected onto a small precolumn, which is subsequently flushed with 20% acetonitrile in water as a clean-up step. The precolumn is switched on-line with the analytical column, and, using 47.5% acetonitrile in water as the eluent, the concentrated zone of iprodione is desorbed and transported to the analytical column. Detection takes place with UV at 229 nm. The resulting chromatograms are free of interferences. The detection limit of the method in 0.02 ppb. Good reproducibility and linear calibration curves are obtained. The results of the method are in agreement with those of capillary GC analysis with cold-on-column injection. The advantage of the method compared to GC is the lower susceptability to errors due to the presence of interferences.

INTRODUCTION

Iprodione (1-isopropylcarbamoyl-3-(3',5'-dichlorophenyl hydantoin) is used on vegetables, fruits and ornamental plants for the control of different fungi, such as Botrytis, Alternaria, Phoma and Rhizoctonia (ref.1). As result of application, iprodione can either by drift or by leaching, enter watersheds. It was considered necessary therefore to develop a method for the determination of iprodione in surface water. The gas chromatographic determination of iprodione is problematic. With packed column GC and electron capture detection quantitation is known to be difficult. The compound shows considerable tailing on 5 out of 6 columns tested in our laboratory (ref.2). Besides, quantitation is hampered by the occurrence of interferences and of matrix and column saturation effects. Capillary column GC eliminates many of these problems but, as will be shown under Results and Discussion, the method is susceptible to interferences. We have recently described an HPLC method for the determination of residues of iprodione and five other fungicides on carrot (ref.3). The method is in principle suitable for application to surface water samples. In this study we have increased the sensitivity and selectivity of the reported HPLC method by making use of on-line preconcentration and clean-up on a short precolumn, packed with C_{18}-modified silica. The principle and application of the method to the analysis of other polar environmental pollutants has been described earlier by one of us (CEG) (ref.4,5). Another important advantage of the method presented is the increased speed of analysis, obtained by using an automated column switching unit.

THEORY

The principle of on-line preconcentration has been extensively described earlier (ref.4,5). On-line preconcentration of aqueous samples followed by HPLC analysis has been shown to be a fast and reliable method for automated sample preparation, which allows the injection of extremely large volumes onto an analytical column, thereby decreasing the detection limits by a factor of 1000 or more. The amount and volume of a sample solution that can be preconcentrated on a precolumn depends on the dimensions of the precolumn, the characteristics of the packing material, the nature of the solvent (usually water) and the flow rate used during sampling. The dimensions of the precolumn are chosen such that a reasonable volume can be sampled without loss of the analyte(s) of interest at a reasonable speed and such that the precolumn does not add considerable band-broadening to the analytical system during the elution step. It has been shown that a good compromise between loadability and performance is obtained with 2-10 mm long, 2-3 mm I.D. precolumns, packed with 5-10 μm particles (ref.6). For most analytical purposes, C_{18} materials have been recommended (ref.4).

Many papers have dealt with the preconcentration of analytes from purely aqueous solutions, such as river water (ref.4,5), soft drinks (ref.7) or serum (ref.8). In order to obtain high concentration factors, the use of a sampling liquid with the weakest possible elution strength towards the applied sorbent is favorable. In case of reversed phase systems, the weakest solvent possible is water. However, if the concentration factors aimed at are below the maximally attainable, and if the analytes of interest are not extremely polar, sampling may take place from organic-aqueous mixtures as well. The possibility of preconcentration from organic-aqueous (or even purely organic) media is extremely important for applications where organic extracts are involved. In the case of surface water analysis, liquid-liquid extraction is often preferred over solid surface extraction since the latter technique can not isolate analytes which are bound to solid matter, present in the sample. A drawback of liquid-liquid extraction as sample preparation for reversed phase HPLC is the incompatibility of the extraction solvent with aqueous eluents. The extraction solvent generally has to be removed and replaced by some water miscible fluid. Besides, the high concentration factors obtained by extraction and evaporation (usually about a factor of 100) are almost completely counterbalanced by the small injection volumes employed in HPLC (generally 10 µl). Since it is not possible to manipulate reproducibility with such small sample volumes, aliquots have to be taken, thereby wasting the bulk of the sample. On-line preconcentration of extracts is an excellent way of increasing the injection volume without causing unacceptable band broadening, thereby increasing the sensitivity of the method. If preconcentration is carried out on a small precolumn, the additional advantages of simultaneous clean-up and guard functions are offered.

Clean-up of extracts using solid surface adsorption is frequently employed in an off-line mode, for instance using Sep Pak-, Baker C_{18}- or similar cartridges. An off-line clean-up procedure for iprodione in carrot extract has been developed by us earlier (ref.3). C_{18} cartridges and step-wise gradient elution with acetonitrile/water mixtures were employed. In order to avoid loss of the analyte due to insolubility or coprecipitation with matrix components, iprodione was applied to the cartridge as a solution in acetonitrile/water (1:1). The off-line method was found to be quite cumbersome. Besides, it leads to a 3-fold dilution of the extract. These drawbacks can be circumvented by making use of on-line preconcentration. Clean-up of the sample can be effected by flushing the precolumn with a weak eluent, before desorption of the analyte onto the analytical column.

The volume and composition of flushing liquid, used for clean-up, should be chosen such that breakthrough of the analyte of interest does not occur.

It has been shown that the behaviour of analytes during preconcentration can be described using simple relationships, derived for liquid chromatography (ref.5).

The volume of a diluted sample solution that can be preconcentrated with 100% recovery on a short precolumn filled with a packing material with high loading capacity, such as C_{18}-silica, can be estimated from its chromatographic behaviour (ref.5). The breakthrough volume from a short precolumn is given by the equation

$$V_B = V_R - 2\,\sigma_v \qquad \{1\}$$

where V_R = retention volume, V_o = dead volume of the precolumn and σ_v = the width of the sample zone of the precolumn. The retention volume, V_R, of the analyte during the sampling or flushing step, can be calculated from:

$$V_R = V_o\,(1 + k_f) \qquad \{2\}$$

where k_f = the capacity factor of the analyte on the precolumn with the flushing solvent as eluent. k_f can be estimated from liquid chromatographic data, assuming a linear relationship between the lorgarithm of the capacity factor, $\ln k$, and the mobile phase volume fraction of organic modifier, $\emptyset$. It has been shown that this assumption is true in first approximation (ref.5). The precolumn dead volume can be calculated from:

$$V_o = \varepsilon^o \cdot Vg \qquad \{3\}$$

where ε^o = precolumn porosity. ε^o is approx. 0.70 for C_{18}-packed precolumns (ref.5). Vg = geometric volume of the precolumn. The bandwidth of the analyte, σ_v, can be calculated from:

$$\sigma_v = \frac{V_o (1 + k_f)}{\sqrt{N}} \qquad \{4\}$$

The plate number, N, of the precolumn is given by:

$$N = L / H \qquad \{5\}$$

where the plate height, H, is assumed to be 3 d_p; d_p is the particle diameter. For a well packed column, H generally is 2-3 times d (ref.9).

The capacity factors of iprodione on the C_{18} material Hypersil ODS in acetonitrile/water (50:50, v/v) and (70:30, v/v) were known from previous work to be 5.19 and 1.10, respectively. From these data the capacity factors of iprodione in different acetonitrile/water compositions used as flushing solvents, can be calculated. Using a 15 x 3.2 mm I.D. precolumn, packed with 7 μm C_{18} particles and assuming a k_f of 282 and 60 for pure water and acetonitrile/water (20:80, v/v) respectively, breakthrough volumes of 22 and 5 ml are found. This would mean that an iprodione extract, dissolved in either water or acetonitrile/water (20:80, v/v), could be sorbed onto the short precolumn, without losses due to breakthrough using resp. 22 or 5 ml of carrier of flushing solvent.

EXPERIMENTAL

Sample extraction

500 ml of surface water was extracted respectively with 100, 50 and 50 ml dichloromethane and the combined extracts were dried over sodium sulphate and concentrated to 10 ml in a Kuderna-Danish apparatus. The extract was evaporated to dryness and, for GC analysis, dissolved in petroleum ether. For HPLC the extracts were taken up in 2 ml acetonitrile. After dissolution, 8 ml water was added. A 2-ml aliquot was injected into the liquid chromatography as described below.

Gas chromatography

Packed column GC was carried out on a Varian 2700 (Palo Alto, CA, USA) gas chromatograph with a Ti^3H electron capture detector under the following conditions:
column: Pyrex, 90 cm x 2 mm (I.D.), packed with 3% OV-210 on Chromosorb W-HP, 100-200 mesh.
temperature: oven 200°C, inlet 220°C, detector 275°C.
carrier gas: nitrogen, 30 ml/min.
injection volume: 5 μl.
Capillary column GC was performed on a Carlo Erba 5160 gas chromatograph with

a Model 400 electron capture detector and automatic cold-on-column injector under the following conditions: column: 25 m x 0.32 mm I.D. fused silica, coated with 0.1 μm SE-52. Injection volume: 1 μl. Temperature program: 45°C (2 min), 25°C/min → 180°C (1 min), 5°C/min → 230°C, 10 min.

Liquid chromatography

The configuration for automated preconcentration and LC analysis is schematically presented in Fig. 1. The apparatus consisted of the following components: Eluent pump (A): Waters M45 (Milford, MS, USA), sample carrier pump (B): Kipp 9208 (Delft, The Netherlands), sample injection valve with 2-ml loop (I): Rheodyne 7125 (Berkely, CA, USA), Rheodyne 7000 switching valve (II) with a 15 x 3.2 mm I.D. Brownlee MPLC New Guard RP18 column (Santa Clara, CA, USA) (P), Analytical column (C): 150 x 4.6 mm I.D. homepacked with 5-μm Hypersil ODS (Shandon, Runcorn, GB), UV detector (D) with 229 nm Cd lamp and filter: Waters 440.

The pneumatic high pressure valves were controlled by a home-made programmable logical controller. Programs were loaded from an Epson HX-20 basic programmable personal computer. Data were processed by a Hewlett Packard 3388A (Waldbronn, GFR) computing integrator. Preconcentration of iprodione was carried out from water or water/acetonitrile (80:20, v/v) and LC analysis took place with 5 ml acetonitrile/water (47.5:52.5, v/v) at a flow rate of 1 ml/min.

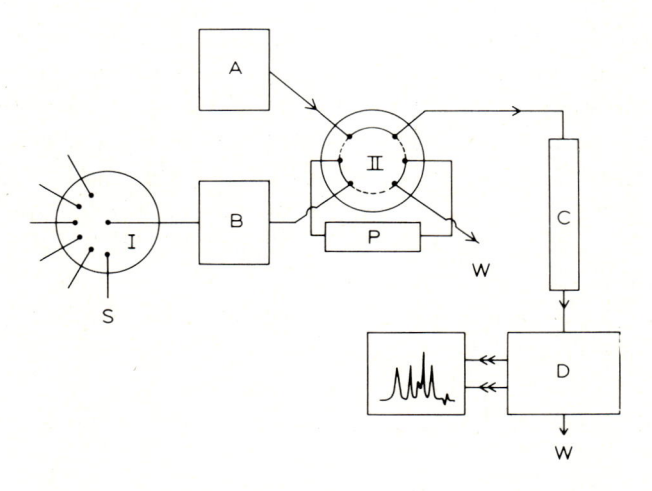

Fig. 1. Confiration of HPLC apparatus for on-line preconcentration and clean-up by column switching. For explanation of symbols: see Experimental.

Chemicals

Dichloromethane was distilled in house to analytical purity. All other solvents used were of analytical grade and obtained from Baker (Deventer, The Netherlands). HPLC quality water was obtained from a Milli-Q (Millipore, Bedford, MS, USA) purification system.

Iprodione was obtained from Rhône Poulenc, Direction des Recherches et du Dévelopement (Centre Nicolas Grillet, France).

RESULTS AND DISCUSSION

The efficiency of the liquid-liquid extraction procedure for iprodione was determined using petroleum ether and dichloromethane as extraction solvents. The recovery of iprodione with petroleum ether was 40-60% only. With dichloromethane the recovery was 95 ± 6% (n=3). This solvent was used for extraction of water samples in all further experiments.

Fig. 2 shows a gas chromatogram of the extract of an iprodione-spiked water sample, on a packed column. Accurate quantitation is not possible due to the broad interfering peak and to matrix and column saturation effects. Due to a matrix-effect recovery of iprodione, as determined by packed column GC, is only 40-60%. The saturation effect is reflected by a steadily increasing response towards a standard injection, which makes accurate calibration impossible.

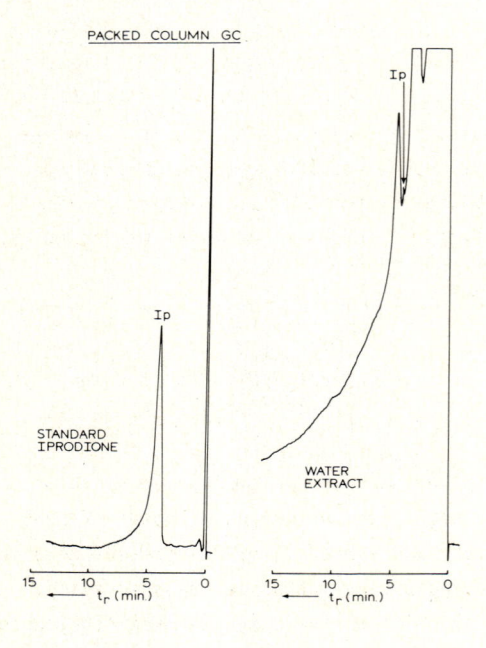

Fig. 2.

Packed column gas chromatography of iprodione (a) standard (0.097 µg/ml); (b) surface water extract. Concentration factor: 100. Column: 3% OV-210. Temp.: 200°C. Further conditions: see Experimental.

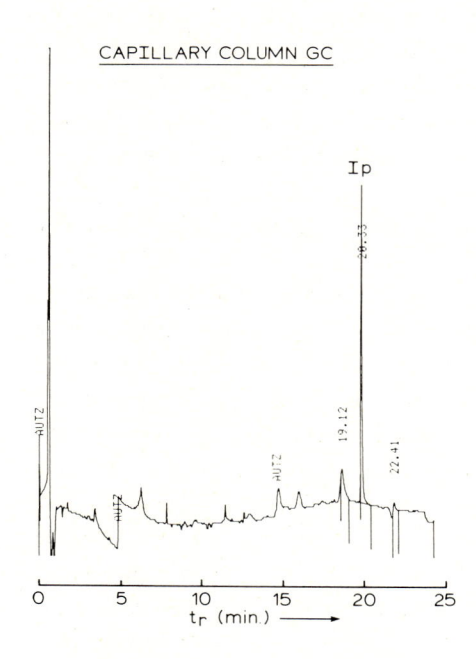

Fig. 3.
Capillary column gas chromatography of iprodione standard (0.025 µg/ml). Column: SE-52. Cold on column injection and temperature programming. Further conditions: see Experimental.

Fig. 3 shows a chromatogram of a water sample analysed by capillary column GC. Although the chromatogram looks much better than that obtained with packed-column GC, and very low detection limits (0.01 ppb) are attainable, it can be seen that many interferences are present in the sample, which may be co-eluted with iprodione. In order to prevent this, the analysis time has to be rather long (20 min). For this reason, and also since capillary column GC with cold-on-column injection is not yet wide-spread, development of an alternative liquid chromatographic procedure for the analysis of iprodione in water was found useful.

The optimal LC conditions for iprodione were known from previous work and are described under Experimental.

With direct loop injection of 50 times concentrated extracts of samples into the HPLC, the detection limits were too high to enable analysis of real life samples.

Further concentration of the extracts by uptake in a smaller volume of solvent was not found attractive as it is known to lead to gross losses of analyte and decreased reproducibility of recovery and since it would increase the background signal. The latter effect would negatively influence the theoretical gain in sensitivity. Preconcentration of 20 ml of the extracts, redissolved in water onto the short precolumn, followed by immediate in-line switching with the analytical column, resulted in chromatograms with many interferences. Therefore it was decided to introduce a flushing step after preconcentration, in order to create an on-line clean-up procedure.

2-ml Aliquots of iprodione standards dissolved in acetonitrile/water (20:80, v/v) were flushed onto the precolumn using both water and acetonitrile/water (20:80, v/v) as carrier solvents. The breakthrough curves were composed by measuring the response for iprodione by HPLC analysis as a function of the flush volume. The results are given in Fig. 5. It can be seen that breakthrough from water does not occur after flushing with up to 20 ml, while with acetonitrile/water (20:80, v/v), breakthrough occurs after flushing with some 6 ml of solvent. These results are in excellent agreement with the data predicted from theory. If the 2-ml sample is flushed onto the precolumn with a too small volume of solvent, e.g. 2.5 ml, the analyte is not completely recovered, due to incomplete loading of the precolumn. With a flush volume of 5 ml acetonitrile/water (20:80, v/v), the recovery of a 2 ml injection of an iprodione standard of 0.142 µg/ml was 94 ± 3%, compared to a 200 µl loop injection of the same amount of analyte directly onto the analytical column.

Next, chromatograms were recorded of 2-ml aliquots of blank water extracts, which were preconcentrated and flushed with different volumes of water and acetonitrile/water (20:80, v/v) prior to in-line switching with the analytical column. Two representative chromatograms of the optimization experiments are given in Fig. 6. It can be seen from Fig. 6a that flushing with 5 ml water is by no means effective for clean-up. In contrast, Fig. 6b shows that flushing the precolumn with 5 ml of acetonitrile/water (20:80, v/v) removes most interferences present in the water extract. The latter procedure was used in all further experiments. It can be seen from Fig. 7a (preconcentrated iprodione standard) that this procedure does not lead to excessive band broadening. Actually, the peak volume after preconcentration from acetonitrile/water (20:80, v/v) was equal to that of a 200 µl loop injection of the analyte, dissolved in the same solvent. If necessary, the peak volume can be further decreased by compression of the sample zone by using a stronger eluent during desorption. This would decrease the detection limit of the method even further.

With this system and procedure, a calibration curve was measured for iprodione standards in acetonitrile/water (20:80, v/v), using 4 concentrations between 0.01 and 0.14 µg/ml. The calibration curve was found to be linear, with a regression coefficient of 0.99995, and to pass through the origin. The repeatability was excellent. For an iprodione concentration of 0.01 µg/ml the mean relative standard deviation was found to 1.9% (n=5). The detection limit, determined as 3 times the peak-to-peak noise level, was 0.5 ng/ml. For surface water samples, spiked with 0.7 and 7.0 ppb of iprodione, the repeatability was: 89 ± 6% (n=4). Fig. 7b shows a chromatogram of a surface water extract, containing

1-2 ppb of iprodione.

A number of water extracts were analysed both by capillary GC and by HPLC using the preconcentration method developed. The results were compared (see Table I) and found to be in agreement by application of Students' t-test.

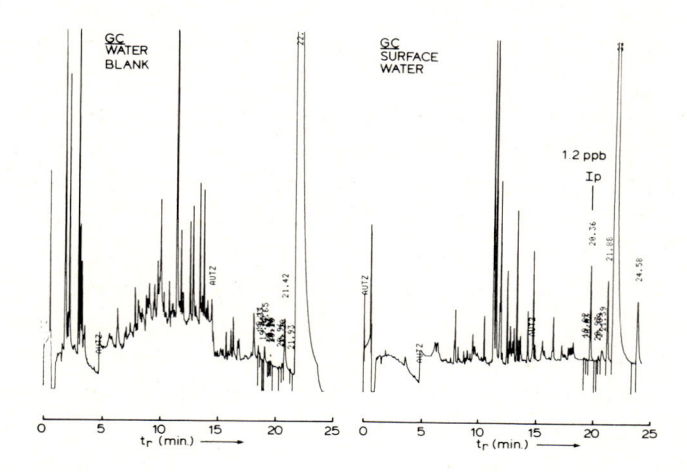

Fig. 4. Capillary column gas chromatography of (a) surface water blank. Concentration factor: 50; (b) surface water extract containing iprodione (1.2 µg/l). Concentration factor: 8. Experimental conditions: see text.

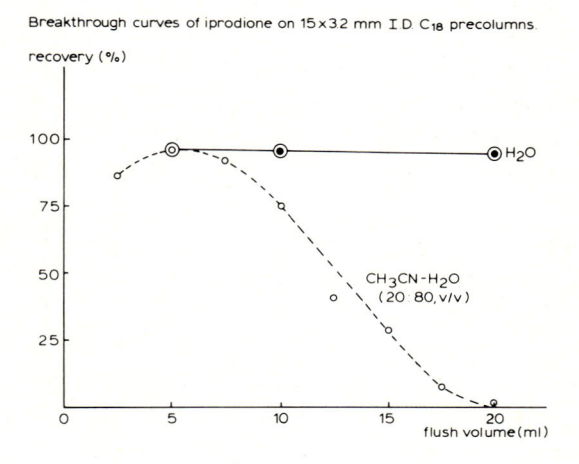

Fig. 5. Breakthrough curve of iprodione from (a) water, (b) acetonitrile/water (20:80, v/v). The response of iprodione (0.142 µg/ml) (arbitrary units) is given as a function of the flush volume. Precolumn: Brownlee MPLC New Guard, 15 x 3.2 mm I.D., packed with 7-µm RP 18. Flow rate during flushing: 1 ml/min. On-line HPLC analysis conditions: see Experimental.

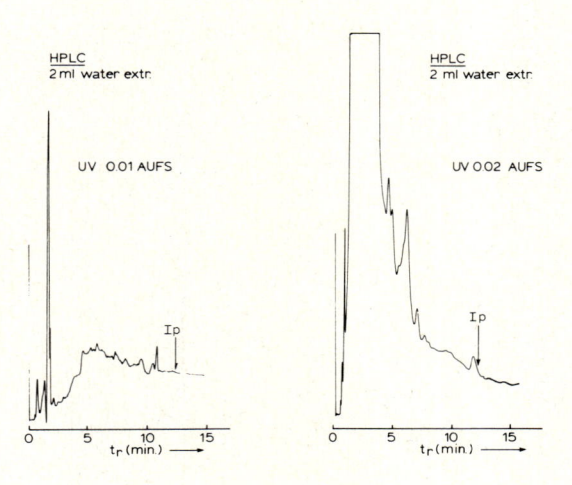

Fig. 6. On-line preconcentration/clean-up and HPLC analysis of 2 ml of a 20-fold concentrated surface water blank extract. Carrier solvent during preconcentration (a) acetonitrile/water (20:80, v/v); (b) water. Sample time: 5 min. Further conditions: see Experimental.

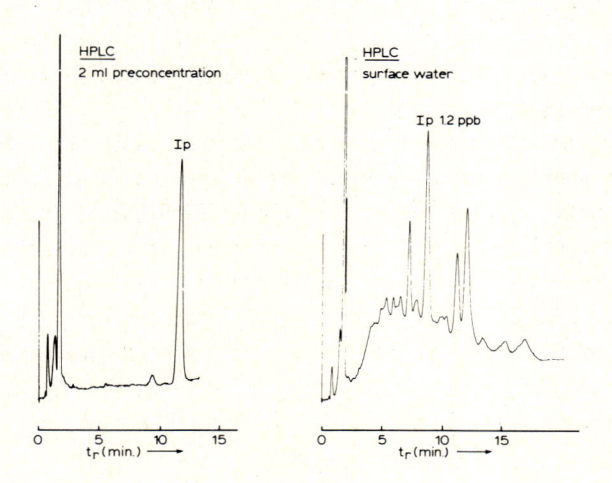

Fig. 7. On-line preconcentration and HPLC analysis of 2 ml of an (a) iprodione standard (0.04 µg/ml); (b) surface water extract containing 1.2 µg/ml of iprodione. Concentration factor: 20. Carrier solvent during preconcentration: acetonitrile/water (20:80, v/v); UV detection at 214 nm; attenuation 0.01 AUFS. Further conditions: see Experimental.

TABLE I

Analysis of surface water extracts.

Sample	Iprodione (µg/l)	
	HPLC	GC
I	0.07	0.02
II	1.20	1.30
III	<0.02	<0.02
IV	0.09	0.08
V	0.10	0.09
VI	<0.02	<0.02
VII	0.15	0.07
VIII	1.10	0.90
IX	4.00	4.10
X	1.00	1.20
XI	0.44	0.51

CONCLUSIONS

On-line preconcentration and clean-up of organic-aqueous water extracts prior to HPLC analysis of iprodione is an attractive alternative to GC analysis. The HPLC method developed is sensitive (detection limit in surface water: 0.02 ppb) and reproducible and can easily be automated. The use of short precolumns, packed with C_{18} material inserted in a high pressure switching valve, allows the application of a flushing step and a heart-cutting procedure, which leads to excellent clean-up, without loss of dilution of the analyte. In this method the precolumn serves simultaneously as guard-, preconcentration and pre-fractionation column. The application of extracted water samples, instead of the untreated surface water samples themselves to the precolumn, as is generally described in literature, has several advantages. The water samples do not need to be filtered before extraction, so bound residues are also collected. The samples can be treated as usual and the extracts divided for other, mostly GC, analyses. The application of extracts instead of uncleaned samples, dramatically increases the lifetime of the precolumn. The cartridge used in this study did not show decreased performance, even after 2 months of extensive use.

ACKNOWLEDGEMENT

We wish to thank Bob Brownlee (Brownlee Labs. Inc.) for the gift of precolumns.

REFERENCES

1. Beschikking Residuen van Bestrijdingsmiddelen, Nederlandse Staatscourant, March 15, 1984
2. P.A.Greve and H.A.G.Heusinkveld, Meded.Fac.Landbouww.Rijksuniv.Gent, 46/1 (1981) 317-324
3. E.A.Hogendoorn, C.E.Goewie, H.H.van den Broek and P.A.Greve, Meded.Fac.Landbouww.Gent, 49/3b (1984) 1219-1230
4. C.E.Goewie, Thesis, Free University Amsterdam (1983)
5. C.E.Werkhoven-Goewie, U.A.Th.Brinkman and R.W.Frei, Anal.Chem. 53 (1981) 2072-2080
6. C.E.Goewie, M.W.F.Nielen, R.W.Frei and U.A.Th.Brinkman, J.Chromatogr. 301 (1984) 325-334
7. H.P.M.van Vliet, Th.C.Bootsman, R.W.Frei and U.A.Th.Brinkman, J.Chromatogr. 185 (1979) 483-489
8. C.E.Werkhoven-Goewie, U.A.Th.Brinkman, R.W.Frei, C.de Ruiter and J.de Vries, J.Chromatogr. 276 (1983) 349-357
9. L.R.Snyder and J.J.Kirkland, Introduction to modern liquid chromatography. Wiley & Sons, New York (1979)

PROGRESS IN THE ISOLATION AND CHARACTERIZATION OF NON-VOLATILE MUTAGENS IN A DRINKING-WATER

P.G. van Rossum

National Institute for Water Research, Council for Scientific and Industrial Research, P.O. Box 395, Pretoria 0001 (South Africa)

ABSTRACT

Using the Ames *Salmonella* mutagenicity test (TA98, without S9), non-volatile mutagenic fractions were obtained by applying a separation scheme on XAD extracts of Pretoria drinking-water. The fractionations obtained were based on solubility, volatility, permeation and adsorption and the mutagenicity yield of each step was estimated. One mutagenic fraction consisted of humic like material, possibly together with polar polynuclear aromatic hydrocarbons.

INTRODUCTION

A survey of Pretoria drinking-water had shown the presence of direct-acting mutagenicity (TA98) (van Rossum, 1982) which could be extracted from water using XAD-resins or methylenedichloride. This mutagenicity was mainly introduced by a waterworks treating river water.

A number of adsorbents were tested for their suitability to extract the mutagens from drinking-water; XAD-7 gave the highest yield and was consequently used in this study (van Rossum, 1984).

This paper describes our attempts to isolate the non-volatile mutagens from Pretoria drinking-water. The isolation scheme applied to a concentrated XAD-7 extract consisted of consecutive fractionation with a LH-20, a silica gel and a porous glass column. HPLC and TLC were used to monitor the purification of the mutagens. The results are useful for the selection and consequent optimization of the most promising of the large number of potential techniques, for the ultimate isolation (preparative or analytical) and finally, for the identification of the mutagens.

EXPERIMENTAL

Care was taken to limit the possibility of the introduction of contaminants and the formation of artifacts. No preparative separations needing acid or alkaline conditions were attempted, very active adsorbents were avoided and evaporation to dryness was done by rotary evaporation at temperatures not higher than 50 °C and the vacuum broken with high purity nitrogen.

Ames tests (Maron, 1983) were done in triplicate (0,1 ml of extract or pure solvent per 2,5 ml of top-agar) and results calculated as mutation ratio. (MR):

$$MR = \frac{\text{Average number of revertants on plates with sample}}{\text{Average number of revertants on plates with solvent}}$$

The 'mutagenicity yield' of a separation step was estimated by comparing the dose-response curves of the extracts before and after the separation step. A part of the material obtained by a separation step was diluted (or concentrated) to the concentration of this material before the separation step and the slope of the dose-response curve between MR = 2 and MR = 3 after the step was expressed as % of the slope before the step.

About 30 m³ of drinking-water was passed through a column (bed volume 2 ℓ) of XAD-7 resin at a flow rate of 2 $\ell.min^{-1}$. The adsorbed compounds were eluted with acetone until the eluate was colourless.

Following evaporation to dryness of the eluate, the mutagenicity of the distillate and the material of the distillation residue soluble in 20 mℓ of MeOH, were determined. During storage at 5 °C, a white precipitate was formed in the methanol solution of the distillation residue. The fraction of the precipitate soluble in acetone was also tested for mutagenicity.

The dark, brown methanol solution of the distillation residue was concentrated to 5 mℓ and placed on top of a LH-20 column, (length 690 mm; dia.: 36 mm; bed volume: 700 mℓ). The column was eluted with methanol and the fractions (50 x 35 mℓ) were analysed for mutagenicity, light absorbance (380 nm; 1 cm cell), total organic halogens (van Steenderen, 1980) and with thin layer chromatography.

Thirty mℓ of the combined mutagenic LH-20 column fractions was used to determine the distribution of the mutagenicity among the strong acids, the weak acids and the amine/neutral compounds in the extract by dissolving the evaporation residue in 30 mℓ of CH_2Cl_2 and extracting 15 mℓ in series with 50 mℓ each of 5% $NaHCO_3$ and 5% NaOH. The aqueous extracts were acidified and extracted with equal amounts of CH_2Cl_2. All the CH_2Cl_2 fractions were evaporated to dryness, the residues dissolved in 15 mℓ of acetone and tested for mutagenicity.

The combined mutagenic LH-20 column fractions were evaporated to dryness and 50 mℓ of benzene added to the residue. After separation, 50 mℓ of acetone was added to the material which did not dissolve in benzene. Methanol (50 mℓ) was added to the separated (sintered glass funnel pore No. 2) material which did not dissolve in acetone. For the Ames test 1 mℓ of each solubility fraction was evaporated to dryness and the residues dissolved in DMSO.

Merck silica gel 60 (14 g, particle size 0,063 - 0,2 mm) was activated by 160 °C for 4 h and partially deactivated by adding purified water (distilled, through XAD resin) to a final water concentration of 10%.

The benzene soluble fraction of the mutagenic LH-20 column eluate was passed through the silica gel column (dia. 14 mm), followed by elution with 50 mℓ each of benzene, dichloromethane, ethylacetate, acetone and methanol and collection of 50 mℓ fractions. The yellow band eluting with ethylacetate was collected separately and made up to 50 mℓ with acetone. For the Ames test 1 mℓ of each fraction was evaporated to dryness and the residue dissolved in 10 mℓ of acetone.

The mutagenic ethylacetate fraction from the silica gel column was evaporated to dryness and the residue dissolved in 1 mℓ of methanol. Exclusion chromatography was done with methanol (0,2 mℓ.min^{-1}) on a controlled porous glass (Corning CPG; pore dia. 235 Å) column (length 500 mm; dia. 6 mm). The fractions of the mutagenic peak were combined and purified water was added until precipitation occurred. The methanol was evaporated under vacuum and the resulting aqueous suspension freeze-dried.

HPLC was done with an Hypersil ODS (5 μm) column (length 30 mm; dia. 6 mm) with a methanol eluent (1,0 mℓ.min.$^{-1}$).

Merck pre-coated silica gel 60 F254 plates were used without further activation. One-dimensional (1-dim.) TLC was done over 10 cm of adsorbent with an eluent which consisted of 90 mℓ of benzene, 8 mℓ of methanol and 4 mℓ of glacial acetic acid. Two-dimensional TLC was done over 6 cm of adsorbent with mixtures of 90% dichloromethane, 10% acetone (1st direction) and 90% benzene, 10% methanol (2nd direction) or with the last eluent in both directions. Spots were detected with 365 and 254 nm ultra-violet light.

Although 2-dim. TLC showed that the freeze-dried material still consisted of a mixture, the following characterization techniques were applied: UV and I.R. (KBr Pellet) and direct-inlet mass spectroscopy, elemental analysis and molecular mass determination by means of vapour pressure osmometry.

RESULTS

The isolation scheme

A brown, tar-like evaporation residue of about 2 g was obtained from the XAD-7 eluate and the estimated mutagenicity yields after evaporation were: Methanol soluble material ≅60% and volatile material ≅50% of the mutagenicity of the XAD eluate (Fig. 1a).

Fractions 15-21 (Fig. 1b) of the Sephadex LH-20 column showed toxicity (dose-response Ames test showed no mutagenicity) with the Ames test and were the only fractions where halogens could be detected. Subsequent GC/MS analysis showed that the main gas chromatographable constituents of these fractions were the herbicides Atrazine and Propazine.

The mutagens eluted after most of the light-absorbing substances and after one bed volume of eluate i.e. they not only permeated the LH-20 but were also

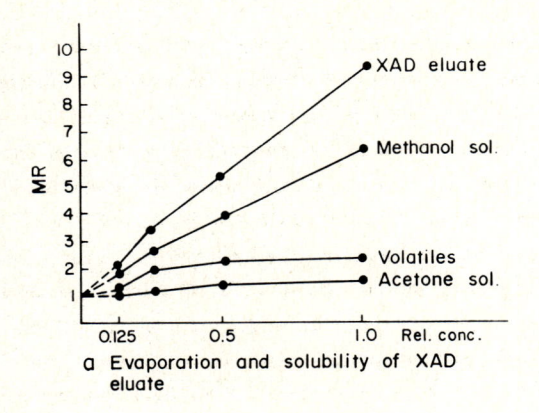

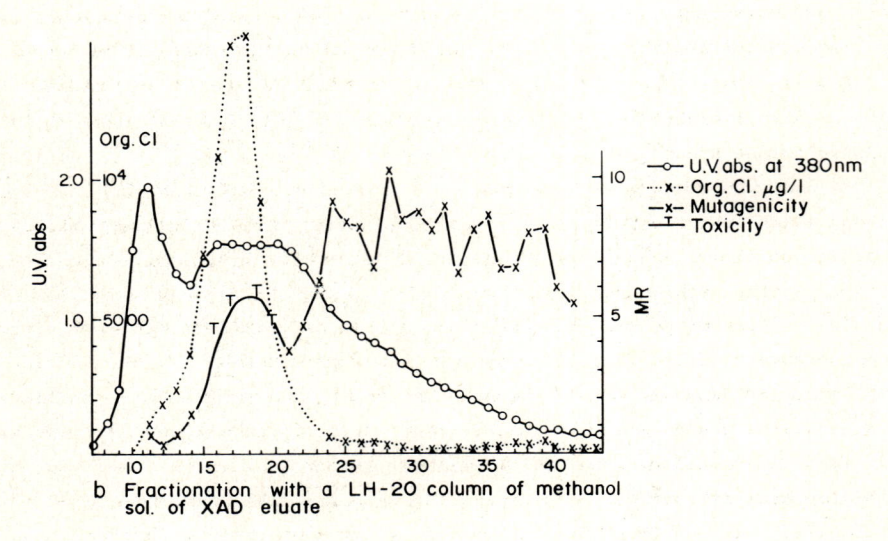

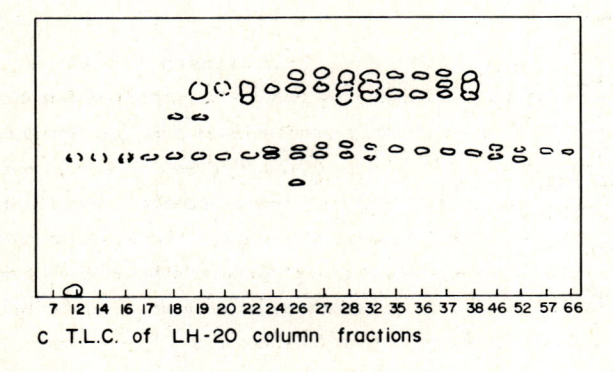

Fig. 1. Separation scheme from XAD eluate to LH-20 fractions.

adsorbed.

Although the mutagenic material always emerged after one bed volume of eluent, the retention times of the MR-peaks of Fig. 1b were non-repeatable. Thin layer chromatography, however, gave repeatable results (Fig. 1c).

Taking the variability of the Ames test into account, the mutation yield due to the separation of non-mutagenic material with LH-20 was 100%.

According to the pH separation experiment, the mutagenic material from the LH-20 column consisted for 60% of base neutrals (Fig. 2a).

The amount of mutagenicity of the dried, combined, mutagenic LH-20 fractions ($\cong$100 mg) soluble in benzene varied between 100% and 60% (Fig. 2b: $\cong$60% soluble in benzene and $\cong$80% soluble in acetone).

With the exception of dichloromethane, all the solvents used, eluted yellow-coloured material from the SiO_2 column. With one-dimensional TLC, the ethyl-acetate fraction (at least 30 mg) showed a band of material between Rf = 0,3 and Rf = 0,6 and recovered $\cong$80% while the benzene fraction recovered $\cong$20% of the mutagenicity that entered the SiO_2-column (Fig. 2c,d). (PAH's like benzo(a)pyrene are at Rf = 0,75.)

A fraction completely excluded and one completely permeated, were obtained with the controlled porous glass column. Their relative proportions depended on the batch of drinking-water extracted but the mutagenicity always resided in the second peak. (Fig. 3a,b). The mutagenicity yield was $\cong$40%.

The purpose of positioning HPLC at the end of the separation scheme was to apply a number of low resolution clean-up steps on the rather large amount of XAD extract. However, after these clean-up steps were applied, no complete fractionation could be achieved with either Hypersil ODS (5 μm) or Sperisorb S5 ODS2 using different combinations of ethanol, methanol, acetonitrile, dioxane, tetrahydrofurane and water. HPLC and Ames tests of the HPLC fractions (Fig. 3c) showed that the mutagens consisted of at least one group of substances with very similar characteristics; i.e. may consist of homologues or isomers.

Two-dim. TLC (same solvent) on the freeze-dried material showed a diagonal band from Rf = 0 to Rf = 1, which indicated a group of substances with varying polarity as well as the absence of dissociating substances (and overloading). Using two eluents a curved fluorescent band with a number of fluorescent blue, yellow and white spots outside or at the boundary of the band were obtained. (Fig. 3d). The ethylacetate fraction from the SiO_2 column as well as the benzene-insoluble (methanol soluble) material of the mutagenic LH-20 fractions gave the same two-dimensional TLC results as the freeze-dried material. The non-mutagenic material, separated with the CPG column, showed fluorescent material at Rf = 1 and a diagonal band up to Rf = 0,5 without distinct fluorescent spots.

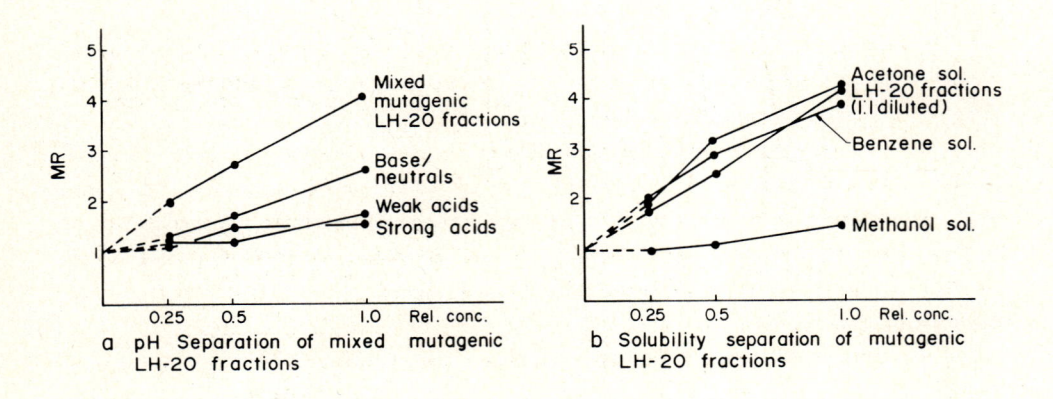

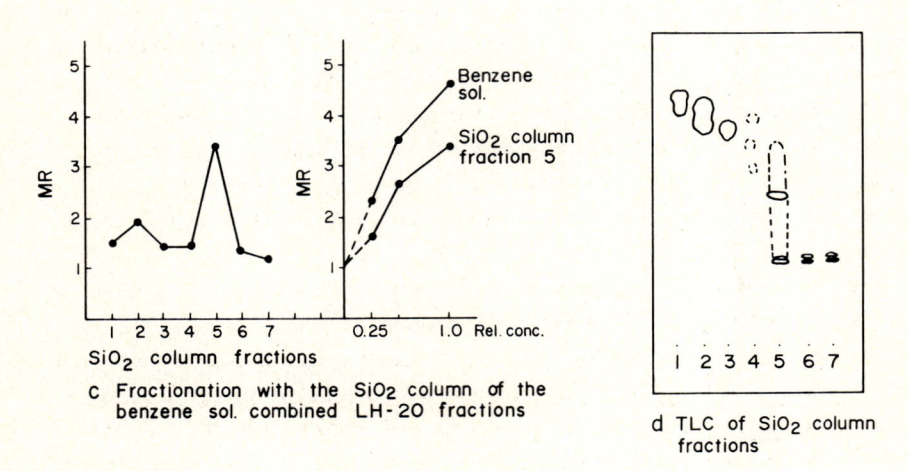

Fig. 2. Separation scheme from LH-20 to SiO₂ fractions.

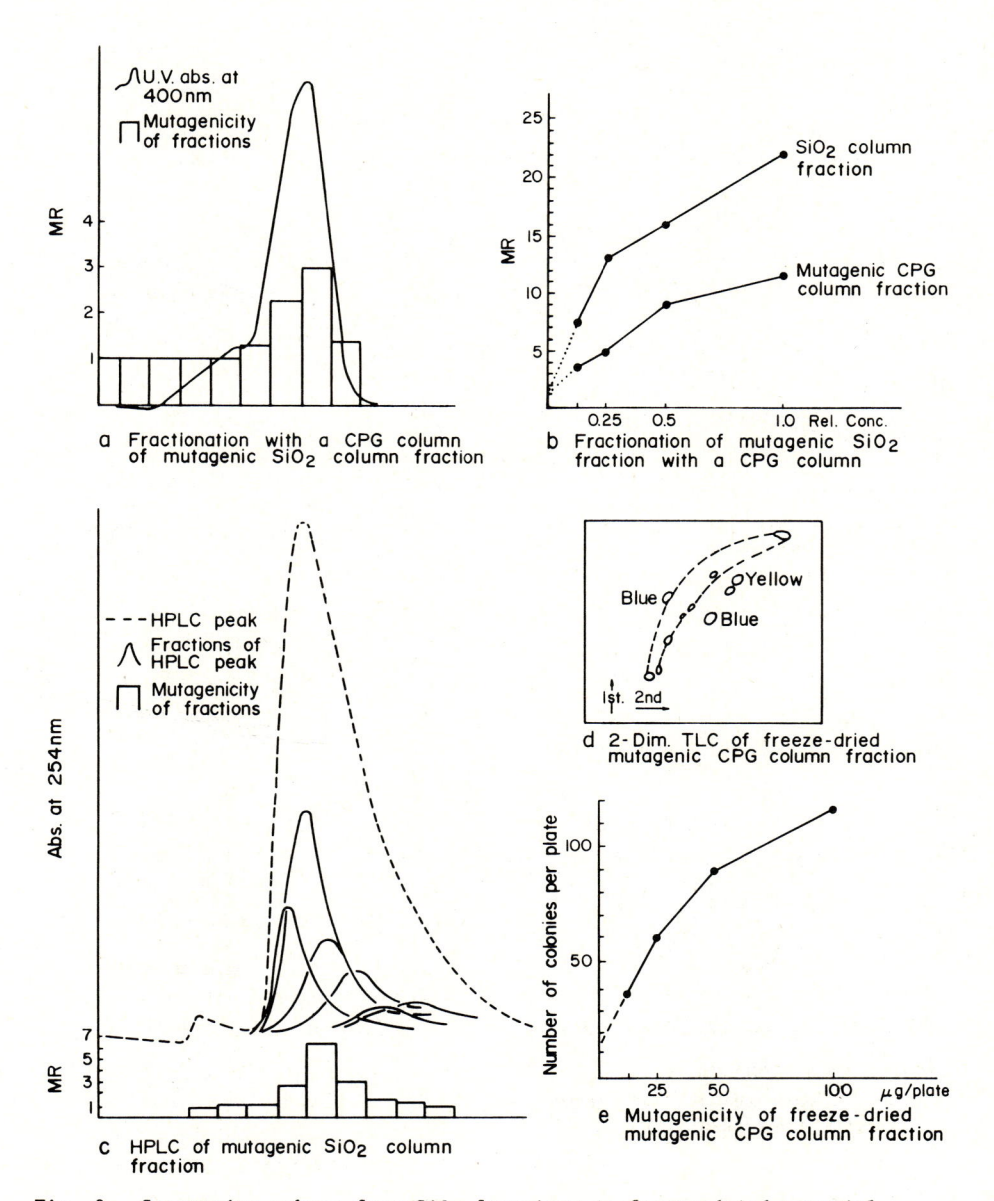

Fig. 3. Separation scheme from SiO₂ fractions to freeze-dried material.

The isolated material

Ultra-violet light absorbance of the freeze-dried material in methanol increased gradually towards the shorter wavelengths, with shoulders at about 236 and 264 nm and caused the material to fluoresce. Infra-red absorption bands appeared at 3400, 2920, 2860, 1690, 1600 and at about 1440 and 1370. A number of absorptions at lower wave numbers were difficult to distinguish from a broad absorption region between 1800 and 900. None of the absorption bands were sharp.

The elementary composition was 46,7% C, 4,7% H and 2,3% N.

With direct-inlet MS, mass spectra were obtained with M/e values at almost every value up to 263. The material evaporated at the highest intensity from the probe at 130 °C. The average molecular mass was 693.

DISCUSSION

The isolation scheme

The ratio between the volume of drinking-water and the volume of the XAD-7 was 10^4. The possibility of replacement of some of the adsorbed material by stronger adsorbing substances cannot be excluded; i.e. the above-mentioned ratio could have influenced the composition of the extract and, consequently, the structure and the results of the applied separation scheme.

Ten mg of mutagenic material was isolated (Fig. 3e) which was obtained from the XAD extract at an overall estimated yield of 11%. Assuming that the mutagens in the XAD extract have equal activity, the drinking-water could contain 3 µg.ℓ^{-1} of non-volatile mutagens.

Although synergism and antagonism (total yield of a separation step >100 %) could have influenced the estimated total yield, the following separation steps need closer attention in order to improve separation and yield:

1. As the XAD eluate was not dried with anh. Na_2SO_4, some mutagens could have been vaporized with the last traces of water when the eluate was taken to dryness. The initial rise of the volatiles' dose/response curve (Fig. 1a) as well as a ≅60% yield of the mutagenicity in the evaporation residue, could confirm this but the mutagens may have been difficult to detect in the distillate due to the presence of toxic substances. GC/MS of the (derivatized) volatile material could be worthwhile.
2. Taking separate mutagenic fractions of the LH-20 column, substituting the HPLC ODS columns with polar ones and eluting with a solvent gradient, could at least produce groups of mutagens with even more similar characteristics.
3. The separation obtained at the solubility fractionation step between the LH-20 and SiO_2 columns could be due to a poor solubilization technique. All the mutagenic material in the dried combined mutagenic LH-20 column fractions

can probably be put onto the SiO_2 column as a dichloromethane solution.

4. Decomposition of mutagens could have occurred on the CPG column (only 40% yield). This is also supported by the appearance of spots on the 2-dim. thin layer chromatogram at Rf >0,7 and Rf <0,3, regions where substances originally present were removed with the SiO_2 column. Permeation chromatography on silanized CPG could prevent decomposition of the mutagens and indicate whether the separated non-mutagenic material was an artifact.

Attempts to improve the % yield of a separation step should be accompanied by an Ames test on the reconstituted fractions in order to establish a synergistic effect.

<u>The isolated material</u>

The results of the characterization of the isolated material indicated that it still consisted of a mixture of marked similarity to humic acids. This observation is supported by the fact that the chlorination of humic material forms mutagens. Acidity could, however, not be established i.e. the main portion of the isolated material can, provisionally, be called 'neutral humic material'.

In this case, decarboxylation could have occurred during chlorination (Larson, 1979). Additionally, attention should be given to the possibility that this neutral humic material is the keto-form (I.R. abs. at 1690) of an enol/keto tautomeric system.

Apart from possible mutagenic neutral humic material, the distinct blue, yellow and other fluorescing spots on the 2-dim. TLC chromatogram could indicate mutagenic polar polynuclear aromatic hydrocarbons. They form a smaller part of the isolated material and no indication of their contribution relative to the total mutagenicity of the isolated material can yet be given.

Although the above-mentioned conclusions in connection with the identity of the mutagens in drinking-water are speculative, they are useful as guides to further attempts at their identification.

It is possible that with today's separation techniques, resolution of the mutagenic material into individual compounds cannot be achieved and attempts will be made to show by analysis rather than by isolation what types of mutagens are present.

The building blocks of the 'neutral humic material' can possibly be established by pyrolysis GC/MS. Furthermore, if pyrolysis brings about the breakdown of a large family of compounds to a few building blocks, temperature-stable polynuclear aromatic hydrocarbons could be better identifiable.

ACKNOWLEDGEMENT

This paper is published by permission of the National Institute for Water

Research. The technical assistance of Christie Engelbrecht in chemical and microbiological work is gratefully acknowledged.

REFERENCES

Larson, R.A. and Rockwell, A.L. Chloroform and chlorophenol production by decarboxylation of natural acids during aqueous chlorination. Env. Sci. & Technol, 13 (1979) 325-329.
Maron, D.M. and Ames, B. Revised methods for the Salmonella mutagenicity test. Mutation Res., 113 (1983) 173-215.
Van Rossum, P.G. et al., Examination of a drinking-water supply for mutagenicity. Wat. Sci. Tech., 14 (1982) 163-173.
Van Rossum, P.G., Determination of the percentage adsorption of unknown organic compounds from water with a multi-column technique. (1984) in press.
Van Steenderen, R.A., The construction of a total organohalogen analyser system. Lab. Pract., 4 (1980) 380-385.

CHARACTERIZATION OF LOW MOLECULAR WEIGHT PRODUCTS DESORBED FROM POLYETHYLENE TUBINGS

C. ANSELME, K. N'GUYEN, A. BRUCHET, J. MALLEVIALLE

Centre de Recherche Lyonnaise des Eaux-Degrémont
38 rue du Président Wilson, 78230 Le Pecq, France

ABSTRACT

A case study is reported which demonstrates the appearance of numerous organic compounds associated with an intense taste and odor problem during the passage of a drinking water through a 300 m polyethylene pipe. Batch experiments show that two principal mechanisms are involved in the occurence of organoleptic changes caused by defective polyethylene tubings ; dissolution of the polymer additives (e.g. alkylphenols...) and oxidation of the internal surface of the pipe during extrusion with subsequent release of the resulting polar compounds (aldehydes, ketones...). Twenty per cent of the pipes tested (n = 264) appeared to be defective and the total release time may exceed several months under low flow rates conditions.

INTRODUCTION

Over the last few years, the use of synthetic pipes (eg polyethylene) for drinking water distribution, has been rapidly increasing (Montiel, 1983). In addition to corrosion considerations, their extensive use in the case of high density polyethylene (HDPE) are motivated by favorable mechanical properties, ease of handling during manufacture and low permeability to external contaminants (Didier, 1984). However in some cases, changes of drinking water organoleptic properties have been noticed. Three principle causes for these changes have been identified in the literature :

- Dissolution of the polymer additives (Flogstad, 1984),

- Oxidation of the internal surface of the pipe and dissolution of the resulting polar compounds (Flogstad, 1984), and

- Migration of external contaminants through the pipe (Mehens, 1984).

This paper reports a case study which demonstrates unambiguously the appearance of numerous organic compounds during the passage of a drinking water through a 300 m polyethylene pipe, in relashionship with an intense taste and odor problem.

MATERIAL AND METHODS

Both laboratory and field conditions were tested. Laboratory samples were obtained by passing tap water through various sections of HDPE pipes with internal diameters respectively of 26 34 and 40 mm. The water velocity was 3 m/mn. Field samples were taken from HDPE portions of distribution systems located in a southern suburb of Paris. Organoleptic qualities of the various samples were evaluated using both a Flavor Profile Analysis (FPA) and the Threshold Taste Number (TTN) according to the standards specified for the European Community (AFNOR, 1983). The FPA technique was developed for the food industry (Arthur D. Little) and has been recently extended to the analysis of tastes and odors in water samples (Krasner, 1983). Erlenmeyer flasks and a temperature of 45°C were adopted for odor evaluations. Samples were collected headspace free and kept refrigerated until analysis. A panel of four or more trained "tasters" evaluated the organoleptic qualities giving a description and an intensity for each taste, odor and sensation observed, with regard to a list of approximately 25 descriptives. The scale for intensity was 0 (absent), 1 (threshold), 12 (strong) with 5 intermediate values.

Chemical analyses initially involved closed loop stripping analysis (CLSA) and was later supplemented by simultaneous distillation extraction (SDE). The CLSA technique used for broad spectrum analysis was similar to that used by Grob (1976). The water bath and carbon trap temperatures were maintained at 45°C and 55°C respectively. The methodology described by Nickerson (1966) was followed in applying the SDE as a complement to the data obtained by CLSA. The SDE technique involves a batch extraction of 3 liters of water. Water and solvent (50 ml methylene chloride) are heated separately, after which they are condensed together. The condensate separates into 2 phases and the solvent is recovered for subsequent concentration (approximately 250 µl) using a set of small Dufton distillation columns. By the nature of the technique, SDE is expected to collect compounds of higher molecular weight and higher polarity than CLSA. Moreover, this analysis is particularly adapted to sensory measurements by the GC/"sniffing" technique.

The resulting extracts were analyzed both by GC/FID (Carbo Erba fractovap 4160, Italy) and GC/MS Ribermag R10-10C, Rueil Malmaison, France). In these two cases, analytic methodology was as follows :

- on column injection of 1 µl on a capillary chromatographic column (chrompack OV1701 50 m long or J and W scientific DB5 30 m long)

- electronic impact at 70 eV

- mass range 20-400.

RESULTS

Case Study

The connection in the South of Paris during summer 1984 of three HDPE sections to a cast steel pipe (Figure 1) resulted one month later in numerous complains due to intense plastic tastes. The low water velocities in this part of the distribution system probably favoured the appearance of this phenomenon. Four water samples were collected as indicated in Figure 1 in each section of the HDPE and in the cast steel pipe, for sensory evaluation and specific organics analysis.

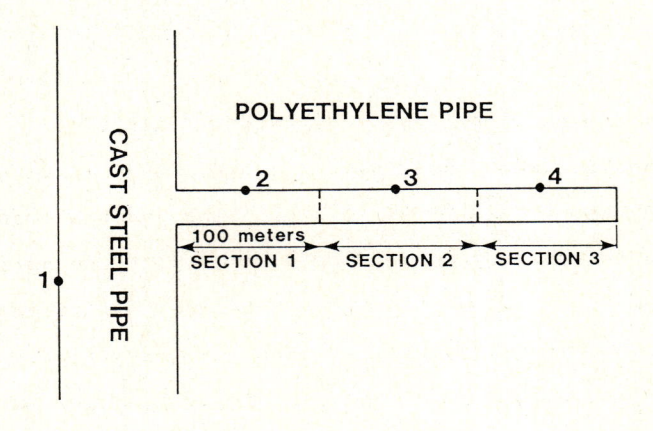

POLYETHYLENE PIPE

CAST STEEL PIPE

100 meters
SECTION 1   SECTION 2   SECTION 3

• Sampling point

Fig.1. Case Study

Sensory evaluation (Table 1) showed that the plastic odor appeared along section 1 and occured at a maximum intensity in section 2.

TABLE 1

Flavor Profile Analysis

	CAST STEEL PIPE	HDPE PIPE		
		SECTION 1	SECTION 2	SECTION 3
TASTE AND ODOR	-	BURNT 4 PLASTIC	BURNT 12 PLASTIC	BURNT 8 PLASTIC
THRESHOLD ODOR NUMBER	1	5	50	30

The CLSA chromatograms obtained for each sampling point (Figure 2) indicate the appearance of numerous organic contaminants during the passage of drinking water along the HDPE pipe, with highest peak intensities detected in sections 2 and 3. As seen in Table 2, the main products identified by GC-MS correspond to polymer additives (Stepek, 1982 ; Di Pasquale, 1984) (lubricants, antioxidants, stabilizing agents) and polar compounds (aldehydes). The quinone detected may be derived from BHT, the well-known phenolic antioxidant (4-methyl-2,6-di-T-butylphenol) reacted with residual chlorine present in the pipe (0.2 ppm). Some of these additives were found at concentrations exceeding 1 µg/l, progressively increasing from section 1 to section 3. Aldehydes are also found at unusually high concentrations in the last two sections.

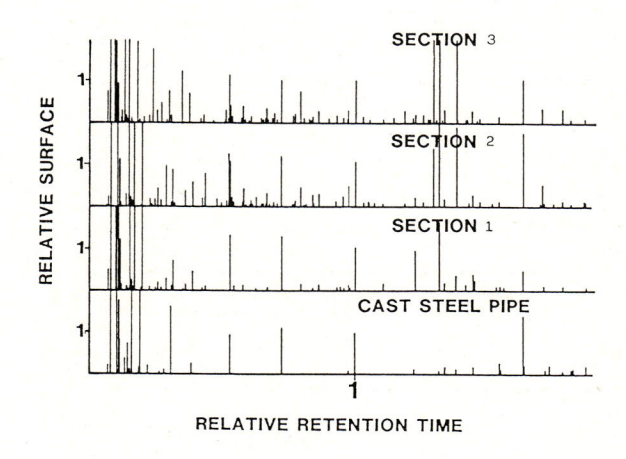

Fig. 2. Analysis of water samples CLSA/GC/FID

TABLE 2

Additive Concentrations in HDPE Pipe

ADDITIVES	CONCENTRATION NG/L			
	CAST STEEL PIPE	HDPE PIPE		
		SECTION 1	SECTION 2	SECTION 3
LUBRICANT				
ALKYL NAPHTALENE	—	—	350	900
ANTIOXIDANTS/STABILIZING AGENTS				
4 ETHYL-2.6-DI-T-BUTYL PHENOL	—	—	70	120
ALKYL THIOPHENE	—	90	500	1800
4 METHYL-2.6-DI-T-BUTYL QUINONE	—	500	2000	3300
ALDEHYDES	50	80	500	600

To get a better understanding of the additives, a HDPE sample of section 2 was obtained and cut into thin slices. Ten grams of these slices were placed in a liter of mineral water (Evian, France) and allowed to soak for 48 hours. Subsequent closed-loop stripping and GC-MS analysis of the soak water (Figure 3 and Table 3) allowed the identification of a broader range of additives : in addition to the compounds previously mentionned, several plasticizers were found and the presence of BHT was confirmed. This product may be directly used as an antioxidant or may be derived from the commonly used thiobisphenols.

Quality control of various HDPE samples

Since the role of HDPE in the degradation of organoleptic properties was clearly evidenced, it seemed necessary to control the quality of the pipes delivered by the French suppliers. A total of 264 hundred meters of HDPE samples were tested according to the following procedure : 2 m sections were rinsed with tap water (flow rate : 700 L/h) during 12 hours and then sealed with

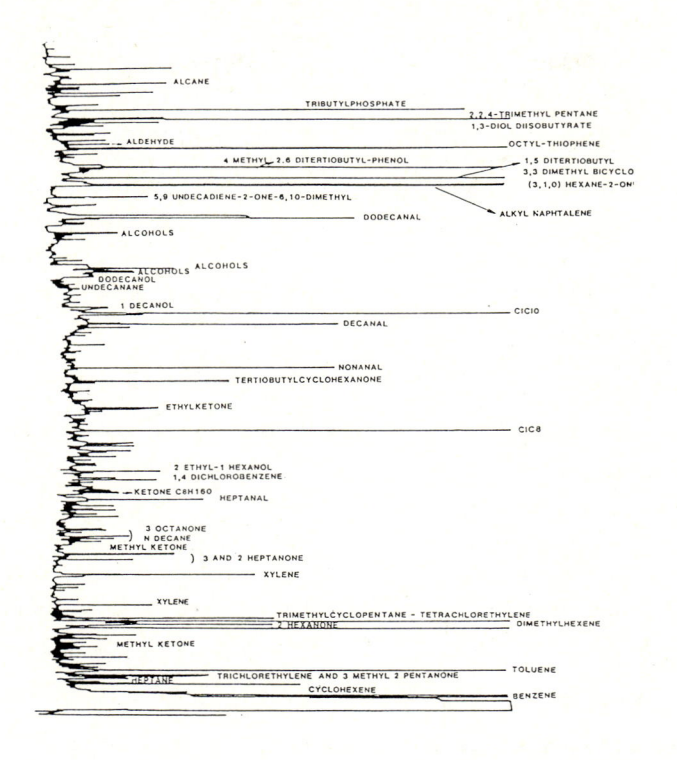

Fig. 3. GC-MS analysis of section 2 soak water

TABLE 3

Additive Concentrations in HDPE Maceration

ADDITIVES	CONCENTRATION NG/l
LUBRICANT	
ALKYL NAPHTALENE	4300
ANTIOXYDANT	
4-METHYL-2.6-DIT-BUTYL PHENOL	2600
4 ETHYL-2.6-DIT-BUTYL PHENOL	70
1,5-DI-T-BUTYL-3.7-DIMETHYL BICYCLO HEXANE-2-ONE	350
ALKYL THIOPHENE	1600
PLASTICIZERS	
2,2,4-TRIMETHYL PENTANE 1,3-DIOL DI ISOBUTYRATE	250
TRIBUTYL PHOSPHATE	160
PHTALATES	660

brass caps and left in contact with tap water over a period of twelve hours. The threshold taste numbers determined for these samples are reported in Table 4. It can be seen that 20% of the pipes tested presented a TTN exceeding the limit fixed by the European communities (threshold 3 at 25°C). Among the defective pipes, more than 50% showed TTN values higher than 30.

TABLE 4

Threshold Taste Number (TTN) of 264 HDPE Samples

TTN	NUMBER OF HDPE SAMPLES	% OF TOTAL
0	212	80
3-10	13	5.0
10-20	2	0.8
20-30	6 } 52	2.3 } 20
30-40	11	4.2
50	20	7.6
TOTAL	264	100

Complementary parameters were measured for three typical samples (high, medium and low taste intensity determined by FPA) and are summarized in Table 5. Nonspecific parameters such as UV absorbance or CH2 index are significantly increased ; the taste intensity does not seem to be directly related with the concentrations of the released additives ; as the total yields of polar compounds (aldehydes, ketones) increase with increased taste intensity, it can be assumed that these polar compounds play a determining role in establishing the final organoleptic sensation.

TABLE 5

Comparative Analyses of Various HDPE Pipes

PIPES	A	B	C
PLASTIC TASTE INTENSITY	2	8	12
CH_2 INDEX µg/L	80	80	300
UV 220 NM	0.16	0.19	0.405
UV 270 NM	0.091	0.171	0.194
ALDEHYDES NG/L	480	570	1400
KETONES NG/L	220	500	1100
ALKYL PHENOLS NG/L	1000	3000	500
ALKYL NAPTHALEN NG/L	530	3150	250
ALKYL QUINONE NG/L	120	520	100

In order to determine the relative influences of the polymer itself and manufacturing process (pipe extrusion), raw granules of polyethylene were left in contact with mineral water (Evian, 10 g/l, 48 h), which was then analyzed by CLSA - GC - MS. Compared with the extract of section 2 soak water, similar concentrations of the same additives were found, but the striking difference was the complete absence of the polar compounds. Indeed this indicates that these compounds are formed by oxidation of the internal surface of the pipe during the cooling phase following the extrusion, in the presence of air (Flogstad, 1984 ; Flogstad, 1984 ; Le Poidevin, 1977 ; Chan, 1967).

Sensory evalution of the additives

A GC - Sniffing method was used as an attempt to identify which specific additives were involved in the odor problem. In this procedure, the effluent of the capillary chromatographic column is split into two parts, 30% supplied to the FID Detector and 70% to the operator nose. Figure 4 represents the GC - Odor chromatogram of a sample of section 2 soak water, extracted by methylene chloride with a steam - distillation - extraction technique. The basic burnt plastic odor detected in the waters studied was found to be due to component number 4 : BHT. Two other aromatic compounds also revealed a plastic odor. However, it must be kept in mind that other molecules found below their odor threshold in the SDE extract, may also have been involved in the final odor detected by the consumers : this could be the case of the 2,2,4 - trimethylpentane - 1,3 - diol - di - isobutyrate (plasticizer), the odorous property of which was checked with a pure standard. Furthermore, as the compounds are smelled separately, the method does not account for the synergistic effects probably acting in such a complex mixture, in particular the polar oxidation by-products.

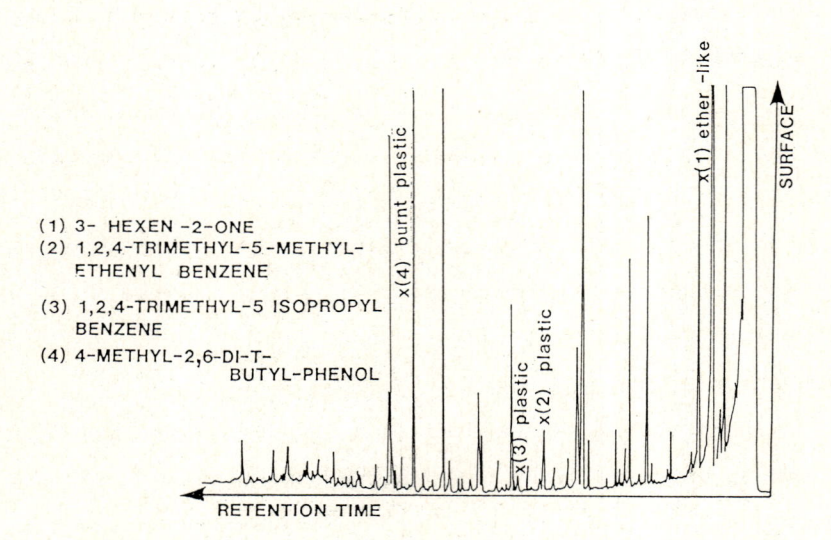

(1) 3- HEXEN -2-ONE
(2) 1,2,4-TRIMETHYL-5-METHYL-ETHENYL BENZENE
(3) 1,2,4-TRIMETHYL-5 ISOPROPYL BENZENE
(4) 4-METHYL-2,6-DI-T-BUTYL-PHENOL

Fig. 4. SDE extract of section 2 soak water GC/odor chromatogram

Evolution of desorption products in a polyethylene pipe

During the case study reported above, the taste and odor problem still remained after four months use (even after long periods of rinsing) which finally led to the replacement of the pipe.

To get information on the duration of the release, a polyethylene pipe (26 mm internal diameter) was flushed with a given volume of water (rinsing volume) and then sealed. After 48 h of soaking with a ratio polymer contact area/soak water volume fixed at 1 cm^{-1}, the TOC and phenol concentrations of the soak water were determined. Figure 5 indicates that the initial peak of TOC observed disappears after the passage of 4 m3, corresponding to a ratio rinsing volume/volume of the pipe equal to 1000. On the other hand, the release of total phenolic additives is more gradual and 20% of the initial release is still observed at a rinsing ratio of 7500. TTN at this point remains very high at about 50 units.

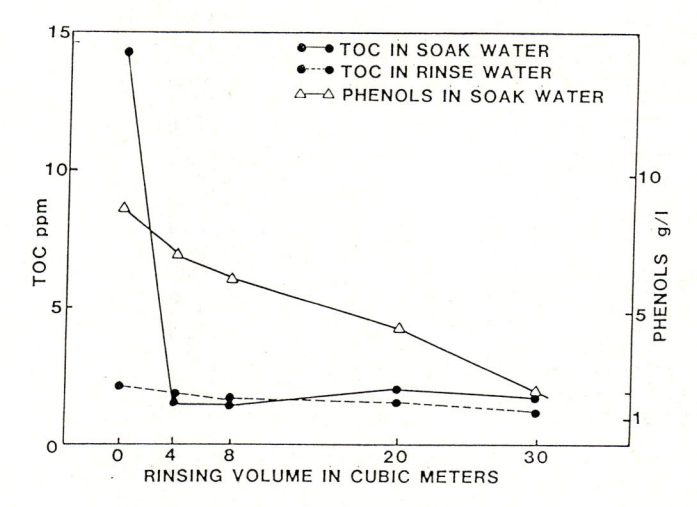

Fig. 5. Evolution of desorption products in a polyethylene pipe

CONCLUSIONS

The case study reported in this paper is a typical example of the degradations occuring in the drinking water distribution systems. Passage through a defective HDPE pipe leads to a nonpotable water with regard to the standard specifications of the European Communities. Besides exceeding values of TTN, this water presents concentrations of phenols approaching the maximum allowed concentration (10 µg/l) which remain undetected by the standard aminoantipyrine method (AFNOR, 1983). Considering the high values and the possible transformation (phenols → quinones) of some desorbed compounds, more toxicological information is required.

The problem sources and their possible remedies are summarized in Table 6.

TABLE 6

ORGANOLEPTIC QUALITY PROBLEM SOURCE	REMEDY
- LOW WATER VELOCITIES IN THE DISTRIBUTION SYSTEM	
- RELEASE OF HDPE ADDITIVES	- USE OF ANTIOXIDANT WITH LOWER SOLUBILITY THAN THIO-BIS-PHENOL
- SURFACE OXIDATION OF HDPE PIPES	- EXTRUSION IN NITROGEN ATMOSPHERE
	- AUTOMATIC CONTROL OF EXTRUDEURS

If the water velocities in the distribution system are difficult to control, the use of an antioxidant with lower solubility than thiobisphenols seems to be the most appropriate solution. It should be noted that the US Food and Drug Administration (FDA) prohibited the antioxidant 4,4 - thiobis(3 - methyl - 6 -

tertiobutylphenol) (Flogstad, 1984). Oxidation during extrusion could be prevented by the use of an inert atmosphere (nitrogen). This latter solution appears, unfortunately, too expensive and thus agreement between French water distributors and HDPE pipes suppliers now provides for organoleptic quality control of each set of manufactured tubings, ensured by the extrudeurs themselves.

REFERENCES

- AFNOR (1), 1983. Norme NF T 90 - 035. Evaluation du goût. Recueil des normes françaises des eaux. Méthodes d'essais. 2ème édition. Paris la Défense.

- AFNOR (2), 1983. Norme NF T 90 - 109. Détermination de l'indice phénol. Recueil des normes françaises des eaux. Méthodes d'essais. 2ème édition. Paris la Défense.

- Arthur D. Little, INC. The Flavor Profile Panel. Internal Report. Cambridge, Massachussetts, USA.

- Chan, M.G. and Lincoln, W., 1967. The Relationship between Rate of Reaction with Oxygen and Chemical Change in Polymers as Indicated by I.R. Analysis. Polymer Engineering and Science. October. pp. 264-268.

- Didier, C., 1984. Le polyéthylène haute densité. Informations Chimie n°256. pp. 193-203

- Di Pasquale, G. and Galli, M., 1984. Determination of Additives in Polyolefins by Capillary Gas Chromatography. Journal of High Resolution Chromatography and Chromatography Communications. Vol 7. pp. 484-486.

- Flogstad, H. (1), 1984. Plastic Pipe Can Transmit Odours. World Water, December. pp. 27.

- FLogstad, H. (2), 1984. Penetration of Plastic Water Pipes by Gases and Solvents. AIDE International Conference. Special Subject n°13. Monastir, Tunisia.

- Grob, K., and Zurcher, F., 1976. Stripping of Trace Organic Substances from Water, Equipment and Procedure. J. Chromatograph. 117. pp. 285-294.

- Krasner, S.W., Mc Guire, M.J., and Fergusson, V.B., 1983. Application of the Flavor Profile Method for Taste and Odor Problems in Drinking Water. Presented at the WQTC Conference, AWWA, Norfolk, Virginia, USA.

- Le Poidevin, G.J., 1977. The Oxidation of Polyethylene in Aqueous Solution. Part 1. Detection and Characterisation of Oxygenated Groups by Infrared Spectroscopy. US Department of

Commerce. National Technical Information Service. Report n°ECRC/N-1062.

- Mehens, J., Peeters, P., and Celens, J., 1984. Diffusion of Water and solvents into HDPE Theoretical and Practical Approach by Simple Laboratory Tests : Case Studies in the Water Distribution System. AIDE International Conference. Special Subject n°13. Monastir, Tunisia.

- Montiel, A., and Mallevialle, J., 1984. Les matières plastiques utilisées lors de la distribution des eaux. AIDE International Conference. Special Subject n°13. Monastir, Tunisia.

- Nickerson, G.B., and Likens, S.T., 1966. Gas Chromatographic Evidence for the Occurence of Hop Oil Components in Beer. J. Chromatograph. Vol 21. pp. 1.

- Stepek, J., and Daoust, H., 1983. Additives for Plastics. Polymers Properties and Applications. Springer - Verlag, New York INC.

CARCINOGENIC AND MUTAGENIC PROPERTIES OF CHEMICALS IN DRINKING WATER

R.J. BULL
College of Pharmacy, Washington State University, Pullman, Washington 99164

ABSTRACT
   Isolated cases of careless handling of industrial and domestic waste has lead to a wide variety of dangerous chemicals being inadvertently introduced into drinking water. However, chemicals with established carcinogenic and mutagenic properties that occur with a high frequency and in multiple locations are limited in number. To date, the chief offenders have been chemicals of relatively low carcinogenic potency. Some of the more common chemicals are formed as by-products of disinfection. The latter process is generally regarded as essential to the production of a "microbiologically safe" drinking water. Consequently, any reductions in what may be a relatively small carcinogenic risk must be balanced against a potential for a higher frequency of waterborne infectious disease.

   The results of recent toxicological investigations will be reviewed to place the potential carcinogenic and mutagenic hazards frequently associated with drinking water into perspective. First, evidence for the carcinogenicity of certain volatile organic compounds such as trichloroethylene, tetrachloroethylene and carbon tetrachloride is considered. Second, the carcinogenic activity that can be ascribed to various by-products of chlorination is reviewed in some detail. Finally, recent evidence that other chemicals derived from the treatment and distribution of drinking water is highlighted as an area requiring move systematic attention.

INTRODUCTION
   A wide variety of chemicals have been found in drinking waters around the world. Modern analytical techniques have identified such a diversity of chemicals that it renders comprehensive review of toxic and potentially carcinogenic properties of each substance virtually impossible. However the sources of toxic chemicals in finished drinking water are derived from three general sources:
   1. Contamination of source water.
   2. By-products of treatment processes.
   3. Leaching from the distribution system.
Examples of contaminants from each source will be discussed in turn.

CONTAMINANTS OF SOURCE
   A great variety of chemicals can contaminate sources of drinking water. By far the most frequent contaminants in this category are trichloroethylene

and tetrachloroethylene because of their large production volumes and widespread use in commerce. The carcinogenicity of trichloroethylene, tetrachloroethylene and to a lesser degree carbon tetrachloride are somewhat more controversial. Consequently, this section will provide an updated review of the data bearing on the carcinogenic properties of these chemicals. Consideration of other contaminants will be for comparative purposes only. A summary of pertinent carcinogenicity and mutagenicity data associated with contaminants of source waters is provided in Tables 1 and 2.

Carbon Tetrachloride

Studies of carbon tetrachloride (Cameron and Karunaratne, 1936; Reuber and Glover 1967a; Reuber and Glover 1967b) have documented the ability of carbon tetrachloride to induce preneoplastic changes in the liver of rats (e.g., hyperplastic nodules and cholangiofibrosis). Subsequent studies (Reuber and Glover, 1970) demonstrated that subcutaneous administration of carbon tetrachloride in corn oil at a dose of 2080 mg/kg twice weekly gave rise to hepatocellular carcinomas in male Japanese rats, Osborne-Mendel rats, and Wistar rats. Black rats and Sprague-Dawley rats failed to survive long enough to develop carcinomas following the same treatment. The authors' called attention to the fact that carcinoma development appeared to be inversely related to the degree of liver cirrhosis across the strains tested. However, this relationship may be fortuitous since those strains developing severe cirrhosis failed to survive. The single level of treatment utilized in this study precludes a clear resolution of this issue. A more recent study (NCI, 1976) of the carcinogenicity of carbon tetrachloride in Osborne-Mendel rats failed to confirm its carcinogenic effects at doses of up to 160 mg CCl_4/kg in corn oil administered 5 times weekly for 78 weeks (47 and 94 mg/kg for males, 80 and 159 mg/kg for females). However, it should be noted that the doses utilized were much lower than used in previous studies.

Eschenbrenner and Miller (1943) demonstrated that carbon tetrachloride was able to induce hepatomas in Strain A/J mice. Tumor yield was generally dose-related, but the effect was enhanced as the spacing between doses was increased from 1 to 5 day intervals to the same total dose of carbon tetrachloride. In subsequent experiments Eschenbrenner & Miller (1946) examined the relationship between liver necrosis and tumor development and found that the liver became resistant to necrotizing doses of carbon tetrachloride with chronic treatment. This observation was most probably related to the inactivation of cytochrome P-450 by carbon tetrachloride observed in modern metabolic studies (Glende et al., 1975). Overall, there appeared to be a relationship between obvious necrosis and the development of hepatomas. However, hepatomas could be induced by doses of carbon tetrachloride that did not induce obvious necrosis in the liver. The authors pointed out that they could not rule out the possibility of a slightly greater rate of liver cell replacement at these lower doses.

They did find evidence of cirrhosis at doses not producing obvious necrosis, indicating some degree of liver injury. Kiplinger and Kensler (1963) also observed the induction of hepatomas following oral administration of carbon tetrachloride in C3H mice relative to historical controls, however, this study did not include concurrent controls. The National Cancer Institute (NCI, 1976) documented a high incidence of hepatocellular carcinomas in the B6C3F1 mice administered at doses of 1250 and 2500 mg/kg by gavage in a corn oil vehicle.

Despite the fact of its recognized carcinogenic effects in rodents there is little convincing evidence that carbon tetrachloride possesses genotoxic properties (IARC, 1979b; USEPA, 1984). On the other hand, the ability of hepatonecrotic doses of carbon tetrachloride to promote the yield of liver tumors in mice is well documented (Pound and McGuire, 1975). Therefore, the bulk of the evidence indicates that the cancer producing activity of carbon tetrachloride are secondary to its hepatotoxic effects and carbon tetrachloride might be considered as a carcinogen that does not act by a "genotoxic" mechanism.

Trichloroethylene

Trichloroethylene (TCE) was first shown to be carcinogenic in a study sponsored by the National Cancer Institute (NCI, 1976). In this study, male B6C3F1 mice were exposed to time-weighted average doses of 1169 and 2339 mg/kg body weight, whereas, females received 869 and 1739 mg/kg of TCE dissolved in corn oil by gavage for a period of 78 weeks. In the same study both male and female Osborne-Mendel rats were administered doses of 549 and 1097 mg/kg TCE by the same means and for the same duration of exposure. Dose-related increases in the incidences of hepatocellular carcinomas were observed in both male and female mice, although the response was much more marked in the males. There were no statistically significant increases in tumor yield in either male or female Osborne-Mendel rats.

Inhalation studies conducted by Henschler et al. (1980) at concentrations of TCE in air of 100 and 500 ppm for 6 h/day, 5 days/week for a period of 18 months demonstrated an increased incidence of malignant lymphoma in female NMRI mice but was without apparent effect in male mice, Wistar rats or Syrian

TABLE 1. Evidence for Carcinogenicity of Carbon Tetrachloride, Trichloroethylene and Tetrachoroethylene

Chemical	Species & Strain (Sex)	Highest Dose	Route	Vehicle	Result	Reference
Carbon Tetrachloride	C3H Mice (M&F)	0.04 ml 2 to 3 times weekly for 8-16 weeks	Oral	Olive Oil	Hepatoma	Edwards, et al., 1941
"	Strain L	0.04 ml 46x over a 4 month period (no	"	"	Hepatoma	Edwards, et al., 1941
"	Strain A Mice (M&F)	2.5g/kg to a total of 30 doses	Oral	Olive Oil	Hepatoma	Eschenbrenner & Miller, 1943
"	B4C3F1 Mice (M&F)	2500/mg/kg 5 X weekly for 78 wks	Oral	2-5% solution in corn oil	Hepatocellular carcinoma	NCI, 1976
"	C3H Mice	1.6 g/kg 3 X weekly for 10 wks	Oral	Corn Oil	Hepatoma	Kiplinger & Kensler, 1963
"	C3H Mice	0.04 ml 2 times weekly for 20-26 weeks	Rectal	Olive Oil	Liver Tumors	Confer & Stenger, 1965
"	Osborne-Mendel Rat (M&F)	160 mg/kg 5 X weekly 78 weeks	Oral	Corn Oil	Negative	NCI, 1976
"	Syrian golden hamsters	12.5 ul/hamster weekly no controls	Oral	Corn Oil	Liver cell carcinomas	Della Porta et.al., 1976
"	Japanese rat (M)	2g/kg bw 2 X weekly	s.c.	Corn Oil	Hepatocellular carcinoma	Reuber & Glover, 1970
"	Osborne-Mendel (M)	"	s.c.	"	Hepatocellular carcinoma	Reuber & Glover, 1970
"	Wistar - (M)	"	s.c.	"	Hepatocellular Carcinoma	Reuber & Glover, 1970

TABLE 1. (Continued)

Chemical	Species & Strain (Sex)	Highest Dose	Route	Vehicle	Result	Reference
Carbon Tetrachloride	Black rat (M)	2g/kg bw 2 X weekly	s.c.	Corn Oil	(Died early)	Rebuer & Glover, 1970
"	Sprague-Dawley (M)	"	"	"	"	Reuber & Glover, 1970
Trichloroethylene	B6C3F1 Mice (M&F)	2339 mg/kg TWA for 78 weeks	Oral	Corn Oil	Hepatocellular carcinoma	NCI, (1976)
"	Osborne-Mendel (M&F)	1097 mg/kg TWA for 78 weeks	Oral	Corn Oil	Negative	NCI, (1976)
"	ICR Mice (F)	0, 150 & 450 ppm in air 7 h	Inh.	Air	Pulmonary adenocarcinoma	Fukuda, et al., 1983
"	Sprague-Dawley rats (F)	0, 150 & 450 ppm in air 7 h per day, 5 days a week	Inh.	Air	Negative	"
"	Mice (M&F)	500 ppm for 18 months	Inh.	Air	Malignant lymphoma	Henschler et al., 1980
"	Rats (M&F)	500 ppm for 18 months	Inh.	Air	Negative	"
"	Hamsters (M&F)	500 ppm for 18 months	Inh.	Air	Negative	"
Tetrachloroethylene	B6C3F1 Mice (M&F)	1072 mg/kg TWA for 78 weeks	Oral	Corn Oil	Hepatocellular carcinoma	NCI, (1977)
"	Osborne-Mendel (M&F)		Oral	Corn Oil	Negative	NCI, (1977)
"	Sprague-Dawley (M&F)	600 ppm in air for 12 months	Inh.	-	Negative	Rampy et al., 1978

TABLE 2. Evidence for Genotoxicity of Trichloroethylene and Tetrochloroethylene

Chemical	Test System	Special Conditions	Highest Dose Tested	Result	Reference
Trichloroethylene	S. Typhimurium	TA100 +/- mouse and rat S-9 fraction in desiccators	1.5 ml in desiccator 2.5 ml cytotoxic	Pos. req. S-9	Simmon et al., 1977
"	S. Typhimurium	TA1950 TA1951 TA1952 TA1535 TA1538 TA100 TA98 Spot test no S-9 used in vitro	0.05 ml/plate	Neg. neg. neg. pos. pos. neg. neg.	Cerna & Kypenova 1977
"	"	TA1950 TA1951 TA1952 Host mediated	LD_{50} & 1/2 LD_{50}	pos. pos. pos.	Cerna & Kypenova 1977
"	E.Coli K12	gal + arg + nad + MTR (forward mutation) +/- mouse microsomes phenobarbital pre-treated	3.3 mM in media	arg + pos. req. metabolic activation	Greim et al., 1975
"	Covalent binding in rat hepatic microsomes	In Vitro		pos.	Van Duuren & Banerjee, 1976
"	Covalent binding to tissue proteins in Wistar rats.	In Vitro	1000 ppm for 5 hours. (approx. the same level of binding seen with vinyl chloride).	pos.	Bolt and Filser 1977

TABLE 2. (Continued)

Chemical	Test System	Special Conditions	Highest Dose Tested	Result	Reference
Trichloroethylene	Covalent binding to protein in rat liver microsomes	In vitro	Unlike vinyl chloride binding is not specific to SH containing protein.	pos.	Bolt and Filser 1977
"	S. Typhimurium TA100	+/- Aroclor 1254 induced rat liver S-9 fraction	100 ug/ml in top agar	neg. with and without S-9	Henschler et al., 1977
"	Saccharomyces cerevisiae XV185-14C	+/- mouse liver S-9 fraction	20 ul/ml	Pos. with S-9 only	Shahin & Von Borstel 1977
"	Fischer Rat F1706 embryo cell system		1.1×10^4 M	Pos.	Price et al., 1977
"	Covalent binding to hepatic proteins in the S.D. Rat	In vivo and In vitro		Pos.	Allemand, et al., 1978
"	Saccharomyces cerevisiae D7	Suspension test, +/- mouse liver 10,000 x g supernatant	40 mM	Pos. point mut. and give conversion required S-10 to be active	Bronzetti, et al., 1978

TABLE 2. (Continued)

Chemical	Test System	Special Conditions	Highest Dose Tested	Result	Reference
Trichloroethylene	Saccharomyces cerevisiae D7	Host mediated	400 mg/kg	pos. ilv reversion + trp conversion	Bronzetti, et al., 1978
=	=	=		Pos. at ade and trp loc.	=
=	D4	=	400 mg/kg		=
=	=	Host mediated			=
=	D4	Host mediated	3700 mg/kg (22 repeated doses of 150 mg/kg + 400 mg/kg day of sacrifice).	pos. trp and ade loci	
	Schizosaccharomyces pombe	P1 +/- mouse and rat liver S-9 fraction pretreated phenobarbital or B-naphthoflavone	22 mM	neg. at ade loci 1,3,4, 5 & 9	Rossi, et al. 1983
=	Schizosaccharomyces pombe	P1 Host mediated	2 g/kg	neg.	Rossi, et al. 1983
=	DNA binding in in vitro & in vivo	Calf thymus, in vitro mouse, in vivo		pos. in vitro neg. in vivo	Bergman, 1983
Tetrachloroethylene	E. Coli K12	gal + arg + nad + MTR +/- Mouse microsomes, phenobarbital pretreated	0.9 mM in media	neg.	Greim et al. 1975

392

TABLE 2. (Continued)

Chemical	Test System	Special Conditions		Highest Dose Tested	Result	Reference
Tetrachloroethylene	S. typhimurium	Spot test-no S-9 used in vitro	TA1950 TA1951 TA1952 TA1535 TA1538 TA100 TA98		neg. neg. neg. neg. neg. pos. neg.	Cerna & Kypenova, 1977
"	S. typhimurium	Host mediated Female ICR mice	TA1950 TA1951 TA1952	LD_{50} and 1/2 LD_{50}	Pos. no dose response evident	Cerna & Kypenova, 1977
"	Macromolecular binding in rat and liver tested in vivo			500 mg/kg (oral dose)	Binding to macro-molecules but no binding detected in purified DNA	Schumann et al. 1980
"	Saccharomyces cerevisiae	Suspension assay +/- mouse liver S-9 Host mediated assay	D7	85 mM 11 g/kg (single dose) 2g/kg x 12 + 4 g/kg	Neg. Neg. Neg.	Bronzetti et al. 1983

hamsters. More recently, Fukuda et al. (1983) were able to show a small
increase in the incidence of pulmonary adenocarcinomas in female ICR Swiss
mice exposed to 150 and 450 ppm TCE (16 and 15%, respectively, versus a
control incidence of 2%) in air 7 h per day, 5 days a week for 104 weeks. No
evidence of carcinogenic activity was observed in female Sprague-Dawley rats
subjected to the same treatment. Consequently, evidence that TCE is
carcinogenic is limited to mice, although three different tumors have been
identified in each of the mouse strains that have been studied.

As a result of the initial observation that TCE was carcinogenic in mice
and because it comes from the same chemical class as vinyl chloride
(chloroethylene), an established human carcinogen, there have been many
studies of the compound's genotoxic properties. Simmon et al. (1977) reported
that TCE increased the reversion rate of S. typhimurium strain TA100 only in
the presence of a 9000 X g supernatant of homogenates made of mouse or rat
liver. The experiments in this study were conducted in a desiccator and
avoided the volatilization of TCE that is encountered in the standard plate
assay of the Ames test when such precautions are not taken. Cerna and
Kypenova (1977) were unable to demonstrate mutagenic activity of TCE in the
spot test utilizing a variety of strains of S. typhimurium in the absence of a
metabolic activation system, but were able to demonstrate positive activity in
host-mediated assays in female ICR mice using strains TA1950, TA1951 and
TA1952. Greim et al. (1975) demonstrated that TCE was capable of more than
doubling the spontaneous mutation rate at the arg+ locus of E. coli strain K12
after incubation with a liver microsomal preparation taken from mice that had
been pretreated with phenobarbital. No activity was attributable to TCE in
the absence of metabolic activation. Henschler et al. (1977) failed to
confirm the mutagenic activity of TCE in S. typhimurium strain TA100 both in
the presence and absence of a metabolic activation system derived from Aroclor
1254 pretreated rats.

In addition to studies in bacterial systems, TCE has been shown to induce
mutations in one yeast test system by two different research groups (Shahin &
von Borstel, 1977; Bronzetti et al., 1978) either in the presence of a
metabolic activation system or in host-mediated assays, but negative under
similar circumstances in another (Rossi et al., 1983). Price et al. (1978)
were able to demonstrate that TCE was capable of transforming Fischer rat
embryo cells. Transformed cells were capable of growth in semisolid agar and
produced undifferentiated fibrosacromas when inoculated into newborn Fischer
rats. Henschler et al., (1977) suggested that the mutagenic activity and the
carcinogenic activity of TCE might be accounted for by the presence of

epichlorohydrin and 1,2-epoxibutane. However, these chemicals produced a mutagenic effect in the absence of a metabolic activation system, an observation that was not observed in prior positive tests of TCE by other authors.

It has been clearly shown that TCE is metabolized to a form capable of covalently interacting with tissue proteins at a level comparable to that observed with vinyl chloride (Van Duuren & Banerjee, 1976; Bolt & Filser, 1977; Allemand et al. 1978). Only very low levels of interaction of TCE with DNA in vivo have been demonstrated (Stott et al., 1982; Bergman, 1983). Under in vitro conditions and in the presence of a metabolic activation system, a higher degree of binding of a TCE metabolite to calf thymus DNA has been observed (Bergman, 1983).

The evidence that TCE is carcinogenic was considered to be limited by IARC (1979d). At that time the evidence that TCE has genotoxic properties was obscured by reports of impurities present in the technical grade solvent (Henschler et al., 1977). However, the mutagenic activity associated with impurities have always shown activity in the absence of metabolic activation, whereas samples of TCE have been routinely negative in the absence of metabolic activation. As indicated above, positive results have been specifically observed under circumstances where metabolic activation was possible in E. coli, S. typhimurium, and Saccharomyces cerevisiae. Moreover demonstration of transformation in Fischer rat embryo cells supports the notion that TCE is able to act as a carcinogen. Negative results using bacterial and yeast systems in vitro appear to involve failure to consider the volatility of TCE or employed lower concentrations than utilized in positive studies. The reasons for the discrepancy in results using two yeast strains in host-mediated assays is not as clear.

The lack of strong evidence of TCE interaction with DNA in vivo (Stott et al., 1982; Bergman, 1983) presents a problem in invoking a genotoxic event as being responsible for the carcinogenic response. The difficulty in demonstrating such an interaction is in sharp contrast to that observed with vinyl chloride, a closely related chemical (Green and Hathway, 1978). On the other hand, it is difficult to completely rule out the possibility that the minor interaction observed (Stott et al., 1982) could be responsible for some of the apparent genotoxic effects of TCE.

The observations of pulmonary adenocarcinomas (Fukuda et al., 1983) and malignant lymphoma (Henschler et al., 1980) in addition to hepatocellular carcinomas previously reported (NCI, 1976), indicate that TCE has some carcinogenic activity in mice. It should be pointed out, however, that these tumors have an appreciable background incidence in the strains of mice utilized. In the liver of mice, doses that produced liver tumors produced definite histological evidence of hepatocellular damage and a dose-related hypertrophic response to trichloroethylene (Stott et al., 1982). These data suggest that liver tumors in long-term studies may have resulted from repeated tissue damage and hypertrophy.

Tetrachloroethylene

To date, the evidence that tetrachloroethylene (perchloroethylene, PCE) is carcinogenic is limited to a study conducted by the National Cancer Institute (1977). The results of this study indicated that time weighted average doses of 536 and 1072 mg/kg to male and 386 and 772 mg/kg given to female B6C3F1 mice by gavage using a corn oil vehicle for 78 weeks resulted in the development of hepatocellular carcinomas. No evidence of increased tumor incidence was observed in Osborne-Mendel rats given time weighted average doses of 471 and 941 mg/kg in males and 474 and 949 mg/kg in females administered under the similar conditions in the same study. Rampy et al. (1978) also failed to observe increased tumor incidences in Sprague-Dawley rats exposed to 600 ppm TCE in air for a period of 12 months.

Evidence that PCE has genotoxic properties is also limited. PCE failed to increase the mutation frequency of E. coli K12 strain (Greim et al. 1975) in the presence or absence of mouse liver microsomes. It was reported (in abstract form) to produce an increase in the reversion rate of S. typhimurium strain TA100 in the absence of metabolic activation in the spot test, but was inactive in all other strains employed (Cerna & Kypenova, 1977). On the other hand, these same authors indicated that PCE was positive in a host-mediated assay utilizing S. typhimurium strains TA1950, TA1951 and TA1952 in female ICR mice at the LD50 and 1/2 of the LD50 in the same abstract. There was no evidence of dose-response. Bronzetti et at. (1983) were unable to demonstrate an effect of PCE using Saccharomyces cerevisiae strain D7 in a suspension assay in the presence or absence of mouse liver S-9 fraction or in a host-mediated assay. It is notable that these same authors had previously demonstrated positive results with TCE under both of these conditions (Bronzetti et al. 1978).

The IARC (1979e) evaluated the evidence that PCE was carcinogenic and found only limited evidence that it was active in mice. There is no substantive basis on which to modify that opinion at the present time.

BY-PRODUCTS OF TREATMENT PROCESSES

The trihalomethanes were the first chemicals that were recognized to be by-products of the chlorination of drinking water (Bellar et al., 1974; Rook, 1974). Trichloromethane (chloroform), dichlorobromomethane, dibromochloromethane, and tribromomethane (bromoform) are the most commonly encountered trihalomethanes found in drinking water. Iodinated derivatives are less commonly observed. A recent, comprehensive review of its carcinogenic, mutagenic, and teratogenic effects has been published by Davidson et al., (1982). The present section will primarily concern itself with data that has become available since this earlier review.

More recently, it has become quite apparent that the trihalomethanes (THM) are only one class of by-products produced in the chlorination of drinking water. Many of the non-THM by-products have been shown to be mutagenic in bacterial systems, but of these only the haloacetonitriles and chlorinated phenols have been shown carcinogenic in experimental animals. Thus, only these two additional classes will be covered in the present review. However, chlorine is known to increase mutagenic activity in drinking water (e.g. Zoeteman et al. 1982; Kool et al. 1982; Meier and Bull, 1985) and in reactions with humic acid, (Meier et al. 1983; Coleman et al., 1984; Meier et al., 1985). On the other hand, it should be remembered that chlorination is the only practice extensively studied. Many by-products remain to be identified with chlorine as well as other alternative modes of disinfection and even those which have been identified have been poorly characterized toxicologically.

In addition to by-products of drinking water disinfection, other treatment processes have the potential for the introduction of carcinogens into drinking water. In this light the carcinogenicity of acrylamide is discussed for illustrative purposes.

Chloroform

A study sponsored by the National Cancer Institute (NCI, 1976) demonstrated that B6C3F1 mice of both sexes developed hepatocellular carcinomas while male Osborne-Mendel rats had increased yields of renal tumors following chronic exposure to chloroform (Table 3). Chloroform was administered to both mice and rats in a corn oil vehicle by stomach tube.

Because of the primary source of chloroform in drinking water is side reactions of chlorination and the necessity for producing microbiologically safe drinking water a more thorough examination of the dose response relationships involved in chloroform-induced cancer was undertaken (Jorgenson et al. 1985). The doses used in this study was extended to a lower range and

TABLE 3. Evidence for Carcinogenicity of Chloroform in Experimental Animals

Species and Strain	Vehicle	Highest Dose Tested	Route	Tumor	References
A Mice	Olive Oil	2350 mg/kg	oral	Liver	Eschenbrenner & Miller, 1945
C57BL Mice	Toothpaste	60 mg/kg/d	oral	none	Roe et al., 1979
CBA Mice	Toothpaste	60 mg/kg/d	oral	none	"
CF/1	Toothpaste	60 mg/kg/d	oral	none	"
ICI Mice	Toothpaste	60 mg/kg/d	oral	Kidney	"
Sprague-Dawley Rat	Toothpaste	60 mg/kg/d	oral	none	"
Beagle Dog	Toothpaste	20 mg/kg/d	oral	none	"
"	"	"	"	Total Neoplasms	Reuber, 1979
Osborne-Mendel Rat	Corn Oil	180 mg/kg/d	oral	Kidney	NCI, 1976
"	"	"	"	Thyroid	Reuber, 1979
"	"	"	"	Cholangiofibromas	"
"	"	"	"	Cholangiocarcinomas	"
B6C3F1 Mice	Corn Oil	477 mg/kg/d	oral	Liver	NCI, 1976
"	"	"	"	Lymphoma	Reuber, 1979
Osborne-Mendel Rat	Drinking Water	130 mg/kg/d	oral	Kidney	EPA-NCI Study.
B6C3F1 Mice	Drinking Water	400 mg/kg/d	oral	None	EPA-NCI Study.

the group sizes were substantially expanded. Chloroform was administered in the drinking water. This design was to avoid problems that may have been associated with gavage dosing and the corn oil vehicle. The expanded group size at low doses would provide a better basis for extrapolation of the results to estimate human risks. Serum and tissue biochemical measurements were made in separate groups of animals on parallel treatments. Additional control groups were utilized for both species in which were restricted in their water consumption to the amount consumed by the high dose groups for each species. This provided a check on the effects that might be secondary to decreased water consumption. Only female B6C3F1 mice and male Osborne-Mendel rats were included in the experiment as these were the sexes which gave the greatest response in the NCI (1976) study.

In the male Osborne-Mendel rat very similar results were observed in the Jorgenson et al. study (1985) as had been obtained in the prior NCI study (1976). In particular, there was a dose-related increase in renal tubular adenomas and adenocarcinomas. The results obtained with the female B6C3F1 mice were, however, substantially different than observed in the original NCI study. There was a complete absence of any dose-related increases in the incidence of hepatocellular carcinomas in mice. This result was particularly significant, since the daily doses of chloroform in the two high exposure groups bracket the low dose which gave rise to an 80% incidence of hepatocellular carcinomas in the earlier NCI study.

These results are interesting in the context of evidence of chloroform carcinogenicity that was obtained prior to the NCI study (Table 3). Eschenbrenner and Miller (1945) noted that hepatocarcinogenic effects of chloroform were confined to doses which produced frank liver necrosis. Studies quoted in Table 3 that failed to produce increases in tumor incidence involved dose levels lower than those used in either the NCI (1976) or Jorgenson et al. (1985) studies. Roe et al. (1979), however, did find increased incidence of renal tumors in ICI mice when chloroform was administered in a toothpaste base at a dose of 60 mg/kg per day. Renal tumors were not produced in the C57BL, CBA, and CF/1 strains at the same dose in the same study. No liver tumors were not observed in any of these mouse strains.

It has been argued that the carcinogenicity was secondary to tissue necrosis followed by regenerative hyperplasia largely because of the early observations of Eschenbrenner and Miller (1945). Chloroform at doses found capable of inducing tumors produces substantial increases of tritiated thymidine incorporation into DNA of both the liver and kidney of the B6C3F1 mouse (Reitz et al., 1982). This result provides evidence of a regenerative response and is consistent with this hypothesis. Along the same line, Moore et al. (1982) provided evidence that doses of 60 mg/kg of chloroform in corn oil resulted in damage to the kidney and evidence of tubular regeneration as measured by tritiated thymidine uptake. These same doses administered in a toothpaste base were without effect. At 240 mg/kg similar effects were seen with both vehicles. This type of response was not strongly apparent in the liver or kidney of Osborne-Mendel rats.

In animals receiving parallel treatments to those used in the Jorgenson et al. study (1985) there was a dose-related increase in the liver fat content apparent in chloroform-treated mice (Jorgenson et al., 1982). This effect was observed only at much higher doses in the rat. Before the changes in liver fat occurred in the rat they were preceded by substantial a depression of serum triglycerides. Data on serum triglycerides of the mice under these conditions are not presently available. Based on the rat data one can speculate that the use of corn oil may have contributed to the development of hepatocellular carcinomas in the first NCI study through some type of interaction with the hepatoxic effects of chloroform.

The difficulty that has been encountered trying to demonstrate that chloroform possesses genotoxic properties (Table 4) indirectly supports the thesis that chloroform induced tumors are the result of overt tissue necrosis. A number of independent workers have failed to show a direct interaction of chloroform or a metabolite with DNA in the mouse or rat liver or kidney (Diaz-Gomez and Castro, 1980; Reitz, et al. 1982; Pereira et al. 1982). On the other hand, chloroform has been shown capable of inducing chromosome breakage and increased sister chromatid exhange frequencies in human lymphocytes in vitro and increased sister chromatid exchange in the bone marrow of mice treated with chloroform in vivo (Morimoto and Koizumi, 1983). It has also been found to enhance the transformation of Syrian Hamster Embryo (SHE) cells by S7 adenovirus (Hatch et al., 1983). It is notable that the concentrations of chloroform tested in these positive experiments tend to be higher than those utilized in similar experiments that had negative results (Kirkland et at., 1981). It must be said, however, that the exact mechanism by which chloroform induces cancer remains elusive. Attempts to directly demonstrate an alternative mechanism for chloroform-induced cancer have produced equivocal data. For example, Pereira et al. (1982) failed to clearly demonstrate either tumor initiating or tumor promoting activity in the rat liver.

In summary, the extent to which chloroform in drinking water represents a carcinogenic hazard to man is still uncertain. There is no doubt that chloroform is capable of producing renal tumors in at least one strain of mice and rats. Chloroform can also induce liver tumors in mice, but this effect seems dependent on the vehicle used. In view of the apparent dependence of the carcinogenic effects of chloroform in the mouse liver on the vehicle in which it was administered, it is suggested that chloroform's carcinogenic activity in the rat may be more appropriate for estimating carcinogenic risk in humans.

Haloacetonitriles

The dihalogenated chlorine and bromine derivatives of acetonitrile have been the principal species identified in chlorinated drinking water. (Trehy & Bieber, 1981; Oliver, 1983). Investigations into the carcinogenicity of these chemicals have been confined to the three members of the chlorinated series, bromochloroacetonitrile and dibromoacetonitrile. Simmon et al. (1977) first documented that dichloroacetonitrile (DCAN) was mutagenic in Salmonella typhimurium. Bull et al. (1985) confirmed the mutagenic properties of DCAN in Salmonella and demonstrated that bromochloroacetonitrile (BCAN) is mutagenic under similar circumstances. Chloroacetonitrile (CAN), trichloroacetonitrile (TCAN) and dibromoacetonitrile (DBAN) were found to be inactive in the presence or absence of a 9000 X g supernatant fraction (S-9) of a liver homogenate taken from rats previously treated with Arochlor 1254. All five haloacetonitriles induced sister chromatid exchange (SCE) in Chinese hamster ovary cells, in vitro, in the absence of rat liver S-9 fraction. None of these same five haloacetonitriles, however, were found capable of inducing mutagenic activity in the mouse micronucleus assay in vivo (5 male and 5 female CD-1 mice in each group) at doses of 0, 12.5, 25 and 50 mg/kg bw for 5 consecutive days in this same study.

Direct evidence of the carcinogenicity of the haloacetonitriles is limited to initiation/promotion experiments in the mouse skin. Experiments employing the oral route of administration (a total dose of 150 and 300 mg/kg bw split between 6 individual doses) followed by a 20 week promotion schedule with 12-O-tetradecanoyl-phorbol-13-acetate (TPA) failed to demonstate a significantly elevated incidence of either benign or malignant skin tumors within a 1 year observation period (Bull et al., 1984). On the other hand, higher doses applied topically (1200, 2400 and 4800 mg/kg, also split between 6 applications), yielded significantly higher tumor incidences with CAN, BCAN and DBAN. DCAN and TCAN produced small increases in cumulative tumor yields, but the changes were not significantly different from controls.

TABLE 4. Evidence for Genotoxic Activity of Chloroform

Test System	Special Conditions	Highest Dose Tested	Result	Reference
S. Typimurium TA1535	+/- Mouse & Rat Liver and kidney S-9	10 mg/plate	neg.	Van Abbe et al. 1982
" TA1537	"	"	"	"
" TA1538	"	"	"	"
" TA98	"	"	"	"
" TA100	"	"	"	"
Syrian Hamster Embryo Cells- Enhanced viral transformation	Applied in sealed chamber	0.5 ml/chamber	pos.	Hatch et al., 1983
V79 cells - 8-azaguanine	Applied in flow through system, then sealed	3% in air	neg.	Sturrock, 1977
Human Lymphocyte - Sister Chromatid Exchange Induction	+/- Rat Liver S-9	400 ug/plate	neg.	Kirkland et al., 1981
Human Lymphocyte Chromosome Damage	+/- Rat Liver S-9	400 ug/plate	neg.	"
E. Coli WP2p and WP2uvr A$^-$p	+/- Rat Liver S-9	10 mg/plate	neg.	"
Human Lymphocyte - Sister Chromatid Exchange, in vitro		6 mg/ml	pos.	Morimoto & Koisumi, 1983
Mouse Bone Marrow - Sister Chromatid Exchange, in vivo		4 x 200 mg/kg/1	pos.	"
Rat Liver, DNA - binding in vivo			neg.	Diaz-Gomez & Castro, 1980

TABLE 4. (Continued)

Test System	Special Conditions	Highest Dose Tested	Result	Reference
Sprague Dawley Rat, DNA binding in vivo	Liver	N.A.	neg.	Pereira et al., 1982
	Kidney	N.A.	neg.	"
B6C3F1 mice, DNA binding,	Liver	N.A.	neg.	"
B6C3F1 mice, DNA binding in vivo	Liver	N.A.	neg.	Reitz et al., 1982

A preliminary investigation of the haloacetonitriles ability to increase the incidence of lung tumors in strain A/J mice was reported by Bull and Robinson (1985). Doses of 10 mg/kg administered three times weekly for 8 weeks significantly increased lung tumor yields at 9 months of age with CAN, TCAN, and BCAN (Bull and Robinson, 1985). However, the tumor incidence was increased to only 32, 28 and 31% respectively. Control incidence of this spontaneous tumor was 10%. Although statistically significant, this small increase in lung tumors is difficult to interpret because of the variable background rate of this tumor in A/J mice.

These data indicate that certain members of the haloacetonitrile class do possess weak carcinogenic properties. This evidence has been confined to the topical route of administration and coupled to the subsequent application of a potent tumor promoting agent, TPA. There is no evidence available concerning the ability of these compounds to induce cancer by a systemic route of administration. Because of their relatively common occurrence as a result of the chlorination of drinking water it is critical that studies examining this issue be initiated.

Chlorinated Phenols

Chlorination of drinking water often gives rise to small quantities of chlorinated phenol derivatives. The principle products are 2-chlorophenol, 2,4-dichlorophenol and 2,4,6-trichlorophenol. There is substantive evidence of carcinogenic properties for only 2,4,6-trichlorophenol. The National Cancer Institute (NCI, 1979) conducted a feeding study of the effects of 5,000 and 10,000 ppm of 2,4,6-trichlorophenol in the diet of F344 rats and 5,214 and 10428 ppm in the diet of B6C3F1 mice. The incidence of lymphomas or leukemias was increased in male rats in a dose-related manner. Hepatocellular carcinomas were increased in both male and female mice and the incidences were dose-dependent.

Chlorinated Aldehyde and Ketone Derivatives

A number of chlorinated aldehydes and ketones have been identified in drinking water or related matrices following chlorination. A number of these by-products have been shown to be mutagenic (Bull and Robinson, 1985). Preliminary results from our studies, indicate that at least two of these chemicals, 2-chloropropenal and 1,3-dichloroacetone are capable of initiating tumors in mouse skin (unpublished results).

Other Treatment Chemicals

A wide variety of chemicals are necessary in the production of a safe and wholesome drinking water. For the most part these chemicals are generally regarded as safe and pose no substantive hazard to the health of the consuming public. However, if the synthesis of some products is improperly controlled they can contain contaminants that are of potential concern. An example of such products is the group of polymeric chemicals that are commonly used clarifying water, the coagulant aids.

The coagulant aids are high molecular weight polymers that are essentially non-toxic and are removed from the finished water by precipitation and filtration. However, the monomeric components used in the synthesis of these polymers are often reactive chemicals that are highly toxic and if sufficiently stable in water could be expected to be in the finished drinking water. One example of such a monomer is acrylamide.

Acrylamide has been recently shown capable of initiating papillomas and carcinomas in the skin of mice whether administered by the oral, intraperitoneal or topical routes of administration (Bull et al., 1984a). It was also shown to be capable of increasing the yield of lung adenomas in strain A/J mice in the same study. More recently, very similar results have been reported with ICR-Swiss mice (Bull 1984b).

Although acrylamide is inactive in the standard plate assay of the Ames' test (Bull et al., 1984a) it does increase the frequency of sister chromatid exchange in Chinese hamster ovary cells, in vitro (Bull et al., 1983). Shiraishi and Yamamoto (1978) and Shiraishi (1978) reported that acrylamide produces aneuploid and polyploid cells in both bone marrow cells and spermatogonia of mice treated with acrylamide by oral or intraperitoneal administration. Bull et al. (1983) observed dose-related increases in spermhead abnormalities in mice with acrylamide. More recently, Vanhorick and Moens (1983) found that acrylamide was a weak inducer of SV40 DNA amplification in SV40-transformed Chinese hamster cells as measured by in situ hybridization techniques. However, acrylamide synergistically enhanced the induction of SV40 DNA synthesis by a number of other carcinogens.

DISTRIBUTION SYSTEM

Contribution of toxic chemicals from the distribution system to finished drinking water can be considerable under certain circumstances. It is beyond the scope of this paper to deal comprehensively with this subject. However, it should be pointed out that certain products used in drinking water distribution are made of monomeric components that are recognized carcinogens. The most obvious example is polyvinyl chloride pipe which is a polymer of the well established human carcinogen vinyl chloride (IARC, 1979). Modern manufacturing techniques have reduced the monomer content of these products to negligible concentrations, but users of such products should carefully avoid substandard materials.

Another product that was widely used in lining of certain water mains, storage tanks and pipes are coal tar and asphaltic-based paints. The high polyaromatic hydrocarbon content of coal tar paints is partially, but not completely responsible for the high carcinogenic activity of these products in the mouse skin (Robinson et al., 1984). Although still containing carcinogenic components, the asphaltic based paints used in contact with potable water are approximately 3 orders of magnitude less potent than their coal tar analogs. Because of the extremely low water solubility of the polyaromatic hydrocarbons that are partially responsible for the carcinogenic activity of coal tar these chemicals would not be expected to be rapidly leached from coatings. However, the extent to which these coatings provide a relatively constant reservoir for the leaching of carcinogen from the surface of the distribution system has not been well studied. However, some information exists to suggest that as these coatings age they can begin to shed fine particulates containing high concentrations of polyaromatic hydrocarbons into the finished drinking water.

DISCUSSION

It is clear from the previous discussion that chemical carcinogens find their way into drinking water through a variety of means. Obviously, industrial contamination is only one and in most cases a minor source of chemicals in drinking water. Evidence that the chlorination by-products 2,4,6-trichlorophenol and the haloacetonitriles have carcinogenic properties indicate that we must look beyond the trihalomethanes as the only potential carcinogenic hazards associated with disinfection. However, whatever the carcinogenic risk these products might represent they must be balanced against the obvious reduction of disease from waterborne infectious disease that results from chlorination. Finally there is a clear need to carefully scrutinize products that are utilized in the treatment and distribution of drinking water since these products can be a major source of harmful chemicals presented at the tap for human consumption.

Despite the need to accept the general notion that carcinogenicity in experimental animals signals cause for concern about human exposures to the same chemical, it must be recognized that considerable uncertainty underlies the extrapolation of that information to man. These uncertainties are both

qualitative and quantitative and have been discussed in detail elsewhere
(IARC, 1982c). For example, if a chemical induces malignant tumors in more
than one species, particularly if it produces tumors that are rare in the
particular strain or species its carcinogenic properties are widely accepted
(IARC, 1982c). On the other hand, a chemical that increases the incidence of
a tumor that has a high spontaneous incidence in the test animal employed are
felt to provide only limited evidence of carcinogenicity. Certain of the low
molecular weight chlorinated hydrocarbons fall into this category. Other
chemicals from the same class, such as vinyl chloride, cause little
controversy because they produce malignant tumors in multiple sites and
different species. It should be recognized that this view does not suggest
that increased incidence of spontaneous tumors is of less concern than
induction of rare tumors. Rather it is a statement related to the wide
variety of factors that can modify the yield of a spontaneous tumor, whereas
induction of a rare tumor probably requires the properties of a complete
carcinogen.

It is ironic that most of the chemicals that occur frequently in drinking
water fall into a class of carcinogens whose carcinogenic effects are
controversial. In general, the controversy involves those chemicals which
induce liver tumors, in particular those whose activity is limited to the
induction of liver tumors in mice. Part of this concern is illustrated by the
fact that the yield of hepatocellular carcinomas in B6C3F1 mice can be
increased substantially by simple performance of a partial hepatectomy
(Newberne et al., 1982). The chemicals most directly involved in this
controversy are trichloroethylene and tetrachloroethylene, but also involve
considerations of the carcinogenicity of carbon tetrachloride and chloroform.

Carbon tetrachloride should in many ways be considered the prototype for
this group, although its carcinogenicity seems somewhat more generally
accepted than the other three compounds. As pointed out above, carbon
tetrachloride is capable of producing liver tumors in a variety of species.
Despite many attempts (many of which have not been published), it has been
virtually impossible to demonstrate any genotoxic activity with carbon
tetrachloride (IARC, 1979b; USEPA, 1984). However, the studies that have
demonstrated carbon tetrachloride induced liver cancer has involved very high
doses, from 1000 to more than 2000 mg/kg administered in an vegetable oil
vehicle 2 to 5 times weekly for extended periods of time.

The experimental results with trichloroethylene in many ways contrast
sharply with that obtained with carbon tetrachloride. Trichloroethylene has
been shown to only induce hepatocellular carcinomas in mice (NCI, 1976).

Although of a rather low potency, the genotoxic properties of trichloroethylene have been demonstrated by a number of investigators. In this respect, evidence of its carcinogenic activity would have to be considered stronger than that of carbon tetrachloride. However, the fact remains that trichloroethylene has only been shown capable of increasing the yield of neoplasms only in strains of mice which have high background incidence of these same neoplasms (Henschler et al., 1980; Fukuda et al. 1983; NCI, 1976b).

Examination of the livers of mice receiving chloroform in drinking water demonstrated a dose-related increase in liver fat content that extend to levels much lower than those utilized for the original NCI corcinogenesis study (Jorgenson et al., 1982). Similar changes were not observed in the livers of Osborne-Mendel rats until much higher levels of chloroform were administered in drinking water. These results suggest that corn oil was in some way involved in the liver pathology that develops in mice treated with chloroform. If so, this would imply that tumors induced by other hepatoxic agents may depend heavily on the corn oil vehicle commonly used in testing of non-polar chemicals. Such an interpretation would suggest that results that have been obtained with chemicals such as trichloroethylene and tetrachloroethylene may not be extrapolated to other species without appropriate qualification. In fact, these data also call into similar question the carcinogenicity of carbon tetrachloride since it has only been shown capable of producing liver tumors when administered in a vegetable oil vehicle.

Tetrachloroethylene increases hepatic DNA synthesis in B6C3F1 mice but not in rats following doses of 1000 mg/kg per day for 11 days (Schumann, et al., 1980). Similarly, Reitz et al. (1982) found substantial increases in DNA synthesis of the liver and kidney of B6C3F1 mice treated with single doses of 60 and 240 mg/kg of chloroform and much smaller increases in the liver and kidney of male Osborne-Mendel rats. These data indicate that the toxic effects of these chemicals in the mouse liver results in substantial tissue regeneration of dose levels utilized in the carcinogenesis bioassays of these chemicals. Consequently, the possibility exists that these chemicals are capable of increasing liver tumor development from spontaneously initiated cells in the same way that carbon tetrachloride has been shown to increase hepatomas in the liver of diethylnitrosamines initiated mice (Pound and McGuire, 1978).

These questions become extremely important when attempting to estimate the risk to human beings consuming drinking water. Presumably the damage that is produced by initiators of cancer (i.e. those that act through damage to DNA) produce a certain degree of irreversible or non-repairable damage that is directly related to the dose of carcinogen. Theoretically, this would lend itself to a linear relationship to dose at low response rates and risks at low concentrations can be reliably estimated from experiments conducted at high doses. Increased tumor incidence due to non-genotoxic effects of chemicals or interactions with nutritional state presumably involve reversible processes. Consequently, the assumption of a linear relationship with dose at low response rates may not be justified. This may substantially affect any estimate of the health hazard associated with the low concentrations of these chemicals that occur in drinking water that is based upon the high doses used in animal experiments.

The data available at present is not suited for projecting the risk to humans from the use of materials in the treatment and distribution of drinking water. Identification of carcinogenic properties of by-products of disinfection other than the trihalomethanes, acryamide and coal tar paints, however, does indicate that the use of such material in contact with potable water requires closer scrutiny. The industry should examine critically time-honored practices in the treatment and distribution of drinking water to gradually replace these materials with safer alternatives.

In summary, evidence that a chemical induces cancer in experimental animals does suggest that there is a potential risk to man regardless of the mechanism involved. Although it is certainly possible that a chemical that produces cancer in experimental animals would not do so in man, it is difficult to develop a prudent course of action on this assumption because the evidence which presently exists would suggest that the reverse is the general case. If tumorigenesis is dependent upon tissue damage, it should be remembered that these agents are also capable of inducing liver damage in humans. Consequently, it should be recognized that the argument presented regarding relationship of tissue damage to tumorigenicity is more of a quantitative consideration in the judgement of risks in man and less of a question of across-species extrapolation than discussions of the issue would ordinarily imply. Therefore, the appearance of all the above chemicals in drinking water is cause for concern and to be avoided wherever possible and practical.

REFERENCES

Allemand, H., Pessayre, D., Descatoire, V., Degott, C., Feldmann, G., and Benhamou, J.P., (1978) Metabolic activation of trichloroethylene into a chemically reactive metabolite toxic to the liver. J. Pharmcol. Exp. Therapeut., 204, 714-723.

Bellar, T.A., Lichtenberg, J.J. and Kroner, R.C., (1974) The occurrence of organohalides in chlorinated drinking waters. J. Am. Water Works Assn., 66, 703-706

Bergman, K., (1983) Interactions of trichloroethylene with DNA in vitro and with RNA and DNA of various mouse tissues in vivo. Arch. Toxicol., 54, 181-193.

Bieber, T.I. and Trehy, M.L., (1983) Dihaloacetonitriles in chlorinated natural waters. In: Jolley, R.L., et al. eds., Water Chlorination: Environmental Impact and Health Effects, 4, Ann Arbor MI: Ann Arbor Science Publi., Inc., pp 85-96.

Bronzetti, G., Zeiger, E., and Frezza, D., (1978) Genetic activity of trichloroethylene in yeast. J. Toxicol. Environ. Health, 1, 411-418.

Bronzetti, G., Bauer, C., Corsi, C., Del Carratore, R., Galli, A., Nieri, R., and Paolini, M., (1983) Genetic and biochemical studies on perchloroethylene in vitro and in vivo. Mutation Research 116, 323-331.

Bolt, H.M. and Filser, J.G., (1977) Irreversible binding of chlorinated ethylenes to macromolecules. Environ. Health Persp. 21, 107-112.

Bull, R.J., Meier, J.R. and Robinson, M. (1983) Evidence of genotoxic effects with acrylamide. Pharmacologist 25, 640.

Bull, R.J., Robinson, M., Laurie, R.D., Stover, G.D., Greisiger, E., Meier, J.R., and Stober, J., (1984a) Carcinogenic effects of acrylamide in sencar and A/J mice. Cancer Res. 44, 107-111.

Bull, R.J., Robinson, M. and Stober, J.A. (1984b) Carcinogenic activity of acrylamide in the skin and lung of Swiss-ICR mice. Cancer Letters 24, 209-212.

Bull, R.J., Meier, J.R., Robinson, M., Ringhand, P., Laurie, D. and Stober, J., (1985) Evaluation of mutagenic and carcinogenic properties of brominated and chlorinated acetonitriles: By-products of chlorination. Fundam. Appl. Toxicol. In press.

Bull, R.J. and Robinson, M. (1985) Carcinogenic activity of haloacetonitriles and haloacetone derivatives in the mouse skin and lung. In: Jolly R.L., et al. eds., Water Chlorination: Environmental Impact and Health Effects, 5, In Press.

Cameron, G.R. and Karunaratne, W.A.E., (1936) Carbon tetrachloride cirrhosis in relation to liver regeneration. J. Pathol. Bacteriol. 42, 1-21.

Cerna, M. and Kypenova, H., (1977) Mutagenic activity of chloroethylenes analyzed by screening system tests. Mutat. Res. 46, 214-215.

Davidson, I.W.F., Sumner, D.D. and Parker, J.C., (1982) A Review of its metabolism, teratogenic, mutagenic and carcinogenic potential. Drug and Chemical Toxicol. 5, 1-87.

Diaz-Gomez, M.I. and Castro, J.A., (1980) Covalent binding of chloroform metabolites to nuclear proteins - no evidence for binding to nuclei acids. Cancer Lett. 9, 213-218.

Eschenbrenner, A.B., and Miller, E., (1943) Studies on hepatomas. I. Size and spacing of multiple doses in the induction of carbon tetrachloride hepatomas. J. Natl. Cancer Inst. 4, 385-388.

Eschenbrenner, A.B., and Miller, E., (1945) Induction of hepatomas in mice by repeated oral administration of chloroform with observations on sex differences. J. Natl. Cancer Inst. 2, 251-255.

Eschenbrenner, A.B., and Miller, E., (1946) Liver neurosis and the induction of carbon tetrachloride hepatomas in strain A mice. J. Natl. Cancer Inst. 6, 325-341.

Fukuda, K., Takemoto, K. and Tsuruta, H., (1983) Inhalation carcinogenicity of trichloroethylene in mice and rats. Industr. Hyg. 21, 243-254.

Glende, E.A., Jr., Hruszkewycz, A., and Recknagel, R.O., (1975) Critical role of lipid peroxidation in carbon tetrachloride-induced loss of aminopyrene demethylase, cytochrome P-450, and glucose-6-phosphatase. Biochem. Pharmacol. 25, 2163-2170.

Green, T. and Hathway, D.E., (1978) Interactions of vinyl chloride with rat-liver DNA in vivo. Chem.-Biol. Interactions. 22, 211-224.

Greim, H., Bonse, G., Radwan, Z., Reichert, D., and Henschler, D., (1975) Mutagenicity in vitro and potential carcinogenicity of chlorinated ethylenes as a function of metabolic oxirane formation. Biochem. Pharmacol. 24, 2013-2017.

Hatch, G.G., Mamay, P.D., Ayer, M.L., Casto, B.C., and Nesnow, S., (1983) Chemical enhancement of viral transformation in Syrian hamster embryo cells by gaseous and volatile chlorinated methanes and ethanes. Cancer Res. 43, 1945-1950.

Hayes, et al., (1984) 90-day studies with carbon tetrachloride in rats via oral gavage. Manuscript in preparation.

Henschler, D., Eder, E., Neudecker, T. and Metzler, M., (1977) Carcinogenicity of trichloroethylene: Fact or artifact? Arch. Toxicol. 37, 233-236.

Henschler, D., Romen, W., Elsasser, H.M., Reichert, D., Eder, E. and Radiran, Z. (1980) Carcinogenicity study of trichloroethylene by long term inhalation in three animal species. Arch. Toxicol. 43, 237-248.

IARC (1977) IARC Monographs on the Evaluation of the Carcinogenic Risk of Chemicals to Humans, 15, Ethylene dibromide, Lyon, pp 195-209.

IARC (1979a) IARC Monographs on the Evaluation of Carcinogenic Risk of Chemicals to Man, 19, Vinyl chloride, Polyvinyl Chloride and Vinyl Chloride - Vinyl Acetate Copolymers. Lyon, pp 377-438.

IARC (1979b) IARC Monographs on the Evaluation of Carcinogenic Risk of Chemicals to Man, 20, Carbon Tetrachloride, Lyon, pp 371-399.

IARC (1979d) IARC Monographs on the Evaluation of Carcinogenic Risk of Chemicals to Man, 20, Trichloroethylene, Lyon, pp 545-572.

IARC (1979e) IARC Monographs on the Evaluation of Carcinogenic Risk of Chemicals to Man, 20, Tetrachloroethylene, Lyon, pp 491-514.

IARC (1979f) IARC Monographs on the Evaluation of Carcinogenic Risk of Chemicals to Man, 20, Chloroform, Lyon, pp 401-427.

IARC (1982) IARC Monographs on the Evaluation of the Carcinogenic Risk of Chemicals to Humans, Suppl. 4, Chemicals, Industrial Processes and Industries Associated with Cancer in Humans, Lyon, pp 7-24.

Jorgenson, T.A., Rushbrook, C.J. and Jones, D.C.L., (1982) Dose-response study of chloroform carcinogenesis in the mouse and rate: Status Report. Environ. Health Persp. 46, 141-149.

Jorgenson, T.A., Meierhenry, E.F., Rushbrook, C.J., Bull, R.J. and Robinson, M. (1985) Carcinogenicity of chloroform in drinking water to male Osborne-Mendel rats and female B6C3F1 Mice. Fundam. Appl. Toxicology, In press.

Kiplinger, G.F., and Kensler, C.J., (1963) Failure of pheonoxybenzamine to prevent formation of hepatomas after chronic carbon tetrachloride administration J. Natl. Cancer Inst., 30, 837-843.

Kirkland, D.J., Smith, K.L., and Van Abbe, N.J., (1981) Failure of chloroform to induce chromosome damage or sister-chromatid exchanges in cultured human lymphocytes and failure to induce reversion in Escherichia coli. Fd. Cosmet. Toxicol., 19, 651-656.

Laskin, S., Sellakumar, A.R., Kuschner, M., Nelson, N., LaMendala, S., Rusch, G.M., Katz, G.V., Dulak, N.C., Albert, R.E., (1980) Inhalation carcinogenicity of epichlorohydrin in non inbred Sprague-Dawley rats., J. Natl. Cancer Inst., 65, 751-757.

Moore, D.H., Chasseaud, L.F., Majeed, S.K., Prentice, D.E., Roe, F.J.C., and Van Abbe, N.J. (1982) The effect of dose and vehicle on early tissue damage and regenerative activity after chloroform administration to mice. Fd. Chem. Toxic., 20, 951-954.

Morimoto, K. and Koizimu, A., (1983) Trihalomethanes induce sister chromatid exchanges in human lymphocytes in vitro and mouse bone marrow cells in vivo. Environ. Research., 32, 72-79.

NCI (1976a) National Cancer Institute Carcinogenesis Bioassay of Chloroform. NTIS No. PB264018/AS.

NCI (1976b) National Cancer Institute Carcinogenesis Bioassay of Trichloroethylene. NCI-CG-TR-2, NIH-76-802.

NCI (1977) National Cancer Institute Bioassay of Tetrachloethylene for Possible Carcinogenicity. (NIH) 77-813.

NCI (1979) Bioassay of 2,4,6-trichlorophenol for possible carcinogenicity. DHEW Publ. No. (NIH) 79-1711.

Newberne, P.M., de Camrargo, J.L.V., and Clark, A. (1982) Chloine deficiency partial hepatectomy and liver tumors in rats and mice. Toxicol. Pathol., 10, 95-109.

Newberne, P.M., Weigert, J., and Kula, N., (1979) Effects of dietary fat on hepatic mixed-function oxidases and hepatocellular carcinoma induced by aflatoxin B_1 in rats. Cancer Res. 39, 3986-3991.

Oliver, B.G., (1983) Dihaloacetonitriles in drinking water: Algae and fulvic acid as precursors. Env. Sci. Technl., 17, No. 2, 80-83.

Pereira, M.A., Lin, L.-H.C., Lippitt, H.M. and Herren, S.L. (1982) Trihalomethanes as initiators and promoters of carcinogenesis. Environ. Hlth. Persp., 46, 151-156.

Pound, A.W. and McGuire, L.J., (1978) Influence of repeated liver regeneration on hepatic carcinogenesis by diethylnitrosamine in mice. Br. J. Cancer, 37, 595-602.

Price, P.J., Hassett, C.m, Mansfield, J.I. (1978) Transforming activity of trichloroethylene and proposed industrial alternatives. In Vitro, 14, 290-293.

Rampy, L.W., Quast, J.F., Balmer, M.F., Leong, B.K.J. and Gehring, P.J. (1978) Results of a long-term inhalation toxicity study in rats of a perchloroethylene (tetrachloroethylene) formulation. Toxicology Research Lab., Dow Chemical Co., Midland, MI 48640.

Reitz, R.H., Fox, T.R. and Quast, J.F., (1982) Mechanistic considerations for carcinogenic risk estimation: chloroform. Environ. Hlth. Persp. 46, 163-168.

Reuber, M.D. and Glover, E.L. (1970) Cirrhosis and carcinoma of the liver male rats given subcutaneous carbon tetrachloride. J. Natl. Cancer Inst., 44, 419-423.

Reuber, M.D. and Glover, E.L., (1976a) Hyperplastic and early neoplastic lesions of the liver in buffalo strain rats of various ages given subcutaneous carbon tetrachloride. J. Natl. Cancer Insti., 38, 891-895.

Reuber, M.D. and Glover, E.L. (1976b) Cholangiofibrosis in the liver of Buffalo strain rats injected with carbon tetrachloride. Exp. Pathol. 48, 319-322.

Reuber, M.D. (1979) Carcinogenicity of chloroform. Environ. Hlth. Persp. 31, 171-182.

Robinson, M., Bull, R.J., Munch, J. and Meier, J. (1984) Comparative carcinogenic and mutagenic activity of coal tarr and petroleum asphalt paints used in potable water supply systems. J. Appl. Toxicol., 4, 49-56.

Roe, F.J.C., Palmer, A.K., Worden, A.N., Van Abbe, N.J., (1979) Safety evaluation of toothpaste containing chloroform. I. Long-term studies in mice. J. Environ. Path. & Toxicol., 2, 799-819.

Rook, J.J. (1974) Formation of haloforms during chlorination of natural water. *Water Treat. Examin.*, 23, 234-243.

Rossi, A.M., Migliore, L., Barale, R. and Loprieno, N., (1983) *In vivo and in vitro* mutagenicity studies of a possible carcinogen, trichloroethylene, and its two stabilizers, epichlorohydrin and 1,2-epoxybutane. *Teratagenesis Carcinogenesis & Mutagenesis*, 3, 75-87.

Shahin, M.M. and von Borstel, R.C. (1977) Mutagenic and lethal effects of -benzene hexachloride, dibutyl phthalate and trichloroethylene in *Saccharonyces cerevisiae*. *Mutat. Res.*, 48, 173-180.

Shiraishi, Y. (1978) Chromosome aberations induced by monomeric acrylamide in bone marrow and germ cells of mice. *Mutat. Res.*, 57, 313-324.

Shiraishi, Y. and Yamamoto, K., (1978) Chromosome aberrations induced by monomeric acrylamide in germ cells of mice. *Proc. Japan Acad.*, 54, Ser. B., 272-276.

Simmon, V.F., Kauhanen, K., Tardiff, R.G. (1977) Mutagenic activity of chemicals identified in drinking water. *Progress in Genetic Toxicology*, pp 249.

Stott, W.T., Quest, J.F and Watanabe, P.G. (1982) The pharmacokinetics and macromolecular intention of trichloroethylene in mice and rats. *Toxicol. Appl. Pharmacol.* 62, 137-151.

Trehy, M.L. and Bieber, T.I., (1981) *Detection, identification and Quantitative analysis of dihaloacetonitriles in chlorinated natural waters.* In: Keith, L.H., ed. Vol. 2, *Advances in the Identification and Analysis of Organic Pollutants in Water*., (Ann Arbor, MI: Ann Arbor Science Publ., Inc.), pp, 941-975.

USEPA (1984) *Research and Development Health Assessment Document for Carbon Tetrachloride*, EPA 600/8-82-001.

Van Duuren, B.L. and Banerjee, S., (1976) Covalent interaction of metabolites of the carcinogen trichloroethylene in rat hepatic microsomes. *Cancer Research*, 3, 2419-2422.

Vanhovick, M. and Moens, W. (1983) Carcinogen-mediated induction of SV40 DNA amplification is enhanced by acrylamide in Chinese hamster CO60 cells. *Carcinogenesis*, 4, 1459-1463.

Weisburger, J.H. and Williams, G.M. (1983) The distinct health risk analysis required for genotoxic carcinogens and promoting agents. *Environ. Hlth. Persp.*, 50, 233-245.

Withey, J.R., Collins, B.T. and Collins, P.G., (1983) Effect of vehicle on the pharmacokinetic and uptake of four halogenated hydrocarbons from the gastrointestinal tract of the rat. *J. Appl. Toxicol.* 3, 249-253.

EFFECT OF TRICHLOROETHYLENE ON THE EXPLORATORY AND LOCOMOTOR ACTIVITY OF RATS EXPOSED DURING DEVELOPMENT

Douglas H. Taylor, Kirk E. Lagory, Daniel J. Zaccaro, Ronald J. Pfohl and R. Dana Laurie. Department of Zoology, Miami University, Oxford, Ohio 45056 and Health Effects Research Laboratory, USEPA, Cincinnati, Ohio 45268 USA

ABSTACT

Trichloroethylene (TCE) is a common contaminant of underground water supplies. To examine the effect of TCE on the developing central nervous system, rats were exposed to TCE throughout gestation until 21 days postpartum via their dams' drinking water. TCE concentrations of 312 mg/l, 625 mg/l and 1250 mg/l were tested. Exploratory behavior was higher in 60- and 90-day old male rats which were exposed to any level of TCE. The effect of TCE-exposure on locomotor activity (running wheel) was also examined in 60-day old males (625 and 1250 ppm exposure groups). Locomotor activity was significantly higher in rats exposed to 1250 ppm TCE. These data suggest that TCE has long-term effects on behaviour.

We thank P.A. Wuest, V. Steel and J.D. Cmielewski for their help throughout this study. This research was supported under cooperative agreement Cr809618 from the Health Effects Research Laboratory, US Environmental Protection Agency, Cincinnati, Ohio. This paper does not necessarily reflect EPA policy.

INTRODUCTION

Trichloroethylene (TCE) is an unsaturated chlorinated hydrocarbon which is commonly used as an industrial solvent (ref. 1). Historically, the most common route of TCE-intake by humans is via inhalation. Consequently, most research has focused on the effects of TCE vapors on physiological and behavioral processes and relatively little attention has been paid to the effects of TCE exposure via drinking water. However, it is now well established that TCE is a common drinking water contaminant (ref. 2, 3, 4, 5, 6).

TCE and its metabolites (e.g., trichloroethanol and trichloroacetic acid) are neurotoxic and affect many parts of the brain (ref. 7, 8, 9, 10). The neurotoxic properties of TCE are presumably the result of its effects on the lipid structure of nerve cell membranes (ref. 11) and its concamitant impairment of membrane permeability (ref. 12). TCE may hinder mobilization of neuro-transmitters and thus have a depressant effect upon conduction of nerve action potentials (ref. 13).

We chose to examine the effects of TCE in rats exposed during development to toxic insult due to their reduced ability to detoxify compounds. In addition, toxicants can have more serious, irreversible consequences if developmental processes are altered. TCE easily crosses the placenta and is found in fetal blood shortly after maternal exposure (ref. 14). Despite this, few studies have investigated the effects of prenatal and early postnatal exposure to TCE on behaviour. In our study, rat pups were exposed via their dam's drinking water during gestation and lactation, the period of the most active CNS development (ref. 14).

METHODS

Female Sprague-Dawley rats were exposed to TCE at several nominal concentrations (312, 625 and 1250 mg/l) in their drinking water. TCE was dissolved in distilled water without the use of a vehicle to avoid the possible confounding effects of the vehicle. Drinking water was provided ad libitum and was changed every 3 days. The degradation of TCE in the water bottles was determined using a gas chromatograph (Hewlett-Packard Model 5792) to estimate the dosage received by the dams. Experimental dams received TCE from 14 days prior to breeding, through gestation, until the pups were weaned at 21 days of age; control dams received distilled water. Thus, experimental pups were indirectly exposed prenatally and postnatally to TCE, or its byproducts. Litters were culled at birth to eight pups. Male rats derived from experimental and control dams were used as the subjects in this study.

Exploratory behaviour during a 15-min. sampling period was measured in an apparatus similar to that described by Berlyne (ref. 15; Fig. 1a). Individual rats were tested once at either 28, 60, or 90 days of age.

A residential cage apparatus (Fig. 1b) was used to monitor wheel-running, feeding, and drinking behaviour in rats for 24 hr/day from 55-60 days of age. Rats were allowed to acclimate to the residential cage for 4 days prior to the onset of data collection.

Statistical analyses were performed using the Statistical Analysis System (ref. 16). One-factor analyses of variance were used to test for treatment effects on the frequency of exploration and the number of beams crossed in the arena at 28, 60, and 90 days of age. Three-factor analyses of variance were used to analyze for treatment, time, and day effects on wheel-running, feeding and drinking in the residential cage. Only the dark part of the photoperiod was analyzed because rats were essentially inactive in the residential cage when the lights were on. Dark hours were pooled into 3-hr periods for analysis to account for ultradian rhythms in rat activity (ref. 17). Probability values ≤ 0.05 were considered significant.

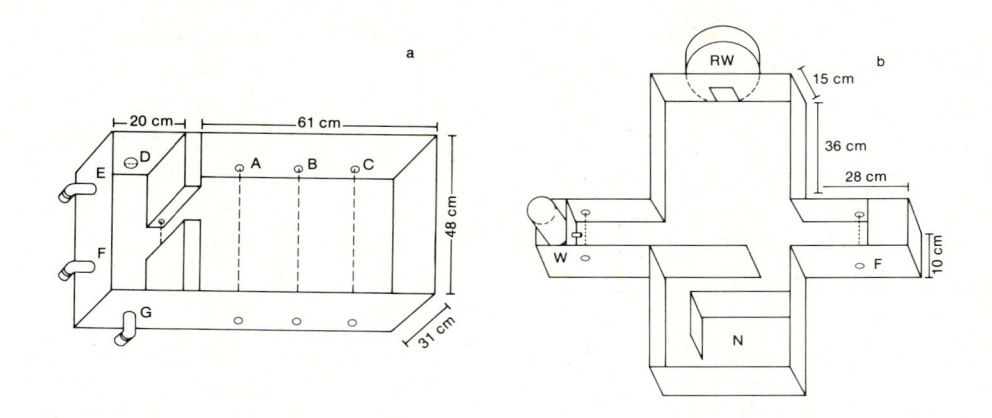

Fig. 1 (a). Exploratory apparatus. The apparatus consisted of 2 chambers: an arena crossed by 3 infrared beams (A-C) and an alcove with 4 exploratory ports bisected by infrared beams (D-G). Cups at the end of these ports contained used bedding from an adult male rat to provide an olfactory stimulus for exploration. A subject explored these ports by inserting its head into the opening.
(b). Residential cage. Infrared beams were positioned to monitor feeding (F) and drinking (W) behaviour; wheel revolutions (RW) were registered electro-magnetically. A nest area (N) was provided. Data from both apparatuses were recorded by a Hewlett-Packard 9825A desktop computer.

RESULTS

Total consumption of TCE during the experiment was calculated from mean water consumption values and the mean concentration of TCE in the drinking water of treated rats. Total consumption values were: 312 mg/l - 646 mg, 625 mg/l - 1102 mg, 1250 mg/l - 1991 mg.

At 28 days of age, no difference in the level of exploratory activity was seen among treatment groups (P = 0.70). However, at 60 and 90 days of age, rats exposed to TCE exhibited increased levels of exploration (see Fig. 2a and 2b). No treatment effects were detected at any age on the number of beams crossed in the arena (28 days: P = 0.68; 60 days: P = 0.24; 90 days: P = 0.29).

From 55-60 days of age, rats exposed to 1250 mg/l TCE were more active on the wheel in the residential cage than were controls or those exposed to 625 mg/l TCE (see Fig. 3). This difference among treatment groups was especially pronounced during the initial 3 h of darkness when rats at this age were most active (dose * time interaction: P = 0.001). No significant differences were detected among treatment groups for the levels or timing of feeding and drinking activities during the night (P 0.05).

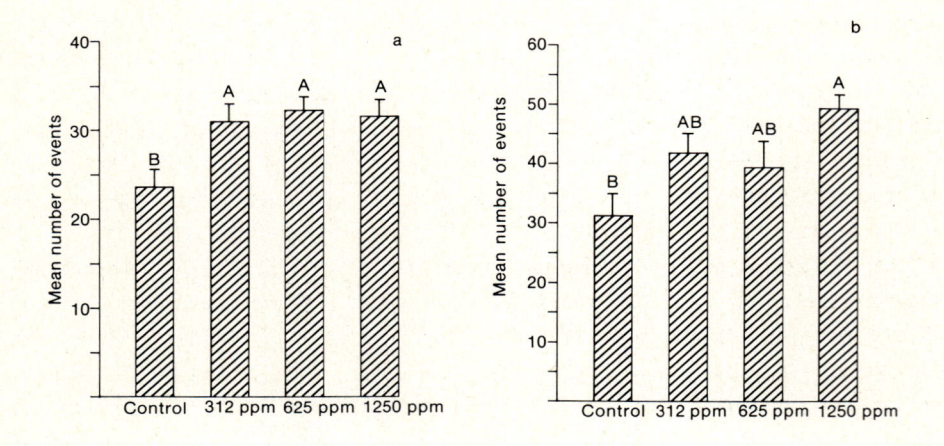

Fig.2. Mean number of exploratory events in 60- (a) and 90- (b) day old rats (+ 2 SE's). Means with the same letter are not significantly different. Treatment effect at 60 days: F = 2.97, P = 0.03; at 90 days: F = 4.7, P = 0.01.

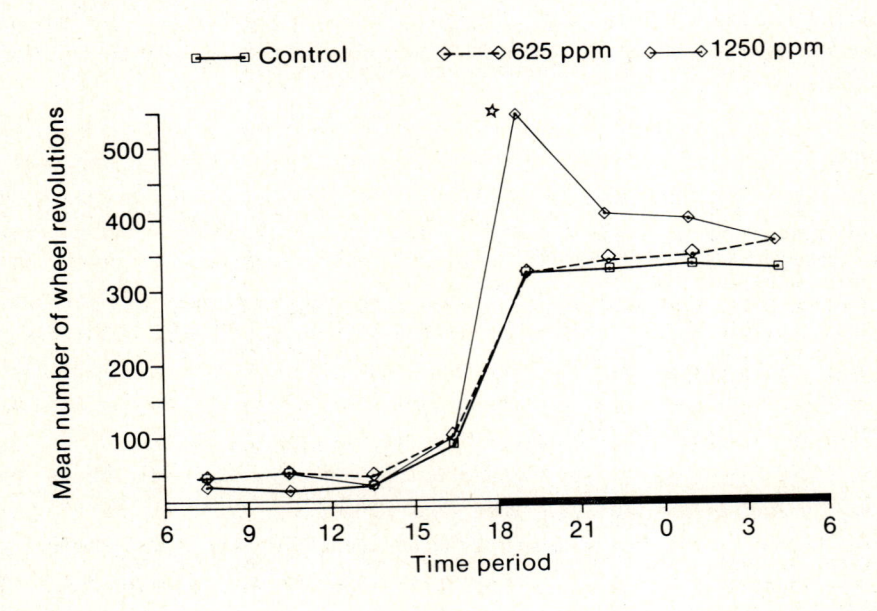

Fig. 3. Wheel-running in the residential cage. Rats exposed to 1250 mg/l TCE were more active during the initial period of darkness (*) than were controls or rats exposed to 625 mg/l.

DISCUSSION

Our data suggest that exposure to TCE during the period of active CNS development can lead to behavioral changes which are present at least 2 months after cessation of exposure. Exploratory behaviour appeared to be more sensitive than locomotor behaviour (wheel-running) to the effects of TCE. This is indicated by the increased levels of exploratory activity in rats exposed to any concentration of TCE 312 mg/l in contrast, wheel-running was altered only in rats exposed to 1250 mg/l TCE.

Other investigators have reported either increased activity or exploratory behavior (ref. 18, 19, 20) possibly as a result of reduction of fear or inhibition. Battig et al (ref. 21) found that adult rats subjected to long-term exposure to TCE vapors (360-420 mg/l for 8 hrs/day, 5 days/week for 44 weeks) exhibited increased levels of exploration relative to controls when tested during the 43rd and 44th week. Exploratory tests performed 5 to 6 days after cessation of exposure, however, indicated no differences between controls and treated animals. These data indicate that the behavioral changes following TCE exposure are reversible if exposure occurs during adulthood.

Behavioral changes after exposure to TCE are presumably linked to the CNS dysfunction documented in animals exposed to TCE. The differences between control and TCE-exposed rats observed in our study long after the termination of exposure indicates that TCE-mediated damage to the CNS during development may be permanent.

REFERENCES

1 E.M. Waters, H.B. Gerstner and J.E. Huff, J. Toxicol. Environ. Hlth., 2 (1977) 671-707.
2 EPA document # 400/5-80-077. Ambient water quality criteria for trichloroethylene, 1980.
3 G. Ziglio, Bull. Environ. Contam. Toxicol., 26 (1981) 131-136.
4 G. Bross, D. DiFranceisco and M.E. Desmond, Toxicology, 28 (1983) 283-294.
5 D.M. Loekle, A.J. Schecter and J.J. Christian, Bull. Environ. Contam. Toxicol., 30 (1983) 199-205.
6 G. Ziglio, G.M. Fara, G. Bettramelli and F. Pregliasco, Arch. Environ. Contam. Toxicol., 12 (1983) 57-64.
7 A.G. Baker, J. Neuropathol. Exp. Neurol., 17 (1958) 649-655.
8 V.J. Bartonicek and A. Brun, Acta. Pharmacol. Toxicol., 28 (1970) 359-369.
9 K.G. Haglid, P. Kjellstrand, L. Rosengren, A. Wronski and C. Briving, Arch. Toxicol., 43 (1980) 187-199.
10 K.G. Haglid, C. Briving, H.A. Hansson, L. Rosengren, P. Kjellstrand, D. Stavron, U. Swedin and A. Wronski, Neurotoxicol., 2 (1981) 659-673.
11 T. Kyrklund, C. Alling, K. Haglid and P. Kjellstrand, Neurotoxicol., 4 (1983) 35-42.
12 P. Kjellstrand, Toxicol. Letters, 14 (1982) 97-101.
13 R.D. Kennedy and A.D. Galindo, Br. J. Anaesth., 47 (1975) 533-540.
14 T.E.J. Healy, T.R. Poole and A. Hopper, Br. J. Anaesth., 54 (1982) 337-341.
15 D.E. Berlyne, J. Camp. Physiol. Physol., 48 (1955) 238-246.
16 SAS Institute, Inc., SAD User's Guide: Statistics, Cary, 1982, 584 pp.

17 L.W. Reiter, G.E. Anderson, M.E. Ash and L.E. Gray, Jr., in H. Zenick and L.W. Reiter (Eds.), Behavioral Toxicology, and Emerging Discipline, EPA 6001 9-77-042:6.1-6.18, 1977.
18 E. Grandjean, Arch. Environ. Hlth., 1 (1960) 106-108.
19 H. Savolainen, P. Pfaffli, M. Tengen and H. Vainio, Arch. Toxicol., 38 (1977) 229-237.
20 A.P. Silverman and H. Williams, Brit. J. Indust. Med. 32 (1975) 308-315.
21 K. Battig and E. Grandjean, Arch. Environ. Hlth., 7 (1963) 80-85.

RESULTS OF A 90-DAY TOXICITY STUDY ON 1,2,3- AND 1,1,2-TRICHLOROPROPANE ADMINISTERED VIA THE DRINKING WATER

D.C. VILLENEUVE, I. CHU, V.E. SECOURS, M.G. COTÉ, G.L. PLAA AND V.E. VALLI

Bureau of Chemical Hazards, Health Protection Branch, Ottawa, Canada

Département de Pharmacologie, Université de Montréal, Montréal, P.Q., Canada

ABSTRACT

Trichloropropanes have been identified as environmental contaminants in sediments of the Great Lakes region of North America. Since these chemicals had the potential to find their way into drinking water, a 90-day feeding study was carried out in order to determine their subchronic toxicity. Groups of 10 male and 10 female weanling Sprague-Dawley rats were supplied drinking water ad libitum, containing 1,2,3- or 1,1,2-trichloropropane at concentrations of 1, 10, 100 or 1000 mg/L for 13 weeks. Emulphor (0.5%) was used to solubilize the chemicals. At the end of the study, the animals were killed and examined for gross and microscopic changes. Heart, liver, brain, kidney and spleen were excised and weighed. Blood was collected and subjected to a comprehensive hematological analysis. Serum was collected and profiled for changes in 12 biochemical parameters and a portion of liver was used to determine mixed function oxidase activity. Although three animals died during the study, their deaths could not be related to treatment. Decreased growth rate was observed in both sexes of the group receiving 1000 mg/L 1,2,3-trichloropropane. There was an increase in liver, kidney and brain weights (relative to body weight) in rats of both sexes fed 1000 mg/L 1,2,3-trichloropropane. Fatty livers were observed in some of the treated animals but a clear dose-relationship was not evident. An elevation in serum cholesterol was observed in female rats fed the highest dose of 1,2,3-trichloropropane. This chemical also induced hepatic aminopyrine demethylase and aniline hydroxylase activities in male rats at the highest dose. Administration of both isomers produced only mild histological changes in the liver, thyroid and kidney of rats at the highest dose. The changes in the liver consisted of an increase in cytoplasmic eosinophilia in the periportal area together with vesication of biliary epithelial nuclei. Morphological changes were characterized by increased anisokaryosis in the proximal epithelium and occasional pyknosis associated with the accumulation of large eosinophilic inclusions. Changes in the thyroid consisted of a mild reduction in follicular size associated with an increased epithelial height. In general, these changes were more severe in the males than females, but were still mild overall. It was concluded that the no-effect level for both chemicals was 100 mg/L (15-20 mg/kg bw/day) and based on effects on growth rate and other changes, the 1,2,3- isomer was judged to be slightly more toxic than the 1,1,2- isomer.

INTRODUCTION

Trichloropropanes (TCP) are byproducts of synthetic processes used to manufacture propylene chlorohydrin and other propylene derivatives (Mark et al., 1979). One of the isomers, 1,2,3-trichloropropane, has been used commercially as a solvent and as a degreasing agent. TCP's have been identified as environmental contaminants in the Lake St. Clair and Detroit River regions of the Great Lakes (Water Quality Board, 1978) and thus have the potential to contaminate drinking water supplies. Since there was a paucity of toxicological information on these chemicals, a 90-day toxicity study was carried out in which two of these chemicals were administered to rats in their drinking water.

EXPERIMENTAL

Materials

1,2,3-trichloropropane was purchased from Aldrich Chemicals (Milwaukee, WI) and had a stated purity of 99%. The identity and purity of this chemical were confirmed by GC, NMR, GC/MS and TLC. 1,1,2-trichloropropane was purchased from Columbia Organic Chemicals (Columbia, SC) and was shown to contain 20% trichloropropene. Catalytic hydrogenation, using hydrogen and platinum oxide at room temperature and atmospheric pressure resulted in a product with greater than 99% purity, as confirmed by GC, NMR, GC/MS and TLC. Emulphor was obtained from Domtar Canada, Montreal.

Methods

Sprague-Dawley rats (60-70g) were randomly assigned to groups containing 10 males or 10 females and were administered 1,2,3- or 1,1,2-TCP in their drinking water at concentrations of 1, 10, 100 or 1000 mg/L. Emulphor was added to all drinking water at a concentration of 0.5% to help solubilize the chemicals. Two control groups per sex were used: one received tap water alone, the other received tap water containing 0.5% Emulphor. The animals were placed on test for 13 weeks and killed at the end of the study. During the study, the animals had free access to food and water. Clinical observations were made daily. The parameters monitored during the study included body weight changes and water intake. After the termination of the study, brain, liver, kidney, heart and spleen were removed and weighed. Blood was taken for hematological evaluation (hemoglobin, packed cell volume, mean corpuscular hemoglobin, mean corpuscular volume, mean corpuscular hemoglobin concentration, total erythrocytes, total and differential counts of leukocytes). Comprehensive biochemical analysis of the serum was done using a Technicon SMA 12/60 analyzer. The parameters measured included sodium, potassium, inorganic phosphorus, total protein, calcium, cholesterol, glutamic oxaloacetic transaminase, total bilirubin, alkaline phosphatase, glucose, uric acid and lactate dehydrogenase. Serum sorbitol dehydrogenase activity was measured by a method described previously (Yagminas and Villeneuve, 1977). Hepatic mixed function oxidase assays included aniline hydroxylase (Fouts, 1963), aminopyrine demethylase (Cochin and Axelrod, 1959) and ethoxyresorufin deethylase (Burke and Mayer, 1974). Tissues were examined histologically as described previously (Villeneuve et al., 1979). Analysis of TCP's was carried out in the drinking water to ensure that the actual concentration was within 10% of the intended level. Statistical analysis of the data was carried out on a Hewlett-Packard 3000 computer using a one-way analysis of variance. When significant differences were observed, the data were subjected to Duncan's Multiple Range Test (Nie et al., 1977) in order to determine which groups were significantly different (P≤0.05) from the controls.

RESULTS

Clinical observations

Three animals died during the study (one female from the 100 mg/L 1,2,3-TCP group, one female from the 100 mg/L 1,1,2-TCP group, and one male from the 1000 mg/L 1,1,2-TCP group). The cause of death could not be determined.

Growth rate and water intake

Body weight gain and water intake data from treatment groups that were significantly different from the vehicle control group are summarized in Table 1.

TABLE 1

Body weight gain and water intake data

Treatment Group	Initial weight (g)	Weight gain (g)	Water Intake (ml/day)	Amount of Chemical Ingested (mg/kg/day)
Control (M)	79	455	53	--
Vehicle control (M)	76	450	45	--
1,2,3-TCP 1000 mg/L (M)	86	309[a,b]	27[a,b]	113
Control (F)	79	192	44	--
Vehicle Control (F)	73	207	36	--
1,2,3-TCP 100 mg/L (F)	79	181	31[a]	17.6
1,2,3-TCP 1000 mg/L (F)	81	140[a,b]	23[a,b]	149

[a] Signifies significant difference from control group (P≤0.05)
[b] Signifies significant difference from vehicle control group (P≤0.05)
M denotes male
F denotes female

The animals from the vehicle control group had a slightly lower water intake than the tap water controls, but their body weight gain was not significantly affected. Only the males fed 1000 mg/L 1,2,3-TCP showed a depressed growth rate which was apparent from week 1 of the study. This group of animals also displayed a decreased water intake throughout the study. Body weight gain in the females was depressed in the group fed 1000 mg/L 1,2,3-TCP. Water intake was reduced in this same group of animals as well as in the group fed 100 mg/L 1,2,3-TCP. Neither body weight gain nor water intake were affected in any of the groups fed 1,1,2-TCP. The intake of animals fed the chemicals at 1000 mg/L ranged from 15-20 mg/kg b.w./day.

Organ weights

Wet organ weights were not altered by treatment in any of the groups. When expressed as a percentage of body weight, the liver was increased in males fed the highest level of each chemical (control, 3.7; vehicle control, 3.7; 1000 mg/L 1,2,3-TCP, 4.5; 1000 mg/L 1,1,2-TCP, 4.0). In females, the animals receiving the two highest levels of 1,2,3-TCP also showed increased liver/body weight ratios (control, 3.6; vehicle control, 3.6; 100 mg/L 1,2,3-TCP, 3.8; 1000 mg/L 1,2,3-TCP, 4.2). Kidney weight, expressed as a pecentage of body weight, was increased in male rats fed the highest dose level of 1,2,3-TCP (control, 0.33; vehicle control, 0.32; 1000 mg/L 1,2,3-TCP, 0.42). Females had increased kidney/body weight ratios in groups fed the two highest dose levels of 1,2,3-TCP (control, 0.36; vehicle control, 0.35; 100 mg/L, 0.40; 1000 mg/L, 0.47). Brain weights, expressed as a percentage of body weight, were increased for both males and females receiving the highest dose level of 1,2,3-TCP (Males: control, 0.38; vehicle control, 0.39; 1000 mg/L 1,2,3-TCP, 0.47. Females: control, 0.71; vehicle control 0.66; 1000 mg/L, 0.81).

Biochemical changes

Serum cholesterol levels were significantly increased compared to both the control and vehicle control groups in female rats fed the highest level of either 1,2,3- or 1,1,2-TCP (control 71 $\pm$ 14 nmoles HCHO/mg protein/hour; vehicle control, 69 $\pm$ 25; 1000 mg/L 1,2,3-TCP, 107 $\pm$ 13; 1000 mg/L 1,1,2-TCP, 96 $\pm$ 14). Hepatic aminopyrine demethylase activity was increased significantly compared to both control groups in males fed 1000 mg/L 1,2,3-TCP (control 50 $\pm$ 11; vehicle control, 53 $\pm$ 11; 1000 mg/L 1,2,3-TCP 69 $\pm$ 9) and in females fed the same chemical at the same level (control, 38 $\pm$ 6; vehicle control 38 $\pm$ 5.1; 1000 mg/L 1,2,3-TCP, 47 $\pm$ 7.8). Aniline hydroxylase activity was significantly increased in only one group, males fed 1000 mg/L 1,2,3-TCP (control, 25 $\pm$ 5 nmoles PAP/mg protein/hour; vehicle control, 23 $\pm$ 4; 1000 mg/L 1,2,3-TCP, 32 $\pm$ 5.5).

Hematological changes

Mild treatment-related changes occurred in only two of the hematological parameters measured. Male rats fed the highest dose of 1,2,3-TCP showed decreased neutrophil counts (control, 1.4 ± 0.84; vehicle control, 1.1 ± 0.61; 1000 mg/L 1,2,3-TCP, 0.65 ± 0.29) and decreased lymphocyte counts (control, 8.6 ± 0.30; vehicle control, 8.9 ± 2.0; 1000 mg/L 1,2,3-TCP, 5.7 ± 1.2). These values, however, were still within the normal ranges established for Sprague-Dawley rats in our laboratories.

Histological changes

The liver, kidneys and thyroid were affected by treatment. In the liver, mild but significant changes were observed in the highest dose levels for both compounds in both sexes. These changes included anisokaryosis, accentuated zonation and the occasional fatty vacuolation. In addition, biliary hyperplasia was noted in female rats fed 1000 mg/L 1,2,3-TCP.

In the kidney, mild changes were also observed in both sexes at the highest dose levels of each compound. These changes consisted of eosinopilic inclusions, pyknosis, nuclear displacement, fine glomerular adhesions and occasionally interstitial reactions and histologic proteinuria. Mild changes occurred in the thyroid of animals of both sexes fed the highest dose level of both chemicals. These changes included angular collapse of some follicles, reduction in colloid density and increased epithelial height. In general, the changes observed in the tissues were milder and less prevalent in animals receiving 1,1,2-TCP and were milder and less prevalent in females compared to males.

DISCUSSION

Neither chemical used in this study caused any overt toxicity. The deaths that did occur could not be related to any specific cause, but since none of the other animals displayed any toxic symptoms, it is unlikely that the deaths were due to the chemicals themselves. 1,2,3-trichloropropane treatment caused a number of changes including growth suppression, liver enlargement, elevated serum cholesterol, increased hepatic mixed-function oxidase activity and histologic alterations in liver, kidney and thyroid. In the present study alterations were generally observed at 1000 mg/L, corresponding to an intake of 15-20 mg/kg bw/day. The change in the liver weight relative to body weight indicates that the chemical can produce liver hypertrophy. The increased brain and kidney weights are probably not real changes since the wet weights of these organs were not affected. Rather, they probably reflect the decreased weight of the animals from that treatment group. Although there were hematological changes present, they were considered mild and were within the normal ranges for Sprague-Dawley rats established in our laboratories. As with the other changes noted in the study, the histological alterations observed in liver, kidney and thyroid were mild in nature and females were less affected than males.

1,1,2-TCP was less toxic and produced fewer changes than 1,2,3-TCP. For instance, 1,1,2-TCP did not cause body weight reduction or changes in organ weights. It did cause elevated cholesterol levels, but did not influence hepatic mixed function oxidase activities. It produced no change in any of the hematological parameters monitored and produced fewer and milder histological changes than 1,2,3-TCP. It was concluded that the no-effect level for both the compounds tested during this study was 100 mg/L and that 1,1,2-TCP was less toxic than 1,2,3-TCP.

ACKNOWLEDGEMENTS

The authors wish to thank Monique Morisset, Marie-Yolaine Thomas, Claudette Lamoureux, Nanette Beament, Barb Reed, Al Yagminas, Jack Kelly, Andre Viau and Mary Beaudette for their technical assistance, and Kathy Nesbitt for typing the manuscript.

REFERENCES

Burke, M.D. and Mayer, R.T., 1974. Ethoxyresorufin: direct fluorimetric assay of a microsomal-o-dealkylation which is preferentially inducible by 3-methylcholanthrene. Drug Metab. Dispos., 2:585-588.

Cochin, J. and Axelrod, J., 1959. Biochemical and pharmacological changes in the rat following chronic administration of morphine, nalorphine and normorphine. J. Pharmacol. Exp. Ther., 125:105-110.

Fouts, J.R., 1963. Factors influencing the metabolism of drugs in liver microsomes. Ann. N.Y. Acad. Sci. 104:975-880.

Great Lakes Water Quality Board Report, 1978. International Joint Commission, Windsor, Ontario.

Mark, H.F., Othmer, D.F., Overberger, C.G. and Seaborg, G.T. (ed), 1979. Encyclopedia of Chemical Technology 5:848-864.

Nie, N.H., Hull, C.H., Jenkins, J.G., Steinbrenner, K. and Bent, D.H., 1977. Statistical Programs for the Social Sciences, Chicago, SPSS, Inc.

Villeneuve, D.C., Valli, V.E., Chu, I., Secours, V., Ritter, L. and Becking, G.C., 1979. Ninety-day toxicity of photomirex in the male rat. Toxicology, 12:235-250.

Yagminas, A.P. and Villeneuve, D.C., 1977. An automated continuous flow assay for serum sorbitol dehydrogenase activity and its use in experimental liver damage. Biochem. Med., 18:117-125.

CARCINOGENICITY STUDY IN RATS WITH A MIXTURE OF ELEVEN VOLATILE HALOGENATED HYDROCARBON DRINKING WATER CONTAMINANTS

P.W.Wester, C.A.van der Heijden, A.Bisschop, G.J.van Esch, R.C.C.Wegman and Th. de Vries
National Institute of Public Health and Environmental Hygiene, P.O.Box 1, 3720 BA Bilthoven, The Netherlands

ABSTRACT
A lifetime carcinogenicity study was carried out in Wistar rats, with a mixture of the following halogenated hydrocarbons: trichloromethane, tetrachloromethane, monobromodichloromethane, trichloroethylene, tetrachloroethylene, 1,2,-dichlorobenzene, 1,3,-dichlorobenzene, 1,4,-dichlorobenzene, 1,2,3,-trichlorobenzene, 1,2,4,-trichlorobenzene, 1,3,5-trichlorobenzene. From this mixture 0.22, 2.2, or 22 mg was added per liter drinking water representing concentrations being three orders of magnitude higher than found in several water wells. Most of the changes found in body weight, hematology and pathology correlated with intercurrent diseases or were in accordance with background pathology. With respect to incidence and time of occurrence of tumors, no significant differences were found between the control and the high dose group when lifespan correction was applied. Thus it is concluded that in the present study no significant toxic or carcinogenic effects are induced by lifetime exposure of rats to a mixture of volatile halogenated hydrocarbons in the drinking water.

INTRODUCTION
From a study performed by the National Institute of Public Health and the National Institute of Water Supply it appeared that 20 out of 232 ground water wells in the Netherlands were contaminated by volatile halogenated hydrocarbons in concentrations over 1 µg/l (Zoeteman, 1979).
For the contamination of ground water several sources can be incriminated, the major sources being disposal of industrial wastes and industrial impoundments and solid waste disposal sites (Speth et al., 1981). Another cause of contamination of drinking water in certain areas is the chlorination of drinking water in order to destroy harmful bacteria, yielding trihalogenated methanes in particular (Deinzer, 1978, Zoeteman, 1979, Crump, 1982). In order to investigate the long term effects with emphasis on carcinogenic properties of these contaminants, a lifetime study has been carried out with a mixture of eleven volatile halogenated hydrocarbons; the choice of these compounds was based upon their occurrence in drinking water wells (Zoeteman, 1979).

EXPERIMENTAL
Animals and maintanance
For the experiment, SPF-derived outbred weanling Wistar rats, (Riv: Tox [M]) were obtained from the Institute's own breeding colony, and allotted to a control group and three dose groups, each group consisting of 60 ♀♀ and 60 ♂♂ rats. They were distributed littermate and aselect over the groups, and were kept under conventional conditions, two per cage according

to sex. The animal rooms were controlled at a temperature of 22 ± 2°C, a 12 hr light-dark cycle and a relative humidity of 40-60%. The air in the animal rooms was continuously refreshed (ventilation fold: 8/hr). To prevent substantial exposure of control animals to the volatile test compounds via the inhalatory route, control and treatment groups were housed separately. The rats received a commercially available semi-synthetic diet (Muracon SSP-Tox standard, Trouw Ltd., Putten, The Netherlands) during the first 33 weeks of the experiment and thereafter a Laboratory Animal Diet (RMH-B Hope Farms, Woerden, The Netherlands). Food and drinking water were provided ad libitum. The diets were regularly analyzed for contaminants, essential elements and vitamins. During weeks 66 and 67, 4 ml Sulfadimidine Na. 33 (Aesculaap BV., Boxtel, The Netherlands) was added per liter drinking water for the treatment of a respiratory tract disease.

Chemicals
The following eleven halogenated hydrocarbons were selected for the mixture under investigation, and were obtained from Fluka unless otherwise stated: trichloromethane (Merck), tetrachloromethane, monobromodichloromethane, trichloroethylene, tetrachloroethylene, 1,2,-dichlorobenzene (Merck), 1,3,-dichlorobenzene, 1,4,-dichlorobenzene, 1,2,3,-trichlorobenzene, 1,2,4,-trichlorobenzene, 1,3,5-trichlorobenzene. Their purity was 97% or higher.

Dosages and route of administration
The rats received drinking water to which 0, 0.22, 2.2 or 22 mg of a mixture consisting of equal quantities of the halogenated hydrocarbons in 1 ml ethanol was added per liter, and which was prepared daily; control animals received 1 ml ethanol/l drinking water. The highest dose was based upon the maximum quantity of this mixture soluble in 1 ml ethanol. The concentrations of the individual halogenated hydrocarbons in the drinking water were measured regularly. From this it appeared that substantial losses had occurred, presumably during preparation procedures, yielding actual concentrations 15 - 50% lower than the intended ones. During a 24-hr period, losses were only minimal, except for tetrachloromethane in groups 2 and 3, and for tetrachloroethylene, 1,3,5- and 1,2,4-trichlorobenzene in group 4. In the tap water, used for drinking water preparation for all groups, minimal concentrations of trichloromethane were measured (0.001 > - 0.019 mg/l). Also, the air in the animal rooms was regularly analyzed, from which it was concluded that the exposure of the animals to chlorinated hydrocarbons via the inhalatory route can be neglected. In addition, the occupational risk for the technicians was considered negligible since the concentrations of those halogenated hydrocarbons of which MAC values are known were several orders of magnitude lower than these limits.

Observations
Health, behaviour and mortality of the animals were checked daily and animals, when moribund or showing large tumor masses, were killed and autopsied. Body weight and water consumption were recorded weekly during the first twelve weeks, and thereafter every four weeks; after one year, water consumption was recorded only every eight weeks.
Hematology was performed by Dr.P.W.Helleman, Unit Clinical Chemistry

and Hematology. Blood samples were collected under ether anesthesia from the retro-orbital plexus after 12 and 24 months of exposure, from 10 animals of each sex and group. The following parameters were determined: hemoglobin concentration , hematocrit value (packed cell volume), erytrocyte, total and differential leucocytes counts. Mean corpuscular volume (MCV), mean corpuscular hemoglobin (MCH) and mean corpuscular hemoglobin concentration (MCHC) were calculated.

After 25 months of exposure, the surviving animals were killed by bleeding from the abdominal aorta under ether anesthesia. All animals, both after scheduled and intercurrent death were subjected to detailed gross inspection. From all animals, the following tissues and organs were collected except when advanced autolysis had occurred: brain, heart, lungs, liver, spleen, kidneys, pituitary, thyroid, thymus, pancreas, adrenals, ovaries, testes, uterus, prostate, mesenteric lymph nodes, submandibular salivary glands, oesophagus, stomach, duodenum, jejunum, ileum, cecum, colon and rectum, urinary bladder, vertebral column, sciatic nerve, quadriceps muscle. In addition, all tumors and lesions suspected of being tumorous were collected. Samples were fixed, embedded in paraffin wax and 5 µm sections were cut and stained with hematoxylin and eosin (HE).

Histopathological examination for cancerous and possible precancerous lesions was carried out on all animals autopsied in the control and high dose group. Also from these groups, ten randomly selected male and female rats surviving two years were subjected to histopathological examination for non-neoplastic lesions.

Except for pathology, experimental data were analysed by the Student's t-test. Pathology data, if indicated, were subjected to statistical analysis according to the prevalence method (Peto et al., 1980).

RESULTS

After four months a high incidence of intercurrent deaths occurred caused by gastrointestinal obstruction from trichobezoars. This condition was almost exclusively observed in females and the incidence was higher in the treatment groups than in the control group. Changing the diet from Muracon SSP-Tox standard to RMH-B was followed by a drop in mortality due to gastro-intestinal obstructions. The latter diet has a higher crude fiber content, which might have exerted a stimulating effect upon gastrointestinal motility and function.

After 60 weeks of exposure, a high morbidity and mortality from a respiratory tract disease was observed in the treatment groups, from which predominantly males were affected. From the lungs of the diseased animals, frequently Bordetella bronchiseptica was isolated among several other microorganisms. Therefore a two-week treatment with sulfadimidine in the drinking water was started for control and treatment groups.

Towards the end of the study age-related clinical signs developed, such as weight loss, poor condition of the fur, dyspnoea, posterior paralysis, chronic peritarsitis. These features occurred equally among control and treatment groups. Except for scheduled sacrifices survival after 12 months was 100% for control males and 96% for males from the top-dose group. Control females showed 82% and top-dose females 74% survival after this period. After 24 months these data were 55% and 38% for males and 49% and 34% for females.

During the first six weeks a significant higher body weight gain occurred in the treatment groups compared with the controls. This phenomenon was not dose-related. During the course of the experiment growth retardation occurred in females of groups 3 and 4 from week 16 onwards and in males from the treatment groups starting at week 60; this feature coincided with the underlying diseases, intestinal obstruction and respiratory infections, respectively.

During the first experimental month, treatment groups showed a higher water consumption than controls, whereas from week 9 a lower water consumption was observed; this dose-related decrease in water consumption persisted in females until the end of the experiment, and returned to the control level in the males after the first experimental year. An increase in water consumption in all groups occurred from week 32 - 36: this coincided with the change of the diet and with the period at which major intercurrent mortality declined.

In hematology after two years significant increases were found in treatment groups for hemoglobin concentration, packed cell volume, erythrocyte counts (females only), MCV and MCH (males only). In differential leucocyte counts, an increase in eosinophils was observed in females from the top dose group after 12 months, whereas this parameter was decreased after 24 months. In the treated males an increase in neutrophils was observed which appeared not to be dose-related. This is supposed to be related to intercurrent respiratory infections during the second year, which were most prominent in these groups.

Gross and histopathological examination of the non-neoplastic lesions showed that, except from the findings in the respiratory tract, these lesions were equally distributed amongst control and treatment groups, and are considered age- and strain related, caused by intercurrent disease or background pathology. The lesions in the respiratory tract included chronic suppurative bronchitis-peribronchitis, lobar bronchiectasis and atelectasis. This condition occurred more pronounced and more frequently in treated animals. A notable finding in two males out of the high dose group was the presence of nodular metaplastic bone structures in the lungs, which were not associated with inflammatory processes. Based upon morphology these lesions were not considered to be neoplastic in nature.

With respect to neoplastic lesions, the absolute number of tumor bearing rats, the total number of tumors, and the total number of tumors according to organ site e.g. pituitary, adrenals, appeared to be lower in the top dose group when compared with the control group. However, when statistical analysis allowing for longevity was applied, no significant differences between the groups was found. In addition, no unusual tumor types for this strain of rats were found after treatment.

DISCUSSION

The present study in which male and female Wistar rats were exposed to a mixture of halogenated hydrocarbons in the drinking water during their lifetime, is in a certain sense an unorthodox study, since not a single compound is under investigation, but a mixture which mimics in exaggerated form the natural exposure of a human population. The composition of this mixture is, in a qualitative sense, representative for the contamination of several ground wells used for the drinking water supply, sampled and analyzed at regular intervals during the period of 1976 - 1978 in the

Netherlands (Zoeteman, 1979). Similar findings are reported from the USA (Speth et al., 1981). Mutagenicity studies carried out with polluted surface or ground water (Loper, 1980; Kool, 1983), revealed mutagenic activity in several test systems; epidemiological studies indicate a slight increase in risk for colon, rectum and bladder cancer, associated with organic contaminants in drinking water (Williamson, 1981 Crump, 1982). However, mutagenicity studies performed at our Institute with this mixture of eleven compounds showed no mutagenic action in five different assays (Voogd et al., 1979), though the concentrations tested were relatively low because of cytotoxicity and poor solubility. Also negative results were obtained with these single components in Salmonella typhimurium TA 100 and TA 98 assays (Voogd and van der Stel, 1979).

Two major intercurrent diseases interfered with this study: intestinal obstruction by trichobezoares and respiratory infections. Trichobezoars could be relieved by a change of diet, resulting in a higher crude fiber intake. The prevalence for this condition in females has not been clearified yet. Infections of the respiratory tract occurred, predominantly in treated males, after 60 weeks of exposure; this predominance can be explained partly to the separate housing regimen. However, it cannot be excluded that impairment of immune functions, as described by Munson et al., (1982) by some of these compounds, has played a role. Since these two intercurrent diseases also occurred in other simultaneous experiments in the laboratory, they are not considered to be caused directly by exposure to the compounds under investigation. Several parameters such as bodyweight gain, mortality and neutrophil counts were affected and coincided with these conditions and are considered to be related to them.

Effects on red blood cell parameters and a dose related decrease in water consumption (predominantly in females) remain unexplained: an aversive taste of the drinking water seems unlikely, since at the start of the experiment even a higher water intake was noted.

From the non-neoplastic histopathological findings after 25 months of exposure, only multiple metaplastic bone formation in the lungs in two out of ten top dose males is considered a notable finding. Since no malignant histological features and not any primary tumor with similar appearance was found, these lesions are considered non-neoplastic. A similar condition is described in several other species, including man, with unknown etiology (Borst, 1976). In the rat however, and in particular the strain used in the present study, no such lesions are as yet described in the literature (Burek, 1978, Kroes, 1981). Whether this condition is fortuitous or is to be attributed to exposure to the test compound, or the intercurrent respiratory infections, remains questionable. The tumor distribution, when corrected for age, proves to be not significantly different between control and top dose group. In addition time of appearance of tumors was not influenced by treatment and no unusual tumor types were observed. It should be noted, however, that due to poor solubility of the test compound in the drinking water, the top dose was not at the MTD (maximum tolerated dose) level.

In conclusion, it can be stated that under the conditions of this study this mixture of eleven halogenated hydrocarbons, in a concentration being three orders of magnitude higher than as found in several water wells, exhibited no evident toxic or carcinogenic effect in the rat.

REFERENCES

Borst, G.H.A., Zwart, P., Mullink, H.W.M.A., and Vroege, C., 1976. Bone structures in avian and mammalian lungs. Vet.Pathol. 13: 98-103.

Burek, J.D., 1978. Pathology of Aging Rats. C.R.C.Press Inc., West Palm Beach, Florida, 230 pp.

Crump, K.S., and Guess, H.A., 1982. Drinking water and cancer: review of recent epidemiological findings and assesment of risks. Ann.Rev.Public Health, 3: 339-357.

Deinzer, M., Schaumburg, F. and Klein, E., 1978. Environmental health sciences center task force review on halogenated organics in drinking water. Environm. Health Perspect. 24: 209-239.

IARC, 1979. Monographs on the Evaluation of the Carcinogenic Risk of Chemicals to Humans: Some Halogenated Hydrocarbons, Volume 20. International Agency for Research on Cancer, Lyon. 609 pp.

IARC, 1982. Monographs on the Evaluation of the Carcinogenic Risk of Chemicals to Humans. Supplement 4. International Agency for Research on Cancer, Lyon. 292 pp.

Kool, H.J., 1983. Organic mutagens and drinking water in the Netherlands. Thesis, Agricultural University, Wageningen.

Kroes, R., Garbis-Berkvens, J.M., de Vries, Th. and van Nesselrooij, H.J., 1981. Histopathological profile of a wistar rat stock including a survey of the literature. J.of Gerontol. 36: 259-279.

Loper, J.C., 1980. Mutagenic effects of organic compounds in drinking water. Mut.Res. 76: 241-268.

Munson, A.E., Sain, L.E., Sanders, V.M., Kaufmann, B.M., White, K.L., Page, P.G., Barnes, D.W. and Borzelleca, J.F., 1982. Toxicology of organic drinking water contaminants: trichloromethane, bromodichloromethane, dibromodichloromethane and trichloromethane. Environm. Health Perspect. 46: 117-127.

Peto, R., Pike, M.C., Day, N.E., Gray, R.G., Lee, P.M., Parish, S., Peto, J., Richards, S. and Wahrendorf, J., 1980. Guidelines for simple, sensitive significance tests for carcinogenic effects in long term animal experiments. Annex in: Long-Term and Short-Term Screening Assays for Carcinogens: A Critical Appraisal. Supplement 2. IARC Monographs on the Evaluation of the Carcinogenic Risk of Chemicals to Humans. pp. 311-426.

Speth, G., Harris, R.H. and Yaru, J., 1981. Contamination of Ground Water by Toxic Organic Chemicals. U.S.Council on Environmental Quality.

Voogd, C.E., Kramers, P.G.N., Knaap, A.G.A.C. and Van Went-de Vries, G.F. 1979. Onderzoek naar de mutagene werking van gechloreerde alifatische en aromatische koolwaterstoffen die in grondwater, bestemd voor drinkwater, voorkomen. Rapport no. 93/79 Tox Farm Chemo, Rijksinstituut voor de Volksgezondheid.

Voogd, C.E. and van der Stel, J.J., 1979. Onderzoek naar de mutagene werking van enkele gehalogeneerde alifatische en aromatische verbindingen, die in drinkwater zijn gevonden. Rapport nr. 174/79 Chemo, Rijksinstituut voor de Volksgezondheid.

Williamson, E.J., 1981. Epidemiological studies on cancer and organic compounds in U.S. drinking waters. Sci.Total Environm. 18: 187-203.

Zoeteman, B.C.J., Piet, G.J., Slingerland, P., Fonds, A.W. and Wegman, R.C.C., 1979. Onderzoek naar de aanwezigheid van vluchtige organische halogeenverbindingen in grondwater bestemd voor de drinkwatervoorziening. Rapp. no. 79-01; Rijksinstituut voor Drinkwatervoorziening; no. 10-79, Rijksinstituut voor de Volksgezondheid.

TARGET ORGAN TOXICOLOGY OF HALOCARBONS COMMONLY FOUND CONTAMINATING DRINKING WATER*

L.W. CONDIE

Toxicology and Microbiology Division, Health Effects Research Laboratory, Environmental Protection Agency, Cincinnati, Ohio 45268 (United States)

ABSTRACT

Some of the most frequent drinking water contaminants are organic halocarbons. This paper will initially summarize the target organ effects of three halocarbons: 1,2-dichloroethane, tetrachloroethylene, and trichloroethylene. Following the brief summaries, a more detailed description of the oral hepatoxicity of carbon tetrachloride is presented. Data are provided that indicate that the hepatotoxicity of carbon tetrachloride is enhanced when administered by corn oil gavage when compared to aqueous suspension gavage.

INTRODUCTION

Large populations are repeatedly exposed to potentially toxic drinking water contaminants in minute amounts over many years or throughout a whole lifetime. When evaluating the potential adverse health effects of a contaminant in drinking water, the magnitude of all adverse effects, whether carcinogenic or non-carcinogenic, should be considered. It is imperative that the most sensitive endpoint of toxicity be identified. It is conceivable that a noncarcinogenic adverse health effect could occur near or within the range of a carcinogenic risk level being considered for a given contaminant. Any attempt to regulate a chemical substance should include an evaluation of adverse health effect levels for the most sensitive noncarcinogenic end-point (i.e., hepatic, behavioral, reproductive).

Hundreds of chemicals have been detected in United States' drinking water, but the majority have been detected infrequently and at low concentrations. The selection of chemicals for revised primary drinking water regulations is based on the following criteria: (1) frequency of occurrence, (2) environmental concentrations detected, (3) size of the exposed populations, and (4) the

---

*The research described in this paper has been peer and administratively reviewed by the U.S. Environmental Protection Agency (EPA) and approved for publication. Mention of trade names or commercial products does not constitute endorsement or recommendation for use.

toxicology of the chemicals. Based on the above criteria, nine volatile synthetic organic chemicals, seven of which are halocarbons, were proposed for subsequent regulation in the United States (EPA, 1984). This paper summarizes the noncarcinogenic toxicology research data for four halocarbons and describes some of the author's latest research findings regarding the hepatotoxicity of carbon tetrachloride. The literature is replete with toxicological data from acute, inhalation experiments conducted with volatile halocarbons. Both inhalation and oral exposure studies are mentioned in this paper, but because of the size limitations of this paper, emphasis is placed on subchronic or chronic animal studies in which the chemical exposure occurred by the oral route of administration.

REVIEW OF TOXICOLOGY OF SELECT HALOCARBONS

1,2-Dichloroethane (DCE)

DCE is widely used in the synthesis of vinyl chloride and other compounds, as an insect fumigant, as a paint and varnish ingredient, as a metal degreaser, and in wetting and penetrating agents. Almost all DCE toxicity studies reported in the literature are based on acute, high-dose inhalation exposures, and little is known of the subtle toxicology of low-level oral exposures to DCE. Chronic inhalation exposures of rats and guinea pigs to DCE at 100 ppm for seven hours daily, five days weekly for up to six months usually produced no deaths and no adverse effects while similar exposures of rats, guinea pigs, rabbits, and monkeys to air containing 400 or 500 ppm DCE generally resulted in high mortality and varying pathological findings such as pulmonary congestion, diffuse myocarditis, and fatty degeneration of the liver, adrenal, kidney, and heart (Heppel et al., 1946; Spencer et al., 1951; Hofmann et al., 1971).

In a low-level feeding study, male and female rats were exposed to DCE for up to 2 years (Alumot and coworkers, 1976). The minimum toxic dietary level of DCE was 1600 ppm as indicated by increases in hepatic total lipids after a five- to seven-week period. No significant alterations in body weight gain or in clinical chemistry values were observed in male or female rats after 2 years exposure to DCE (250 and 500 ppm). In addition, no alteration of fertility or reproduction was detected in the rats. In a separate study Lane et al. (1982) administered DCE at concentrations up to 1000 mg/kg/day in the drinking water of mice and found no adverse reproductive effects. The limited studies indicate that DCE is not teratogenic to mice or rats.

Tetrachloroethylene or perchloroethylene (PCE)

PCE is primarily used as a chemical solvent, heat transfer medium, and in the manufacture of fluorohydrocarbons. Investigations into the target organ toxicity of PCE have primarily involved inhalation exposures. The most characteristic effect of acute, inhalation exposure to PCE is central nervous system depression. At higher doses, centrilobular hepatic infiltration occurs which is followed by hepatic necrosis. Renal damage is seen at higher doses of PCE than at which initial liver damage occurs. Subchronic inhalation exposure effects are also manifested as liver and kidney damage.

The National Cancer Institute (1977) carcinogenesis bioassay of PCE demonstrated a high incidence of nephropathy in both male and female mice and rats that were orally gavaged with doses of PCE ranging from 386 to 1070 mg/kg/day for as long as 78 weeks. While there was little evidence of hepatotoxicity at those exposure conditions, there was a high incidence of hepatocellular carcinoma in the mice only. In a separate study, B6C3F1 mice and Sprague-Dawley rats were orally administered PCE for 11 consecutive days at levels of 100, 250, and 1000 mg/kg. Histological changes were seen in rat livers at the highest dose and in mouse livers at all dose levels (Schumman et al., 1980).

In a recent study sponsored by EPA, a subchronic toxicology study with PCE in drinking water was conducted (Hayes et al., 1985a). Male and female rats received approximate daily doses of 14, 400, and 1400 PCE mg/kg/day for 90 consecutive days. Body weights were significantly lower in male and female rats at the higher dose. 5'Nucleotidase activity was increased in a dose dependent manner, suggesting possible hepatotoxicity; other blood chemistry parameters were unaffected by the treatment. Liver and kidney organ to body weight ratios were elevated at the higher doses.

Trichloroethylene (TCE)

TCE is utilized as a metal degreaser, dry cleaning solvent, starting material in organic synthesis, fumigant, and refrigerant. Acute inhalation experiments indicate the major effects of TCE, like other chlorinated aliphatic hydrocarbons, are liver and kidney damage, and central nervous system depression. Depressed myocardial contractility has also been noted. Mild toxicological effects have been noted in subchronic and chronic experiments with TCE. Studies in which mice were orally administered TCE by gavage five days weekly for three weeks indicated that 250 mg/kg caused minimal hepatoxic effects which were evident only upon histopathological examination of the liver (Stott et al., 1982); doses of 500 mg/kg or greater caused increases in liver weight and centrilobular hypertrophy. A subchronic drinking water study was conducted in which mice were exposed to TCE at concentrations up to 5000 ppm for periods up to six months (Tucker et al., 1982). The following minimal toxicological effects were demonstrated: increases in liver and kidney weights, changes in hepatic microsomal drug-metabolizing enzymes, and alterations in urinary protein and ketone levels.

The only long-term mouse and rat toxicology studies employing oral administration of TCE were performed as part of the investigation of TCE's carcinogenic potential (National Cancer Institute, 1976). Dose-related reduction in body weight gain and decreased survival time were manifested in male and female rats throughout the study. Although no noncarcinogenic histopathological changes were detected in the livers of mice or rats, slight to moderate degeneration of renal proximal tubular epithelium was present in the low and high dose groups of both male and female mice and rats.

HEPATOTOXICITY OF CARBON TETRACHLORIDE (CT)

CT has been widely used in the manufacture of chlorofluorcarbons, as grain fumigant, in various solvents, and in numerous cleaning agents. CT is a relatively stable environmental contaminant, and environmental exposure to man can occur via the air, water, and food. Acute oral and inhalation studies with CT indicate that the primary target organ is the liver. In fact, CT is one of the most intensively investigated hepatotoxic agents with respect to its short-term toxicity and mechanism of toxicity. The current view is that the molecular target of CT is various cellular membranes, especially the hepatic endoplasmic reticulum. It is thought that the initial biochemical events in CT hepatotoxicity are the covalent binding of CT reactive metabolites to microsomal lipids and proteins, lipid peroxidation of polyunsaturated fatty acids in membranes and the inhibition of microsomal calcium pump activity (Recknagel and Glende, 1973; Recknagel, 1983; Mitchell et al., 1984; Reynolds and Moslen, 1974). It is not currently possible to state which of these events or combination of events is ultimately responsible for the hepatotoxicity of CT.

Humans are exposed to CT via the environment and the workplace. It has become evident that CT is potentially toxic to humans at low doses by either the oral or inhalation routes of exposure (Louria and Bogden, 1980). Individuals with previous hepatic injury, such as alcohol induced cirrhosis, are more sensitive to the hepatotoxicity of CT. Although there have been a number of subchronic and chronic inhalation studies with CT, none have adequately defined the oral dose response relationships of CT hepatotoxicity. Alumot and coworkers (1976) fed rats diets containing CT which resulted in a daily exposure of up to 18 mg/kg body weight for two years duration. Unfortunately, no biochemical, clinical, or histological changes could be detected. The studies were also hampered by chronic respiratory disease and poor survival of the rats. Prompted by the need for subchronic toxicity data for risk assessment of the noncarcinogenic effects of ingested CT, the EPA sponsored two subchronic toxicity studies in which CT was administered by oral gavage (Bruckner et al., 1985; Hayes et al., 1985b).

In one subchronic study adult, male Sprague-Dawley rats were administered CT by gavage at dose levels of 1, 10, and 33 mg/kg/day in a total volume of 1.0 ml corn oil per dose (Bruckner et al., 1985). The rats were dosed five times weekly for a total of 12 weeks after which time nine rats per group were allowed to recover from the CT exposure for 13 additional days before being sacrificed.

A dose response relationship was seen in the subchronic rat study. Oral administration of 1 mg/kg was without effect on clinical chemistry values, liver and kidney morphology, and body weight gain. Animals ingesting 10 mg/kg CT exhibited slight alterations in some of the parameters that were measured. Sorbitol dehydrogenase showed a two- to three-fold increase over controls during the three-month period. Mild centrilobular vacuolization was seen in the liver of each animal examined at sacrifice and there was no evidence of other serious degenerative changes.

Repeated oral administration of 33 mg/kg CT resulted in substantial toxicity. Several serum enzymes that are indicators of hepatic damage were substantially increased. The livers of these rats exhibited extensive lesions indicative of cirrhosis. Vacuolization, periportal fibrosis and bile duct hyperplasia were present. Hepatocytes with multiple or hyperchromatic nuclei were commonly seen; hyperplastic nodules were also found. Many of the structural lesions persisted after the 13-day recovery period.

A second subchronic toxicity study was conducted with mice and consisted of 20 animals of each sex per group (Hayes et al., 1985b). The groups consisted of naive untreated mice, corn oil vehicle control, and four exposure groups (12, 120, 540, and 1200 mg/kg/day). Animals were dosed by corn oil gavage for 90 consecutive days.

In both male and female animals several serum enzyme activities were increased at all dosage levels in an apparent dose-dependent manner. Serum bilirubin values increased in a dose-dependent manner at the three higher doses. Blood glucose levels were decreased in a dose-dependent manner in both sexes. Serum protein and calcium levels were significantly increased at the high dose. Other increased serum parameters were cholesterol, phosphorus, chloride, and creatinine. Liver and spleen weights and organ to body weight ratios were increased in a dose-dependent fashion in both sexes. Thymus and brain weights and organ to body weight ratios were increased at the 540 and 1200 mg/kg doses.

Histopathological examination indicated that there was no evidence of kidney toxicity. Hepatotoxicity was prominent in the treated animals and the severity was dose dependent. Focal necrosis around central veins and inflammation was evident at the lowest dose in both males and females. Fat deposition and hepatocytomegaly was more severe in males than females at this dose. At 120 mg/kg, hepatoxic lesions were more severe than at the low dose, necrotic hepatitis was common and the inflammatory infiltrate often involved the portal area. Fat deposition and large foam cells were present. Two males at the 120 mg/kg dose, two males at the 540 mg/kg dose, and one female at the high dose had circumscribed nodules in the liver with a line of demarcation from the surrounding tissue (neoplastic nodules). At the two highest doses, necrosis was less apparent because regeneration of liver cells had occurred. Mitoses and karyomegaly were also common in all treatment groups. Significant findings from the two CT subchronic studies are presented in Table 1.

TABLE 1
Subchronic hepatotoxicity of carbon tetrachloride.

	CD-1 Mice[a]	Sprague-Dawley Rats[b]
Study duration	90 days	12 weeks 2 weeks recovery
Group size	20 males 20 females	15-16 males
Dosage	0, 12, 120, 540, 1200 mg/kg Gavage 7/week	0, 1, 10, 33 mg/kg Gavage 5/week
Vehicle	1.0 ml corn oil/100 gm	1.0 ml corn oil/dose
Effects	Effects were seen at all dose levels Increased liver weight Marked increases in serum enzymes Histopathological changes	1 mg/kg No adverse effect level 10 mg/kg Slight increases in sorbitol dehydrogenase activity Mild hepatic centrilobular vacuolization 33 mg/kg Marked hepatotoxicity Elevated serum enzyme levels (reversible) Cirrhosis present (irreversible)

[a]Hayes et al., 1985b.
[b]Bruckner et al., 1985.

Since humans are exposed to CT by several different routes, subchronic toxicity studies were needed in employing orally administered CT in order to provide data for risk assessment from drinking water exposure. These studies should have ideally employed a vehicle similar to the one by which humans are exposed. Exposure of CT via the food is impractical because of its high volatility, while the low water solubility of CT prevents utilization of high doses by this route. Therefore, the oral gavage route of amdinistration with corn oil as the vehicle was employed in the two subchronic CT toxicity studies (Bruckner et al., 1985; Hayes et al., 1985b).

However, a recently emphasized concern is whether administration of test substances such as CT by gavage introduces additional factors that may decrease the relevancy of the experimental data. Corn oil gavage of a test substance could modify the pharmacokinetics of that test substance significantly which might change its absorption, target organ concentration, metabolism, or elimination. In fact, the use of corn oil gavage delivers a bolus of high concentration of test substance and also a high caloric fat load to the animal which may alter the nutritional status of the experimental animal. Therefore, a study was conducted to determine the potential for the corn oil vehicle to alter the hepatotoxicity of CT (Condie et al., 1985).

Male and female CD-1 mice, 12 animals/group, were administered CT by oral gavage at dose levels of 0, 1.2, 12, and 120 mg/kg in either corn oil or aqueous suspension (1% Tween 60). The mice were dosed five times weekly for total of 90 days duration. Hepatotoxicity of CT was determined primarily by measuring serum enzyme activities (see Table 2) and by histopathological evaluation.

TABLE 2

Summary of serum enzyme activities from male and female mice exposed to carbon tetrachloride by either corn oil or aqueous gavage.[a,b]

Dose	Male Mice		Female Mice	
	Corn oil	1% Tween 60	Corn oil	1% Tween 60
1.2 mg/kg	NC	NC	NC	NC
12 mg/kg	+	NC	+	NC
120 mg/kg	+++	++	+++	++

[a] Condie et al., 1985
[b] NC = no change; + = slight increase; ++ = moderate increase; +++ = marked increase

Histopathological findings confirmed the no observed adverse effect level determined by serum enzyme activities. There were no significant differences between the low-dose (1.2 mg/kg CT) corn oil and aqueous suspension gavage groups of mice except for increased fatty deposits in the corn oil low-dose group. In the mid-dose exposure group, hepatocellular cytomegaly was prominent in the corn oil vehicle female and male groups but absent in the corresponding aqueous suspension vehicle groups. Hepatic necrosis was present in nine out of ten male, mid-dose corn oil vehicle animals and was absent in the corresponding aqueous suspension gavage male mice. In the high-dose exposure groups, hepatic necrosis was evident in all exposure groups but the incidence was 100% for the male and female corn oil vehicle animals as compared to 50% for the male and female aqueous suspension vehicle mice. These data indicate that under these experimental conditions the corn oil vehicle lowered the no observed adverse effect level from CT exposure by an order of magnitude and that the hepatotoxicity in the high exposure groups of animals was greatly enhanced by the administration of CT in corn oil as compared to the aqueous medium.

CONCLUSIONS

Despite wide gaps in our knowledge of the metabolism and ultimate fate of chemicals in humans, properly conducted animal experiments can improve our estimates of risk to human populations from environmental exposures to chemical substances. The experimental tests on laboratory animal are most valuable when the chemical agents are administered to the animals in a manner similar to human exposure. The bulk of toxicological data from experiments with halocarbons commonly found contaminating drinking water has come from inhalation studies. Since the target organ toxicity of many halocarbons appears to be greater from the inhalation route of administration as compared to the oral route, caution must be taken when extrapolating experimental data from the inhalation route of administration to the oral route. The EPA has conducted or sponsored numerous toxicology studies by the oral route of exposure to minimize certain events that can greatly influence the toxicological impact of a chemical. These events include absorption, distribution, metabolism, excretion, arrival at the site of action, and interaction with other substances.

While CT has been one of the most extensively investigated hepatotoxins, most of the studies have been short-term studies employing high doses of CT by various routes of administration. Since humans are exposed to CT by several sources, subchronic toxicity studies were needed which employed daily oral administration to provide data for hazard assessment. The experimental studies (Bruckner et al., 1985; Hayes et al., 1985b) indicate that mice and rats are quite sensitive to the continuous low-level exposure of CT. These studies should be useful in the assessment of short- and long-term risks of CT ingestion. However, there are potential confounding factors with these two subchronic studies since the CT was given by corn oil gavage. Scientific evidence indicates that untoward physiological and nutritional changes may be induced in rodents when vegetable oils are used as the vehicle for the administration of chemical substances (Nutrition Foundation, 1983). Studies conducted by the author (Condie et al., 1985) indicate that the corn oil gavage enchances the oral toxicity of CT. Perhaps additional studies are warranted by administering micro-encapsulated CT in the animal feed.

Clearly, additional research is needed to better define the toxicological properties of the halocarbons that are commonly found in drinking water. Because of the many uncertainties that exist in connection with the extrapolation of data from the reported animal experiments, better methodologies are needed to more clearly define the extent that humans are at risk from the low-level exposures to organic halocarbons in drinking water. For example, studies of the comparative metabolism of chemical substances of laboratory animals and man are urgently needed. There is also a need for better toxicological data for many drinking water contaminants which cannot be evaluated because of lack of toxicological experimentation. Additional toxicological information can be obtained about numerous halocarbons as well as other drinking water contaminants from a National Research Council series (1977, 1980, 1983).

REFERENCES

Alumot, E., Nachtomi, E., Mandel, E. and Holstein, P., 1976. Tolerance and acceptable daily intake of chlorinated fumigants in the rat diet. Food Cosmet. Toxicol., 14: 105-110.

Bruckner, J.V., Muralidhara, S., Luthra, R., Kyle, G.M., MacKenzie, W.F. and Acosta, D., 1985. Oral toxicity of carbon tetrachloride: acute, subacute and subchronic studies in rats. Fundam. Appl. Toxicol., in press.

Condie, L.W., Laurie, R.D., Robinson, M. and Bercz, J.P., 1985. Influence of corn oil gavage vehicle on hepatotoxicity of carbon tetrachloride. Submitted for publication.

Environmental Protection Agency, 1984. National primary drinking water regulations; volatile synthetic organic chemicals. Fed. Reg., 49: 24300-24355.

Hayes, J.R., Condie, L.W. and Borzelleca, J.F., 1985a. Acute, 14-day repeated dosing and 90-day subchronic toxicity studies of carbon tetrachloride in CD-1 mice. Submitted for publication.

Hayes, J.R., Condie, L.W. and Borzelleca, J.F., 1985b. The subchronic toxicity in rats of tetrachloroethylene (perchloroethylene) in drinking water. Submitted for publication.

Heppel, L.A., Neal. P.A., Perrin, T.L., Endicott, K.M. and Porterfield, V.T., 1946. The toxicology of 1,2-dichloroethane (ethylene dichloride). J. Ind. Hyg. Toxicol., 28: 113-120.

Hofmann, H.T., Birnsteil, H. and Jobst. P., 1971. Zur inhalationstoxicitat von 1,1-und 1,2-dichlorathan. Arch. Toxikol., 27: 248-265.

Lane, R.W., Riddle, B.L. and Borzelleca, J.F., 1982. Effects of 1,2-dichloroethane and 1,1,1-trichloroethane in drinking water on reproduction and development in mice. Toxicol. Appl. Pharmacol., 63: 409-421.

Louria, D.B. and Bogden, J.D., 1980. The dangers from limited exposure to carbon tetrachloride. CRC Crit. Rev. Toxicol. Aug.: 177-188.

Mitchell, J.R., Smith, C.V., Lauterburg, B.H., Hughes, H., Corcoran, G.B. and Horning, E.C., 1984. Reactive metabolites and the pathophysiology of acute lethal cell injury. In: J.R. Mitchell and M.G. Horning (Editors), Drug Metabolism and Drug Toxicity. Raven Press, New York, pp. 301-319.

National Cancer Institute, 1976. Carcinogenesis bioassay of trichloroethylene. NCI-CG-TR-2. DHEW Publ. No. (NIH) 76-802. Department of Health, Education and Welfare, Washington. D.C.

National Cancer Institute, 1977. Bioassay of tetrachloroethylene for possible carcinogenicity. NCI-CG-TR-13. DHEW Publ. No. (NIH) 77-813. Department of Health, Education and Welfare, Washington, D.C., 47 pp.

National Research Council, 1977. Drinking water and health. Report of the Safe Drinking Water Committee, Advisory Center on Toxicology, Assembly of Life Sciences. National Academy of Sciences, Washington, D.C., 489-856.

National Research Council, 1980. Drinking water and health. Report of the Safe Drinking Water Committee, Board on Toxicology and Environmental Health Hazards, Assembly of Life Sciences. National Academy Press, Washington, D.C., 67-263.

National Research Council, 1983. Drinking water and health. Report of the Safe Drinking Water Committee, Board on Toxicology and Environmental Health Hazards, Commission of Life Sciences. National Academy Press, Washington, D.C., 9-117.

Nutrition Foundation, 1983. Report of the ad hoc working group on oil/gavage in toxicology. The Nutrition Foundation, Washington, D.C., 45 pp.

Recknagel, R.O., 1983. A new direction in the study of carbon tetrachloride hepatotoxicity. Life Sci., 33: 401-408.

Recknagel, R.O. and Glende, E.A., 1973. Carbon tetrachloride hepatotoxicity: an example of lethal cleavage. CRC Crit. Rev. Toxicol. Nov.: 263-297.

Reynolds, E.S. and Moslen, M.T., 1974. Chemical modulation of early carbon tetrachloride liver injury. Toxicol. Appl. Pharmacol., 29: 377-388.

Schumman, A.M., Quast. J.F. and Watanabe, P.G., 1980. The pharmacokinetics and macromolecular interactions of perchloroethylene in mice and rats as related to oncogenicity. Toxicol. Appl. Pharmacol., 55: 207-219.

Spencer, H.C., Rowe, V.K., Adams, E.M., McCollister, D.D. and Irish, D.D., 1951. Vapor toxicity of ethylene dichloride determined by experiments on laboratory animals. Arch. Ind. Hyg. Occup. Med., 4: 482-493.

Stott, W.T., Quast, J.F. and Watanabe, P.G., 1982. The pharmacokinetics and macromolecular interactions of trichloroethylene in mice and rats. Toxicol. Appl. Pharmacol., 62: 137-151.

Tucker. A.N., Sanders, V.M., Barnes, D.W., Bradshaw, T.J., White, K.L., Sain, L.E., Borzelleca, J.F. and Munson, A.E., 1982. Toxicology of trichloroethylene in the mouse. Toxicol. Appl. Pharmacol., 62: 351-357.

INHALATION EXPOSURE IN THE HOME TO VOLATILE ORGANIC CONTAMINANTS OF DRINKING WATER

JULIAN B. ANDELMAN

Center for Environmental Epidemiology, Graduate School of Public Health, University of Pittsburgh, Pittsburgh, Pennsylvania 15261 (USA)

ABSTRACT

Our field studies show that indoor air concentrations of volatilized trichloroethylene (TCE) can be substantial when TCE-contaminated water is used domestically. Using a model shower, increases in TCE water concentrations, water temperature and drop path (time) increased the steady-state air TCE concentrations. Volatilization was incomplete and the rates were comparable to predicted ones. Indoor air models show that the inhalation route of exposure for such chemicals has the potential for being much greater than by direct ingestion. This should be considered in developing regulations to limit adverse health impacts from contaminants of potable water.

INTRODUCTION

There is a growing awareness of the presence of a wide range of chemicals that can occur in the home environment and cause a continuous, although normally low-level human exposure. The regulation of the chemical quality of public water supplies has been based on the assumption that the ingestion of such water constitutes the only or at least principal route by which an adverse human exposure can occur. The question arises, can potable waters brought into the home contribute significantly to these exposures by routes other than ingestion, specifically as a result of the volatilization of chemicals into the indoor air, resulting in an inhalation dose; also due to the absorption of these chemicals through the skin, primarily during bathing when a large portion of the body surface comes in contact with the water for substantial periods of time? Second, if such exposures can be expected to occur and are significant, how do they compare quantitatively to those from the direct ingestion of water?

In recent years there has been a large number of studies and assessments of indoor air pollution, such as that by a committee of the National Research Council (1981), and more recently by Spengler and Sexton (1983). These list a variety of sources of indoor air pollutants, including consumer products, materials of construction, combustion sources and tobacco smoke, but rarely refer to potable water as a possible vehicle or source, other than for radon. Concern about indoor radon exposure from water has reached the point of discussion by the U.S. Environmental Protection Agency (1983) in an advance notice of proposed rulemaking (ANPR) concerning its possible revision of primary drinking water regulations. It is noted there that "airborne exposure from radon released into the home from water might be more significant than direct ingestion from drinking water;" also, that "it appears that radon may contribute one of the most significant cancer risks of any substance in drinking water." They indicate that maximum contaminant limits are under consideration and request comments on the need for such regulation of radon.

The U.S. Environmental Protection Agency (1982) noted similarly in an ANPR that it was considering the need to regulate volatile synthetic organic chemicals in drinking water. However, neither in the 1982 or 1983 ANPR did the discussion consider the possible inhalation exposure route in the home for these volatile chemicals, other than for radon. Similarly, dermal exposures were not discussed. Also these routes of exposure have been considered elsewhere only rarely.

The purpose of this paper is to consider the likely extent of the inhalation routes of exposure to chemicals in potable water brought into the home. Skin absorption of such chemicals during bathing has been discussed by others (Brown et al., 1984), indicating that this route of exposure may be comparable to that from the direct ingestion of water.

HUMAN EXPOSURE PARAMETERS AND AIR AND WATER USES

In order to assess human exposure to and intake of chemicals from water within a home it is necessary to have an understanding of the quantities and the ways in which such water is used. Both of these are highly variable. Table 1 shows a range of such typical daily per capita domestic uses in the U.S. These values are normally multiplied by the number of people living within a home to estimate the quantity of water brought into the home on a daily basis. Using these data for a family of four, the range is about 160 to 280 gallons or 600 to

TABLE 1

TYPICAL DAILY U.S. DOMESTIC WATER USE PER CAPITA (BOND ET AL., 1973)

Use	Quantity per capita per day	
	Gallons	Liters
Drinking and cooking	1-2	4-8
Dishwashing	1-4	4-15
Garbage disposal unit	0-4	0-15
Laundering, cleaning	3-9	11-34
Bathing	10-25	38-95
Toilet	24	91
Total	39-68	148-257

1000 liters per day. It is apparent from Table 1 that the quantities of water used in the home that most directly relate to ingestion (drinking and cooking) are a very small part of the total water usage. Much larger quantities of water are used that can and do result in volatilization of chemicals in the home and their subsequent inhalation.

Just as the water uses within a home are highly variable, so are the daily quantities of water ingested by humans. The International Commission for Radiological Protection (ICRP) has assessed various reports on the ranges of such intakes measured for adults and children under a variety of conditions such as ambient air temperature and nature of human activity (ICRP, 1975). Direct intake of tap water itself, in contrast to water intake from milk or water-based drinks (tea, coffee, bottled beverages), can be small. The ICRP reports that under normal conditions the direct intake of tap water ranges from 45 to 730 mL per day for adults, compared to a total fluid (water) intake range of 1000 to 2400 mL. More recently the variability of drinking water intakes by adults from studies in Canada, Great Britain, the Netherlands and New Zealand has been reported (Gillies and Paulin, 1983). Among these studies the mean daily intakes, including those for beverages made with tap water, ranged from 0.96 to 1.30 liters per day. In one of the New Zealand studies the relative standard deviation of such intakes for about 2000 adults was 570 mL per day, the mean being 960 mL per day. In order to have a standardized basis for assessing the quantities of contaminants in potable water that may be ingested, the ICRP has developed a fluid (water) intake table for reference man (Table 2). For reference adult man, adult woman and child the tap water reference intake is in each case less than 15 percent of the total fluid (water) intake.

TABLE 2

DAILY FLUID INTAKE FOR REFERENCE MAN (ICRP, 1975)

Source	Fluid intake (mL per capita per day)		
	Adult man	Adult woman	Child (10 yr)
Milk	300	200	450
Tap water	150	100	200
Other	1,500	1,100	750
Total fluid	1,950	1,400	1,400

The 1950 mL per capita total fluid intake for reference adult man is approximately equal to the 2-liter quantity frequently used to establish standards for and assess health effects from exposures to chemicals and other constituents in public water supplies. It is conservative to do so (in a protective sense). Nevertheless, to assess more accurately such impacts, particularly in relation to other human inhalation exposures to chemicals in water, the types of waters ingested and their variabilities should be considered.

Although respiratory rates can also be quite variable, the ICRP has developed a set of rates and volumes for reference man air intake, as shown in Table 3. Although the 10 m^3 per day value corresponding to the respiratory volume for adult man of 9.6 m^3 in an 8-hour working period is often used in assessing daily exposures in the workplace, for air exposures within a home a different value should be used normally, depending on the fraction of the day spent in the home and the activities undertaken. Thus, for example, an adult not leaving the home might appropriately be considered to have a daily respiratory volume of about 20 m^3, twice the amount normally used for an 8-hour work exposure period. At the same time, where there is a shorter period of unusual air exposure, such as in a shower or bathroom contaminated with a volatile chemical, a short-term respiratory rate is of interest. From Table 3, a value of 1 to 1.2 m^3 per hour is a good estimate for adult reference man or woman in this regard.

TABLE 3

RESPIRATORY VOLUMES FOR REFERENCE MAN, CUBIC METERS OF AIR BREATHED (ICRP, 1975)

Activity	Respiratory volume (m^3)			
	Adult man	Adult woman	Child (10 yr)	Infant (1 yr)
8-hr working light activity	9.6	9.1	6.2	2.5 (10 hr)
8-hr non-occupational activity	9.6	9.1	6.2	-
8-hr resting	3.6	2.9	2.3	1.3 (14 hr)
Total (m^3/day)	23	21	15	3.8

Such estimates for water and air exposures do not address the question of the absorption of chemicals from the air and water media taken in via ingestion or inhalation. The relative absorption factors for a given intake by these two routes of exposure can be quite variable and specific to each chemical. These will not be addressed. Our principal focus will be exposure, which we define as the quantity of a chemical contained in a given amount of inhaled air or ingested water.

PHYSICO-CHEMICAL ASPECTS

The volatilization of chemicals from water used within buildings has both equilibrium and time (rate) components that should be considered in estimating the indoor air concentrations that can ensue. These in turn will interact with the air movement factors, such as the rates of forced air ventilation, infiltration and exfiltration to establish the air concentrations to which people will be exposed.

The equilibrium and kinetic aspects of such volatilization have been assessed for a variety of low solubility chemicals that may be encountered as contaminants of natural and treated fresh waters (MacKay and Wolkoff, 1973). They showed that a convenient expression for the equilibrium of the volatile chemical between the air and water phases is

$$P_i = C_i P_{is}/C_{is} \tag{1}$$

where P_i is the air concentration (atm) of the compound "i" in equilibrium with its water concentration C_i (mol/m^3, essentially millimol/L); P_{is} is the vapor pressure of the pure compound and C_{is} its solubility in water. The ratio P_{is}/C_{is} is the well known Henry's law constant H_i. Thus

$$H_i = P_{is}/C_{is} \tag{2}$$

and therefore

$$P_i = H_i C_i \tag{3}$$

This is Henry's law which states simply that the equilibrium vapor pressure of a volatile constituent above an aqueous solution is directly proportional to its concentration in that aqueous solution. Although the Henry's law constant has been measured for some anthropogenic chemicals, Mackay and Wolkoff's analysis is useful in that it is a tool to predict values of H_i for chemicals of limited water solubility from other measured quantities, namely the vapor pressure of the pure chemical and its solubility in water.

Temperature usually affects equilibrium constants such as H_i. The vapor pressure of pure compounds will increase with temperature. Solubilities of chemicals in water can either decrease or increase with temperature. Over the temperature range of 20 to 30 °C carbon tetrachloride solubility in water is relatively constant at about 800 mg/L, while chloroform decreases in solubility slightly with increased temperature from about 8.1 to 7.6 g/L over the same temperature range (National Academy of Sciences, 1978). Yet over that temperature range the vapor pressures of the pure liquids increase substantially with temperature, from 90 to 140 mm mercury for carbon tetrachloride, and from 155 to 246 mm mercury for chloroform. Thus Equation 2 indicates that for each of these chemicals the net effect is that H_i will increase over this temperature range; and for a given concentration in water the equilibrium vapor pressure for each of these constituents will similarly increase. This effect can be important in establishing the extent to which such chemicals in water brought into a home may volatilize and be affected by the temperature of the water use.

Mackay and Wolkoff (1973) further discuss how the <u>rate</u> of evaporation of such chemicals from dilute aqueous solution can be expressed by a first order differential equation:

$$dC_i/dt = kM_iH_iC_i \qquad (4)$$

where k is a proportionality constant including such factors as the rate of evaporation of water and its vapor pressure; M_i is the molecular weight of the evaporating chemical; and t is time. When the volatilizing chemical in the air above the water in a home is below the equilibrium vapor pressure, there is a "driving force" to volatilize. As the air concentration of the chemical builds up and approaches equilibrium, the reverse process becomes important, so that the net evaporation rate decreases.

These concepts have been utilized in interpreting the results of experimental studies of the transfer of six organic chemicals (listed in Table 4) from aqueous solution to the atmosphere, using oxygen as a reference compound (Roberts and Dandliker, 1983). The chemicals listed there are in order, dichloro-difluoromethane, carbon tetrachloride, tetrachloroethylene, 1,1,1-trichloroethane, trichloroethylene (TCE) and chloroform, most of which are of

TABLE 4

AIR-WATER EQUILIBRIUM CONSTANTS AND TRANSFER RATE COEFFICIENTS FOR SEVERAL VOLATILE ORGANIC CONTAMINANTS (ROBERTS AND DANDLIKER, 1983)

Compound	Henry's const. (atm x m^3/mol)	Mass trans. prop. coeff.[a]
CCl_2F_2	1.5	0.66
CCl_4	2.4×10^{-2}	0.62
$CCl_2=CCl_2$	1.5×10^{-2}	0.61
CH_3CCl_3	1.5×10^{-2}	0.61
$CHCl=CCl_2$	9.9×10^{-3}	0.62
$CHCl_3$	5.3×10^{-3}	0.56

[a]Compared to oxygen; see text

interest in that they have been found to be present in many groundwaters and public water supplies.

As shown in Table 4, aside from the first compound, the Henry's law constants, H_i, do not vary by more than a factor of about five. However, a more varied group of compounds of limited water solubility can have a wide range of values of H_i (Mackay and Leinonen, 1975). For example, they reported values of 2×10^{-7}, 4×10^{-5}, 6×10^{-4}, and 7×10^{-3} for dieldrin, DDT, Aroclor 1242 (a PCB) and toluene, respectively, at 25 °C. Thus in order to assess a possible equilibrium situation that might be encountered, such as in a home, the H_i value for the specific compound must be utilized, rather than assuming that it might lie in the relatively narrow range of values listed in Table 4.

Roberts and Dandliker (1983) evaluated their evaporation rate experiments in terms of a traditional mass-transfer rate expression:

$$dC_i/dt = K_L A C_i \qquad (5)$$

where K_L is the overall mass-transfer rate coefficient and A is the specific surface area (m^2) per volume of aqueous solution from which the solute is evaporating. This is similar to Equation 4, but states explicitly that the rate of evaporation will necessarily increase in proportion to the surface area of the solution. Roberts and Dandliker note that the mass transfer rate coefficient K_L may be influenced by both the liquid- and gas-phase resistances (to diffusion) at the air-water interface. Following the approach of Mackay and Leinonen (1975), they estimated that for all compounds with H_i greater than 4.8×10^{-3}, which includes all of those listed in Table 4, liquid phase resistance is rate limiting. Interpreting their experimental results in terms of Equation 5, they defined a mass transfer proportionality constant essentially as the ratio of K_L for each compound divided by that for oxygen. In Table 4 it is seen that for each compound this has a value of about 0.6, consistent with the fact that for each of these compounds the diffusivity in water is about 0.6 times that of oxygen (Roberts and Dandliker, 1983), and therefore that the rate controlling step in these experiments is liquid phase diffusion. This then provides the basis for relating mass transfer rates from water surfaces to the diffusion coefficients (diffusivity) in the aqueous phase. However, the authors note that additional work is needed to assess the possible influence of gas-phase resistance in practical applications.

INDOOR AIR MODELING TO ESTIMATE RELATIVE AIR AND WATER EXPOSURES

Various models have been used to estimate expected indoor pollutant concentrations, including one- and multi-compartment models, as well as empirical models based on statistical evaluation of concurrent indoor and outdoor air concentrations and other relevant terms (Wadden and Scheff, 1983). The single compartment model has been used widely in analyzing indoor air concentrations, as described by Wadden and Scheff. Two air concentration terms are used, C_i and C_o, the indoor and outdoor values, respectively. The source term S is the quantity of pollutant generated indoors per unit time (e.g., milligrams per hour). The model can encompass a pollutant decay factor R, often treated as a first-order rate proportional to C_i, and a mixing factor k, which reflects a deviation from ideal mixing within the home (k = 1) and which has been found to range from 0.1 to 0.33. This mixing factor is important in that, to the extent that there is "dead space" within a home, the effective indoor air volume in this model will decrease and the steady-state indoor air concentration from a pollutant source will increase.

The outdoor air concentration C_o is of importance in the model in that it will manifest itself as an indoor air concentration from makeup air and infiltration. Also, because of loss to the outdoors by exfiltration and exhaust, even if there is no decay mechanism for the pollutant or removal by filtration in air recirculation, there will be a finite buildup with time for a continuously generated pollutant.

This single compartment model in simplified form has been used to estimate the steady-state indoor air concentration of a continuously generated pollutant (Wadden and Scheff, 1983). The parameters are as follows: V, home air volume, 450 m^3; q_1, indoor air recirculation rate, 1350 m^3/hr (this is equivalent to three room air changes per hour); q_2, infiltration rate, 338 m^2/hr (this was taken also to be the exfiltration rate, q_3); and k, the mixing factor, 0.15. They further assumed that the outdoor air concentration of the pollutant C_o was zero and that there was no removal of gaseous pollutants by the filters. For certain chemicals, such as formaldehyde and carbon monoxide, they assumed that there were no decay mechanisms within the home.

For this set of assumptions the steady-state form of the equation based upon the model takes the simple form:

$$C_i = S/(kq_2) \tag{6}$$

Thus the steady-state indoor air concentration is directly proportional to the source term S and inversely to the mixing factor and infiltration rate. Using the values cited above for the latter terms

$$C_i = 0.02S \tag{7}$$

where C_i has the units of mass/m^3 and the source term S, mass/hr.

It should be noted that the air recirculation rate q_1 does not appear in Equation 6, nor does the indoor air volume, because it was assumed that there were no removal nor decay mechanisms within the home. When these occur, the final form of the equation becomes more complex.

Such an approach was used to relate measurements of radon within buildings to the uses of water from which the radon volatilizes (Prichard and Gesell, 1981). They utilized a simple single compartment model which they expressed in the form

$$C_i = (C_w/24q_2)\sum_j e_j W_j \tag{8}$$

where q_2 is the volume of outdoor air passed through the dwelling per hour; "j" refers to each different use of water in the home that can result in volatilization of radon; e_j is the transfer efficiency (fraction of radon present that volatilizes) in that use; W_j is the average daily quantity of water in such use; C_w is the concentration of radon in the water brought into the home; and C_i is the calculated indoor air radon concentration. The symbols for the terms in the Equation 8 have been modified to be consistent with those used in this paper. Equations 6 and 8 are essentially the same in form in that $C_w \sum_j e_j w_j$ is equivalent to the source term S of Equation 6 and $24q_2$ is the daily air flow through the home, equivalent to kq_2 of Equation 6. The Prichard and Gesell model does not, however, utilize a mixing factor term k.

Equation 8 predicts that the steady-state indoor air radon concentration from water will be proportional to its concentration in the water when the other parameters are constant. Table 5 shows the result of their calculation of the

TABLE 5

RADON RELEASE FROM DOMESTIC USE OF WATER CONTAINING 1000 pCi RADON PER LITER (PRICHARD AND GESELL, 1981)

Use	Daily consumption (liters)	Transfer efficiency (%)	Radon liberated (pCi)
Showers	150	63	94,500
Tub baths	150	47	70,500
Toilet	365	30	109,500
Laundry	130	90	117,000
Dishwasher	55	90	49,500
Drinking and kitchen	30	30	9,000
Cleaning	10	90	9,000
Total	890	-	459,000

quantities of radon released (from water containing 1000 pC$_i$ radon/L) for each water use estimated for a typical family of four, using transfer efficiencies determined experimentally. The transfer efficiencies shown there vary from 30 to 90 percent, but the average efficiency weighted by water use is about 50 percent; that is, 50 percent of all radon brought into the home in these water uses is volatilized within the home, the remainder presumably leaving in the domestic wastewater. They note that since radon is not chemically reactive, the removal patterns by air flow and otherwise follows the same pattern as the loss of any other relatively non-reactive pollutant.

They also estimated the variability in steady-state concentrations in air that could be encountered, depending on the type of building and air exchange rates. These estimates indicate that the air radon concentrations from water can be expected to vary considerably, the extremes of the range differing by a factor of 50. Finally they note that the exposure models were corroborated by a series of measurements of radon in air, the results suporting the models within experimental error.

The likely extent of volatilization or transfer efficiency in indoor water uses will be affected by the Henry's Law constant. For example, using for chloroform the H value of 5.3×10^{-3} shown in Table 4, one can calculate at 25 °C using Equation 3 and the ideal gas law that at equilibrium

$$C_A = 0.2 C_W \tag{9}$$

where C_A and C_W are the respective concentrations in air and water in units of mass per volume, such as mg/L. For a typical U.S. domestic water use of 30 L/hr and an air flow rate of 3×10^5 L/hr, the ratio of air to water use V_A/V_W is 10^4. At Henry's Law equilibrium, in such uses one would predict that the relative masses of a volatile constituent that would be distributed between the air and water, as expressed by M_A/M_W would be

$$M_A/M_W = (C_A/C_W)(V_A/V_W) \tag{10}$$

For chloroform, using Equation 9 in conjunction with Equation 10 and the value of 10^4 for V_A/V_W, one calculates a value for M_A/M_W equal to 2000. This indicates that the equilibrium for chloroform is displaced in the direction of essentially complete volatilization, as it would be for many volatile constituents. In such cases the principal factors that would affect the achievement of equilibrium are the time of water exposure to air in the home and the area of the water-air interface. The latter in turn would be influenced substantially by the hydrodynamic regime, such as the geometry and turbulence in a clothes washer.

The simple indoor air model that leads to Equations 6 and 7 can be used to generate an estimate of the maximum indoor air concentrations that are likely to result from the volatilization of chemicals from water brought into a home. For example, one can use the data from Table 1 which indicate that the per capita total daily indoor use of water is in the range of 39 to 68 gallons per day or 592 to 1028 liters per day for a family of four. Taking a value of 720 liters per day, this is equivalent to 30 liters per hour. If one assumes further that all of the volatile constituent at concentration C_w (mass per liter) in the water volatilizes into the indoor air (100 percent transfer efficiency), the source term S becomes

$$S = 30 C_w \tag{11}$$

which when combined with Equation 7 leads to

$$C_i = 0.6 C_w \tag{12}$$

recognizing that the concentration units are different (C_i, mass per m^3 in air; C_w, mass per L in water).

The next question is that of the relative exposures to a volatile pollutant from the ingestion and inhalation routes. Using the data from Tables 2 and 3 for reference man, and taking adult man as an example, one can use a daily respiratory volume of 20 m^3 and a daily water intake of either 2 liters (total fluid) or 0.15 liter (tap water). The daily exposure from water to a volatile constituent is then calculated simply by multiplying the concentration in water C_w (mg/L) by either 2 or 0.15. The daily exposure from air is the product of 20 times C_i, using Equation 12 to determine the latter. These are presented in Table 6 for the water concentration of a constituent, C_w, and a specific example

TABLE 6

ESTIMATED INDOOR INHALATION AND INGESTION EXPOSURES FOR AN ADULT MALE FROM VOLATILE CHEMICALS IN WATER USING A ONE-COMPARTMENT INDOOR AIR QUALITY MODEL

Water conc. (mg/L)	Exposure, mg/day		
	Air	Water 2 L intake	Water 0.15 L intake
C_w	12 C_w	2 C_w	0.15 C_w
0.01	0.12	0.02	0.0015

of the latter, namely 0.01 mg/L. It is apparent from Table 6 that exposure from the air route is substantially greater than that from water ingestion using this single compartment air model and the assumptions stated above. If a daily ingestion of two liters is assumed, the air exposure is higher than that from water by a factor of 6. This ratio becomes much larger, namely 80, if the smaller estimate of 150 mL per day is used for water ingestion. One should be cautious, however, in using these ratios as more than an example of the possible relative exposures that can occur via the drinking water and air inhalation routes. Nevertheless, they do indicate the possibility of substantially greater air exposures to volatilized constituents from water used within the home and, therefore, the need to consider this route of exposure in assessing health effects from such contaminants in potable water supplies.

TRICHLOROETHYLENE IN FIELD AND LABORATORY STUDIES

We investigated the possible volatilization of trichloroethylene (TCE) into indoor air within buildings in a small community using individual wells obtaining water from an aquifier measured to contain about 40 mg TCE/L (Andelman et al., 1985). Using a continuous real-time monitor with an infra-red detector, measurements were taken in closed rooms in two homes and a small municipal building on one day in July 1983. Prior to turning on water in bathrooms, no TCE could be detected in the indoor air above the detection limit for the instrument, namely 0.5 mg/m^3. However, TCE was readily detected in the bathrooms with water running. As expected, the air concentration levels increased with time as shown in Table 7. In home B, the highest air concentration of TCE measured after 17 minutes of the shower running was 81 mg/m^3, approximately one-third of the eight-hour time-weighted threshold limit value of 270 mg/m^3 for the work environment (American Conference of Governmental Industrial Hygienists, 1984). To estimate the possible inhalation exposures within these homes from these shower uses, one can take a value of about 40 mg/m^3 as the average air concentration during the shower period. Assuming one-hour per week in the shower and an air breathing volume of 1.2 m^3 in that hour, this would correspond to a dose of 48 mg TCE per week (neglecting any question of the fraction of the dose that is absorbed). If one also assumes an ingestion of the contaminated water at 150 mL/day (1.05 L/week), the water dose is 42 mg TCE per week, indicating that the shower air and ingestion doses may be reasonably comparable. One would also expect an increased air dose from exposure elsewhere in the home.

To investigate further the factors that influence TCE air concentrations resulting from showers, a scaled-down model shower was constructed and operated with known concentrations of TCE injected continuously into the inlet water, typically in the range of 1.5-2.9 mg TCE/L. The experimental details and

TABLE 7

INDOOR AIR CONCENTRATIONS OF TRICHLOROETHYLENE (TCE) IN BUILDINGS USING WELL WATER CONTAINING APPROXIMATELY 40 mg TCE PER LITER (ANDELMAN ET AL., 1985)

Location	Time	Conc., mg/m^3
Municipal Bldg.		
Ladies rest room, water running	12:20	35
Home A		
Kitchen	1:34	ND[a]
Bathroom, shower on	1:53	32
	1:56	48
	2:01	67
	2:03	72
Home B		
Bathroom upstairs, shower on	2:20	ND
	2:27	19
	2:30	35
	2:33	40
	2:37	81
Bathroom downstairs, shower on	3:08	ND
	3:17	64
	3:24	67

[a]ND - not detected above detection limit of 0.5 mg/m^3

complete results are reported elsewhere (Andelman et al., 1985). The shower chamber volume was 0.1 m^3 and the regulated air throughput 0.005 m^3/min, equivalent to three air changes per hour. Based on the diameter of the shower head orifices, it was estimated that a spherical droplet emitted from the shower head would have a diameter of approximately 0.25 mm. The shower water flow was 0.3 L/min at a temperature which was varied between 23 and 41 $^{\circ}$C, the drop path being adjusted from 25 to 75 cm.

Continuous monitoring of air concentrations of TCE in the shower chamber was performed by pumping air from it to either an infra-red or flame ionization detection system. Steady-state air concentrations were achieved typically within 60 minutes, after which TCE injection was discontinued. The decay in TCE air concentration was also monitored.

Increases in TCE inlet water concentration, water temperature and drop path height all increased the steady-state air TCE concentrations, as expected. Figs. 1 and 2 are typical buildup and decay curves showing the effects of the two latter parameters. In all experiments the measured air TCE concentrations were considerably lower than those predicted by Henry's Law for the corresponding water TCE concentrations. At 25 $^{\circ}$C one can calculate such an

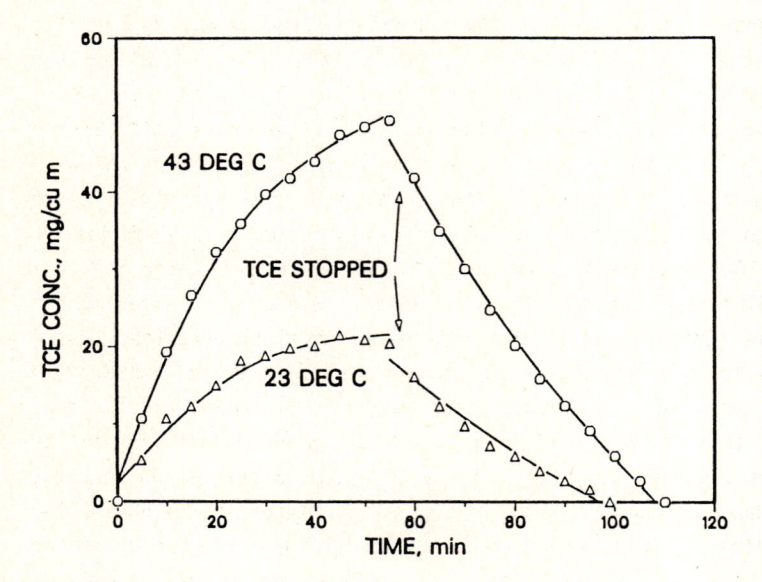

Fig. 1. Effect of water temperature on TCE air concentration in model shower.

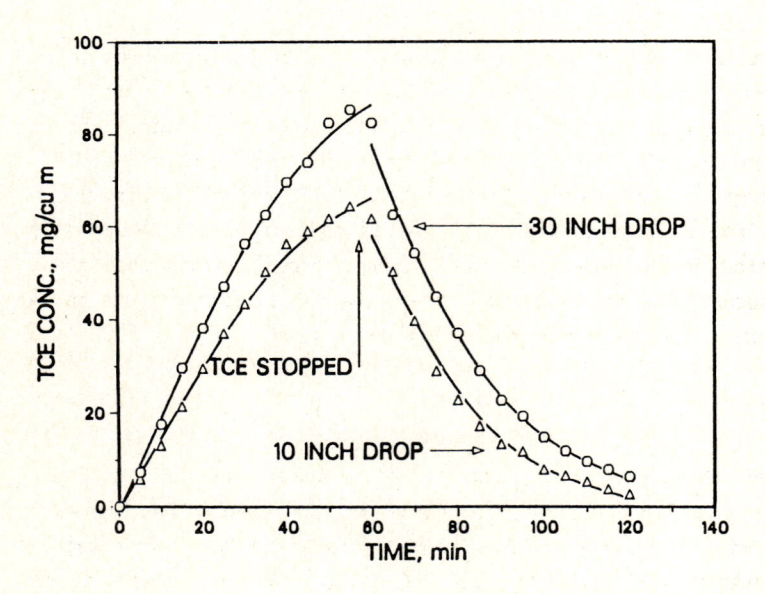

Fig. 2. Effect of drop path height on TCE air concentration in model shower.

equilibrium relationship for TCE comparable to that for chloroform in Equation 9. For TCE at this temperature C_A/C_W would thus equal 0.4. For the typical shower water concentration of 2 mg TCE/L used in these experiments, the equilibrium air concentration expected at 25 $^\circ$C is then 0.8 mg TCE/L or 800 mg/m^3, well above the approximately 20 mg/m^3 shown in Fig. 1 for the 23 $^\circ$C curve. Thus one can conclude that the rate of volatilization is sufficiently slow so that Henry's Law equilibrium is not attained and that the interplay between the kinetics of volatilization and the transport of TCE from the chamber by air passed through it establishes the TCE air concentration at a given time.

Measurements of TCE in the inlet and effluent shower chamber water indicate that typically 40-60 percent of the TCE volatilized, depending on experimental conditions. This is generally consistent with theoretical expectations, based on a model of gas transfer across an air-water interface (Liss and Slater, 1974). Assuming a probable 1.0 m/sec droplet velocity for the likely droplet size in our system, their model predicts about 20-45 percent volatilization for the 0.25-0.75 second drop time expected for the 10 inch (25 cm) and 30 inch (75 cm) droplet paths shown in Fig. 2. For many volatile solutes the rate-limiting step in the volatilization process is diffusion across the liquid film at the water-air interface. The substantial increase with temperature in the rate of TCE volatilization shown in Fig. 1 is consistent with the well known effect of temperature on diffusion of solutes in aqueous solution, including across such a liquid film.

Based on the fact that the measured air concentrations of TCE were well below those expected at equilibrium, and that 40-60 percent of the TCE volatilized from the water, one can write a mass-balance equation using a constant rate of input, K, for TCE volatilizing from the shower water per unit time. This K in fact represents a complex process of a first-order rate of volatilization which gradually decreases for a given drop of water as it moves from the shower head and reaches the bottom of the shower chamber. Nevertheless, within that short time (less than a second) one can reasonably treat the volatilization rate on an integrated average basis. The mass-balance equation is then

$$V(dC_A/dt) = K - FC_A \tag{13}$$

where V is the shower chamber volume, C_A the air concentration of TCE, and F the rate of flow of air through the chamber. This equation assumes that there is rapid and complete mixing within the chamber. (Although we have experimental evidence to the contrary, we have not yet quantified the incompleteness of mixing.) As the TCE air concentration increases, eventually the rate of removal will equal the rate of volatilization and a steady-state will be attained. At that point dC_A/dt will be equal to zero and, from Equation 13,

$$C_A(\text{steady-state}) = K/F \qquad (14)$$

The function describing the build up to steady-state is obtained by integrating Equation 13 to yield

$$\ln(1 - FC_A/K) = -(F/V)t \qquad (15)$$

Equation 15 describes the build-up portion of the experimental curves of Figs. 1 and 2 to the point where the TCE injection is stopped. In several of our experiments, such as the 23 °C curve of Fig. 1, it appears that steady-state was reached by the time the TCE injection was stopped. The function for the time required to attain a given fraction f of the steady-state concentration is obtained by combining Equations 14 and 15 to give

$$\ln(1 - f) = -(F/V)t_f \qquad (16)$$

Taking the shower system parameters of F and V equal to 0.005 m^3/min and 0.1 m^3, respectively, and using Equation 16, $t_{0.9}$ has a value of 46 minutes, somewhat smaller than the approximately 55 minute time when the TCE injection was normally discontinued, and consistent with a leveling off of the TCE air concentration around that time.

The decay portion of the curve after discontinuing TCE injection is represented simply by Equation 13 with K equal to zero, the integrated form being

$$\ln(C_o/C_t) = (F/V)t \qquad (17)$$

Although not shown here, the decay curves did follow this relationship with constant values for the slopes when the F and V parameters were not changed.

Extending these results from the scaled-down model shower system to full-size shower systems with their considerably longer droplet paths, one can expect that a large fraction of TCE and other highly volatile contaminants of potable water would volatilize in shower usage. This can result in an inhalation exposure, both to the user of the shower and to other inhabitants of the home as the shower air is distributed through it.

SUMMARY AND CONCLUSIONS

Air and water measurements in homes have shown the presence of trichloroethylene (TCE) and other volatile chemicals, such as radon, due to contamination of groundwater systems. Rate and equilibria relationships for such constituents volatilizing into air from water have been studied. These can be used to estimate the extent to which this process occurs indoors.

Indoor air quality models have been developed and used to predict the steady-state air concentrations of a variety of volatile constituents originating from materials used in the home. These can be utilized for volatile chemicals and radon-222 in potable water. Using such indoor air models, calculations of inhalation exposures indicate that this route can be substantial in comparison with direct ingestion of the contaminated water.

In a model shower system using water containing TCE, the air in the vicinity of the shower water was found to increase in TCE concentration with time. The air concentrations of TCE were dependent as expected on several factors, including water temperature and height of the drop path. These experiments indicate that the resulting human inhalation exposures can be substantial for highly contaminated waters, and this has been shown in field mesurements as well. Such exposures should be considered in the development of regulations to limit the possible adverse health impacts from the use of contaminated potable water.

ACKNOWLEDGEMENTS

This research was supported in part by Cooperative Agreement CR-811173 between the U.S. Environmental Protection Agency (EPA) and the Center for Environmental Epidemiology of the University of Pittsburgh. Since this manuscript has not been subjected to EPA peer and administrative review policy, it does not necessarily reflect views of the Agency and no official endorsement should be inferred.

REFERENCES

American Conference of Governmental Industrial Hygienists, 1984. TLVs Threshold Limit Values for Chemical Substances and Physical Agents in the Work Environment and Biological Exposure Indices with Intended Changes for 1984-85. ACGIH, Cincinnati, p. 32.

Andelman, J.B., Couch, A. and Thurston, W., 1985. Inhalation exposure in indoor air to volatile constituents in potable water. In preparation.

Bond, R.G., Straub, C.P. and Prober, R., 1973. Handbook of Environmental Control. Volume III: Water Supply and Treatment, CRC Press, Cleveland, p. 155.

Brown, H.S., Bishop, D.R. and Rowan, C.A., 1984. The role of skin absorption as a route of exposure for volatile organic compounds (VOCs) in drinking water. Amer. J. Public Health, 74: 479-484.

Gillies, M.E. and Paulin, H.V., 1983. Variability of mineral intakes from relationship of water quality to cardiovascular disease. Intern. J. Epid., 12: 45-50.

Liss, P.S. and Slater, P.G., 1974. Flux of gases across the air-sea interface. Nature, 247: 181-184.

International Commission for Radiological Protection (ICRP), 1975. Report of the Task Group on Reference Man, Edition no. 23. Pergamon Press, New York.

Mackay, D., and Wolkoff, A.W., 1973. Rate of evaporation of low-solubility contaminants from water bodies to atmosphere. Environ. Sci. Technol., 7: 611-614.

Mackay, D., and Leinonen, P.J., 1975. Rate of evaporation of low-solubility contaminants from water bodies to atmosphere. Environ. Sci. Technol., 9: 1178-1180.

National Academy of Sciences, 1978. Chloroform, Carbon Tetrachloride and Other Halomethanes. National Academy Press, Washington, D.C., 294 pp.

National Research Council, 1981. Indoor Pollutants. National Academy Press, Washington, D.C., 537 pp.

Prichard, H.M. and Gesell, T.F., 1981. An estimate of population exposures due to radon in public water supplies in the area of Houston, Texas. Health Physics, 41: 599-606.

Roberts, P.V. and Dandliker, P.G., 1983. Mass transfer of volatile organic contaminants from aqueous solution to the atmosphere during surface aeration. Environ. Sci. Technol., 17: 484-489.

Spengler, J.D., and Sexton, K., 1983. Indoor air pollution: A public health perspective. Science, 221: 9-16.

U.S. Environmental Protection Agency, 1983. National Revised Primary Drinking Water Regulations; Advance Notice of Proposed Rulemaking. Federal Register 48 (194, October 5, 1983), 45502-45521.

U.S. Environmental Protection Agency, 1982. National Revised Primary Drinking Water Regulations, Volatile Synthetic Organic Chemicals in Drinking Water; Advanced Notice of Proposed Rulemaking. Federal Register 47 (43, March 4, 1982), 9350-9358.

Wadden, R.A. and Scheff, P.A., 1983. Indoor Air Pollution. John Wiley and Sons, New York, 213 pp.

EPIDEMIOLOGIC STUDIES OF ORGANIC MICROPOLLUTANTS IN DRINKING WATER

GUNTHER F. CRAUN

U.S. Environmental Protection Agency, Health Effects Research Laboratory, 26 West Saint Clair, Cincinnati, Ohio (U.S.A.)

ABSTRACT

Epidemiologic studies have been conducted in order to make a quantitative statement about associations between drinking water contaminants and disease. The basic measures of the association are a rate ratio or relative risk and rate difference or attributable risk. The appropriateness of this measure is dependent on components of study design, data collection, and the analysis of epidemiologic data, and these must be evaluated for each study to determine precision (lack of random error) and validity (lack of systematic error). Internal validity includes considerations for preventing selection bias, minimizing observation bias, and assessing, preventing, and controlling confounding bias within a particular study. No single epidemiologic study is likely to provide a definitive answer, and the results of epidemiologic studies must be interpreted in the context of other scientific information. Epidemiologic studies of organic micropollutants in drinking water have been reviewed and are summarized based on these considerations.

INTRODUCTION

Over the past decade, concern has increased over the possible adverse effects of organic micropollutants found in drinking water. In 1974 the U.S. Environmental Protection agency (EPA) reported finding some 100 organic compounds in drinking water supplies in the United States. At present, some 1000 organic micropollutants have been identified in water (EPA, 1980). Many of these compounds are organic solvents and industrial chemicals from discharges to surface water, but agricultural practices, domestic waste sources, surface runoff, toxic waste disposal, and leaking storage tanks also contribute organic pollutants to water. Groundwater supplies have been found to contain volatile synthetic organic compounds, such as tetrachloroethylene, trichloroethylene, and 1,1,1-trichloroethane, often in concentrations much higher than are found in the most contaminated surface waters (Craun, 1984). Some of the organic micropollutants have been identified as animal or human carcinogens, but the vast majority have not been tested for carcinogenicity.

The use of chlorine as a disinfectant to prevent the waterborne transmission of infectious diseases has also been implicated as contributing organic micropollutants to drinking water. Under certain conditions, the reaction of free chlorine with selected precursor substances in drinking water produces a group of halogen-substituted single carbon compounds refered to as trihalomethanes (THMs). The predominant chlorination by-products are chloroform and bromodichloromethane, but dibromochloromethane and bromoform are also frequently found (Rook, 1974; Bellar et al., 1974). The concentration of the THMs is dependent upon the presence and concentration of the necessary precursors, chlorine dose and contact time, water temperature, and pH (Symons et al., 1981). The THMs have been found in both groundwater and surface water supplies, but up to an order-of-magnitude higher levels of THMs have been reported in surface water supplies (Burke et al., 1983). Under typical circumstances chloroform is the compound most often formed and is found at concentrations generally exceeding those of other organic micropollutants by a substantial margin (Wilkins and Comstock, 1981). Chloroform has been shown to be a carcinogen in animal studies at high dose levels, and the other THMs are mutagenic in bacterial tests (Craun, 1984; Hoel and Crump, 1981). Chloroform and the other THMs have likely been present in water supplies as long as chlorine has been used as disinfectant, since the important precursors which react with free chlorine to produce these compounds are naturally occurring aquatic humic substances, such as humic and fulvic acids and bromide rather than synthetic organic contaminants. The use of chlorine as a water disinfectant is widespread, and some 170,000,000 people in the United states regularly consume water which has been chlorinated (EPA, 1981).

Since 1974 a number of statistical and epidemiologic studies have been conducted in the United States to assess the relationship between cancer and drinking water quality. These studies differ markedly in their design and in what they can reveal about these possible relationships. Because of the widespread exposure to chlorination by-products, most of these studies have focused on the relationship between cancer and the chlorination of drinking water. Historical information on water sources and chlorination practices have primarily been used as surrogate measures of exposure to THMs, but in some instances current levels of THMs have been used in conjunction with various operational parameters to estimate historical exposure levels. In order to properly interpret these epidemiologic studies, it is important to have a basic understanding of the various types of epidemiologic studies and a framework for their evaluation.

TYPES OF EPIDEMIOLOGY STUDIES

I find Monson's (1980) classification of various types of epidemiologic studies to be most useful (see Table 1).

TABLE 1

Types of epidemiologic studies*

    I. Experimental
   II. Nonexperimental

       A. Descriptive
       B. Analytic

           1. Longitudinal

              a. Cohort or Follow-up

                 (1) Prospective
                 (2) Retrospective

              b. Case-comparison

           2. Cross-sectional

---

*Reprinted with permission from Monson, R.R., Occupational Epidemiology. Copyright CRC Press, Inc., Boca Raton, FL (1980).

Since the investigator has no control over exposure and does not randomly assign exposure to the study participants, the epidemiologic studies of water chlorination exposures are considered to be nonexperimental in nature. Many of the early studies were descriptive, and information was avilable on exposure and disease for groups of persons or only on disease. The epidemiology subcommittee of the National Academy of Sciences Safe Drinking Water Committee (NRC, 1980) reviewed all of these descriptive studies conducted through 1978. Nine of ten descriptive studies found associations between chlorinated surface water and either cancer incidence or mortality, and three additional studies found associations between current levels of THM and cancer mortality. These studies are felt to be of limited value, as doubts remain as to whether the observed associations are a result of exposure to chlorinated water or due to potential confounding characteristics. Although some demographic variables were considered in most of these studies, data are not available for individuals, and it is impossible to adequately control for the effects of potential confounders in descriptive epidemiologic studies. The National Academy of Sciences' review noted that "the bladder, stomach, large intestine and rectum, which were cancer sites identified in a number of geographic areas, warrant further study" and recommended analytic epidemiologic studies be conducted so a quantitative measure of the association could be obtained. In an analytic study, information on exposures, disease, and potential confounding characteristics are available for each study participant, and confounding bias can be assessed and controlled. Analytic studies can be either longitudinal or cross-sectional. In a longitudinal study, the time sequence between exposure and disease can be inferred, but in a cross-sectional study, the data on exposure and disease relate to the same point in time. Longitudinal epidemiologic studies include case-comparison and cohort designs. In a case-comparison study, individuals enter the study on the basis of disease status, and various exposures are determined. In a cohort study, individuals enter the study on the basis of exposure status, and each individual is followed to determine morbidity or mortality. In a prospective cohort study, the disease has not occurred at the time the exposed and nonexposed groups are defined. In a retrospective cohort study, the disease has occurred at the time the exposed and nonexposed groups are defined. Both case-comparison and cohort studies of associations between water chlorination and cancer require an assessment of previous or historical exposures. Prospective cohort studies, as well as retrospective cohort studies, require an assessment of both historical and current exposures, since it is unlikely that a cohort under 45-50 years of age would be assembled and followed prospectively for cancer incidence or mortality.

EVALUATION OF EPIDEMIOLOGIC STUDIES

Analytic studies allow the epidemiologist to make a quantitative statement about an observed association between water chlorination and cancer by comparing rates of similar types. The basic measures of this association are the rate ratio or relative risk and rate difference or attributable risk. In a cohort study, incidence rates can be determined and the attributable risk calculated. In a case-comparison study, the proportion of cases and comparison subjects is arbitrarily determined by the investigator, and since disease rates are meaningless, the exposure-odds ratio is determined. The exposure-odds ratio is equivalent to the rate ratio or relative risk for rare diseases (Monson, 1980). A rate ratio of unity (1.0) indicates no association; any other rate ratio indicates some association. Based on Monson's (1980) experience, ranges of the rate ratio may be used to judge the strength of the observed association (see Table 2).

TABLE 2
A guide to strength of association[*]

Rate Ratio	Strength
1.0 - 1.2	None
1.2 - 1.5	Weak
1.5 - 3.0	Moderate
3.0 - 10.0	Strong
>10.0	Infinite

*Reprinted with permission from Monson, R.R., Occupational Epidemiology. Copyright CRC Press Inc., Boca Raton, FL (1980).

Any rate ratio below 1.2 indicates essentially no association, as one or more uncontrolled or unknown confounding factors could lead to a weak association, and the epidemiologist is limited in his ability to identify and measure such weak confounding factors. It is generally accepted that if some confounding factor exists which accounts for a larger rate ratio, its detection should be relatively simple (Monson, 1980). The size of a rate ratio, however, has little to do with the possibility that an association could be due to possible systematic bias or random variability. This must be assessed separately (see Table 3).

TABLE 3
Accuracy of quantitative statement about association between exposure and disease

---

I. Precision (lack of random error)

    A. Size of study population
    B. Efficiency of information

II. Validity (lack of systematic error)

    A. Internal validity

        1. Selection bias
        2. Observation bias
        3. Confounding bias
        4. Misclassification

    B. External validity

        1. Scientific generalization
        2. Interpreted in context of other information

Precision is influenced primarily by the size of the study population and efficiency of the collection of information for individuals in the study. The association should be based on reasonably large numbers of individuals, but even with large numbers random variation may occasionally lead to an observed rate ratio of 1.0 when the true rate ratio is larger (1.2). Assessing the validity of epidemiologic associations requires a search for potential sources of systematic bias. Internal validity of a particular study must be ensured before considering external validity (Monson, 1980). If the criteria used to enroll subjects in a study are not comparable, the data cannot be used to measure an association between exposure and disease because of selection bias. The selection criteria between the two groups must be different in order for selection bias to occur. An inaccurate definition of disease or of exposure that applies equally to the two groups results in random misclassification, and this can only alter the results of a study toward no association between exposure and disease. Selection bias must be prevented; it cannot be controlled. If data are collected on two groups using methods which are not comparable, the data contain incorrect information on the association between exposure and disease because of observation bias. Observation bias can be prevented in a cohort study if the exposure status of study participants is not known when information is obtained on disease status, and in a case-comparison study no observation bias is possible if neither the patient nor data collector know the diagnosis when information on exposure is collected. To minimize observation bias in situations where the interviewer or patient knows case status, objectivity is sought in obtaining information. If a characteristic exists which is a cause of the disease and is also associated with both exposure and disease in any particular study, an appearance of association may be conveyed because of confounding bias. Confounding bias does not result from any error of the investigator. It is a basic characteristic of all epidemiologic studies and must be considered as the possible explanation for any observed association. Information can be collected on known or suspected confounding characteristics to prevent or control this bias. If a characteristic can be made or demonstrated to have no association with exposure or with disease in a particular study, that characteristic cannot be confounding. Matching is a technique generally employed in the study design to prevent confounding, and stratification or multivariate techniques are employed to assess and control confounding at the time of data analysis. External validity concerns extending the results of several epidemiologic studies to the target population and is generally referred to as scientific generalization. The results of epidemiologic studies must be interpreted in the context of other information, and no single study is likely to provide a definitive answer. Conscientious investigators will discuss methods used to prevent selection bias, minimize observation bias, and assess, prevent, or control confounding bias, and how these are likely to have influenced the interpretation of the results.

Recently, results have been reported for a cohort study and several case-comparison studies of the relationship between water chlorination or current levels of THMs and cancer in the United States. I have reviewed the cohort study and the case-comparison studies reported since 1981 (Craun, 1985) and others (Cantor, 1983; Crump and Guess, 1982; Hoel and Crump, 1981; Shy and Struba, 1980) have reviewed the earlier studies. The reader is referred to these references for a more extensive review than is presented here. Wilkins and Comstock (1981) studied three historical cohorts, each distinguished by a different degree of exposure to chloroform and other chlorination by-products, and reported incidence rates for cancer of the bladder among men and cancer of the liver among women to be nearly two-fold higher in the drinking water cohort supplied with chlorinated surface water at home when compared with the cohort supplied with unchlorinated groundwater. The results of this study cannot be interpreted as a true association because the reported rates are unstable and subject to random variation. Few cancer deaths occurred in the cohort, and the confidence intervals reported for the rate ratios are wide and include 1.0. Because the confidence intervals include 1.0, neither can the null hypothesis of no association be rejected.

A variation of the traditional case-comparison study using data available from death certificates for decedent cases and comparison subjects has been used to obtain information on associations of cancer and water quality. Since no interviews are conducted in this type study, advantages of this approach include considerable cost savings, but the disadvantages of this approach include the loss of important information on residence history, exposures, and confounding factors. A review (Crump and Guess, 1982) of five case-comparison mortality studies completed through 1981 concluded that information provided by the studies strengthened the evidence for an association between rectal, colon, and bladder cancer and water chlorination but were not sufficient to establish a causal relationship between chlorinated by-products in drinking water and cancer. Because of methodological limitations and the small increased risks observed in these studies, it is not possible to separate possible associations from potential confounding characteristics which could not be assessed or controlled. The two most recently reported studies have employed traditional case-comparison study designs where individuals were interviewed (Cragle et al., 1985; Cantor et al., 1985). These offer the potential for establishing a casual relationship between water chlorination and cancer and are reviewed in this article.

ASSOCIATION BETWEEN WATER CHLORINATION AND COLON CANCER

Cragle et al. (1985) investigated the relationship between water chlorination and colon cancer using 200 incident cases of colon cancer from seven hospitals in North Carolina and 407 hospital-based comparison subjects without evidence of cancer and no history of familial polyposis, ulcerative colitis, adenomatus polyposis, or any other major chronic intestinal disorder. Both cases and comparison subjects were required to be residents of the state for at least ten years to be included in the study. Comparison subjects were matched on age, race, gender, vital status, and hospital. Additional information on various potential confounders including alcohol consumption, genetic risk (number of first-degree relatives with cancer), diet, geographic region, urbanicity, education, and number of pregnancies, was obtained by either mailed questionnaire or telephone interview. Approximately 71 percent of the eligible population was included in the study. Water exposures were verified for each address and trichotomized on the following: 1) groundwater, no chlorination; 2) groundwater, chlorination; 3) surface water, chlorination. Because less than seven percent of the subjects were found to have consumed chlorinated groundwater at their residence, these study subjects were included in the chlorinated surface water group and the analysis confined to chlorinated versus nonchlorinated water. Logistic regression analysis showed genetic risk, a product term between alcohol consumption and high fat diet, and an interaction term between age and chlorination to be positively associated with colon cancer. The relationship between chlorinated water and colon cancer was found to be highly dependent upon age. Rate ratios for persons who drank chlorinated water at their residence for 16 or more years were consistently higher than those exposed to chlorinated water less than 16 years, but a statistically significant association between water chlorination and colon cancer, controlling for possible confounders, was found only for those above age 60 (see Table 4).

TABLE 4
Chlorination odds ratios for colon cancer reported by Cragle et al. (1985) in the North Carolina study

Age	Odds Ratio (95% C.I.) 1-15 years exposure	>15 years exposure
60-69	1.18 (0.94, 1.47)	1.38 (1.10, 1.72)
70-79	1.47 (1.16, 1.84)	2.15 (1.70, 2.69)
80-89	1.83 (1.32, 2.53)	3.36 (2.41, 4.61)

ASSOCIATION BETWEEN WATER CHLORINATION AND BLADDER CANCER

Cantor et al. (1985) recently reported results from a collaborative EPA-National Cancer Institute (NCI) study of the association between water chlorination and bladder cancer. Included were a total of 2982 persons (73 percent of those eligible to participate) between the ages of 21 and 84 diagnosed with cancer of the urinary bladder in 1978 and residing in ten areas of the United States (Connecticut, Iowa, New Jersey, New Mexico, Utah, and the metropolitan areas of Atlanta, Detroit, New Orleans, San Francisco, and Seattle) and 5782 population-based comparison subjects, randomly selected and frequency matched on gender, age, and study area. Subjects were interviewed at home by a trained interviewer, and data were collected for a number of possible confounders including smoking, occupation, artificial sweetner use, coffee and tea consumption, and use of hair dyes. A complete residence history was obtained to categorize individuals according to water sources and chlorination status on a year by year basis, and information was obtained on use of bottled water and fluid consumption. Of the 587,568 person-years lived by all residents since 1940, 76 percent were at a known water source. Logistic regression analysis was used to control for potential confounders. Among persons in all study areas combined, relative risk was not elevated in those respondents living in areas with chlorinated water supplies for 20, 20-39, 40-59, and 60 or more years. However, it should be noted that this study was originally designed to determine if saccrhin was a human carcinogen rather than to determine cancer risks associated with water chlorination. The study areas were not selected to provide the optimal variabilty of water sources and treatment, and the statistical power of the study is less than suggested by the large number of individuals studied, as the five metropolitan areas are served primarily by chlorinated water supplies. Among the 10 study areas, participants from the three states with agricultural land use did show elevated risk for bladder cancer with the number of years at a surface source, but the number of participants in these areas was small compared with the other areas. Among nonsmokers who were never employed in a high-risk occupation (a group otherwise at low risk for bladder cancer) the risk was elevated among those served by chlorinated surface sources with evidence of a duration of exposure-response relationship (see Table 5).

TABLE 5

Relative risks for bladder cancer in nonsmokers served by a chlorinated surface water source (Cantor et al., 1985)

Years at a Residence Served by Chlorinated Water	Relative Risk (95% C.I.)
1-19	1.3 (0.7, 2.2)
20-39	1.5 (0.9, 2.4)
40-59	1.4 (0.9, 2.3)
60+	2.3 (1.3, 4.2)

FUTURE STUDIES

It is planned to conduct another collaborative case-comparison study with NCI beginning in late summer 1985 to further study the suggestion of an association between chlorinated water and bladder cancer incidence in agricultural areas. Approximately 2500 incident cases of cancer of the colon, rectum, bladder, brain and pancreas (liver and kidney if feasible) and approximately 1500 population-based controls matched for race, gender, age will be interviewed for information on lifetime residential history, smoking, occupation, medical history, diet, socio-economic status to determine cancer risks associated with chlorinated drinking water and agricultural runoff.

Young et al. (1981) reported an association between colon cancer mortality in Wisconsin and THM exposures in drinking water as estimated by the average daily chlorine dosage of water over a twenty-year period. The study included 8,029 cancer deaths and 8,029 noncancer deaths in white females matched on county of residence, year of death, and age. Death certificates provided information on urbanicity, maritial status, and occupation, and these were considered as potential confounders. This association was further pursued in an interview study of 370 incident cases of colon cancer and 1450 population-based comparison subjects (Young and Kanarek, 1984). This case-comparison study found no association between estimated THM exposures and colon cancer incidence, but colon cancer cases were found to more frequently consume water from municipal groundwater supplies. Since groundwaters in the United States have been shown to be contaminated with synthetic votalite organic compounds (Craun, 1984), the possibility must be considered that misclassification of exposure occurred and was responsible for not observing an association in this study. We are currently planning to collaborate with these investigators at the University of Wisconsin in a study of the exposures of this population to organic contaminants in groundwater. The data obtained from this epidemiology study will be reanalyzed considering these exposures.

To obtain data on incidence rates of cancer associated with water chlorination requires a cohort study, however, the populations studied must be sufficiently large in order to have the statistical power to demonstrate an association or properly interpret a study where no association is found. Only one cohort study (Wilkins and Comstock, 1981) has been attempted to study associations between water chlorination and cancer, and this was of limited size and power. Currently, we are collaborating with investigators at Mount Sinai School of Medicine and the American Cancer Society (ACS) to determine the feasibility of conducting a large cohort study. The ASC Cancer Prevention Study I includes a cohort of 1,086,000 individuals assembled in 1959 and followed annually for 12 years; the ACS Cancer Prevention Study II includes a cohort of 1,200,000 individuals assembled in 1982 to be followed for at least ten years. A questionnaire on personal characteristics, including age, gender, race, occupation, place of birth, education, residence history, medical history, exercise, diet, smoking, alcohol consumption, medications, and household sources of water, was administered in the earlier study and will be administered in the newly initiated study. Both studies will require historical data on water exposures based on residence histories, and feasibility considerations include costs of obtaining these data and ensuring that sufficient populations are available in areas with variability in water sources, treatment, and quality.

SUMMARY

The two most recently reported case-comparison epidemiologic studies (Cragle et al.,1985; Cantor et al., 1985) have provided an indication of the magnitude of the cancer risk associated with chlorinated water for colon and bladder cancer. Based on Monson's (1980) guide to interpreting the strength of an association, a weak to moderate association between water chlorination and colon cancer was observed in an elderly population. This association was found to be stronger among the elderly who had been exposed to chlorinated water for more than 15 years. A moderately strong association (RR = 2.3) between chlorinated water and bladder cancer was observed in an otherwise low-risk population, nonsmokers, who had received chlorinated surface water for 60 or more years. Additional case-comparison and cohort studies are required to confirm these findings and are being planned in the United States. Any interpretation as to the causal association suggested by these studies should await the completion of these planned studies and other relevant research in this area.

REFERENCES

Bellar, T.A., Lichtenberg, J.J., and Kroner, R.C., 1974. The occurrence of organohalides in chlorinated dirnking water. J. Am. Water Works Assoc., 66:703-6

Burke, T.A., Amsel, J. and Cantor, K.P., 1983. Trihalomethane variation in public drinking water supplies. In: R.L. Jolley, W.A. Brungs, J.A. Cotruvo, R.B. Cumming, J.S. Mattice, and V.A. Jacobs (Editors), Water Chlorination: Environmental Impact and Health Effects, Volume 4, Ann Arbor Science Publishers, Inc., Ann Arbor, MI., pp. 1343-51.

Cantor, K.P., 1983. Epidemiologic studies of chlorination by-products in drinking water: An overview. In: R.L. Jolley, W.A. Brungs, J.A. Cotruvo, R.B. Cumming, J.S. Mattice, and V.A. Jacobs (Editors), Water Chlorination: Environmental Impact and Health Effects, Volume 4, Ann Arbor Science Publishers, Inc., Ann Arbor, MI., pp. 1381-97.

Cantor, K.P., Hoover, R., Hartge, P., Mason, T.J., Silverman, D.T., and Levin, L.I., 1985. Drinking water source and risk of bladder cancer: a case-control study. In: R.L. Jolley, R.J. Bull, W.P. Davis, S. Kate, M.H. Roberts, Jr., and V.A. Jacobs (Editors), Water Chlorination: Chemistry, Environmental Impact, and Health Effects, Volume 5, Lewis Publishers, Inc., Chelsea, MI., pp. 143-49.

Cragle, D.L., Shy, C.M., Struba, R.J., and Siff, E.J., 1985. A case-control study of colon cancer and water chlorination in North Carolina. In: R.L. Jolley, R.J. Bull, W.P. Davis, S. Katz, M.H. Roberts, Jr., and V.A. Jacobs (Editors), Water Chlorination: Chemistry, Environmental Impact, and Health Effects, Volume 5, Lewis Publishers, Inc., Chelsea, MI., pp. 151-57.

Craun, G.F., 1984. Health aspects of groundwater pollution. In: G. Bitton and C. Gerba (Editors), Groundwater Pollution Microbiology, John Wiley and Sons, Inc., New York, pp. 135-79.

Craun, G.F., 1985. Epidemiologic considerations for evaluating associations between the disinfection of drinking water and cancer in humans. In: R.L. Jolley, R.J. Bull, W.P. Davis, S. Katz, M.H. Roberts, Jr., and V.A. Jacobs (Editors), Water Chlorination: Chemistry, Environmental Impact, and Health Effects, Volume 5, Lewis Publishers, Inc., Chelsea, MI., pp. 131-41.

Crump, K.S. and Guess, H.A., 1982. Drinking water and cancer: Review of recent epidemiologic findings and assessment of risks. Ann. Rev. Public Health, 33:339-57.

Environmental Protection Agency, 1980. Briefing: The occurrence of volatile organics in drinking water, Office of Drinking Water, Washington.

Environmental Protection Agency, 1981. Number of facilities and population served by various treatment processes. Federal Reporting Data System, Office of Drinking Water, Washington.

Hoel, D.G. and Crump, K.S., 1981. Waterborne carcinogens: A scientific view. In: R.W. Crandall and L.B. Love (Editors), The Scientific Basis of Health and Safety Regulation, Brookings Inst., Washington, pp. 1973-95.

Monson, R.R., 1980. Occupational Epidemiology. CRC Press, Inc., Boca Raton, FL., 219 pp.

National Research Council, 1980. Epidemiology studies. Drinking Water and Health, Volume 3, National Academy Press, Washington, pp. 5-21.

Rook, J.J., 1974. Formation of haloforms during chlorination of natural waters. J. Soc. Water Treat. Exam., 23:234-43.

Shy, C.M. and Struba, R.J., 1980. Epidemiologic evidence for human cancer risk associated with organics in drinking water. In: R.L. Jolley, W.A. Brungs, and R.B. Cumming (Editors), Ann Arbor Sciences, Inc., Ann Arbor, MI., pp. 1029-1042.

Symons, J.M., Stevens, A.A., Clark, R.M., Geldreich, E.E., Love, Jr., O.T., and DeMarco, J., 1981. Treatment Techniques for Controlling Trihalomethanes in Drinking Water, Environmental Protection Agency, Cincinnati, OH., 289 pp.

Young, T.B., Kanarek, M.S., and Tsiatis, A.A., 1981. Epidemiologic study of drinking water chlorination and Wisconsin female cancer mortality. J. Nat. Cancer Inst., 67:1191-98.

Young, T.B. and Kanarek, M.S., 1984. Incidence case-control study of colon cancer in Wisconsin. International Epidemiologic Assoc., Vancouver, B.C., August.

Wilkins, III, J.R. and Comstock, G.W., 1981. Source of drinking water at home and site-specific cancer incidence in Washington County, Maryland. Am. J. Epidemiol., 114:178-190.

METABOLITES OF CHLORINATED SOLVENTS IN BLOOD AND URINE OF SUBJECTS EXPOSED AT ENVIRONMENTAL LEVEL

G. ZIGLIO, G. BELTRAMELLI, F. PREGLIASCO and G. FERRARI

Istituto di Igiene, Università, Via F. Sforza 35, 20122 Milan, Italy

ABSTRACT

After a two-level selection, 141 blood donors living in Milan, Italy, were analyzed for their content of plasma and urinary trichloroacetic acid (TCA) and trichloroethanol (TCE). Environmental levels of exposure to trichloroethylene (TRI) and tetrachloroethylene (PER) through drinking water and air were also measured. The plasma TCA levels were in the range of previously found concentrations. Relationships among plasma and urinary metabolites were found and discussed.

INTRODUCTION

Blood levels of trichloroacetic acid (TCA), a common metabolite of trichloroethylene (TRI) and tetrachloroethylene (PER) have been previously studied in popultion groups exposed to these compounds at environmental levels (Ziglio et al., 1983; Ziglio et al., 1984a,b).

The major findings of these studies are summarized as follows:
1. The presence of plasma trichloroacetic acid (TCA) demonstrates that these compounds are taken up in the blood, absorbed and metabolized.
2. TCA is detectable also in subjects not exposed through drinking water.
3. TCA levels in people supplied with TRI/PER contaminated drinking water are statistically higher than levels in people supplied with uncontaminated water.
4. TCA does not seem be related to sex, age and body mass.

A new study was undertaken in a group of subjects better controlled as regards confounding from short-term, high-level exposure at home (use of spot removers, some hobbies). Urinary TCA and trichloroetanol (TCE, metabolite of TRI, not of PER) were also measured in an attempt to discriminate the exposure to TRI from that to PER.

METHODS

All blood donors of Maggiore Hospital Center of Milan were selected following a computerized procedure which took into consideration age, zip code, profession and attendance. Among the respondents, only those classified as exposed to TRI/PER through only drinking water and air were chosen.

From May to October 1984, 141 subjects were controlled for their levels of plasma TCA and urinary (spot sample) TCA and TCE. Plasma TCA was determined by GLC-ECD following the procedure of Ziglio et al. (1984c). Urinary TCA and TCE were determined simultaneously combining, with some modifications, the gaschromatographic procedures of Breimer et al. (1973), Monster and Boersma (1975), and Humbert and Fernandez (1976).

Drinking water concentrations (wells and family taps) of TRI/PER were determined by the head-space technique (GLC-ECD). Atmospheric levels (24hours average) were measured at a single station five times per month and each day before specimens collection. The compounds were concentrated on a GAC filter and, after desorption by toluene, analyzed by GLC-ECD (NIOSH, 1975).

RESULTS

Of the 141 participants, 84(60%) were males. The age (M+F) ranged from 18 to 61 years (70% under 41). Because TCA levels are not related to sex and age variables (Ziglio et al., 1983), no breakdown is provided here. The 141 subjects were categorized as exposed to TRI/PER via drinking water and air as shown in Table 1.

TABLE 1

Categorization of the exposure to TRI and PER through drinking water and air (141 subjects, M+F).

Drinking water				Air	
Subjects (M+F)	64	45	32	Subjects (M+F)	141
TRI, ug/L	(12-30)[a]	(35-69)	(70-123)	TRI, ug/m^3	(1.7-26.9)[a] 6.2[b]
PER, ug/L	(2-33)	(14-68)	(20-46)	PER, ug/m^3	(2.9-40.0) 10.0

[a] Range
[b] Mean value on 25 samples

The percentile distribution of TCA plasma levels are depicted in Table 2.

TABLE 2

Percentile distributions of plasma TCA (ug/L), urinary TCA (ug/g cr) and urinary TCE (ug/g cr).

Percentile	TCA (pl)	TCA (ur)	TCE (ur)
5	8.1	6.2	14.4
10	10.0	8.2	17.6
50	17.9	19.4	40.6
90	39.0	44.8	88.9
95	44.9	57.5	98.9
98	60.0	72.0	130.2

The same table shows the percentile distribution of the urinary metabolites expressed as ug/g creatinine for balancing differences in diuresis.

In the urine samples, the concentrations of TCE (metabolite of TRI alone) were higher than the corresponding concentrations of TCA in more than 50% of the cases (see Table 3).

TABLE 3

Percentile distribution of Ratio values between TCE and TCA in the urine.

Percentile	R. value	Percentile	R. value
5	0.4	55	1.2
10	0.5	70	1.5
15	0.6	85	1.9
30	0.8	90	2.2
45	1.0	95	2.6
50	1.1	98	4.9

Utilizing a polynomial regression model, only 16-28% of the observed variability of urinary metabolites (as a single or total) were explained by TCA plasma concentration (See Table 4). No improvement was obtained including higher level terms.

TABLE 4

Functional relationships between plasma TCA as independent variable and TCA, TCE, and TCA + TCE (ur): only first order analysis shown.

Dependent Var.	TCA (ur)	TCE (ur)	TCA+TCE (ur)
N. Observations	141	141	141
Intercept	10.66	9.96	20.63
Slope	0.58	0.72	1.30
F-Value	27.665**	51.376**	54.349**
R^2	0.166	0.270	0.281

** $P < 0.01$

DISCUSSION

TCA plasma concentrations were similar to the range found in other studies implicating people continuously exposed.

No detailed pharmacokinetic studies on animals or humans exist showing percent of absorbtion, and rates of metabolism and elimination of TRI and PER, alone or in combination, when exposed at environmental levels. In people with long-term exposure to TRI at occupational levels, the ratio between urinary TCE and TCA ranges from 1 to 2 (WHO, 1981). In our study, all 141 subjects are continuously exposed to TRI and PER, although at different levels and with different patterns. More than 40% of these subjects present TCE/TCA ratios between 1 and 2. Considering that TCE has a substantially higher rate of elimination than TCA and that TCA is the only metabolite common to TRI and PER, the major differences from these values could be interpreted as: recent increase in TRI exposure (ratios > 4) or an important contribution of PER exposure (ratios < 0.5). Therefore, the most important contribution to TCA levels is due to TRI. This is also in agreement with the small percent of biotransformation of PER. Based on the differences between concentrations in air and drinking water and considering the substantial contribution to the global exposure resulting from home uses of the water (Ziglio et al., 1983; Andelman, 1985), drinking water should be considered as the most important source of TRI.

REFERENCES

Andelman, J.B., 1985. Inhalation exposure in the home to volatile organic contaminants of drinking water. This Journal.
Breimer, D.D., Ketelaars, H.C.J. and Van Rossum, J.M., 1973. Gaschromatographic determination of chloral hydrate, trichloroethanol and trichloroacetic acid in blood and urine employing head-space analysis. J. Chromatography, 88: 55-63.
Humbert, B.E. and Fernandez, J.G., 1976. Simultaneous determination of trichloroacetic acid and trichloroethanol by gas chromatography. Int. Arch. Occup. Environ. Hlth., 36: 235-241.
Monster, A.C. and Boersma, G., 1975. Simultaneous determination of trichloroethylene and metabolites in blood and exhaled air by gas chromatography. Int. Arch. Occup. Environ. Hlth., 35: 155-163.
NIOSH, 1975. Manual of analytical methods. NIOSH New Publication N° 75-121.
WHO, 1981. Rapport Techniques, 664: 61-83.
Ziglio, G., Fara, G.M., Beltramelli, G. and Pregliasco, F., 1983. Human environmental exposure to trichloro and tetrachloroethylene from water and air in Milan, Italy. Arch. Environ. Contam. Toxicol., 12: 57-64.
Ziglio, G., Beltramelli, G., Pregliasco, F., Arosio, D. and De Donato, S., 1984a. Esposizione ambientale a solventi clorurati in popolazioni studentesche di un Comune del Nord Italia. Ig. Mod., 82: 133-161.
Ziglio, G., Beltramelli, G., Pregliasco, F. and Mazzocchi, M.A., 1984b. Esposizione per via idrica a tricloroetilene in alcuni nuclei familiari del Comune di Porto Mantovano (MN). Ig. Mod., 82: 591-605.
Ziglio, G., Beltramelli, G. and Pregliasco, F., 1984c. A procedure for determining plasmatic trichloroacetic acid in human subjects exposed to chlorinated solvents at environmental level. Arch. Environ. Contam. Toxicol., 13: 129-134.

CRITICAL CONSIDERATIONS ON THE SIGNIFICANCE OF CARCINOGENIC AND MUTAGENIC COMPOUNDS IN DRINKING WATER

C.A. VAN DER HEIJDEN AND C.F. VAN KREIJL
National Institute of Public Health and Environmental Hygiene,
Laboratory for Carcinogenesis and Mutagenesis, P.O. Box 1,
3720 BA Bilthoven, The Netherlands

ABSTRACT

Public and scientific concern has been expressed on the possible hazards of trace amounts of organic compounds with carcinogenic and mutagenic properties, identified in drinking water. For a number of these compounds, the carcinogenicity is well established according to IARC criteria, but the extremely low concentrations (< 1 µg/l) indicate a neglectable risk to humans. Some compounds, mainly volatile halogenated alkylated hydrocarbons, may be present at higher concentrations, but for these the weight of evidence for carcinogenicity often is very poor, being demonstrated in mouse liver only. The relevance of mouse liver tumours may be seriously questioned, especially after exposure to hepatotoxic doses and in the absence of sufficient evidence for genotoxicity. It is therefore not justified, to use a non-threshold approach in the toxicological evaluation of these compounds. More or less similar conclusions can be derived for the organic "mutagens" identified in water, that is either their concentration is extremely low or sufficient evidence for genotoxicity is lacking. It is concluded therefore, that, at the present time, drinking water in the Western world can be regarded in general as "chemically safe".

INTRODUCTION

Since present methods of water treatment in western societies are very well capable of preventing contamination of drinking water by pathogenic organisms, the interest in the drinking water field has shifted over the past two decades towards the occurrence of chemical contaminants. This shift in interest was further strengthened by the increased use of chemically contaminated surface waters as source for drinking water preparation.

Subsequent chemical surveys carried out in the 1970's have shown the presence of low concentrations of large numbers of inorganic and in particular organic compounds in drinking water in the Netherlands as well as in other countries. Among these so-called micropollutants, several compounds known or suspected of having carcinogenic and/or mutagenic properties were also found to be present. This knowledge has led to public and scientific concern about the possibility of adverse health effects, especially in relation to the contribution of drinking water to human cancer.

CRITERIA TO CONSIDER A COMPOUND AS HUMAN CARCINOGEN

Before an assessment of the cancer risk associated with exposure to a particular chemical substance can be made, the following fundamental questions should be answered first: "what is the weight of evidence to consider a substance as carcinogenic" and even more important "what is the level of evidence needed to consider a substance as a human carcinogen"?

The International Agency for Research on Cancer (IARC) considers three lines of evidence as relevant for the assessment of carcinogenicity of chemical compounds (IARC, 1982). These are data from human epidemiological studies, animal experiments and short-term tests for genotoxicity. For each of these three lines criteria have been set to judge the data as either sufficient evidence, limited evidence or inadequate evidence. Furthermore, based on the combined evidence, a classification in three categories is made regarding the weigth of evidence to consider a chemical as a human carcinogen:

1. Carcinogenic to humans

    This category includes substances for which there is sufficient evidence from epidemiological studies to support a causal association between exposure and human cancer.

2. Probably carcinogenic to humans

    This category includes substances for which the epidemiological

evidence of human carcinogenicity ranged from <u>limited</u> to <u>inadequate</u> and for which there is <u>sufficient</u> evidence in animals.
3. <u>Cannot be classified as to its carcinogenicity to humans</u>

It is important to emphasize that the key terms sufficient, limited and inadequate underlying the above classification, relate only to the relationship between exposure to a chemical and an increased tumour incidence, or, for the short-term tests, the occurrence of genotoxic effects and not to the extent of the activity nor to the mechanisms involved.

MECHANISMS OF CHEMICAL CARCINOGENESIS AND RISK ASSESMENT

It has become increasingly evident that all chemical carcinogens do not act via the same mechanism of carcinogenesis. Most theories regarding the mechanisms of chemical carcinogenesis may be generally classified as "genetic" or "epigenetic" (non-genetic) in nature.

The most common interpretation of a genetic mechanism of carcinogenesis is embodied in the somatic mutation theory. The base of this theory is that the direct interaction of a chemical with DNA (e.g. by alkylation, intercalation) can result in a somatic cell mutation, which may ultimately lead to a transformed or neoplastic cell. How these initiated cells with altered genetic information result in a cancer is very poorly understood, but results from recent molecular biological studies in the rapidly expanding field of oncogene research demonstrate the requirement of at least two but probably more genetic changes. In addition to the initiation stage there are two further operational stages, the so-called promotion and progression stages.

With regard to the epigenetic mechanisms it appeared that several findings cannot be reconciled with a mutagenic base of carcinogenesis. These include observations that not all carcinogens are mutagens, that a number of carcinogens do not possess initiating properties in vivo, that the implantation of inert plastic - or metal film does induce tumours, and that tumours do occur after recurrent cytotoxicity in the target organ.

The assumption that a threshold does not exist for carcinogens is mainly based on the somatic mutation theory, which implies that theoretically, even one molecule of a genotoxic carcinogen has the capacity to initiate a single cell, which subsequently may give rise to a neoplasm. As a consequence there is general agreement among scientists involved in regulation, that risk assessment of carcinogens that act via a genetic mechanism should be based on a non-threshold approach. For such chemicals an absolute safe exposure level will be zero. However when exposure to low levels of such a substance does occur - for instance due to practical infeasibility to remove them or for economic reasons - exposure values can be calculated which are associated with a calculated risk of one additional cancer per life-time per 10^6 or 10^5 exposed people; such a risk is generally regarded as neglectable. For these low dose extrapolations a lineair non-threshold model is commonly used in the Netherlands.

In contrast, for those chemicals which demonstrate an epigenetic, non-mutational, mechanism of carcinogenesis and which do not show clear evidence of genetic activity in short-term tests a non-threshold approach is often not justified. For these substances it is assumed that an exposure level does exist at which no adverse effects will occur. This means that for risk assessment general toxicological criteria can be applied and that an acceptable daily intake can be calculated, making use of a no-effect level and a safety factor.

ORGANIC CARCINOGENS AND MUTAGENS IDENTIFIED IN DRINKING WATER

From the hundreds of organic compounds detected in drinking water as published in the past years, only a very small number fulfill the IARC criteria of having sufficient evidence for carcinogenicity (Kool et al., 1982; Noordam, 1983, 1985). Of these, only two compounds - benzene and hexachlorobenzene - can actually be considered as human carcinogens (IARC class 1), and eight compounds - benzo(a)pyrene, O-toluidine, polychlorinated biphenyl (PCB's), carbontetrachloride, chloroform, DDT (p,p'), 1,2-dibromoethane, 2,4,6-trichlorophenol - can be classified as probably carcinogenic to man (IARC class 2), whereas three compounds - benzo(b)fluoranthene, 1,2-dichloroethane, indeno (1,2,3-cd) pyrene - do not meet the requirements for IARC class 1 or 2, but have sufficient evidence for carcinogenicity in animals. In addition, about twenty substances were detected which have only limited evidence according to the IARC criteria;

among these are trichloroethene, tetrachloroethene (both IARC class 3) and a few chlorinated ethanes (no IARC classification).

Except for some halogenated aliphatic hydrocarbons, which will be discussed below, none of the above mentioned compounds is observed in drinking water in concentrations above 1 µg per liter. It appeared that, even when the most conservative, lineair non-threshold model is used to calculate the carcinogenic risk for the population, trace amounts of these compounds in drinking water have a neglectable contribution to cancer based on a life time water consumption; therefore their presence in drinking water, given the low concentration, is considered to be acceptable.

Some halogenated aliphatic hydrocarbons e.g. carbontetrachloride, trichloroethene, may be present at levels above 1 µg per liter. It should be emphasized here that for this group of chemicals the weight of evidence is very much questioned because the effects seem to be restricted mainly to the liver of rodents, especially that of mice.

In the scientific community the relevance of mouse liver tumours as an indication for carcinogenic effects is much in debate (The Nutrition Foundation, 1983). Whereas there is general agreement that the observed changes in the liver are appropriately diagnosed, there appeared to be many problems with respect to interpretation and biological meaning. Apart from these technical difficulties the relevance of liver tumours is questioned and this for the following, main, reasons. 1) The high spontaneous incidence of liver tumours in mice, which even for one strain of mice appeared to be quite heterogeneous, both within as between various laboratories. 2) The results of a number of bioassays of the U.S. National Toxicology Program, which showed that, although most of the agents which were shown to be carcinogenic in rodents, induced neoplasms in more than one species and at more than one site, a number of agents were found to induce only mouse hepatomas. While most of the compounds that are carcinogenic in more than one species and at more than one site also appeared mutagenic, the majority of the agents that affects mainly the mouse liver do have far from sufficient evidence for mutagenicity in short-term tests.

Based on the experimental data it can be concluded that trichloroethene, tetrachloroethene, carbontetrachloride and to a certain extent also chloroform, fall in this special group of liver carcinogens. Although the mechanism whereby such compounds induce this specific neoplastic response is not fully understood, it is likely to be related to the observation that at high doses, which do produce this response, the compounds are also very toxic to their target organs, which is the liver and sometimes the kidney (in the case of chloroform and trichloroethene); in other words, the cancer producing activity of these compounds seemed to be secondary to their tissue damaging effect.

Since the above considerations do not justify that these substances are placed in the category of carcinogens for which a non-threshold approach for risk assessment is indicated, general toxicological criteria should be applied to derive an acceptable daily intake, making use of a no-effect level and a safety factor. In doing so, it appeared that also for those compounds in drinking water, which have been found at concentrations above 1 ug per liter, no adverse effects are to be expected.

In drinking water also a great number of compounds with mutagenic properties have been identified. Again the majority of them were detected at trace amounts only and it might be stated that in general the genetic risk involved is extremely low, if there is a risk at all. For those compounds which may be present at higher concentrations, the evidence for genotoxicity is rather poor, being mostly marginal results from bacterial tests only or equivocal results in other testsystems. Thus the assumption that these compounds have a risk due to their genetic activity, either a genetic risk by itself or a possible carcinogenic risk based on the non-threshold model, is therefore as yet unwarranted.

In conclusion, it may be stated that, although (extremely) low concentrations of organic carcinogens and mutagens have been identified in drinking water, there are at the present time no toxicological reasons to consider the drinking water as carrying a hazard for human health; therefore the drinking water in the Western world can be regarded in general as "chemically safe".

ACKNOWLEDGEMENT

The authors want to acknowledge the critical reading of the manuscript by their colleague Dr.A.G.A.C.Knaap.

REFERENCES

IARC, WHO-International Agency for Research on Cancer., 1982. Supplement 4; Chemicals, industrial process and industries associated with cancer in humans, volumes 1-29, Lyon, France

Kool, H.J., van Kreijl, C.F. and Zoeteman, B.C.J., 1982. Toxicology assesment of organic compounds in drinking water. Crit.Rev.Toxicol., 12, 307-357.

Noordam, P.C., 1983. Toxicologische beoordeling van een aantal zowel in het water van Rijn en/of IJsselmeer en/of Maas als in het hieruit bereide drinkwater aangetroffen xenobiotische organische verbindingen. KIWA-report SWO-83.224.

Noordam, P.C., 1985. Toxicologische aspecten in: Drinkwater uit oevergrondwater (Ed: D.van der Kooy). KIWA-mededeling nr.89.

The Nutrition Foundation, 1983. The relevance of mouse liver hepatoma to human carcinogenic risk. A report of the international expert advisory committee to the Nutrition Foundation. The Nutrition Foundation, Inc., 1983 ISBN 0-935368-37-X.

Drinking water and health hazards in environmental perspective

B.C.J. Zoeteman, Director, National Institute for Public Health and Environmental Hygiene, Bilthoven, The Netherlands

Summary

Among the present environmental issues drinking water quality and more specifically organic micropollutants receive not the highest priority.
The long tradition of potable water quality assurance and the sophisticated evaluation methodologies provide a very useful approach which has great potential for wider application in environmental research and policy making.

Water consumption patterns and the relative importance of the drinking water exposure route show that inorganic water contaminants generally contribute much more to the total daily intake than organic micropollutants. An exception is chloroform and probably the group of typical chlorination by-products. Among the carcinogenic organic pollutants in drinking water only chlorination by-products may potentially increase the health risk. Treatment should therefore be designed to reduce chemical oxidant application as much as possible.

It is expected that in the beginning of next century organic micropollutants will receive much less attention and that the present focus on treatment by-products will shift to distribution problems. Within the total context of water quality monitoring microbiological tests will grow in relative importance and might once again dominate chemical analysis the next century.

As disinfection is the central issue of the present water treatment practice the search for the ideal disinfection procedure will continue and might result in a further reduction in the use of chemical oxidants.

Our biggest problems

Is the health of organic contaminants in drinking water our biggest problem, one may ask oneself. The answer is: no. Is the quality of drinking water a major factor in human health could be the next question. In the developed world the answer is fortunately enough also: no, although still about 3000 cases of waterborne desease have been estimated to occur annually in the most powerful nation of this world (Cotruvo, 1984).

This is, however, neglectible compared to the recent WHO estimate that last year 15 million children died in the developing nations due to waterborne diseases.

If we were political responsible for the world as a whole and if we had to make a choice between reducing drinking water organics in the western world or contribute with the same effort to a safe water supply in the developing nations the decision would be easy to make. However, we are no politicians, let us stick to our own countries. Most of us work in organisations where public water supply is part of the agency that is responsible for environmental protection.

Where would I locate the health risks of drinking water in comparison with the other environmental issues, I asked myself.

Presently the main topics on the political agenda in the EC-countries are:
- acid deposition abatement
- introduction of cleaner and less noisy vehicles
- disposal of toxic chemical wastes
- nuclear energy or renewable energy sources
- clean-up of contaminated soil and sediment

while in the next decade other issues will probably be added to this priority list, such as:
- greenhouse effect due to CO_2
- changes in the ozone layer
- indoor-air pollution
- the hypothesis of nuclear winter.

All these problems need the direct attention of our decision-makers and include vast amounts of money. Against this background drinking water obtains a low priority, maybe a too low priority (figure 1).

The ultimate environmental calamity is the use of nuclear weapons followed by a further environmental catastrophy. Estimates by Sagan (1983/1984) show temperature drops varying from 5-50°C in the Northern Hemisphere during periods of 4-12 months, depending on the severity of the nuclear exchange.

Such a calamity would of course by far outweigh all other environmental issues mentioned before and for the first time environment starts to become a significant factor of strategic defense planning. Our existence on this planet would actually be at risk. We feel threatened externally by toxic vapours, acid rain and eventually a fatal darkening of the sky and internally by coronary heart disease, cancer and AIDS.

Within this context we have to look at our problem of today as society feels all these stresses and has to decide to what extent money will be allocated on this particular problem.

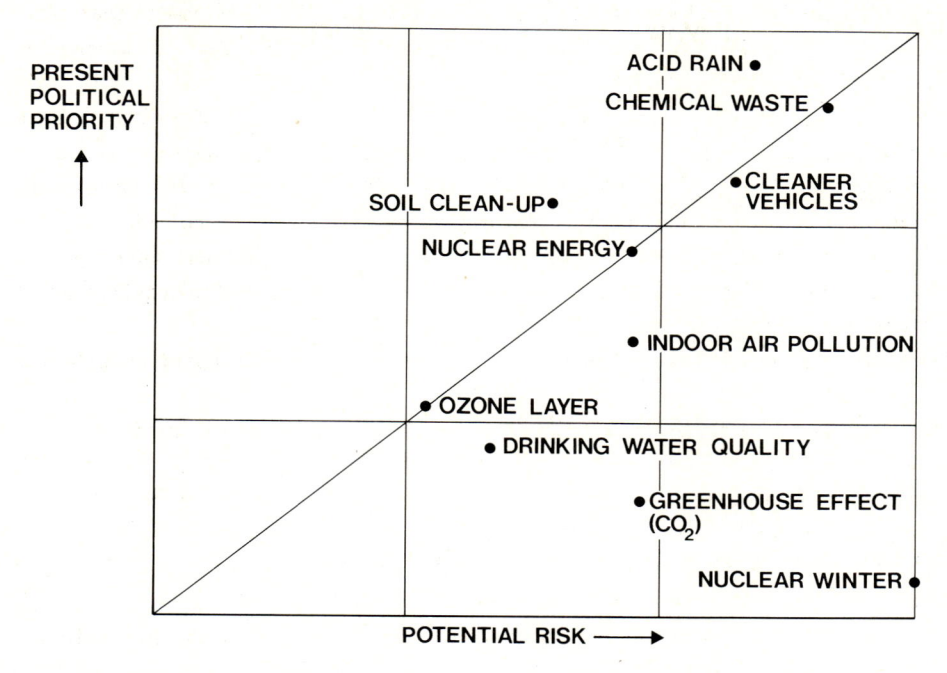

Figure 1. Rating of drinking water quality and other environmental issues according to present political priority and potential risk.

In comparison with other environmental issues much money has been spent on the study of drinking water organics. This is due to the recent discovery of the occurrence of these compounds in the drinking water, to our wish that drinking water must be absolutely safe and to the traditionally excellent organization of the water supply industry. We see that advanced analytical techniques and extremely sensitive epidemiological and toxicological methodologies have been developed to study drinking water quality.

Drinking water organics have been handled in a way that may seem somewhat overdone to those dealing with air or soil contamination but it also provides the most sophisticated examples of how environmental problems can be tackled. The study of drinking water quality is the most advanced branch of environmental research. It is with this in mind that I like to draw some conclusions from the experiences gained during the past decades and especially during this symposium.

Water consumption changes

A primary factor for all regulations to limit exposure of the population to environmental contaminants is assessment of the relative importance of the exposure route. In The Netherlands RIVM started in 1983 the production of so-called basic documents for priority pollutants in which all relevant aspects for standard setting, including the importance of exposure routes, are described. These documents are ment to promote a balanced set of standards for water, air, soil, drinking water and eventually nutritional and occupational exposures.

In our context the first question should be: How much tapwater is actually consumed and in which form? In The Netherlands the average daily intake was in 1978 1.3 litres per head (Zoeteman, 1980), while similar values are found in the UK (Hopkins and Ellis, 1980) and Canada (Health and Welfare, 1981).

More important is that in The Netherlands relative water consumption is roughly speaking 10 times higher for babies (0.15 l/kg) than for adults (0.016 l/kg). Moreover large differences occur in individual water consumption patterns. This is partly due to the presence of taste and odour affecting contaminants in drinking water (Zoeteman, 1980). Although consumtion of tea showed to be independent of tea taste the consumption of water as such dropped from 0.32 l/day to 0.15 l/day in case tapwater had an offensive taste and odour.

About 0.5% of the Dutch population consumes more than 4 litres a day (Haring et al., 1979). Why is this group of consumers not taken as the critical group of the population on which standard setting is based, one may ask.

On the other hand water consumption patterns are changing. Few people consume tap water. It is often replaced by bottled water. In several countries bottled water sales trippled in the past years (Zoeteman, 1981).

In case these mineral waters like Spa, Evian and Perrier replace alcoholic or high caloric beverages this is probably the maximum contribution drinking water can make to health in our affluent society.

Contribution of drinking water to total daily intake

Knowing the quality of water consumed, the next factor to assess is the relative importance of the drinking water exposure route.

The recently published WHO guidelines for drinking water quality (1984) provide in Volume 2 supporting information which can be used to derive for most of the substances mentioned the relative contribution of drinking water to the total daily intake. Based on this WHO document and some additional publications a survey has been composed and presented in Appendix 1. Table 1 summarizes the data and shows that practically all known organic micropollutants in drinking water contribute less than 1% to the total daily intake of these compounds. The inorganic contaminants seem to be of much greater interest for human health, particularly fluoride, lead and magnesium. The only exception among the many organics is chloroform (Table 2).

Table 1. Survey of the relative contribution of drinking water contaminants to their mean daily intake by man

Contribution range (%) of drinking water contaminants to the mean daily intake

< 0.1	0.1 - 1.0	1.0 - 10	≥ 10
	Al, As, Be, Fe, Se, Ag	Ba, Cd, Cr, Mn, Hg, Ni, Na, SO_4, Cl	Ca, F, Pb, Mg, NO_3
Vinylchloride	Carbontetrachloride	Trichloroethene	Chloroform
Aldrin/Dieldrin	1,2 Dichloroethane		
Chlordane	Tetrachloroethene		
DDT	Benzo(a)pyrene		
Hexachlorobenzene			
Heptachlor(epoxide)			
Lindane			
Benzene			

Table 2. Main exposure routes for some drinking water contaminants

Substance	Drinking water	Food	Air	Smoking
	% contribution to total intake			
Fluoride	50	50	< 1	-
Lead	32	65	3	-
Magnesium	29	71	< 1	-
Calcium	16	83	< 1	-
Chloroform	15	77	8	-
Nitrate	14	85	< 1	-
Trichloroethene	1	5	94	-
Benzo(a)pyrene	1	87	4	8
DDT	< 1	100	< 1	-
Vinylchloride	< 1	5	95	-
Benzene	< 1	56	44	-

Generally speaking those compounds seem to be of interest where man is manipulating water quality either by chemical treatment or by distributing it through piping materials that release compounds. Lead is a good example of the latter and chloroform is an indicator for the total group of halogenated by-products, such as halophenols, halo-acids, halo-acetonitrites etc. which are probably mainly ingested via the drinking water route.

A closer look at Table 2 also shows that with the exception of the volatile halogenated organics for which air is the major exposure route, food is always the most important contributor to the daily intake. Exposure to inhalation can also be traced back for some organics to indoor tap water use (Andelman, 1985). This shows the need to look at these problems in an integrated way. Since adequate water treatment techniques have been developed and applied the past decades industrial organic micropollutants present in raw water sources are generally sufficiently removed to make the drinking water exposure neglectable.

Toxicity of drinking water organics: likelihood and evidence

I will not summarize in detail the findings of the rapidly expanding literature dealing with toxicological and epidemiological evidence for possible health effects of organic micropollutants in drinking water.

It is, however, clear that the risk resulting from drinking water exposure is generally very small, that the mixture of organic contaminants is complex and changing in time and that available methodology is often not sensitive enough to identify and quantify the specific hazards with sufficient likelihood. This may be illustrated by a number of statements of the speakers at this symposium. Craun (1985) showed that "no single epidemiologic study is likely to provide a definitive answer in this question". Kool (1983), after reviewing
methodologies and shortcomings of animal carcinogenicity studies on concentrates of drinking water organics, concluded that "the contribution of drinking water, if there is a contribution at all, should be relatively small (less than 1%)". Van der Heijden (1985) stated that "the experimental evidence for the vast majority of compounds is very limited and generally based on positive results in a bacterial test system only, often at very high doses. At present no methods are available to quantify the genetic risk after exposure to mutagens". These uncertainties and difficulties may confuse us.

On the other hand, they may be balanced by a number of certainties. If epidemiological studies indicate an increased risk of cancer due to organic contamintants this is always associated with the application of water chlorination (Williamson, 1981; Zoeteman e.a., 1982; Craun, 1985).

Most of the mutagenic activity is also likely to be a product of chlorination (Loper, 1980; Kool and Van Kreijl, 1984). Mutagenic activity can be removed by activated carbon (Monarca e.a., 1983) and can be reduced by dechlorinating agents (Cheh e.a., 1980) but not by bioling (Kool, 1983). The obvious hypothesis that a number of halogenated polar compounds is formed during chlorination has been confirmed in different studies.

Christman et al. (1985) have shown the production of large quantities of chlorinated acetic acids besides the well-known trihalomethanes and carcinogenic halo-acentonitrites and chlorinated phenols (Trehy and Bieber, 1981; Zoeteman et al., 1982). Bull (1985) indicated the relevance of some carcinogenic halogenated propanones and propanals, while other expect that halogenated nitro compounds might be the main causes of increased mutagenicity of chlorination (Kool et al., 1985).

The likelihood that a substance is of no health concern has been increased by the application of risk estimation models in combination with the results of low dose exposures of animals to carcinogens.

Table 3. Comparison between typical tapwater concentrations of carcinogens (see appendix 1) and concentrations corresponding to a calculated cancer risk level of 10^{-5} at the upper bound* (Anderson, 1983)

Compound	Typical tapwater concentration ($\mu g/l$)	Risk level of 10^{-5} at upper bound ($\mu g/l$)	Ratio concentration/ risk level
Aldrin	1×10^{-3}	7.4×10^{-4}	1
Benzene	2×10^{-1}	7	3×10^{-2}
DDT	1×10^{-3}	2×10^{-4}	5
1,2 Dichloroethane	1×10^{-1}	9	1×10^{-2}
Carbontetrachloride	1×10^{-1}	4	3×10^{-2}
Chlordane	1×10^{-3}	5×10^{-3}	2×10^{-1}
Chloroform	20	2	10
HCB	1×10^{-3}	7×10^{-3}	1×10^{-1}
Heptachlor	1×10^{-3}	3×10^{-3}	3×10^{-3}
Lindane	1×10^{-1}	1×10^{-1}	1
Tetrachloroethene	3×10^{-1}	8	4×10^{-2}
Trichloroethene	1	27	4×10^{-2}
Vinylchloride	1×10^{-1}	20	5×10^{-3}

* linearized multistage model

In Table 3 the level at which the calculated excess lifetime cancer risk is one in 100.000 (at the upper bound of uncertainty and using a linearized multistage model) is compared with typical drinking water concentrations for 13 contaminants. Once again this table shows that particularly the chlorination by-product chloroform is of serious interest, while practically all contaminants of industrial nature remain below this risk level.

These certainties make it very likely that drinking water toxicology should concentrate on by-products formed during treatment with Cl_2 or ClO_2. All other issues seem to be neglictible or are of far less importance.

That chloroform is a relative weak carcinogen and not belonging to the most relevant class of chlorination by-products indicates that:
- further research should concentrate on the identity of the polar by-products of chlorination and on their biological effects,
- treatment should be optimized to obtain disinfection at the lowest dose of oxidant possible.

The latter goal should be achieved without waiting for the identification of the toxicants formed during chlorination and in spite of the probable fact that chlorination by-products do not pose a substantive carcinogenic hazard to man. This hazard should be avoided as much as possible in line with the general principle to eliminate the introduction of persistant toxic compounds in our food and into the environment.

Speculations on future trends

The art of predicting the future is partly the art of extrapolating past trends. As long-term predictions can only be generalizations we firs have to change the microscope we used to assess water toxicity for the telescope to look back several decades in water supply history.

The first major problem in relation to organic micropollutants was the presence of phenols in river water which created taste and odour problems after the widespread introduction of chlorination during the thirties. Increasing the chlorine dose reduced the taste problem while the disappearance of gas factories and coaltar plants after the second world war resulted in deminished concentrations of phenols and aromatic hydrocarbons in the surface waters.

The problem of foaming followed two decades later due to the use of non-biodegradable alkylbenzene sulphonates as detergents. Legislation in the beginning of the sixties (Bock and Wickbold, 1966) rapidly reduced the problem to acceptable proportions. This actually was one of the first successful operations to solve an environmental water problem.

The next major problem resulted from the rapid growth of the chemical and refinery industry and the use of pesticides in agriculture. These extremely toxic compounds caused world-wide concern due to their acute toxicity,

WAVES OF PUBLIC CONCERN IN RELATION TO DRINKING WATER CONTAMINATION BY ORGANIC MICROPOLLUTANTS

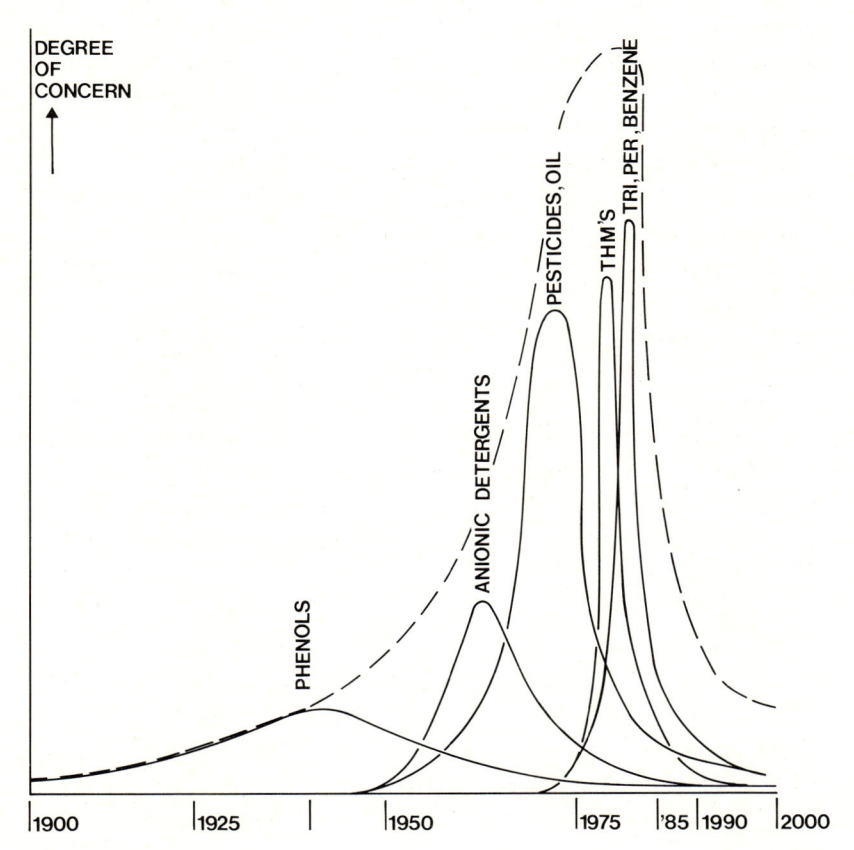

Figure 2

bioaccumulation and persistence. In Europe the endosulphan incident of May 1969 caused by Hoechst resulted in the closure of several water intakes along the Rhine and public statements of our Prime Minister for the press. It was the starting point for the revitalization of the Rhine Convention, EC legislation and a growing public awareness of the link between environmental pollution and potable water supply (Greve, 1971). Pesticides were the first to appear on so-called black lists of substances for which the discharge in water should be eliminated. Within a few years their concentration dropped considerably and nowadays pesticides in drinking water are generally well below acceptable levels. In the same period oil pollution was recognized and controlled.

The fourth wave of concern started in the middle of the seventies with the discovery of the production of trihalomethanes during water chlorination. In

Western-Europe it never reached the peak in publicity which occurred in the U.S.A. This was probably due to the fact, that chlorine never obtained in Europe such a central position in drinking water preparation as in North-America. Before 1980 the use of chlorine had dropped in The Netherlands to 50% of it's use in 1975 and a major research effort which occupies us still today was the result. In The Netherlands the top of public environmental concern was initiated by the fifth wave in the early eighties due to the discovery of organic micropollutants in groundwater used for drinking water production. Trichloroethylene and benzene were on the headlines of our newspapers and within one or two years the most costly national environmental program to clean soil pollution sites was established. What started as an amazing drinking water problem ended up into the most costly environmental operation of our country and maybe of this century. It's effects on potable water supply were limited to incidental additional treatment and safeguarding of adequate protection zones.

SCHEMATIC ILLUSTRATION OF CHANGES IN WATER QUALITY SURVEILLANCE EFFORT OF WATER SUPPLY INDUSTRY DURING THE SECOND PART OF THIS CENTURY (CATAGORIES SOURCES, TREATMENT, DISTRIBUTION)

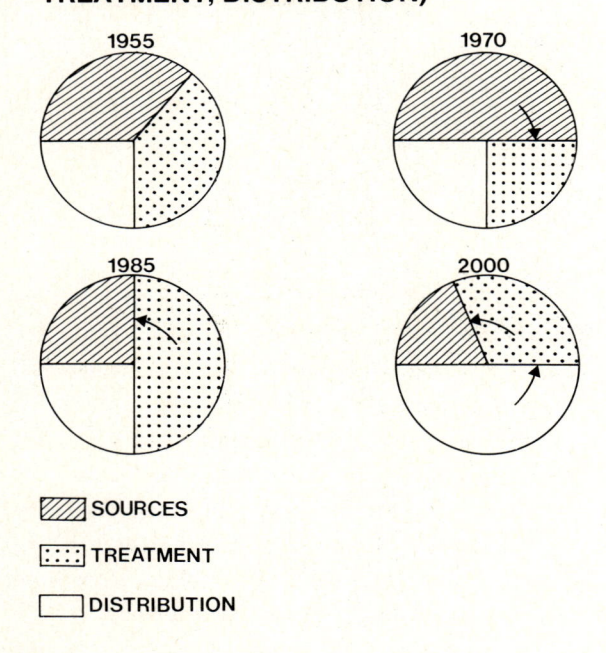

Figure 3

What is next? Presently we are concerned with other by-products of water chlorination than THM's such as haloacetonitriles and haloacetic acids. I do not expect that source contaminants will once again cause such a public concern as we have seen the past two decades.

Our sources are becoming cleaner (Slooff et al., 1984). We will better control and avoid the formation of treatment by-products. This leaves distribution as the potential important source of contamination in the future. Therefore I see three trends. The first is that the degree of public concern will deminish the next decades, now and than interrupted by peaks cuased by new discoveries of organic contaminants in tapwater, as indicated in figure 2.

Secondly, the origin of the micropollutants of concern will change.
Till the sixties the contamination of the source was in the centre of our interest, while presently we are focussing on treatment by-products. Finally, say at the beginning of next century, it might well be that our attention will be given to micropollutants released by coatings, plastic pipes and bacterial after-growth in the distribution systems as illustrated in figure 3.

SCHEMATIC PRESENTATION OF TRENDS IN THE NATURE OF DRINKING WATER QUALITY MONITORING DURING THE 20TH CENTURY

☐ ORGANIC CHEMISTRY
⁙ INORGANIC CHEMISTRY
▨ MICROBIOLOGY
▧ TOXICOLOGY

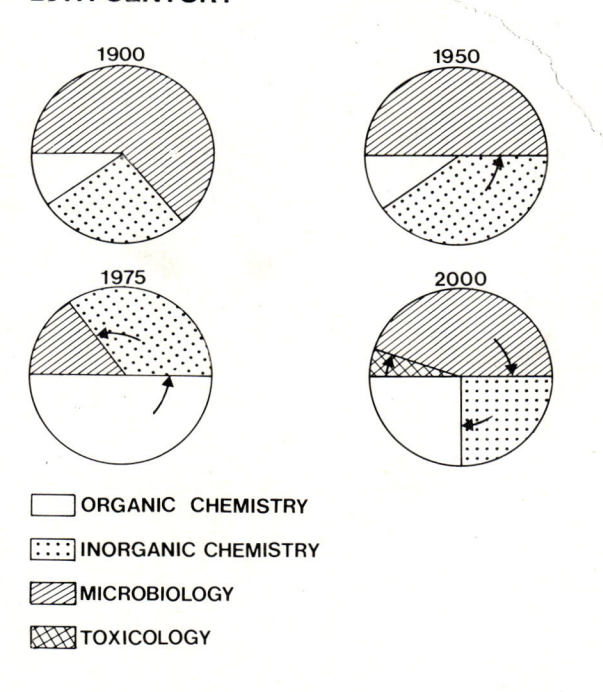

Figure 4

The third trend deals with the type of water quality aspects that is of major concern (figure 4). Water supply started last century to guarantee drinking water that was bacteriologically safe. During this century the bacteriological aspects were more and more put out of sight by inorganic and later organic chemical aspects of water quality. it is my expectation that the better we clean our water sources and optimize water production and water treatment the more relevant becomes once again the inorganic and finally the microbiological quality of drinking water.

So I expect that at the start of the next century water distribution and (micro)biological water quality will be again the central issues in drinking water quality assurance.

Drinking water approaches with wider application potential

As promised in the introduction I now come to the approaches in public water supply which merit a wider application in environmental policies. Due to the mature status of the water supply industry most of its problems can be handled in the quiet atmosphere of solid cost-benefit optimization studies. A good example (figure 5) was recently presented by Cotruvo (1984) in relation to the cost of chloroform reduction and the benefit of less cancer treatment costs. This type of objectivation is still scarce in the setting of standards for environmental quality, but will become more common in the next decades, also in Europe.

Standard setting procedures and inclusion of other exposure routes were practiced for drinking water quality in an early stage. It is amazing that the most advanced risk assessment methodologies have been developed for the relative small risks associated with drinking water. Nowadays they obtain already a wider application in the assessment of air pollutants and soil pollutants. Similar trends can be described for the handling of the exposure to radioactive materials, which subject obtains much attention but generally results in smaller risks than those caused by recent environmental problems. In both cases it is the large existing organization that more or less autonomously creates further refinements in the scientific approaches. One of the major benefits of this achievement will be the use of the water supply experience for other environmental problems that have nowadays a high priority. Water supply experts can therefore move to other area's of environmental research and policy making.

The case of water chlorination has shown the validity of the rule that pollution should be treated as close as possible to it's source. Waste water containing enteric bacteria and virusses should therefore be mainly treated before discharge into our water sources.

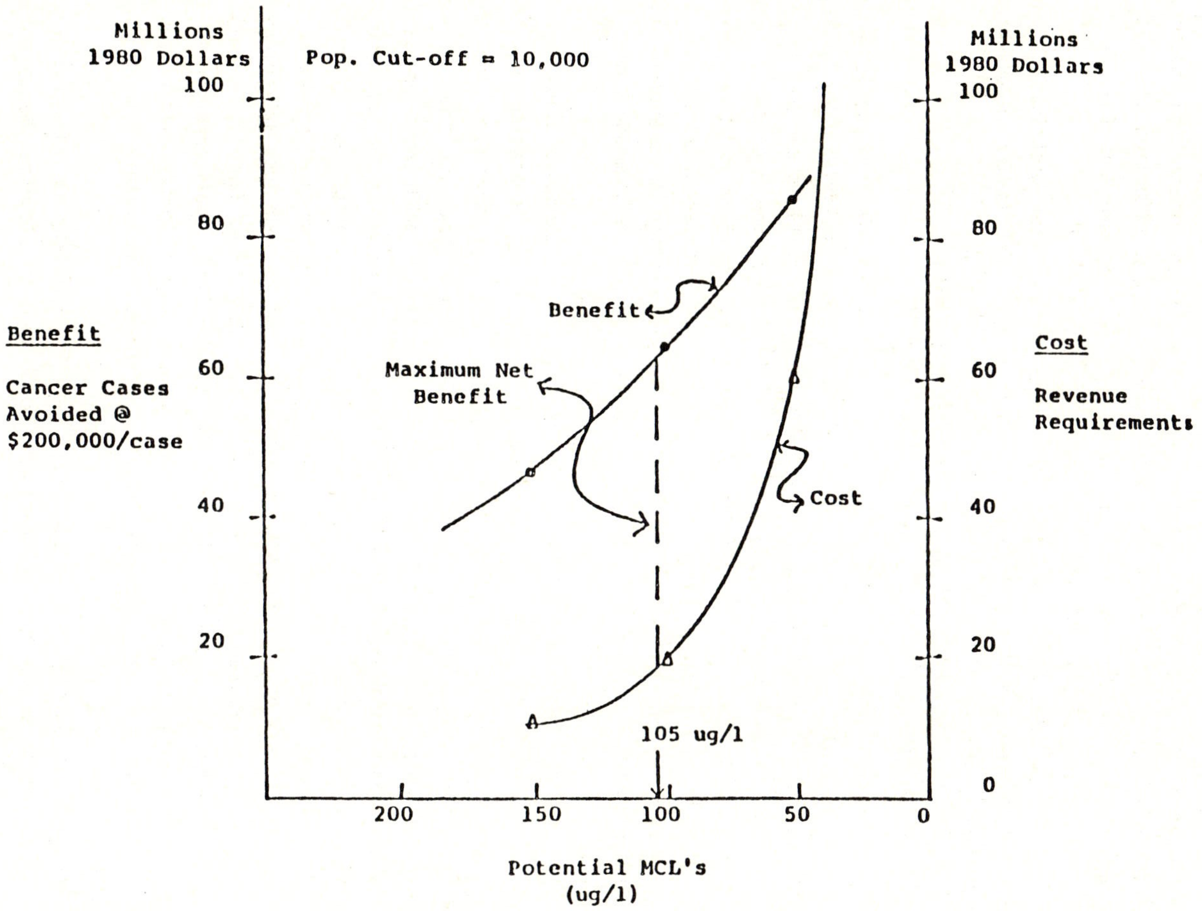

Figure 5. Benefit vs. Cost for TTHM Regulation (Cotruvo, 1984)

Finally very few public sectors have so much experience with the continuous effort aiming at a high quality product, that may cause acute disease in case of failures and that is delivered to the consumer in an only partly controlled packaging material of extremely large dimensions.

Disinfection after the chlorine era

This long experience I just described will be needed in the next decades when the glamour of public interest will fade away but the risk of chemical and microbiological contamination remains continuously present (figure 6). In this context research to further improve disinfection is certainly not at a final stage. As Fiessenger (1985) stated the "ideal disinfectant still remains to be found, which will be persistant, cause no taste and odour, will not react with water constituents and is harmless to man". I believe that the solution is easier than the chemical engineer working in an industrialized region imagines. Het should start to ask himself how he would design a treatment system in case he was not allowed to use a chemical oxidant at all. And sooner or later he would find that the ideal disinfectant has not a chemical nature, but is simply allowing the water enough time by applying for instance soil infiltration, open

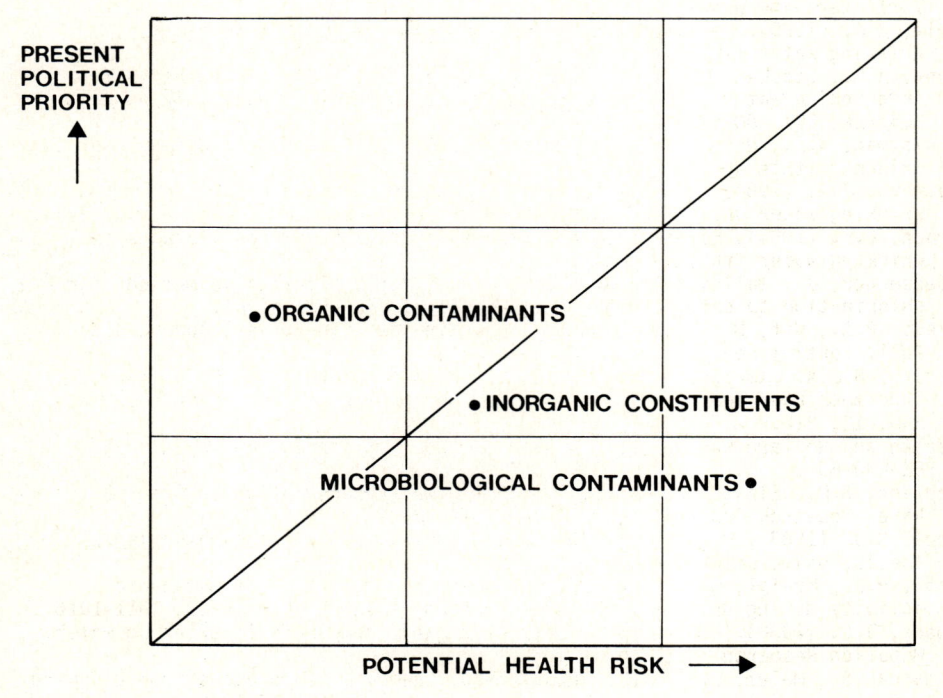

Figure 6
RATING OF DRINKING WATER CONSTITUENTS ACCORDING TO PRESENT POLITICAL PRIORITY AND POTENTIAL RISK

storage or slow sand filtration. Of course this solution is not meant for regions in developing countries or in remote areas where water is scarce and thechnical control is difficult to achieve.

I can summarize this paper by saying that the essential issue we have been discussing is the health hazard of chemical oxidation by-products and that this hazard can be avoided to a large extend by applying the same principle as is used for the production of good wine: disinfect drinking water with time!

Acknowledgement

The author gratefully acknowledges the help of Dr. A. Minderhoud and Ir. J. Hrubec in the preparation of this paper.

REFERENCES

Andelman, J.B. (1985). Inhalation exposure in the home to volatile organic contaminants of drinking water (this symposium).
Anderson, E.L. (1983). Quantitative approaches in use to assess cancer risk, Risk Analysis, 3, No. 4.
Bock, K.J., Wickbold, R. (1966). Auswirkungen der Umstellung auf leicht abbaubare Waschrohstoffe in einer grosstechnischen Kläranlage und im Vorfluter, Vom Wasser, 33, 242-253.
Bull, R.J. (1985). Carcinogenic and mutagenic properties of chemicals in drinking water (this symposium).
Chek, A.M., Stockedopole, J., Koski, P., Cole, L. (1980). Non-volatile mutagens in drinking water: production by chlorination and destruction by sulphite, Science, 207, 80-92.
Christman, R.F., Nordwood, D.L., Johnson, J.D. (1985). By-products of chemical oxidants (this symposium).
Cotruvo, J.A. (1984). Risk assessment and control decisions for protecting drinking water quality, IARC Workshop Drinking Water and Health (in press).
Craun, G.F. (1985). Epidemiological studies of organic micropollutants in drinking water (this symposium).
Fiessenger, J., Mallevaille, J., Rook, J.J. (1985). Alternative methods for chlorination to safeguard hygienic quality (this symposium).
Greve, P.A., Wit, S.L. (1971). Endosulphan in the Rhine river, Journ. Water Poll. Control Fed., 43, No. 12, 2338-2348
Haring, B.J.A., Karres, J.J.C., Poel, P. van der, Zoeteman, B.C.J. (1979). Onderzoek naar de gebruiksgewoonten bij drinkwaterconsumptie in Nederland, H$_2$O, 12, 212.
Health and Welfare Canada (1981). Tapwater Consumption in Canada, Report 82-EHD-80.
Hopkins, S.M., Ellis, J.C. (1980). Drinking water consumption in Great Brittain Water Research Centre, Techn. Report, TR 137.
Kool, H.J. (1983). Organic mutagens and drinking water in The Netherlands, Thesis, University of Wageningen, 76
Kool, H.J., Kreijl, C.F. van (1984). Formation and removal of mutagenic activity during drinking water preparation, Water Research, 18, 1011-1016
Loper, J.C. (1980). Mutagenic effects of organic compounds in drinking water, Mutation Research, 76, 241.
Monarca, S., Meier, J.R., Bull, R.J. (1983). Removal of mutagens from drinking water by granular activated carbon, Water Res., 17, 1015-1026.
Sagan, C. (1983/1984). Nuclear war and climatic catastrope: some political implications, Foreign Affairs, No. 62202, Winter 1983-1984, 257.

Singh, H.B., Salas, L.J., Smith, A.J. and Shigeishi, H. (1981). Measurements of some potentially hazardous organic chemicals in urban environments, Atmosph. Envir., 15, 601-611.
Slooff, W., Kreijl, C.F. van, Zwart, D. de (1984). Biologische parameters en oppervlaktewater (meetnetten), H$_2$O, 17.
Trehy, M.L., Bieber, T.I. (1981). Detection, identification and quantitative analysis of dihaloacetonitriles in chlorinated natural waters, In: Keith, L.H. ed., Advances in the identification and analysis of organic pollutants in water (Ann. Arbor, MI, Ann. Arbor Science Publ.), 2, 941-975.
Williamson, S.J. (1981). Epidemiological studies on cancer and organic compounds in U.S.-drinking waters, Sci. Tot. Env., 18, 187-203.
World Health Organization (1984). Guidelines for drinking water qualtity, 2, Geneva.
Zoeteman, B.C.J. (1980). Sensory assessment of water quality, Pergamon Press, Oxford, 96
Zoeteman, B.C.J. (1981). Onderzoek naar de relatie drinkwater en gezondheid, H$_2$O, 13.
Zoeteman, B.C.J., Hrubec, J., Greef, E. de, Kool, H.J. (1982). Mutagenic activity associated with by-products of drinking water disinfection by chlorine, chlorine dioxide, ozone and UV-irradiation, Env. Health. Persp., 46, 197-205.

Appendix 1

Estimated mean daily intake of 34 substances by man (based on WHO, 1984)

Substance	Unit	Drinking Water*	Food	Air**	Other (incl. smoking)	Total	% Drinking water contribution
1	2	3	4	5	6	7	8
Aluminium	mg	0.1	90	-	-	90	0.1
Arsenic	µg	5	2100	4	-	2100	0.2
Barium	µg	80	800	0.0001	-	880	9
Beryllium	µg	0.4	50	0.02	-	50	0.8
Cadmium	µg	1	40	0.05	1	42	2
Calcium	mg	200	1000	-	-	1200	20
Chloride	mg	100	6000	-	-	6100	2
Chromium	µg	10	200	0.02	1	210	5
Fluoride	mg	2	1	0.02	-	2	50
Iron	mg	0.1	20	-	-	20	0.5
Lead	µg	100	200	10	-	310	30
Magnesium	mg	100	250	-	-	350	30
Manganese	mg	0.05	5	-	-	5	1
Mercury	µg	0.1	10	1	-	11	1
Nickel	µg	10	400	4	-	410	2
Nitrate	mgN	10	60	0.1	-	70	10
Selenium	µg	1	150	-	-	150	0.7
Silver	µg	0.1	50	1	-	51	0.2
Sodium	mg	100	4000	-	-	4100	2
Sulphate	mg	50	500	-	-	550	9
Benzene	µg	0.2	250	200	-	450	0.04
Benzo(a)pyrene	µg	0.01	1	0.05	0.1	1.1	0.9
Carbontetrachloride	µg	0.1	5	20	-	26	0.4
Chlorobenzene	µg	0.1	-	20	-	20	0.5
Chloroform	µg	20	100***	10	-	130	10
1,2 Dichloroethane	µg	0.1	-	20	-	20	0.5
Tetrachloroethene	µg	0.3	5***	100	-	105	0.3
Trichloroethene	µg	1	5***	100	-	106	1
Vinylchloride	µg	0.1	10***	200	-	210	0.05
Pesticides:							
Aldrin/Dieldrin	µg	0.001	2	0.001	-	2	0.05
Chlordane	µg	0.001	10***	0.02	-	10	0.01
DDT	µg	0.001	10	0.02	-	10	0.01
Heptachlor(epoxide)	µg	0.001	1***	0.01	-	1	0.1
Hexachlorobenzene	µg	0.001	1	-	-	1	0.1
Lindane	µg	0.1	100	0.1	-	100	0.1

* see also Zoeteman (1980); values are often lower than indicated
** see also Singh et al (1981)
*** rough estimate

Author Index

Andelman, J.B. 443
Anselme, C. 371

Backlund, P. 257
Baxter, K.M. 93
Beltramelli, G. 473
Bisschop, A. 427
Borén, H. 265
Bouanga, F. 115
Brauch, H.-J. 27
Bruchet, A. 371
Bull, R.J. 385

Carlberg, G.E. 265
Chen, A.S.C. 155
Christman, R.F. 195
Chu, I. 421
Condie, L.W. 433
Coté, M.G. 421
Cotruvo, J.A. 7
Craun, G.F. 461
Croue, J.P. 217

De Kruijf, H.A.M. 59
De Laat, J. 115
De Leer, E.W.B. 211
De Vries, Th. 427
Dore, M. 115, 217, 223
Duguet, J.P. 299

Erkelens, C. 211

Fawell, J.K. 317
Ferrari, G. 473
Fielding, M. 317
Fiessinger, F. 299

Goewie, C.E. 349
Grimvall, A. 265

Hogendoorn, E.A. 349
Hoigné, J. 169
Holmbom, B. 343
Hrubec, J. 229

Janssen, D.B. 121
Janssens, J. 187
Johnson, J.D. 195

Kool, H.J. 229
Kronberg, L. 257, 343
Kruithof, J.C. 137
Kühn, W. 27

Lagory, K.E. 415
Laurie, R.D. 415
Legube, B. 217
Lewis, W.M. 83

Mallevialle, J. 371
Merlet, N. 223
Möller, M. 265

N'Guyen, K. 371
Noordsij, A. 273
Norwood, D.L. 195

Pensar, G. 257
Pfohl, R.J. 415
Piet, G.J. 229
Plaa, G.L. 421
Pregliasco, F. 473
Puijker, L.M. 137, 273

Rittmann, B.E. 99
Rook, J.J. 299

Secours, V.E. 421
Snoeyink, V.L. 155
Sontheimer, H. 27

Taylor, D.H. 415
Thibaud, H. 223
Tikkanen, L. 257, 343
Toft, P. 45

Valli, V.E. 421
Van de Kerkhoff, J.F.J. 293
Van der Gaag, M.A. 137, 273, 293
Van der Heijden, C.A. 427, 479
Van Dijck, H. 187
Van Dijk-Looijaard, A.M. 59
Van Esch, G.J. 427
Van Hoof, F. 187
Van Kreijl, C.F. 229, 479
Van Rossum, P.G. 361
Villeneuve, D.C. 421

Wegman, R.C.C. 427
Wester, P.W. 427
Wigilius, B. 265
Witholt, B. 121

Zaccaro, D.J. 415
Ziglio, G. 473
Zoeteman, B.C.J. 487

Subject Index

Adsorption, method for removal of micropollutants 155
Alternative chlorination methods 299
Ames Salmonella mutagenicity test, used to isolate non-volatile mutagens 361

Bacterial activity, on granular activated carbon 115
Bank-filtered river water, drinking water quality 273
Biodegradation, of organic micropollutants 99
Biological contamination, of drinking water 7
Biological removal, of biodegradable compounds 115
Biotechnology, application to drinking water preparation 121
Blood level of trichloroacetic acid and trichloroethanol from drinking water 473

Carcinogenic and mutagenic properties of chemicals, in drinking water 385
Carcinogenic and mutagenic compounds in drinking water, critical considerations 479
Carcinogenicity study, with halogenated hydrocarbons 427
Chlorination of humic substances in aqueous solution 217
Chlorination, effect on mutagenic activity 299
Chlorination, its influence on water quality 137
Chlorination, of drinking water 27
Chlorine oxide, used for removal of micropollutants 169
Chloropicrin, see trichloronitromethane 223
Coagulation, method for removal of micropollutants 155
Control of organics in drinking water, standards, legislation and practice 45

Drinking water and health hazards, environmental perspective 487
Drinking water quality regulations, in The Netherlands 59

Epidemiologic studies, of organic micropollutants 461

Factors affecting the removal of organic micropollutants 155
Fungicide determination, in surface water 349

Granular activated carbon filters, influence of microbial activity on removal of organic compounds 115
Granular activated carbon filtration, effect on water quality with and without ozonation 137
Ground water pollution 59
Groundwater, contamination by landfill leachates 93

Halocarbons, target organ effects 433
Halogenated hydrocarbons, effect on carcinogenicity 427
Hazardous compounds in drinking water, identification and assessment 317
Humic substances, chlorination in aqueous solution 217
Humic water, fractionation of mutagenic compounds formed during chlorination 343

Inhalation exposure, to volatile organic compounds 443
International Standards for Drinking Water, WHO guidelines on organic micropollutants 83
Iprodione, identification in drinking water by liquid chromatography 349

Landfill leachates, effect on groundwater quality 93
Linear aldehydes, formation by ozonization 187

Mutagenic activity, in drinking water treated with various oxidizing agents 257
Mutagenic activity, effect of different treatment processes 229
Mutagenic compounds, formed during chlorination of water 343

Mutagenicity testing of water, with fish 293

Neurotoxic effects, of trichloroethylene 415
Non-volatile mutagens, in drinking water 361

Organic chemicals in drinking water, identification and assessment 317
Organic micropollutants in drinking water, an overview 7
Organochlorine compounds, in river water 273
Organoleptic changes, caused by defective polyethylene tubings 371
Oxidant by-products, identified by new analytical techniques 195
Ozone, used for removal of micropollutants 169

Pentachlororesorcinol, intermediate in chloroform production 211
Polyethylene, low molecular weight products desorbed from 371

Risk assessment/management decision model, as an aid to control decisions for contamination problems 7

Secondary oxidants, produced during the decomposition of ozone 169
Sister chromatid exchange, for the detection of mutagens in water 293

Subchronic toxicity, of trichloropropanes 421
Surface water pollution 59

Target organ effects, of halocarbons 433
Techniques for concentrating mutagenic compounds, a comparison 265
Toxic chemicals, in drinking water 385
Toxic xenobiotics, removal by microorganisms 121
Trace organic quality, of chalk groundwater 93
Treatment processes, evaluation with respect to mutagenic activity in drinking water 229
Trichloroethylene, effect on the exploratory and locomotor activity of rats 415
Trichloroethylene, inhalation exposure 443
Trichloronitromethane, produced by chlorination of water 223
Trichloropropanes, subchronic toxicity 421
Trihalomethanes, formation during chlorination 27

Urine level of trichloroacetic acid and trichloroethanol from drinking water 473

Volatile organic micropollutants, in groundwater 93

Waste treatment processes, using microorganisms 121